University of Liverpool

se r

y

we

Nutritional Neuroscience

NUTRITION, BRAIN, AND BEHAVIOR

Series Editor: Chandan Prasad, PhD

Professor and Vice Chairman (Research)
Department o f Medicine
LSU Health Sciences Center
New Orleans, LA, USA

Published Titles:

DHEA and the Brain
Edited by Robert Morfin
ISBN 0-415-27585-7

Coffee, Tea, Chocolate, and the Brain
Edited by Astrid Nehlig
ISBN 0-415-30691-4

Nutritional Neuroscience

Edited by

Harris R. Lieberman

Military Nutrition Division
U.S. Army Research Institute of Environmental Medicine
Natick, MA

Robin B. Kanarek

Tufts University
Department of Psychology
Medford, MA

Chandan Prasad

Louisiana State University Health Sciences Center
Department of Medicine
New Orleans, LA

Taylor & Francis
Taylor & Francis Group

Boca Raton   London   New York   Singapore

A CRC title, part of the Taylor & Francis imprint, a member of the
Taylor & Francis Group, the academic division of T&F Informa plc.

Published in 2005 by
CRC Press
Taylor & Francis Group
6000 Broken Sound Parkway NW, Suite 300
Boca Raton, FL 33487-2742

International Standard Book Number-10: 0-415-31599-9 (Hardcover)
International Standard Book Number-13: 978-0-4153-1599-9 (Hardcover)
Library of Congress Card Number 2004058492

Library of Congress Cataloging-in-Publication Data

Nutritional neuroscience / edited by Harris R. Lieberman,
    Robin B. Kanarek, Chandan Prasad.
        p. ; cm. -- (Nutrition, brain, and behavior)
    Includes bibliographical references and index.
    ISBN 0-415-31599-9 (alk. paper)
    1. Brain. 2. Nutrition. I. Lieberman, Harris R. II. Kanarek, Robin B. III. Prasad,
Chandan, 1942- IV. Series.
    [DNLM: 1. Brain--physiology. 2. Nutrition. 3. Behavior. 4. Diet. WL 300 N9769 2005]

QP376.N865 2005
612.8'2--dc22   2004058492

Taylor & Francis Group
is the Academic Division of T&F Informa plc.

Visit the Taylor & Francis Web site at
http://www.taylorandfrancis.com

and the CRC Press Web site at
http://www.crcpress.com

Preface

Nutritional neuroscience is an emerging, interdisciplinary field that relates directly to many health care and quality-of-life issues at the forefront of modern society. Scientific and commercial interest in nutritional neuroscience, and in the general areas of diet, nutrition, weight loss, and dietary supplements, has grown dramatically in the last 10 years. In 1994, as part of the Dietary Supplement Health and Education Act, the Congress of the United States authorized the establishment of the Office of Dietary Supplements at the National Institutes of Health. The journal *Nutritional Neuroscience*, which is devoted to the relationships between nutrition and brain function, was started in 1998 and is now indexed in Medline. Symposia, conferences, and other meetings addressing the relationships between brain function and diet are held regularly

The regulatory status and availability of dietary supplements and similar products vary considerably from nation to nation, but in most countries a wide variety of these compounds can be obtained easily. In the United States, an extremely broad and diverse range of dietary supplements can be found not only at specialty health food stores, but at virtually all pharmacies, supermarkets and general merchandise outlets. Other products such as foods for weight loss, functional foods, nutraceuticals, and medical foods are also marketed, although in many cases the category to which a product is assigned may not be clearly defined. Many of these products are marketed for their effects on behavior or brain function, the focus of the field of nutritional neuroscience. Recommendations of diets for general health, weight loss, or specific medical conditions can be found in many popular publications, often with little validated scientific support.

This book will be of interest to a wide variety of readers who have backgrounds in nutrition, psychology, neuroscience, or a related clinical field such as medicine, clinical dietetics, nursing, or clinical psychology. The chapter authors were asked to make their respective contributions accessible to a more general audience than for a typical edited scientific volume written for specialists in the field. The contents of this volume are therefore quite diverse, and include material on methodological issues, as well as chapters addressing the effects of a wide range of foods, specific nutrients, food constituents, food additives, and dietary supplements on brain function and behavior.

We anticipate that this book will prove useful as an advanced undergraduate or graduate/professional textbook in courses that introduce the field of nutritional neuroscience. It also will be of great value to health care professionals who are considering the use of specific diets or dietary supplements in their practices, or who are frequently asked about nutrition, diet, and supplements by their patients. We also hope this volume will provide scientists who work in the field of nutritional neuroscience with a source of consolidated information in their own specialty, and allow them to become acquainted with subject matter areas with which they are not familiar. Even experts in this field, because of its interdisciplinary nature, often find it difficult to locate all the relevant literature on a particular topic using standard search methodologies.

In closing we would like to thank the authors of every chapter for their dedication and scholarly efforts that made this volume possible.

Harris R. Lieberman, Ph.D.
Robin B. Kanarek, Ph.D.
Chandan Prasad, Ph.D.

Editors

Harris R. Lieberman is a research psychologist in the Military Nutrition Division of the U.S. Army Research Institute of Environmental Medicine (USARIEM) in Natick, Massachusetts. Dr. Lieberman is an internationally recognized expert in the area of nutrition and behavior and has published more than 100 original full-length papers in scientific journals and edited books. He has been an invited lecturer at numerous national and international conferences, government research laboratories, and universities.

Dr. Lieberman received his Ph.D. in physiological psychology in 1977 from the University of Florida. On completing his graduate training, he was awarded an NIH fellowship to conduct postdoctoral research at the Department of Psychology and Brain Science at the Massachusetts Institute of Technology (MIT). In 1980, he was appointed to the research staff at MIT and established an interdisciplinary research program in the Department of Brain and Cognitive Sciences to examine the effects of food constituents and drugs on human behavior and brain function. Key accomplishments of the laboratory included development of methods for assessing the effects of food constituents and environmental factors on human brain function and determination that specific foods and hormones reliably altered human performance and mood.

In 1990, Dr. Lieberman joined the civilian research staff of USARIEM, where he has continued his work in nutrition, behavior, and stress. From 1994 to 2000, he was chief or deputy chief of the Military Nutrition program at USARIEM. His recent research has addressed the effects of various nutritional factors, diets, and environmental stress on animal and human performance, brain function, and behavior. He holds two patents for novel technologies to assess and enhance cognitive performance. Dr. Lieberman currently chairs an International Defense Panel on Cognitive and Ergogenic Aids.

Robin B. Kanarek received a B.A. in biology from Antioch College in Yellow Springs, Ohio, and an M.S. and a Ph.D. in psychology from Rutgers University in New Brunswick, New Jersey. She is currently the dean of the Graduate School of Arts and Sciences and professor of psychology and nutrition at Tufts University in Medford, Massachusetts. Her primary research interests are in the area of nutrition and behavior. She has conducted research on the effects of nutritional variables on the development of obesity, the physiological and behavioral factors influencing diet selection in experimental animals and humans, the role of nutrients in determining the consequences of psychoactive drugs, and the importance of nutrition for cognitive behavior in children and adults. She has authored or coauthored more than 100 books, book chapters, and articles and has presented her research at numerous international and national conferences. Her research has been funded consistently for the last 25 years by the National Institutes of Health (NIH) as well as by other government agencies and private companies. Dr. Kanarek has been actively involved in graduate education and teaching throughout her time at Tufts, serving as the mentor for more than 15 Ph.D. students. In 2000, she was named John Wade Professor and received the Tufts University Senate Professor of the Year award.

Dr. Kanarek's experience includes research fellow, Division of Endocrinology, University of California, Los Angeles (UCLA) School of Medicine, and research fellow in nutrition at Harvard University. She is a member of the editorial boards of *Physiology and Behavior, Nutritional Neuroscience,* and the *Tufts Diet and Nutrition Newsletter* and is a past editor-in-chief of *Nutrition and Behavior.* In addition, she regularly reviews articles for peer-reviewed journals, including *Science, Brain Research Bulletin, Pharmacology Biochemistry and Behavior, Brain Research,*

Journal of Nutrition, *American Journal of Clinical Nutrition*, and *Annals of Internal Medicine*. From 1995 to 2001, she was a member of the National Academy of Sciences, Committee on Military Nutrition Research. Dr. Kanarek also has served on review committees for the National Science Foundation, NIH, and USDA Nutrition Research and as a member of the Program Committee of the Eastern Psychological Association. She is a fellow of the International Society for Behavioral Neuroscience. Her other professional memberships include the Society for the Study of Ingestive Behavior and Society for Neurosciences.

Chandan Prasad graduated from Louisiana State University in 1970 with a Ph.D. in microbiology/biochemistry. After 8 years at the NIH in Bethesda, Maryland, as Fogerty fellow and senior staff fellow, he returned to New Orleans to join the faculty of the LSU School of Medicine. He is currently professor (medicine and neuroscience) and vice chairman (research) in the Department of Medicine at the LSU Health Sciences Center in New Orleans. The current focus of Dr. Prasad's research is on adipocyte biology and the role of dietary supplements in obesity, diabetes, and heart diseases. He has authored more than 200 papers in the area of appetite regulation, obesity, and nutrition. He holds four U.S. and international patents for treatment of obesity and alcoholism. He serves as editor-in-chief of *Nutritional Neuroscience* and associate editor of *Current Topics in Nutraceutical Research*. He also serves as series editor for *Nutrition, Brain, and Behavior*. He is married to Shail Gupta, M.A., and has three sons — Anand Prasad, B.S., M.D.; Amit Prasad, B.S., M.D. (student); and Anoop Prasad, B.S., J.D. (student).

Contributors

David Benton
Department of Psychology
University of Wales
Swansea, Wales, United Kingdom

Charles J. Billington
Minnesota Obesity Center
Minneapolis VA Medical Center
and
Department of Medicine
University of Minnesota
Minneapolis, Minnesota

John E. Blundell
School of Psychology
University of Leeds
Leeds, United Kingdom

Tammy M. Bray
College of Health and Human Sciences
Oregon State University
Corvallis, Oregon

Amanda Carey
U.S. Department of Agriculture —
  Agricultural Research Service
Human Nutrition Research Center at
  Tufts University
Boston, Massachusetts

James R. Connor
College of Medicine
Pennsylvania State University
Hershey, Pennsylvania

R. Todd Coy
Department of Psychology
Tufts University
Medford, Massachusetts

Kristen E. D'Anci
Department of Psychology
Tufts University
Medford, Massachusetts

Jan Berend Deijen
Department of Clinical Neuropsychology
Vrije Universiteit
Amsterdam, The Netherlands

Edzard Ernst
Peninsula Medical School
Universities of Exeter and Plymouth
Exeter, United Kingdom

Rachel L. Galli
Department of Psychology
Simmons College
Boston, Massachusetts

Dorothy W. Gietzen
School of Veterinary Medicine
University of California
Davis, California

Rubem Carlos Araújo Guedes
Department of Nutrition
Universidade Federal de Pernambuco
Recife, PE, Brazil

Jason C.G. Halford
Department of Psychology
University of Liverpool
Liverpool, United Kingdom

Jürg Haller
Human Nutrition and Health
F. Hoffman-LaRoche Ltd.
Basel, Switzerland

Ruth B.S. Harris
Department of Foods and Nutrition
University of Georgia
Athens, Georgia

W. Thomas Johnson
U.S. Department of Agriculture —
  Agricultural Research Service
Grand Forks Human Nutrition
  Resource Center
Grand Forks, North Dakota

James A. Joseph
U.S. Department of Agriculture —
  Agricultural Research Service
Human Nutrition Research Center at
  Tufts University
Boston, Massachusetts

Robin B. Kanarek
Department of Psychology
Tufts University
Medford, Massachusetts

Thomas J. Koehnle
Department of Neuroscience
University of Pittsburgh
Pittsburgh, Pennsylvania

Catherine M. Kotz
Minnesota Obesity Center
Minneapolis VA Medical Center
and
Department of Food Science and Nutrition
University of Minnesota
Minneapolis, Minnesota

John H. Lazarus
College of Medicine
University of Wales
Cardiff, Wales, United Kingdom

Monica Leibovici
Department of Psychology
Tufts University
Medford, Massachusetts

Allen S. Levine
Minnesota Obesity Center
Minneapolis VA Medical Center
and
Department of Psychiatry
Department of Food Science and Nutrition
Department of Medicine
University of Minnesota
Minneapolis, Minnesota

Mark A. Levy
College of Health and Human Sciences
Oregon State University
Corvallis, Oregon

Harris R. Lieberman
Military Nutrition Division
U.S. Army Research Institute of
  Environmental Medicine
Natick, Massachusetts

Caroline R. Mahoney
Department of Psychology
Tufts University
Medford, Massachusetts
and
U.S. Army Soldier Systems Center
Natick, Massachusetts

Danica Martin
Department of Health Studies and
  Gerontology
University of Waterloo
Waterloo, Ontario, Canada

Wendy Foulds Mathes
Department of Psychology
Tufts University
Medford, Massachusetts

Pawel K. Olszewski
Minnesota Obesity Center
Minneapolis VA Medical Center
and
Department of Medicine
University of Minnesota
and
College of Veterinary Medicine
Minneapolis, Minnesota

Domingo J. Piñero
Steinhardt School of Education
New York University
New York, New York

Chandan Prasad
Department of Medicine
Louisiana State University
  Health Sciences Center
New Orleans, Louisiana

Barbara Shukitt-Hale
U.S. Department of Agriculture —
  Agricultural Research Service
Human Nutrition Research Center on Aging
  at Tufts University
Boston, Massachusetts

Andrew Smith
School of Psychology
Cardiff University
Cardiff, United Kingdom

Holly A. Taylor
Department of Psychology
Tufts University
Medford, Massachusetts

Patricia E. Wainwright
Department of Health Studies and
  Gerontology
University of Waterloo
Waterloo, Ontario, Canada

Chiho Watanabe
Graduate School of International Health
University of Tokyo
Tokyo, Japan

Rinah Yamamoto
Department of Psychology
Tufts University
Medford, Massachusetts

Simon N. Young
Department of Psychiatry
McGill University
Montreal, Quebec, Canada

Table of Contents

PART III   Micronutrients, Brain Function, and Behavior

PART IV   Foods and Supplements that Modulate Brain Function

Part I

Fundamental Issues and Methods in Nutritional Neuroscience

1 Human Nutritional Neuroscience: Fundamental Issues

Harris R. Lieberman

CONTENTS

1.1 INTRODUCTION

Scientific examination of the relationships between nutrition and human behavior is difficult. Our ability to assess human behavior is limited, and many key issues in the field of human behavior assessment are controversial. In addition, knowledge of the critical nutritional factors that modulate human behavior is inadequate, especially with regard to the effects, if any, of chronic nutritional manipulations. Nevertheless, interest in both the acute and chronic effects of diet on behavior permeates our society. Dietary supplements that ostensibly will enhance cognitive performance are widely advertised, and other supplements that promise weight loss by modifying behavior are popular. Nutritional products claiming to relieve stress and depression or alleviate memory loss are common. A number of supplements and foods intended to increase energy are also widely available. Scientific evidence validating such claims is often weak and sometimes nonexistent. Therefore, a key objective of the field of nutrition and behavior should be to provide methods that can reliably prove or disprove the claims made by manufacturers, vendors, and the popular press. It is essential that scientists agree on the appropriate methods and interpretation of data so that a consensus on key issues can be achieved and evidence-based policy recommendations provided to government agencies and consumers. If commercial organizations make behavioral claims for their products, such as the ability to improve particular aspects of cognitive performance and mood, including memory, mental energy, stress relief, depression, and anxiety, then the critical evidence must be behavioral. Secondary endpoints, such as biochemical or physiological markers, although of substantial scientific value, cannot be considered suitable substitutes for functional outcomes if the claims are for changes in human behavior.

In the U.S., dietary supplements are regulated by the Dietary Health and Education Act of 1994. The requirements for marketing a dietary supplement, unlike those for drugs, including over-the-counter (OTC) drugs among which dietary supplements are sold, do not require the manufacturer to prove efficacy and safety. This lack of regulation has substantially limited development of scientific methods to assess dietary factors, because the economic incentives present for pharmaceutical development do not exist. Fortunately, in the U.S., a National Institutes of Health (NIH) component, the Office of Dietary Supplements (ODS), does support considerable research in this area. In addition, some supplement manufacturers have conducted or supported research on their products (Alford et al., 2001). However, in the U.S., the strict regulatory process that governs research on drugs does not exist to review and approve the findings of studies on nutritional interventions such as diets or dietary supplements. Unsubstantiated behavioral claims by those marketing dietary supplements are particularly common, perhaps because they are difficult to challenge.

1.2 FOOD CONSTITUENTS THAT AFFECT HUMAN BEHAVIOR

Human studies on the behavioral effects of nutrients, food constituents, and dietary supplements have focused on four classes of substances: (1) the macronutrients, protein, carbohydrate, and fat; (2) micronutrients such as vitamins and minerals; (3) dietary constituents such as caffeine; and (4) dietary supplements such as *Ginkgo biloba*, St. John's Wort, and melatonin. The methods used to study these compounds differ in some respects, but share many common elements. Also, many of the general principles in the area of human nutritional neuroscience are not unique to the field but are drawn from related fields such as psychopharmacology and experimental psychology.

It is widely recognized that the magnitude of the effects of food constituents on behavior is less dramatic than that of the effects of drugs that influence similar functions (Lieberman et al., 1982; Wurtman, 1982). The relative lack of potency of food constituents on behavior is not surprising because plant and animal products that have potent effects on the brain are considered to be drugs, substances of abuse, or poisons. Examples of such naturally occurring substances include scopolamine from the nightshade plant, opium from the poppy, and tetrodotoxin from the fugu fish.

The subtle effects of dietary constituents on the brain and behavior complicate the task of scientists working in this area. The modest size of effects of nutritional factors on behavior has resulted in more scientific disagreements on fundamental issues than in related behavioral fields such as psychopharmacology, in which scientific consensus on major issues is readily obtained. For example, psychopharmacologists agree that benzodiazepines and related compounds are hypnotics (Curran, 1991). The amino acid tryptophan, when administered in sufficient doses, also has hypnotic-like properties, but its effects, although less potent than those of hypnotic drugs, are the subject of considerable controversy (Lieberman, 2003). Use of appropriate and highly rigorous behavioral methods in nutritional neuroscience is therefore particularly critical, so that scientists can reliably replicate the results of their peers and reach agreement on key issues.

There are few individual food constituents that unquestionably affect human behavior. Based on recent literature reviews, it appears that caffeine is the best example of a dietary constituent with clear behavioral effects (Bellisle et al., 1998; Smith, 2002; Lieberman, 2003). Although many other dietary constituents and supplements appear to have behavioral effects, as discussed in detail in various chapters of this book, unqualified scientific agreement on these substances does not exist. Therefore, with the exception of caffeine, which is a stimulant, we lack established, generally accepted examples of dietary factors that affect functions such as perception, learning, memory, and sleep and mood states such as depression and anxiety.

1.3 EXPERIMENTAL CONTROL

The importance of experimental control in the design and conduct of human behavioral research on nutritional factors cannot be overemphasized. Because, as noted previously, the effects of dietary treatments tend to be relatively modest, exacting control of all extraneous parameters is essential. The key elements of experimental control in such nutritional studies include double-blind procedures, use of placebo treatments, and standardization and control of all aspects of the testing situation. Other critical factors include ensuring that an adequate sample size is obtained, selecting appropriate behavioral tests, having a conservative approach to data analysis and interpretation, and replication of results across and within laboratories.

1.3.1 USE OF DOUBLE-BLIND PROCEDURES AND PLACEBOES

The use of double-blind testing procedures is desirable not only in drug trials but also in any study of human behavior in which the investigator has an opportunity to influence the volunteers' behavior, including studies of dietary constituents. The term *double-blind* means that neither the investigator responsible for testing the volunteer nor the volunteer is aware of the experimental condition being tested. It is well known that the influence of the investigator can be exerted unintentionally and that the most subtle of cues can significantly affect the behavior of humans. It is easy for an investigator to accidentally influence the volunteer to provide responses the investigator desires. Similarly, if volunteers become aware of their treatment condition and have even the most rudimentary knowledge of the desired outcome of the study, their behavior could be altered. This can occur even if the subjects are not informed of the precise hypothesis being tested, because volunteers may have preconceptions regarding the study outcome. Implementation of foolproof double-blind procedures can be complex and demanding, but it is an essential component of study design.

To ensure that the volunteer is blind to the treatment condition, it is critical that experimental treatments are indistinguishable. In drug studies and in studies of dietary supplements provided in pill or capsule form, this level of control includes not only an identical color, weight, and size of dosage formulation but also taste, because trace amounts of certain compounds on the exterior surface of pills or capsules can be potent sensory cues.

In studies of dietary factors, especially macronutrients, the formulation of identical treatments can be physically unattainable. For example, it is not possible to formulate a nonnutritive placebo for a meal-sized treatment because consumption of such a placebo would result in gastrointestinal distress. The availability of nonnutritive food substitutes, particularly for carbohydrates and lipids, provides some opportunity for formulating adequate placebo treatments, but absolute placebo control is not usually possible. In such instances, the investigator can attempt to ensure that the volunteer is unaware of the experimental condition by concealing the identity of the treatment. This can sometimes be accomplished by "misdirection" with regard to the critical independent variable being tested, within the boundaries of ethical treatment of volunteers. For example, if a volunteer mistakenly believes an important element of a study is to evaluate the effects of the color or taste of food on behavior because these are the most obvious properties of the foods presented, attention is drawn to these aspects of the foods and away from other differences in their composition that are the key independent variables being tested. Treatments can be formulated with salient colors, textures, or flavors that are independent of the parameter under investigation, and which therefore mask the critical differences between the treatments under investigation. Controlling for any unanticipated effects of such manipulations is essential. The use of such techniques can involve little or no deception of the volunteers, because they can be informed of the real purpose of the study but are unable to distinguish one treatment condition from another, which is the purpose of using placeboes and blind testing conditions. Some investigators have obviated the need for placebo control by administering nutrients directly into the circulation or gut, but such heroic efforts can only be employed in limited circumstances (Wells et al., 1995). Furthermore, such techniques have their own shortcomings because the normal process of digestion is bypassed in such studies, thereby limiting the generalizability of the study. In situations in which identical treatments cannot be formulated, it is essential that the investigators responsible for testing volunteers be blind to the treatment. This can be accomplished by separating the functions of treatment administration from behavioral testing. To the extent possible, the individuals administering the treatments should also have as little knowledge as possible of its composition.

1.3.2 DOSE–RESPONSE DESIGNS

In one of the first papers to consider methodological issues relating to the effects of dietary constituents on behavior, it was suggested that studies should be designed to generate dose–response functions (Dews, 1982). This is a standard design used in pharmacological studies and provides a robust form of internal control. In addition to yielding information regarding the parametric effects of a particular nutrient or supplement, use of a dose–response design can demonstrate internal consistency in a study, a valuable form of validation. If a treatment generates a dose-dependent pattern of behavioral change, the study has internally replicated its key findings. Dose–response studies are not always appropriate and can be misinterpreted. For example, when conducting preliminary studies of dietary factors (particularly if the dose of the constituent is defined by its naturally occurring level), a dose–response study might not be justified because only a very narrow range of doses is of interest. The relatively moderate size of the effects of dietary constituents on behavior makes conducting dose–response studies difficult, as large sample sizes are required to distinguish the effects of different doses. For example, in a between-subject design study that distinguished the behavioral effects of two doses of a carbohydrate beverage on cognitive performance and mood, a total of 143 subjects were tested (Lieberman et al., 2002a). In addition, interpreting dose–response studies can be difficult, because monotonic dose–response functions are not always observed. For example, when caffeine is administered in high doses, it has some adverse effects on mood. When it is given in the moderate doses typically found in foods, there are beneficial effects (Kaplan et al., 1997; Smith, 2002; Lieberman et al., 2002b). Many drugs have similar dose–response properties, with desired effects observed at a lower range of doses but undesirable side effects occurring at higher doses.

1.3.3 CONSISTENCY AND CONTROL OF THE TEST ENVIRONMENT

Because humans are so easily influenced by subtle aspects of their surroundings, particularly the demeanor of other humans, standardization of the physical and social aspects of the testing environment is critical in studies of nutritional factors. Investigators must be carefully trained to behave in a consistent, professional manner and must deal identically with any testing issues that arise. Instructions provided to volunteers should be standardized. If possible, the same investigator or team of investigators should be responsible for testing all the subjects in a single study. The environment should be constant with factors such as time of day, temperature, humidity, and light levels under experimental control. Testing equipment should be identical for all subjects.

1.3.4 VOLUNTEER MOTIVATION

One aspect of the experimental environment that is particularly difficult to control is volunteer motivation. Depending on local Human Use Regulations, various incentives for optimal performance can sometimes be employed. For example, small awards for the best performance among volunteers participating in studies can be provided. Sometimes, competition among volunteers or teams of volunteers can be encouraged. The consequences of using such incentives in behavioral studies of nutritional compounds are not known. However, many investigators have observed that poorly motivated subjects often appear as outliers when data are analyzed. A subtle but important aspect of subject motivation is the demeanor of the investigators. Another important motivational factor can be the adequacy and comprehensibility of the rationale provided to volunteers participating in research studies.

1.3.5 CONFOUNDING FACTORS

A number of factors, if not properly controlled, can obscure the results of what would otherwise appear to be well-designed studies of the behavioral effects of food constituents. Examples of such confounding factors can be found in the literature on behavioral effects of caffeine. Because caffeine is usually present in the diet, failure to control for its use among volunteers in studies of this substance can mask its effects. Individuals who normally use large amounts caffeine will be less sensitive to its effects than those who normally consume little caffeine (Goldstein et al., 1969). Also, failure to control use of caffeine by volunteers before testing can be a confounding variable, not only in studies of caffeine, but also in studies of other food constituents. Finally, some personal habits can influence responsiveness to caffeine. For example, regular use of tobacco can significantly alter caffeine's pharmacokinetics (Murphy et al., 1988). Similar factors might influence the metabolism of other dietary constituents, particularly those that are not nutrients themselves.

1.4 REPLICATION

A fundamental rule of experimental science is that studies must be replicated within and across laboratories before they become accepted as valid. Given the many difficulties discussed previously associated with the study of dietary constituents, replication is particularly critical in this field. Pressures on investigators to produce positive results by some funding agencies and to publish positive results for career advancement can sometimes create subtle biases in the work of even the most careful scientist. The predisposition of some journals to publish positive results and reject negative findings exacerbates this problem. Because dietary supplements are so readily available in the U.S., it is important that claims for beneficial effects of particular dietary supplements or nutrients be proven rigorously.

1.5 SELECTION OF APPROPRIATE TESTS

1.5.1 COGNITIVE PERFORMANCE ASSESSMENT

Selection of appropriate behavioral tasks to detect the effects of food constituents on human behavior is another crucial element in the design of such studies (Lieberman et al., 1986). Unfortunately, although many standardized behavioral tasks are available, the choice of the most appropriate task to assess a specific behavioral parameter, such as memory, vigilance, or learning, is a matter of considerable controversy (e.g., see Gawron, 2000; Wetherell, 1996). In addition, tests typically employed to assess a particular cognitive function, such as learning, invariably have a multifactorial substrate. For example, nearly all learning tasks require sensory processing, memory, and motor output. To further complicate matters, even subtle variations in the manner in which a particular test is employed, such as the number of trials administered, can be important with regard to the likelihood of detecting an effect (Lieberman et al., 1986; Lieberman, 1989).

When selecting tests to evaluate certain food constituents, such as caffeine, sufficient information is available in the literature to allow the investigator to determine optimal tests to use (Lieberman, 2001; Smith, 2002). However, for most food constituents, a coherent body of literature to assist the investigator in choosing the optimal tests is not available. In such a situation, one must rely on the literature in closely related experimental fields, usually psychopharmacology, to select appropriate cognitive tests. Use of tests from clinical fields may not always be appropriate. Such tests are typically designed to detect individuals who exhibit abnormal behaviors, as opposed to detecting the effects of treatments on population samples. Also, many cognitive tests are designed to evaluate a very specific theoretical construct and are not intended for use in a more general context. There are many elegant tests of virtually every aspect of memory, but most are so specific that they are not very useful for the study of dietary constituents (Poon, 1987). Although helpful, the strategy of seeking direction from psychopharmacology and related fields can be of limited use if there is no drug or other independent variable known to alter cognitive performance in a manner that is similar to the nutritional factor under investigation. An example of a nutritional compound with putative behavioral effects but no corresponding drug is the amino acid tyrosine, the subject of a chapter in this volume. Tyrosine appears to reduce the adverse effects of acute stress on cognitive performance, a unique property (Banderet and Lieberman, 1989; Deijen et al., 1999).

Another key issue when tests are selected is their susceptibility to practice effects, particularly when a crossover design is employed. In general, practice on behavioral tests before their use in the actual study is essential. This reduces between- and within-subject variability. Appropriate use of statistical techniques, such as carefully counterbalancing the order of treatment conditions, is also critical to ensure that any residual practice effects not eliminated by pretest practice are controlled.

1.5.2 APPROPRIATE USE OF MOOD QUESTIONNAIRES

Mood questionnaires are frequently used to assess the effects of dietary constituents on behavioral state. They are widely accepted as valid measures of mental states and can be administered in a few minutes, unlike cognitive test batteries. One of most widely employed mood questionnaire in nutrition behavior research is the Profile of Mood States (POMS). This questionnaire provides six individual subscales: vigor, fatigue, tension, depression, anxiety, and anger, as well as an overall indication of mood state (McNair et al., 1971). The POMS and other mood questionnaires are used in various fields, including experimental psychology, exercise physiology, and psychiatry. Depending on the particular mood assessed, they can be highly correlated with tests of cognitive function (Glenville and Broughton, 1978; Nicholson and Stone, 1986; Bolmont et al., 2000). Although mood questionnaires do not have the cachet of tests of cognitive performance because they are not considered to be "objective," they are useful for documenting and explaining the effects of dietary constituents on human behavior. The distinction between an objective and a subjective test is more a matter of appearance than reality. Cognitive performance tests seem to be objective because they apparently bypass the emotional content of the behavior assessed. Standardized mood question-naires are no less objective than cognitive tests — both are reliable measures of particular aspects of human behavior. Each type of test requires the subject to make a response to a particular sequence of standardized stimuli. Mood questionnaires can be particularly valuable in studies of dietary constituents because they explain and validate the results of cognitive performance tests. Further-more, they can provide evidence that a compound has effects not readily detected by cognitive tests. For example, caffeine alters a variety of mood states in a manner that would not necessarily be expected given its stimulant-like actions. Mood questionnaires have shown that in normal volunteers given moderate doses of caffeine, depression and hostility are reduced and clarity of mind, imagination, and energy increased (Leathwood and Pollet, 1982; Amendola et al., 1998).

In general, if consistent effects are observed by using different measurement techniques, such as performance tests and mood questionnaires, then the results of a study are more likely to be valid. Caffeine consistently affects tests of vigilance and closely related mood states such as vigor and fatigue (Smith, 2002; Lieberman, 2003). Drugs that are stimulants, such as amphetamine, also have consistent effects on these factors (Magill et al., 2003). Use of mood questionnaires can also provide information on confounding factors, such as unanticipated changes in fatigue, anxiety, or depression of volunteers during testing.

1.6 CONCLUSION

As discussed previously, there are many obstacles associated with studying the relationships between nutrition and human behavior. However, use of appropriate methods can greatly expedite progress in the field. As societal pressures for resolution of the various issues increase, scientists must be able to reach consensus with regard to the optimal paradigms to study the relationship between nutrition and behavior. Fortunately, sound scientific methodology exists for addressing most of the issues in this emerging field.

ACKNOWLEDGMENTS

This work was supported by the U.S. Army Medical Research and Materiel Command (USAM-RMC). Approved for public release, distribution is unlimited. The views, opinions and/or findings in this report are those of the authors and should not be construed as an official Department of the Army position, policy or decision unless so designated by other official documentation. Citation of commercial organization and trade names in this report do not constitute an official Department of the Army endorsement or approval of the products or services of these organizations.

REFERENCES

Alford, C., Cox, H., and Wescott, R., The effects of Red Bull energy drink on human performance and mood, *Amino Acids,* 21(2), 139, 2001.

Amendola, C.A., Gabrieli, J.D.E., and Lieberman, H.R., Caffeine's effects on performance and mood are independent of age and gender, *Nutr. Neurosci.,* 1, 269, 1998.

Banderet, L.E. and Lieberman, H.R., Treatment with tyrosine, a neurotransmitter precursor, reduces environmental stress in humans, *Brain Res. Bull.,* 22, 759, 1989.

Bellisle, F. et al., Functional food science and behaviour and psychological functions, *Br. J. Nutr.,* 80, S173, 1998.

Bolmont, B., Thullier, F., and Abraini, J.H., Relationships between mood states and performances in reaction time, psychomotor ability, and mental efficiency during a 31-day gradual decompression in a hypobaric chamber from sea level to 8848 m equivalent altitude, *Physiol. Behav.,* 71(5), 469, 2000.

Curran, H.V., Benzodiazepines, memory and mood: a review, *Psychopharmacology (Berl.).,* 105(1), 1, 1991.

Deijen, J.B. et al., Tyrosine improves cognitive performance and reduces blood pressure in cadets after one week of a combat training course, *Brain Res. Bull.,* 48(2), 203, 1999.

Dews, P.B., Comments on some major methodologic issues affecting analysis of the behavioral effects of foods and nutrients, *J. Psychiatr. Res.*, 17(2), 223, 1982.

Gawron, V.J., *Human Performance Measures Handbook*, Lawrence Erlbaum Associates, Hillsdale, NJ, 2000.

Glenville, M. and Broughton, R., Reliability of the Stanford Sleepiness Scale compared to short duration performance tests and the Wilkinson Auditory Vigilance Task, *Adv. Biosci.,* 21, 235, 1978.

Goldstein, A., Kaizer, S., and Whitby, O., Psychotropic effects of caffeine in man. IV. Quantitative and qualitative differences associated with habituation to coffee, *Clin. Pharmacol. Ther.,* 10(4), 489, 1969.

Kaplan, G.B. et al., Dose-dependent pharmacokinetics and psychomotor effects of caffeine in humans, *J. Clin. Pharmacol.,* 37, 693, 1997.

Leathwood, P.D. and Pollet, P., Diet-induced mood changes in normal populations, *J. Psychiatr. Res.,* 17(2), 147, 1982.

Lieberman, H.R., Cognitive effects of various food constituents. In *Handbook of the Psychophysiology of Human Eating,* Shepherd, R., Ed., John Wiley & Sons, Chichester, U.K., 1989, pp. 251–270.

Lieberman, H.R., The effects of ginseng, ephedrine and caffeine on cognitive performance, mood and energy, *Nutr. Rev.,* 59(4), 91, 2001.

Lieberman, H.R., Nutrition, brain function and cognitive performance, *Appetite,* 40(3), 245, 2003.

Lieberman, H.R. et al., Mood, performance and pain sensitivity: changes induced by food constituents, *J. Psychiatr. Res.*, 17, 135, 1982.

Lieberman, H.R., Falco, C.M., and Slade, S.S., Carbohydrate administration during a day of sustained aerobic activity improves vigilance, as assessed by a novel ambulatory monitoring device, and mood, *Am. J. Clin. Nutr.,* 76(1), 120, 2002a.

Lieberman, H.R. et al., Effects of caffeine, sleep loss, and stress on cognitive performance and mood during U.S. Navy SEAL training. Sea-Air-Land, *Psychopharmacology (Berl.),* 164(3), 250, 2002b.

Lieberman, H.R., Spring, B., and Garfield, G.S., The behavioral effects of food constituents: strategies used in studies of amino acids, protein, carbohydrate and caffeine. *Diet and Behavior: A Multidisciplinary Evaluation, Nutr. Rev.,* 44, 61, 1986.

Magill, R.A. et al., Effects of tyrosine, phentermine, caffeine, D-amphetamine, and placebo on cognitive and motor performance deficits during sleep deprivation, *Nutr. Neurosci.,* 6(4), 237, 2003.

McNair, D.M., Lorr, M., and Droppleman, L.F., *Profile of Mood States Manual*, Educational and Industrial Testing Service, San Diego, CA, 1971.

Murphy, T.L. et al., The effect of smoking on caffeine elimination: implications for its use as a semiquantitative test of liver function, *Clin. Exp. Pharmacol. Physiol.,* 15(1), 9, 1988.

Nicholson, A.N. and Stone, B.M., Antihistamines: impaired performance and the tendency to sleep, *Eur. J. Clin. Pharmacol.,* 30(1), 27, 1986.

Poon, L. (Ed), *Handbook for Clinical Memory Assessment of Older Adults*, American Psychological Association, Washington, D.C., 1987.

Smith, A., Effects of caffeine on human behavior, *Food Chem. Toxicol.,* 40(9), 1243, 2002.

Wells, A.S., Read, N.W., and Craig, A., Influences of dietary and intraduodenal lipid on alertness, mood, and sustained concentration, *Br. J. Nutr.,* 74(1), 115, 1995.

Wetherell, A., Performance tests, *Environ. Health Perspect.,* 104 (Suppl. 2), 247, 1996.

Wurtman, R.J., Introduction, *J. Psychiatr. Res.,* 17(2), 103, 1982.

2 Consideration of Experimental Design for Studies in Nutritional Neuroscience

Ruth B.S. Harris

CONTENTS

2.1 INTRODUCTION

The objective of this chapter is to outline a series of procedures that can be used to test a food, ingredient, or dietary supplement for beneficial effects on mental performance. Nutrient requirements for optimal development and maintenance of health are well established, as evidenced by the recent publication of new dietary reference intakes for macronutrients (Food and Nutrition Board, 2002). In addition, the need to consume adequate levels of specific nutrients for normal development of the central nervous system has been recognized for many years (Imura and Okada, 1998; Wainwright, 2002). More recently, there has been a growing interest in using nutrition to improve mental function in adults (Bland, 1995; Jorissen and Riedel, 2002). In this situation, the improved function cannot be attributed to growth and development of neural tissue, but is related to preservation of functionality of existing neurons or to changes in the rates of release or uptake of neurotransmitters that modify mood, acuity, or cognition. Many cultures around the world, especially Asian, have a tradition of using foods or herbs for their perceived health benefits or medicinal qualities, but it is only recently that the westernized society has accepted the concept of delivery of health benefits by specific dietary components. This has led to the development of a nutraceutical, or functional food, industry that sells foods or dietary supplements reported to deliver a physical or functional benefit (Jones, 2002). The role of functional foods in clinical nutrition has recently been reviewed by Jones (2002) and he has provided the following definitions for functional foods, nutraceuticals, and natural health products. Functional foods are foods that provide a physiological benefit to the consumer beyond that of simple nutrition, such as reducing risk for chronic disease, but are not intended for disease treatment. Functional foods contain bioactive ingredients that are integral to the food and can easily be incorporated into the diet. Nutraceuticals possess bioactive components that are independent of food matrices and are consumed in pills, capsules, or ampoules. Capsules containing linoleic acid are an example of a nutraceutical (Jones, 2002). In contrast, natural health products have been consumed for a long time and can be considered medicinal foods that may be used to treat disease. St. John's Wort is reported to be an effective antidepressant (Bilia et al., 2002) and is an example of a natural health product (Jones, 2002).

Health claims made on food products are regulated and have to be supported by scientific evidence but there is little regulation of claims made for nutritional supplements (Hathcock, 2001), and, in the U.S., it is within this category that some products report benefits that include improved memory, relaxation, or decreased stress. Although many of these claims may prove to be invalid, some of the products may enhance a specific aspect of mental function. This chapter describes the steps that have to be taken to scientifically validate the relation between consumption of a specific nutrient or dietary supplement and improved mental function.

2.2 ASSOCIATION OF CONSUMPTION WITH BEHAVIOR

The first steps in the process of associating a cognitive or behavioral benefit with a specific nutrient or dietary ingredient are to identify the source of the information that initiated the claim, evaluate its scientific validity, and determine whether the bioactive component of the food has been identified. With claims that have the least scientific validity, the only information available on the potential performance benefit of a particular product may be promotional literature provided by the manufacturer. In this situation, it is important to determine whether any scientific evidence that is cited is published and whether it was performed, or sponsored by, the company selling the product. If the research was supported by the manufacturer, then it will be limited to one specific preparation and positive outcomes may not be applicable to other manufacturing conditions. In the opposite extreme, there is a relatively large volume of literature for foods, such as garlic (Amagase et al., 2001; Kik et al., 2001), that have been reported to have multiple health benefits and that also are widely used culinary ingredients. The scientific literature on these foods or ingredients will range from epidemiology (Fleischauer and Arab, 2001) to the basic chemistry of the active compound (Bartholomeus Kuettner et al., 2002) and will incorporate information on multiple preparations and sources (Imai et al., 1994). In this situation, it is necessary to critically review the information, determine its relevance to the interest at hand, and identify issues that remain to be resolved.

Once it is established that there is a need to demonstrate a direct relation between consumption of a specific food or ingredient and an improvement in behavior or performance, it is necessary to identify the appropriate animal model for behavioral measures. The model will be determined both by the behavior that needs to be quantified and by evidence that the animal behavior is representative of the anticipated human behavioral response. If the literature associating behavior and consumption of the food is minimal, or nonexistent, then the first series of studies will be on a small animal model, usually a rodent. The benefits of conducting studies with rodents rather than primates or human subjects include cost, number of replicates, and the ability to conduct invasive tests, such as measurement of concentrations of neurotransmitters in specific brain nuclei. Obviously, the primary criteria for selecting the experimental model are the ability to measure the aspect of performance or behavior that is influenced by the food being tested and a demonstration that the behavioral response in the animal correlates with that in humans. Because it is anticipated that consumption of the test food or supplement will demonstrate a gain of function, the conditions of the behavioral test might have to be modified to make it possible to detect an incremental improvement in performance beyond that of control animals.

There are many validated behavioral tests for measuring performance in rats and mice, many of which have recently been described in detail by Crawley (2000). The behavioral tests include measures of fear, anxiety (Blanchard et al., 2001; Holmes, 2001; Richardson, 2000), learning and memory (D'Hooge and De Deyn, 2001), discrimination (De Beun, 1999), and motivation

(Koob, 2000). It is vital that the investigator be fully cognizant of both the limitations of each behavioral measure and the principles behind the measure, because small changes in procedure can result in an inappropriate test, a misinterpretation of the outcome of the test (Gerlai, 2001), or a change in the response of the animal (Peeters et al., 1992). In addition, there might be conflicting interpretations of the same behavior. For example, immobility of a rat in a Porsolt swim test is interpreted both as a coping strategy in stressed animals (Butler et al., 1990) and as defeat in depressed animals (Hilakivi et al., 1989). Many investigators recommend performing a series of different tests for each behavior in order to measure different aspects of a complex behavior and to obtain reliable information on the behavior of the animal (Crawley, 1985; Ivens et al., 1998; van Gaalen and Steckler, 2000).

Another aspect of behavioral tests that should be considered is whether the intervention that has been applied to the experimental group has introduced physical limitations on the animals that would prevent appropriate performance in a behavioral test. In our studies with stressed animals, we found that the mixed physical and psychological stress of restraint combined with water immersion induced a motor deficit in the rats, which prevented testing of their motivational behavior in a paradigm that required lever pressing (Youngblood et al., 1997). Similarly, sucrose preference has been used by many investigators as a measure of anxiety in rats exposed to chronic mild stress. In our hands, we were only able to induce changes in sucrose preference if the rats were water deprived for 12 h before the test. There was no effect of stress on preference if the rats were fully hydrated (Youngblood et al., 1997). A similar consideration is whether the induced change in behavior can be misinterpreted. For example, locomotion and exploration are commonly used measures of anxiety in rats and mice. Stimulants such as amphetamine may lead to an increase in locomotor activity, which has the potential to be misinterpreted as a decreased level of anxiety (Crawley, 1985; Holmes, 2001).

Once it has been determined which animal model is to be used, a final consideration is to select the appropriate strain of animal. There is a significant amount of evidence that different strains of rats and mice have different levels of baseline performance in various behavioral tests. The differences between mouse strains have recently been reviewed by Crawley (2000). C57BL/6J mice have a very low level of anxiety-type behavior compared with A/J mice (Crawley, 1996). In a study that compared performance of four different strains of mice in a battery of anxiety tests, A/J mice consistently showed a high level of anxiety-type behaviors, but the order rankings of the different strains (C57BL/6ChR, C57BL/6J, Swiss Webster/J, and A/J) varied between the different tests (van Gaalen and Steckler, 2000). In a Morris water maze test of acquisition, DBA/2J, C57BL/6J, and CD-1 mice learned the response but BALB/cByJ mice performed poorly (Francis et al., 1995) and FVB/N mice show a significant impairment of learning in the Morris water maze (Royle et al., 1999). These genetic differences in performance in a variety of behavioral tests have now been recognized as an issue in the phenotyping of transgenic mice, in which the background strain of the mouse may have a greater impact on behavior than the absence, or overexpression, of a specific gene (Crawley, 2000; Wilson and Mogil, 2001). In rats, aged Fisher 344 rats perform less well in a Morris water maze than aged Long Evans rats (Lindner and Schallert, 1988) whereas Sprague Dawley rats perform better than Wistar-Kyoto rats (Diana et al., 1994). In addition, Fawn-Hooded rats show more anxiety-type behaviors than Wistar rats (Hall et al., 2000). Examination of the neuroendocrine and behavioral response to stress has shown that Wistar-Kyoto rats have an attenuated noradrenergic and arousal response but a greater adrenal response to stress than Sprague Dawley rats do. In contrast, Lewis rats have an attenuated endocrine response to stress compared with Sprague Dawley rats, but show similar levels of arousal (Pardon et al., 2002). The differences in stress responsiveness and emotionality of rat strains are now being used to advantage to identify the genetic basis of the neuroendocrine response to stress (Mormede et al., 2002).

2.3 VALIDATION OF THE RELATIONSHIP BETWEEN NUTRIENT AND MENTAL FUNCTION

It is essential to clearly demonstrate a direct cause-and-effect relation between consumption of a bioactive ingredient and an improvement in mental function. Selection of the appropriate animal model and behavioral test for evaluation of the anticipated benefit is critical. The efficacy of the beneficial effect of the nutritional supplement on behavior can be evaluated in a variety of ways that are determined by how much is known about the active ingredient, the effective dose, and the site of action. If little information is available, then the first experiment would be a dose–response study with multiple time points.

The method of administration of the test substance depends on its form. If the substance can be dissolved and sterilized, then different doses can be injected peripherally or centrally. Because the focus of this chapter is nutrients that deliver a performance benefit, it would be more appropriate to concentrate on delivering the compound into the gastrointestinal (GI) tract either directly by gavage as a defined dose of solubilized test material or by voluntary consumption as part of a diet. Gavage would be most appropriate for foods or supplements that are expected to be consumed as a snack or a beverage that will deliver an immediate benefit, whereas foods that are expected to improve a specific aspect of mental function on a continuous basis should be incorporated into the diet. The ingredient or supplement would be included in the diet at increasing levels, with the lowest dose delivering a daily intake that is equivalent to a normal daily intake for human subjects (calculated as milligram per kilogram body weight) and may increase up to 1000-fold this intake. Measures of food intake would be used to calculate actual consumption of the ingredient; therefore, it is important to ensure that the active ingredient is incorporated uniformly throughout the diet and that it cannot be easily separated from the rest of the diet. If the ingredient has a strong flavor or induces a specific mouth-feel, it is possible that the highest doses will be unattractive to the animal and food intake will be decreased. If intake declines, then the specificity of the response to sensory properties, rather than induction of sickness or malaise, should be confirmed by carrying out a conditioned taste aversion test (Scalera, 2002). In this test, the control diet and the supplemented diet are each offered to each rat on different days. Each diet is paired with a flavor, presented as a flavored solution at the same time as the food is available. After the rats have learned to associate a specific flavor with each diet, they are allowed to choose between the two flavors that have been paired with the different diets. If the supplemented diet induced malaise, or other negative physiological effects, the animal will avoid the flavor that was paired with that diet. If the reduction in intake was because of low palatability of the food, there will not be any aversion to the flavor that was paired with the supplemented diet. Ideally, a third treatment group will be included as a positive control for induction of malaise, and will receive a compound such as lithium chloride that is known to induce a conditioned aversion response (Rushing et al., 2002). Similar conditioned taste aversion tests can also be performed if the test ingredient is gavaged and it appears that the behavior of the animals is impaired following the highest treatment doses.

The timing and duration of treatment with the bioactive substance will be determined by the temporal aspects of the anticipated benefits. If the ingredient is expected to give an acute boost to mental performance, then behavior would be tested within a short time of the animals being given the ingredient. In this situation, a crossover design study, in which performance is measured both following consumption of the test material and following consumption of a control material, allows each animal to serve as its own control and improves the statistical power of the test. Several time points may have to be evaluated to identify the optimal interval between consumption of the ingredient and the improved mental performance; for example, moderate doses of caffeine improve certain types of memory (Smith et al., 1994), mood, attention, and reaction time (Lieberman, 2001) even in subjects that have not abstained from caffeine consumption before the performance testing (Warburton et al., 2001). The time interval between consumption of caffeine and circulating concentrations of caffeine reaching a plateau is approximately 30 min (Wang and Lau, 1998) and the half-life of caffeine is approximately 3 h (Wang and Lau, 1998). Therefore, the optimal time for testing the effects of caffeine on performance is between 30 and 180 min after oral or gastric administration. In contrast, the time course of testing ingredients that are fed as part of the basal diet and are expected to produce a chronic improvement in mood or cognition would be extended over periods of days or weeks. For example, studies that have evaluated the protective effects of antioxidants on neural function lasted for 1 month in rats (Cotter et al., 1995) and for 24 weeks in older humans (Rai et al., 1991). The extended duration of these studies allows identification of the minimum period of exposure needed to induce a beneficial effect on cognitive function, the time required to obtain the peak benefit, and whether chronic consumption of the ingredient leads to adaptation, resulting in only a transient improvement in mental performance.

2.4 SITE AND MECHANISMS OF ACTION

Once a beneficial effect of a nutrient on mental performance has been convincingly demonstrated in an animal or human study, the site and mode of action need to be identified. It is important to understand the mechanism of the response, as this can highlight potential strategies for enhancing the benefit. Alternatively, it can raise awareness of potential negative side effects that have to be considered. The ingredient may improve mental performance directly by modifying neuronal function within a specific brain nucleus or may act indirectly by influencing the availability of other nutrients or neurochemical precursors.

If the ingredient is thought to act directly and the beneficial behavioral effect has been identified and confirmed, then existing knowledge of the function of specific brain sites may help localize the site that mediates the improvement in performance. If the behavior is complex, or if there is no obvious candidate site for the mediation of a change in a simple behavior, then it may be possible to localize the responsive areas by measuring the expression of early response genes, such as *c-fos* (Wang et al., 2002), in animals treated with the test ingredient. This methodology was used to demonstrate that the stimulants amphetamine, caffeine, and cocaine each produce individual patterns of *c-fos* activation in the rat brain (Johansson et al., 1994), which implies that they work through different mechanisms. In contrast, it may not be possible to define the site of action of some nutrients that produce a global improvement in performance and mood. For example, there are many claims that carbohydrate improves mood and may enhance memory and mental performance. These responses have been attributed to increased glucose availability in neural tissue (Owens et al., 1997), the release of β-endorphins in response to consumption of preferred or "rewarding" foods (Drewnowski et al., 1992), and changes in serotonin metabolism due to increased transport of tryptophan across the blood-brain barrier (Yokogoshi and Wurtman, 1986). The scientific validity of claims for the beneficial effect of carbohydrates on mood has recently been reviewed by Benton (2002). Irrespective of whether or not all the claims are valid, this may be considered an example of a complex response to a single macronutrient. It would be impossible to provide a simple association between consumption of carbohydrate and a change in function in a defined brain area that leads to a specific beneficial change in mental performance or behavior.

If the change in mental performance following consumption of a specific nutrient is associated with a change of function at one specific site, then this can be confirmed by injecting the active factor into the appropriate brain area and monitoring the performance of the animal. Although this type of study is relatively straightforward, it is essential that appropriate controls be incorporated to account for nonspecific responses to the compound and that small volumes of material be applied to reduce the possibility of nonspecific activity in areas surrounding the site of interest. A dose–response study would have to be performed if the previous evidence has been derived from animals consuming the bioactive ingredient and in which the concentration of active fraction reaching the brain has not been determined. It is likely that very small doses of the active compound will be required for an effect following direct application into the brain compared with those needed for an effect following ingestion. Subsequent experiments would investigate the mechanism by which the active fraction modifies the neurochemical activity in that area of the brain. These studies may be performed *in vivo* in an animal model or *in vitro* by using either tissue preparations or cell lines that are representative of the cells that respond to the active compound.

If it is hypothesized that the nutrient or supplement being tested changes the availability of precursors for neurotransmitters, then the concentration of the precursor and the transmitter can be measured in specific areas of the brain. This can be done by dissecting a large block of tissue that includes the nuclei of interest and measuring the concentrations of the precursor and transmitter by high-pressure liquid chromatography (HPLC; Youngblood et al., 1999) or it may be measured directly by microdialysis (Kennedy et al., 2002). There are benefits and limitations to each procedure. Measurement of tissue neurotransmitter concentrations by HPLC has the limitation of using relatively large pieces of tissue, and thereby site specificity is lost. In contrast, microdialysis measures the temporal pattern of release of neurochemicals in specific sites in a conscious animal. Microdialysis involves the placement of a semipermeable membrane in a specific area of the brain. Fluid from the extracellular space diffuses into the probe and is collected and analyzed. This methodology permits analysis of temporal changes in neurotransmitter release in response to a specific treatment (Smagin et al., 1997) or during the performance of a specific task (Fried et al., 2001). For example, microdialysis was used to demonstrate that oral consumption of caffeine increased hippocampal acetylcholine release by antagonism of adenosine receptors (Carter et al., 1995). One limitation of the procedure is that collection of the neurotransmitter is dependent on diffusion down a concentration gradient into the probe. Therefore, it has to be assumed that diffusion is independent of everything except the amount of neurotransmitter released into the extracellular compartment. A second practical issue is that the sample volume is very small, and therefore eluates have to be pooled over periods of 10 to 20 min (Smagin et al., 1997; Bjorvatn et al., 2002; Pain et al., 2002), which limits the temporal specificity of analysis made by traditional methods such as HPLC. The recent evolution of capillary electrophoresis with laser-induced fluorescent analysis has largely resolved this problem as extremely small samples of eluate are required for analysis and frequent time points can be measured (Kennedy et al., 2002). Microdialysis also is labor intensive and only a small number of animals, or subjects, can be measured at one time. For both methodologies, the interpretation of changes in concentration of a specific neurotransmitter, or precursor, is limited because it may result from an increased rate of synthesis or a decreased rate of degradation of the product.

Additional studies may determine whether the factor of interest exaggerates or inhibits the activity of a specific neurotransmitter. For example, slices of rat cortex were used to demonstrate a biphasic effect of caffeine on acetylcholine release (Pedata et al., 1984) and synaptosomes prepared from rat brain tissues were used to demonstrate that the activity of different stimulants, including ephedrine, was associated with release of norepinephrine rather than dopamine or serotonin (Rothman et al., 2001). *In vitro* systems also may provide information on the functions of bioactive compounds that are not directly dependent on the modification of neurotransmitter activity. A recent study examined the neuroprotective and neurotrophic activity of plant-derived phytoestrogens in primary cultures of fetal rat hippocampal cells (Zhao et al., 2002). The results demonstrate that the phytoestrogens have neuroprotective effects but are much less effective than endogenous estrogens (Zhao et al., 2002). In another *in vitro* study, the protective effects of plant-derived antioxidants against neural cell death caused by free radicals were demonstrated by using a neurite cell line (Mazzio et al., 2001).

Recent advances in genetic techniques have provided additional tools for determining the central mechanism that facilitates a change in behavior. Holmes (2001) reviewed evidence from transgenic and gene knockout mice, which has contributed to understanding the neurotransmitters involved in mediating anxiety-type behaviors. Similarly, Silva et al. (1997) have described recent insights provided by gene targeting techniques into the mechanisms that mediate learning and memory processes in the hippocampus. Depletion of single components of a receptor-induced signaling pathway makes it possible to determine not only which neurotransmitters and receptors are essential for the improvement in behavior or mental performance but also which proteins downstream of the receptor are essential for the response. For example, Linskog et al. (2002) recently demonstrated a critical role for DARPP-32 (dopamine- and cyclic AMP-regulated phosphoprotein) in caffeine-induced stimulation of locomotor activity in mice. The stimulatory effect of caffeine in mice that were deficient in DARPP-32 was greatly inhibited and equivalent to that seen in mice treated with an adenosine receptor antagonist (Lindskog et al., 2002).

There is some potential for false negative results in mice in which a protein is either deleted or overexpressed in all cells of a developing animal. Many systems have functional redundancy such that deletion of one protein does not result in the expected phenotype of a knockout mouse (Crawley, 1996). Alternatively, deletion of a specific protein in a developing animal may be compensated for by an increased level of expression of an associated protein (Crawley, 1996). This potentially confounding aspect of gene deletion can be overcome by using tissue-specific or timed induction of the genetic mutation. Site specificity of gene deletion can be achieved by flanking the gene of interest with *loxP* sequences in one line of animals. A second line of mice carries the enzyme Cre recombinase linked to a promoter that is specific for the tissue of interest. Cre deletes the gene flanked by *loxP*; therefore, crossing the two mutants results in progeny in which a gene is deleted in a specific tissue type. Gene deletion in adult animals overcomes the confounds introduced by developmental compensation for gene mutation and can be achieved by linking the promoter for a gene that is to be overexpressed, or Cre for gene deletion, to systems that are induced by tetracycline or interferon (Kuhn et al., 1995). Combination of site-specific promoters with a tetracycline-sensitive system provides control over both the time and site of gene mutation, providing a powerful tool for examining the functional importance of specific proteins. Antisense technology provides an alternative means of inducing site- and time-specific knockdown of proteins (Ho and Hartig, 1999; Roth and Yarmush, 1999). Repeated application of an antisense oligonucleotide can inhibit mRNA expression for a specific protein. Specificity of the response is controlled by including animals treated with missense oligonucleotide or with mixed-base oligonucleotide (Liebsch et al., 1999). The advantages of this procedure are that it is much less expensive and faster than developing lines of transgenic mice and it can be used in rats, which are rarely used for transgenic studies. One disadvantage is that the efficiency of mRNA knockdown is usually 10 to 40% (Heinrichs et al., 1997; Smagin et al., 1998), compared with total deletion of a gene in a knockout mouse.

2.5 SOURCE OF NUTRIENT

Early in the process of evaluating the performance-enhancing attributes of a food or ingredient it becomes essential to identify the active compound in the food. This is important for a number of reasons, including having the ability to titrate dose against response and to allow comparison of efficacy of different preparations of the same food. For foods that have a long history of association of delivering benefits, there will be a substantial literature on the nature of the active ingredient and assays for its detection may be well established. For example, the active components in garlic are sulfur-containing compounds. In intact cloves of garlic, one of the major sulfur containing compounds is alliin, an *s*-alk(en)yl-L-cysteine sulfoxide (Amagase et al., 2001). When the cell walls of the garlic are broken, the enzyme alliinase is released and lyses the alliin to alliicin, which immediately degrades to diallyl sulfides, diallyl disulfides, dially trisulfides, and ajoene (Amagase et al., 2001). The bioactivity of each of the sulfur compounds in garlic has not been clearly defined, but assays have been developed for alliin and for many of its metabolites, making it possible to evaluate and compare the composition of different garlic preparations and to associate the health benefits of garlic with tissue and serum concentrations of specific compounds (Amagase et al., 2001). *Ginkgo biloba* is reported to improve peripheral and brain blood flow (Chung et al., 1999; Zhang et al., 2000), to improve learning and memory in animal (Topic et al., 2002) and human subjects (Kennedy et al., 2001), and to slow the rate of mental degeneration in Alzheimer's patients

(Oken et al., 1998). Although the bioactive ingredients in *Ginkgo biloba* have not been fully identified (Bedir et al., 2002), the ability to precisely measure the concentration of flavonoids and terpenoids in ginkgo allows a quantitative comparison among different preparations (Li et al., 2002).

If there is no information available on the nature of the active compound in an ingredient or supplement being tested, it may be difficult to develop an appropriate chemical assay that correlates with the bioactivity of the material. In this situation, the behavioral test being used to demonstrate the performance enhancement also may have to be used as a bioassay. For example, different sources of garlic have been tested for their anticarcinogenic effect in a rat model of chemically induced mammary tumors (Ip and Lisk, 1994). Bioassays tend to include high levels of within- and between-assay variability, are imprecise, and, at best, are semiquantitative. If the change in mental performance is associated with a measurable change at the cellular level, such as an increased release of a neurotransmitter, a change in cellular metabolism, or inhibition of ligand binding to a specific receptor, then it may be possible to develop an *in vitro* bioassay. For example, primary culture of fetal neuronal cells has been used to demonstrate the neuroprotective effects of the antioxidant α-lipoic acid (Zhang et al., 2001). *In vitro* procedures will be inherently less variable than an assay that requires behavioral testing in animals, and, if the assay uses a cell line rather than primary cell culture, the interassay variability also will be minimized. The bioactivity of extracts from feverfew has been compared by using a leukocyte bioassay (Brown et al., 1997).

If the source of bioactivity in the food or ingredient is not known, then at some point it will be necessary to commit a significant effort to identifying the active compound. This can be done by subjecting the food to a series of extraction procedures with progressive removal of different classes of compounds followed by testing each extracted fraction for its bioactivity. The various traditional and nontraditional methods for extracting bioactive ingredients from herbs and plant materials have recently been reviewed by Vinatoru et al. (1997, 2001), and the procedures for extracting antitumor polysaccharides from medicinal plants has been described by Wong et al. (1994). The task of identifying the active compound can become complicated if the compound is unstable when subjected to the physical and chemical procedures used for extraction or if bioactivity results from a combination of factors and loss of one of these factors results in a significant reduction in functionality. When the bioactive fraction is unknown and its concentration in the ingredient is unknown, the extracts should be tested in amounts that are proportionally equivalent to the amount of the intact ingredient that was functional in the bioassay. Once the active component has been identified, it is possible to develop a quantitative assay that can be used to test different sources of ingredient for the presence of the bioactive material. It is essential that the chemical or biochemical assay detects either the bioactive form of the compound or a compound that is present in absolute proportion to the active ingredient, because measurement of unrelated compounds will result in a poor prediction of the functional benefits of different sources of the ingredient or supplement.

2.6 OPTIMIZING THE RELATION BETWEEN CONSUMPTION AND ENHANCEMENT OF PERFORMANCE

Once the site and mechanisms of action of the active compound have been determined, other dietary and physiological factors can be modulated to optimize the beneficial effect that is delivered by consumption. Because we are considering the evaluation of ingredients or foods that are consumed, the digestibility and bioavailability of the ingredient become an important issue. If the factor is active in the brain, the bioavailability is determined both by absorption from the GI tract and by the rate of transport across the blood-brain barrier. Absorption from the GI tract will be determined by the digestibility of the food itself and possibly by other components of the diet that are consumed at the same time. Dietary fat content will influence the efficiency of absorption of fat-soluble substances; in contrast, high-fiber diets may lower the absorption of some factors due both to a decrease in intestinal transit time and a potential for the fiber to complex the active compound and inhibit digestion and absorption.

Because this book focuses on nutrients that enhance mental performance, the mechanisms by which the ingredient or supplement influences brain function also have to be considered. Some factors will cross the blood-brain barrier, where they modify neurotransmitter release or uptake or may agonize or antagonize specific receptors. For example, it has been shown that a diet containing a high concentration of vitamin E results in increased levels of vitamin E in various areas of the brain, including the hippocampus (Joseph et al., 1998). The increased vitamin E concentration is associated with decreased indices of oxidative stress in brain tissue and improved performance in a Morris water maze task (Joseph et al., 1998). Other factors may influence brain function indirectly by inhibiting or enhancing the transport of precursors of neurotransmitters across the blood-brain barrier. For example, increasing the concentration of valine in the diet can reduce the concentration of tryptophan in certain areas of the brain by competing for transport across the blood-brain barrier (Youngblood et al., 1999). Whether or not the bioactive ingredient crosses the blood-brain barrier, it will be important to identify the nutritional status that optimizes activity of the compound.

One factor that may influence the response is the time of day that the ingredient is consumed. Many hormones and neurotransmitters have a circadian or pulsatile pattern of release. If one of these endogenous proteins either enhances or inhibits the action of the bioactive ingredient, then careful synchronization of consumption with the peak of a synergistic factor, or trough of an antagonist, has the potential to enhance activity. For example, transient elevations of glucocorticoids may facilitate development of long-term memory (Sandi et al., 1997). There is also some evidence of diurnal variation in certain types of mental function, including cognition, that appears to be associated with arousal state (Blake, 1967) and may be secondary to hormonal status. A more important issue is to determine the frequency with which the bioactive compound should be consumed. If the factor acts centrally and has to cross the blood-brain barrier, its concentration in the circulation needs to be maintained at, or above, a level that ensures efficient transport into the brain, due to either passive diffusion down a concentration gradient or by active transport.

2.7   TRANSFER TO CLINICAL TRIALS

Once a reliable relation between consumption of a particular nutrient or supplement and improvement in mental function has been established, the optimal ingredient preparation, frequency, and time of dosing have been identified, and the mechanism of action has, at least partially, been elucidated in an animal model, the ingredient will have to be tested in human subjects. There are well-established guidelines for demonstrating efficacy and safety of new drugs and defined criteria that have to be met before a health claim can be made for a food product. In contrast, dietary supplements, which include nutraceuticals and natural health products, are subject to much less stringent legal control (Hathcock, 2001). A more recent concern is that nutraceuticals with a demonstrated health benefit may have the unintended side effect of modifying activity of other prescribed medications. For example, St. John's Wort is accepted by many as being an effective antidepressant, but it also appears to activate the P450 enzyme system, which promotes clearance of other drugs (Bilia et al., 2002).

Irrespective of the legal and commercial aspects of demonstrating a relation between consumption of a food or supplement and improvement of mental performance, clinical trials will be essential for any product that is intended to be used to effect by humans. These trials will not only confirm the results from animal studies that show a beneficial effect of the food or ingredient on mental performance, but also determine the most effective dose and frequency of consumption of the ingredient in subjects consuming their normal diet. Testing in humans in a free-living condition will account for any unanticipated interactions among diet, lifestyle, and improvement of mental function.

REFERENCES

Amagase, H., Petesch, B.L., Matsuura, H., Kasuga, S., Itakura, Y. (2001) Intake of garlic and its bioactive components. *Journal of Nutrition,* 131, 955S–962S.

Bartholomeus Kuettner, E., Hilgenfeld, R., Weiss, M.S. (2002) Purification, characterization, and crystallization of alliinase from garlic. *Archives of Biochemistry and Biophysics,* 402, 192–200.

Bedir, E., Tatli, I.I., Khan, R.A., Zhao, J., Takamatsu, S., Walker, L.A., Goldman, P., Khan, I.A. (2002) Biologically active secondary metabolites from *Ginkgo biloba. Journal of Agricultural Food Chemistry,* 50, 3150–3155.

Benton, D. (2002) Carbohydrate ingestion, blood glucose and mood. *Neuroscience and Biobehavioral Reviews,* 26, 293–308.

Bilia, A.R., Gallori, S., Vincieri, F.F. (2002) St. John's Wort and depression: efficacy, safety and tolerability-an update. *Life Sciences,* 70, 3077–3096.

Bjorvatn, B., Gronli, J., Hamre, F., Sorensen, E., Fiske, E., Bjorkum, A.A., Portas, C.M., Ursin, R. (2002) Effects of sleep deprivation on extracellular serotonin in hippocampus and frontal cortex of the rat. *Neuroscience,* 113, 323–330.

Blake, M. (1967) Time of day effects on performance in a range of tasks. *Psychonomic Science,* 9, 349–350.

Blanchard, D.C., Griebel, G., Blanchard, R.J. (2001) Mouse defensive behaviors: pharmacological and behavioral assays for anxiety and panic. *Neuroscience and Biobehavioral Reviews,* 25, 205–218.

Bland, J.S. (1995) Psychoneuro-nutritional medicine: an advancing paradigm. *Alternate Therapies in Health and Medicine,* 1, 22–27.

Brown, A.M., Edwards, C.M., Davey, M.R., Power, J.B., Lowe, K.C. (1997) Pharmacological activity of feverfew (*Tanacetum parthenium* (L.) Schultz-Bip.): assessment by inhibition of human polymorphonuclear leukocyte chemiluminescence in-vitro. *Journal of Pharmacy and Pharmacology,* 49, 558–561.

Butler, P.D., Weiss, J.M., Stout, J.C., Nemeroff, C.B. (1990) Corticotropin-releasing factor produces fear-enhancing and behavioral activating effects following infusion into the locus coeruleus. *Journal of Neuroscience,* 10, 176–183.

Carter, A.J., O'Connor, W.T., Carter, M.J., Ungerstedt, U. (1995) Caffeine enhances acetylcholine release in the hippocampus in vivo by a selective interaction with adenosine a1 receptors. *Journal of Pharmacology and Experimental Therapeutics,* 273, 637–642.

Chung, H.S., Harris, A., Kristinsson, J.K., Ciulla, T.A., Kagemann, C., Ritch, R. (1999) *Ginkgo biloba* extract increases ocular blood flow velocity. *Journal of Ocular Pharmacological and Therapeutics,* 15, 233–240.

Cotter, M.A., Love, A., Watt, M.J., Cameron, N.E., Dines, K.C. (1995) Effects of natural free radical scavengers on peripheral nerve and neurovascular function in diabetic rats. *Diabetologia,* 38, 1285–1294.

Crawley, J.N. (1985) Exploratory behavior models of anxiety in mice. *Neuroscience and Biobehavioral Reviews,* 9, 37–44.

Crawley, J.N. (1996) Unusual behavioral phenotypes of inbred mouse strains. *Trends in Neuroscience,* 19, 181–182.

Crawley, J.N. (2000) *What's Wrong with My Mouse? Behavioral Phenotyping of Transgenic and Knockout Mice.* Wiley-Liss: New York.

De Beun, R. (1999) Hormones of the hypothalamo-pituitary-gonadal axis in drug discrimination learning. *Pharmacology Biochemistry and Behavior,* 64, 311–317.

D'Hooge, R., De Deyn, P.P. (2001) Applications of the Morris water maze in the study of learning and memory. *Brain Research: Brain Research Reviews,* 36, 60–90.

Diana, G., Domenici, M.R., Loizzo, A., Scotti de Carolis, A., Sagratella, S. (1994) Age and strain differences in rat place learning and hippocampal dentate gyrus frequency-potentiation. *Neuroscience Letters,* 171, 113–116.

Drewnowski, A., Krahn, D.D., Demitrack, M.A., Nairn, K., Gosnell, B.A. (1992) Taste responses and preferences for sweet high-fat foods: evidence for opioid involvement. *Physiology & Behavior,* 51, 371–379.

Fleischauer, A.T., Arab, L. (2001) Garlic and cancer: a critical review of the epidemiologic literature. *Journal of Nutrition,* 131, 1032S–1040S.

Food and Nutrition Board, IOM (2002) *Dietary Reference Intakes for Energy, Carbohydrates, Fiber, Fat, Protein and Amino Acids (Macronutrients).* The National Academies Press: Washington, D.C.

Francis, D.D., Zaharia, M.D., Shanks, N., Anisman, H. (1995) Stress-induced disturbances in Morris water-maze performance: interstrain variability. *Physiology & Behavior,* 58, 57–65.

Fried, I., Wilson, C.L., Morrow, J.W., Cameron, K.A., Behnke, E.D., Ackerson, L.C., Maidment, N.T. (2001) Increased dopamine release in the human amygdala during performance of cognitive tasks. *Nature Neuroscience,* 4, 201–206.

Gerlai, R. (2001) Behavioral tests of hippocampal function: simple paradigms complex problems. *Behavioral Brain Research,* 125, 269–277.

Hall, F.S., Huang, S., Fong, G.W., Sundstrom, J.M., Pert, A. (2000) Differential basis of strain and rearing effects on open-field behavior in Fawn Hooded and Wistar rats. *Physiology & Behavior,* 71, 525–532.

Hathcock, J. (2001) Dietary supplements: how they are used and regulated. *Journal of Nutrition,* 131, 1114S–1117S.

Heinrichs, S.C., Lapsansky, J., Lovenberg, T.W., De Souza, E.B., Chalmers, D.T. (1997) Corticotropin-releasing factor CRF1, but not CRF2, receptors mediate anxiogenic-like behavior. *Regulatory Peptides,* 71, 15–21.

Hilakivi, L.A., Ota, M., Lister, R.G. (1989) Effect of isolation on brain monoamines and the behavior of mice in tests of exploration, locomotion, anxiety and behavioral "despair." *Pharmacology Biochemistry and Behavior,* 33, 371–374.

Ho, S.P., Hartig, P.R. (1999) Antisense oligonucleotides for target validation in the CNS. *Current Opinion in Molecular Therapeutics,* 1, 336–343.

Holmes, A. (2001) Targeted gene mutation approaches to the study of anxiety-like behavior in mice. *Neuroscience and Biobehavioral Reviews,* 25, 261–273.

Imai, J., Ide, N., Nagae, S., Moriguchi, T., Matsuura, H., Itakura, Y. (1994) Antioxidant and radical scavenging effects of aged garlic extract and its constituents. *Planta Medica,* 60, 417–420.

Imura, K., Okada, A. (1998) Amino acid metabolism in pediatric patients. *Nutrition,* 14, 143–148.

Ip, C., Lisk, D.J. (1994) Enrichment of selenium in allium vegetables for cancer prevention. *Carcinogenesis,* 15, 1881–1885.

Ivens, I.A., Schmuck, G., Machemer, L. (1998) Learning and memory of rats after long-term administration of low doses of parathion. *Toxicological Sciences,* 46, 101–111.

Johansson, B., Lindstrom, K., Fredholm, B.B. (1994) Differences in the regional and cellular localization of c-fos messenger RNA induced by amphetamine, cocaine and caffeine in the rat. *Neuroscience,* 59, 837–849.

Jones, P.J. (2002) Clinical nutrition. 7. Functional foods — more than just nutrition. *Canadian Medical Association Journal,* 166, 1555–1563.

Jorissen, B.L., Riedel, W.J. (2002) Nutrients, age and cognition. *Clinical Nutrition,* 21, 89–95.

Joseph, J.A., Shukitt-Hale, B., Denisova, N.A., Prior, R.L., Cao, G., Martin, A., Taglialatela, G., Bickford, P.C. (1998) Long-term dietary strawberry, spinach, or vitamin E supplementation retards the onset of age-related neuronal signal-transduction and cognitive behavioral deficits. *Journal of Neuroscience,* 18, 8047–8055.

Kennedy, D.O., Scholey, A.B., Wesnes, K.A. (2001) Differential, dose dependent changes in cognitive performance following acute administration of a *Ginkgo biloba/Panax ginseng* combination to healthy young volunteers. *Nutritional Neuroscience,* 4, 399–412.

Kennedy, R.T., Watson, C.J., Haskins, W.E., Powell, D.H., Strecker, R.E. (2002) In vivo neurochemical monitoring by microdialysis and capillary separations. *Current Opinion in Chemical Biology,* 6, 659–665.

Kik, C., Kahane, R., Gebhardt, R. (2001) Garlic and health. *Nutrition, Metabolism and Cardiovascular Diseases,* 11, 57–65.

Koob, G.F. (2000) Animal models of craving for ethanol. *Addiction,* 95, S73–S81.

Kuhn, R., Schwenk, F., Aguet, M., Rajewsky, K. (1995) Inducible gene targeting in mice. *Science,* 269, 1427–1429.

Li, X.F., Ma, M., Scherban, K., Tam, Y.K. (2002) Liquid chromatography-electrospray mass spectrometric studies of ginkgolides and bilobalide using simultaneous monitoring of proton, ammonium and sodium adducts. *Analyst,* 127, 641–646.

Lieberman, H.R. (2001) The effects of ginseng, ephedrine, and caffeine on cognitive performance, mood and energy. *Nutrition Reviews,* 59, 91–102.

Liebsch, G., Landgraf, R., Engelmann, M., Lorscher, P., Holsboer, F. (1999) Differential behavioural effects of chronic infusion of CRH 1 and CRH 2 receptor antisense oligonucleotides into the rat brain. *Journal of Psychiatric Research,* 33, 153–163.

Lindner, M.D., Schallert, T. (1988) Aging and atropine effects on spatial navigation in the Morris Water Task. *Behavioral Neuroscience,* 102, 621–634.

Lindskog, M., Svenningsson, P., Pozzi, L., Kim, Y., Fienberg, A.A., Bibb, J.A., Fredholm, B.B., Nairn, A.C., Greingard, P., Fisone, G. (2002) Involvement of darpp-32 phosphorylation in the stimulant action of caffeine. *Nature,* 418, 774–778.

Mazzio, E., Huber, J., Darling, S., Harris, N., Soliman, K.F. (2001) Effect of antioxidants on L-glutamate and *N*-methyl-4-phenylpyridinium ion induced-neurotoxicity in PC12 cells. *Neurotoxicology,* 22, 283–288.

Mormede, P., Courvoisier, H., Ramos, A., Marissal-Arvy, N., Ousova, O., Desautes, C., Duclos, M., Chaouloff, F., Moisan, M.P. (2002) Molecular genetic approaches to investigate individual variations in behavioral and neuroendocrine stress responses. *Psychoneuroendocrinology,* 27, 563–583.

Oken, B.S., Storzbach, D.M., Kaye, J.A. (1998) The efficacy of *Ginkgo biloba* on cognitive function in Alzheimer disease. *Archives of Neurology,* 55, 1409–1415.

Owens, D.S., Parker, P.Y., Benton, D. (1997) Blood glucose and subjective energy following cognitive demand. *Physiology & Behavior,* 62, 471–478.

Pain, L., Gobaille, S., Schleef, C., Aunis, D., Oberling, P. (2002) In vivo dopamine measurements in the nucleus accumbens after nonanesthetic and anesthetic doses of propofol in rats. *Anesthesia and Analgesia,* 95, 915–919.

Pardon, M.C., Gould, G.G., Garcia, A., Phillips, L., Cook, M.C., Miller, S.A., Mason, P.A., Morilak, D.A. (2002) Stress reactivity of the brain noradrenergic system in three rat strains differing in their neuroendocrine and behavioral responses to stress: implications for susceptibility to stress-related neuropsychiatric disorders. *Neuroscience,* 115, 229–242.

Pedata, F., Pepeu, G., Spignoli, G. (1984) Biphasic effect of methylxanthines on acetylcholine release from electrically stimulated brain slices. *British Journal of Pharmacology,* 83, 69–73.

Peeters, B.W., Smets, R.J, Broekkamp, C.L. (1992) The involvement of glucocorticoids in the acquired immobility response is dependent on the water temperature. *Physiology & Behavior,* 51, 127–129.

Rai, G.S., Shovlin, C., Wesnes, K.A. (1991) A double-blind, placebo controlled study of *Ginkgo biloba* extract ("tanakan") in elderly outpatients with mild to moderate memory impairment. *Current Medical Research and Opinion,* 12, 350–355.

Richardson, R. (2000) Shock sensitization of startle: learned or unlearned fear? *Behavioral Brain Research,* 110, 109–117.

Roth, C.M., Yarmush, M.L. (1999) Nucleic acid biotechnology. *Annual Review of Biomedical Engineering,* 1, 265–297.

Rothman, R.B., Baumann, M.H., Dersch, C.M., Romero, D.V., Rice, K.C., Carroll, F.I., Partilla, J.S. (2001) Amphetamine-type central nervous system stimulants release norepinephrine more potently than they release dopamine and serotonin. *Synapse,* 39, 32–41.

Royle, S.J., Collins, F.C., Rupniak, H.T., Barnes, J.C., Anderson, R. (1999) Behavioural analysis and susceptibility to CNS injury of four inbred strains of mice. *Brain Research,* 816, 337–349.

Rushing, P.A., Seeley, R.J., Air, E.L., Lutz, T.A., Woods, S.C. (2002) Acute 3rd-ventricular amylin infusion potently reduces food intake but does not produce aversive consequences. *Peptides,* 23, 985–988.

Sandi, C., Loscertales, M., Guaza, C. (1997) Experience-dependent facilitating effect of corticosterone on spatial memory formation in the water maze. *European Journal of Neuroscience,* 9, 637–642.

Scalera, G. (2002) Effects of conditioned food aversions on nutritional behavior in humans. *Nutritional Neuroscience,* 5, 159–188.

Silva, A.J., Smith, A.M., Giese, K.P. (1997) Gene targeting and the biology of learning and memory. *Annual Review of Genetics,* 31, 527–546.

Smagin, G.N., Howell, L.A., Ryan, D.H., De Souza, E.B., Harris, R.B. (1998) The role of CRF2 receptors in corticotropin-releasing factor- and urocortin-induced anorexia. *Neuroreport,* 9, 1601–1606.

Smagin, G.N., Zhou, J., Harris, R.B., Ryan, D.H. (1997) CRF receptor antagonist attenuates immobilization stress-induced norepinephrine release in the prefrontal cortex in rats. *Brain Research Bulletin,* 42, 431–434.

Smith, A., Kendrick, A., Maben, A., Salmon, J. (1994) Effects of breakfast and caffeine on cognitive performance, mood and cardiovascular functioning. *Appetite,* 22, 39–55.

Topic, B., Tani, E., Tsiakitzis, K., Kourounakis, P.N., Dere, E., Hasenohrl, R.U., Hacker, R., Mattern, C.M., Huston, J.P. (2002) Enhanced maze performance and reduced oxidative stress by combined extracts of *Zingiber officinale* and *Ginkgo biloba* in the aged rat. *Neurobiology of Aging,* 23, 135–143.

van Gaalen, M.M., Steckler, T. (2000) Behavioural analysis of four mouse strains in an anxiety test battery. *Behavioral Brain Research,* 115, 95–106.

Vinatoru, M. (2001) An overview of the ultrasonically assisted extraction of bioactive principles from herbs. *Ultrasonics Sonochemistry,* 8, 303–313.

Vinatoru, M., Toma, M., Radu, O., Filip, P.I., Lazurca, D., Mason, T.J. (1997) The use of ultrasound for the extraction of bioactive principles from plant materials. *Ultrasonics Sonochemistry,* 4, 135–139.

Wainwright, P.E. (2002) Dietary essential fatty acids and brain function: a developmental perspective on mechanisms. *Proceedings of the Nutrition Society,* 61, 61–69.

Wang, H., Shu, S.Y., Bao, X.M., Yang, W.K. (2002) Expression of immediate-early genes c-fos and c-jun in the marginal division of striatum during learning and memory. *Academic Journal of the First Medical College of PLA,* 22, 9–12.

Wang, Y., Lau, C.E. (1998) Caffeine has similar pharmacokinetics and behavioral effects via the IP and PO routes of administration. *Pharmacology Biochemistry and Behavior,* 60, 271–278.

Warburton, D.M., Bersellini, E., Sweeney, E. (2001) An evaluation of a caffeinated taurine drink on mood, memory and information processing in healthy volunteers without caffeine abstinence. *Psychopharmacology,* 158, 322–328.

Wilson, S.G., Mogil, J.S. (2001) Measuring pain in the (knockout) mouse: big challenges in a small mammal. *Behavioral Brain Research,* 125, 65–73.

Wong, C.K., Leung, K.N., Fung, K.P., Choy, Y.M. (1994) Immunomodulatory and anti-tumour polysaccharides from medicinal plants. *Journal of International Medical Research,* 22, 299–312.

Yokogoshi, H., Wurtman, R.J. (1986) Meal composition and plasma amino acid ratios: effect of various proteins or carbohydrates, and of various protein concentrations. *Metabolism,* 35, 837–842.

Youngblood, B.D., Ryan, D.H., Harris, R.B. (1997) Appetitive operant behavior and free-feeding in rats exposed to acute stress. *Physiology & Behavior,* 62, 827–830.

Youngblood, B.D., Smagin, G.N., Elkins, P.D., Ryan, D.H., Harris, R.B. (1999) The effects of paradoxical sleep deprivation and valine on spatial learning and brain 5-HT metabolism. *Physiology & Behavior,* 67, 643–649.

Zhang, J., Fu, S., Liu, S., Mao, T., Xiu, R. (2000) The therapeutic effect of *Ginkgo biloba* extract in shr rats and its possible mechanisms based on cerebral microvascular flow and vasomotion. *Clinical Hemorheology and Microcirculation,* 23, 133–138.

Zhang, L., Xing, G.Q., Barker, J.L., Chang, Y., Maric, D., Ma, W., Li, B.S., Rubinow, D.R. (2001) Alpha-lipoic acid protects rat cortical neurons against cell death induced by amyloid and hydrogen peroxide through the AKT signalling pathway. *Neuroscience Letters,* 312, 125–128.

Zhao, L., Chen, Q., Diaz, B.R. (2002) Neuroprotective and neurotrophic efficacy of phytoestrogens in cultured hippocampal neurons. *Experimental Biology and Medicine,* 227, 509–519.

3 Assessment of Animal Behavior

Rachel L. Galli

CONTENTS

3.1 INTRODUCTION

Nutritional neuroscience, with its focus on the workings of the nervous system, is often concerned with behavioral outcomes. Assessment of animal behavior encompasses a huge variety of methodologies and species. [For a review see Waddell and Desai (1981).] This chapter provides a general overview of some of the most frequently used paradigms with the most commonly used animal model, the rat. Conditioned and unconditioned behavior and animal models of human disorders are addressed as to methodology, variables, and experimental considerations.

Many laboratory animals have been used in neuroscience research, ranging from worms and flies to hedgehogs and primates; however, the vast majority of behavioral studies are performed with rodents. Using rats for experimental purposes has many advantages. Rats are relatively easy and inexpensive to obtain and house. They are hardy, reproduce readily, and their lifespan of 2 to 3 years facilitates longitudinal studies. There is a large body of literature available regarding their dietary requirements, genetics, anatomy, physiology, and metabolism. They are smart and docile when handled correctly. Rats are omnivorous, continue to gain weight throughout their lifespan, and have the same amino acid requirements as humans do, particular benefits for nutrition research.

When selecting laboratory animals or interpreting results, it can be important to consider strain differences. Of the more than 200 known strains of rat, the following five are most frequently used in research: Fischer 344, Sprague-Dawley, Long-Evans, Wistar, and Brown Norway (Weindruch and Masoro, 1991; van der Staay, 2000). Experiments have demonstrated that along with age, sex, and health status, strain differences can affect physiological and behavioral parameters. [For a review see Kacew et al. (1998).] The use of inbred strains limits genetic variability but can introduce strain-related differences in performance, in response to treatments, and in vulnerability to pathologies.

Housing conditions and feeding regimens can also affect central nervous system (CNS) development and behavior patterns. In evaluating research, it can be important to note whether the animals have been housed individually in standard laboratory cages as is typical or whether they have been housed in groups or in enriched environments. Enriched environments generally include relatively large multioccupant chambers with opportunities for exercise and with a changing array of novel stimuli for the animals to interact with. The presence of cage mates, novel stimuli, and the opportunity for exercise (i.e., a running wheel) can lead to enhanced neuronal complexity (Greenough and Bailey, 1998). The standard laboratory feeding regimen for rats is *ad libitum* (free) access to food and water. Nutritionally balanced rat chows are commercially available, and specially designed diets can be ordered. Although some behavioral tasks maintain animals on restricted food or water intake in order to use food or water as a reward, this is generally applied only for a limited time period before and during testing. As discussed later, in nutritional research it is often desirable to use other forms of motivation where possible.

Experimental considerations extend beyond choice of strain, housing, and feeding regimen to include time of day, lighting, temperature, noise level, odors, and other conditions (Clark et al., 1997). Rats must be accustomed to human contact by repeated exposure to handling before behavioral testing in a process referred to as gentling. Petting and holding the animals for a few minutes each day for several days have a notable effect. The animals quickly become comfortable with being handled and this decreases the stress response, which might otherwise confound results (Schmitt and Hiemke, 1998; Vigas, 1989) Successful analysis of animal behavior requires investigators to have extensive training and practice in methodology, knowledge of testing considerations specific to each task (i.e., sample size, room size, environmental cues), and an understanding of national and institutional regulations regarding the use of laboratory animals. (Information on regulations can be found at www.nal.usda.gov/awic/legislat/awicregs.htm.)

In the study of CNS function, nutrition–brain, and brain–behavior relationships, it is often useful to employ a battery of behavioral tests. This chapter provides an overview of a number of commonly used methods, including information on the interpretation of variables and important testing considerations. First, several examples of unconditioned behavior are discussed, followed by sections on classical conditioning, operant conditioning, maze tasks, and animal models of disorders. Although the behavioral tests are discussed as they apply to rats, all the tasks discussed have been used with mice, as illustrated in the final example. This chapter is not intended as a stand-alone guide to the implementation of behavioral measures. (For more detailed methodological information consult the references cited.) It is intended to provide a general understanding of many of the methodologies in use at present in the behavioral neurosciences.

3.2 UNCONDITIONED BEHAVIOR

This section focuses on the analysis of motor skills and spontaneously emitted behaviors.

3.2.1 GENERAL ACTIVITY

The advent of photo beam activity monitoring cages connected to a computerized recording system now allows for the automated and continuous measure of general motor activity in home cages or novel environments. Depending on the number and placement of the photo beams and the sophistication of the software programming, further divisions of activity into fine movement, ambulatory movement, and rearing can be made based on photo beam interruptions (Casadesus et al., 2001; Yu et al., 1985). Increases or decreases in the level of motor activity and changes in the pattern of activity can provide general but useful information on the effects of nutritional, pharmacological, and environmental manipulations.

3.2.2 EXPLORATORY ACTIVITY

Measurement of exploratory behavior can take place in a variety of environments. [For reviews see Alleva and Sorace (2000) and Kelley et al. (1989).] Most commonly used is the open field. For rats, this is a 1 m² white open box with the floor marked by a grid dividing the region into inner and outer units. Animals are placed individually into the open field and their movement (unit crossings), rearings, and time spent grooming and being inactive are measured either manually or by a computer-assisted image analysis system for a time period of 2 to 30 min. Activity in the open field is a complex behavior interpreted as a measure of motor activity, emotional reactivity, and exploration. Novel objects can be introduced into the field to increase the opportunities for exploration. Behavior patterns in the open field can be affected by circadian rhythm, age, hunger, level of arousal, and food, drug, or neurotransmitter manipulations (Kelley et al., 1989; Schmitt and Hiemke, 1998).

3.2.3 MOTOR SKILLS

A variety of tests can be used to assess motor performance. The four tasks described here are simple in design and have demonstrated reliability for the measure of strength, balance, and coordination (Ingram, 1983; Shukitt-Hale, 1999). Grip strength is measured by the wire hang, in which the rat grips a thick insulated wire and latency to fall to a cushioned surface is timed. Balance and coordination are measured with a rat balance beam. Animals are placed on the center of a narrow plank or a cylindrical rod and the distance that they locomote and the time they remain balanced are recorded. Muscle tone, strength, and balance are tested by using an inclined screen. Each rat is placed in a compartment of wire mesh that is tilted 80° to the horizontal plane of the table. Latency to slide on the screen is timed. The rotorod is a thick, textured rod that slowly rotates at a gradually accelerating pace. The rat is placed onto the rotorod and, like a lumberjack, must balance and adjust its position while the rod rotates. The length of time that the animal retains its position on the accelerating rod before falling to a cushioned surface is recorded as a measure of motor coordination, agility, and balance. Performance on these tasks declines with age and has been shown to respond to dietary supplementation (Galli et al., 2002; Joseph et al., 1999; Shukitt-Hale, 1999).

3.3 CONDITIONED BEHAVIOR

In the study of animal behavior, the terms *conditioned* and *conditioning* are synonymous with *learned* and *learning*.

3.3.1 CLASSICAL CONDITIONING

In classical conditioning, also called Pavlovian conditioning, a neutral stimulus is paired with an unconditioned stimulus, one that automatically elicits an unconditioned (unlearned) response. For example, presenting a food stimulus will elicit salivation: this is a naturally occurring or unlearned response. In classical conditioning, presentation of a neutral stimulus, say the sound of a bell, is paired with presentation of the unconditioned stimulus, in this example, food. After repeated pairings, the sound of the bell will elicit the response of salivation even in the absence of food. Learning has occurred and the sound of bell has become a conditioned stimulus eliciting a learned response; salivation following the bell ring is now a conditioned response. Compared with more complex forms of learning, the neurophysiological substrates that support classical conditioning are fairly well understood, making the process a useful tool to investigate the effects of treatments on cognitive behavior.

Depending on the nature and timing of a dietary manipulation, it may be possible to distinguish treatment effects on learning and memory. Treatments given before and during conditioning may affect acquisition of the learned response. Giving a treatment immediately following the initial training session may affect consolidation, the process by which learned associations become established in long-term memory. Finally, treatments given before retest trials may affect retention.

Recently, there has been a resurgence in the use of one classical conditioning paradigm in particular, the conditioned emotional response (Alleva and Sorace, 2000). A classically conditioned emotional response occurs when a neutral stimulus is repeatedly paired with an aversive stimulus until the initially neutral stimulus alone elicits the same fearful response as the aversive stimulus. The fear response in rats includes increases in heart rate, breathing, and stress hormone levels and a species-typical posture called freezing. Usually, a brief foot shock is used as the aversive stimulus evoking the automatic emotional response and a tone is used as the initially neutral stimulus. After several pairing of the tone with the painful stimulus, the tone alone elicits the conditioned emotional response (Pare, 1969).

One way to assess the effects of a nutritional treatment on conditioning is to measure the strength of the conditioned response after a period of time ranging from minutes to days following the establishment of the response. One can also test the number of trials it takes to extinguish the conditioned response. If the conditioned stimulus, the tone, is repeatedly presented without any further pairing with the unconditioned stimulus, the foot shock, the conditioned emotional response decreases and eventually ends and is said to be extinguished. The number of trials taken to completely extinguish the conditioned emotional response can be used as a measure of how well the association was learned and remembered.

Conditioned place preference is a form of classical conditioning that can be used to assess a rodent's preference between two stimuli (Spiteri et al., 2000; Stohr et al., 1998). The rat is placed in the center of a chamber with three compartments. The small center start compartment has access to two side compartments equivalent in size but with distinctive appearances provided by patterns and textures on the walls and floors and by the lighting. The stimuli to be tested are placed in separate compartments. In the first stage of testing, the rat is placed in the central start area and allowed access only to one side of the apparatus with its stimulus. The process is repeated several times for each stimulus–compartment pair. In the next stage, the rat is placed in the center start area and allowed open access to both sides of the apparatus. The relative amount of time the animal spends in each compartment is seen as indicative of its preference for the stimulus presented in that location. It is important to counterbalance the pairing of stimuli with compartment, and the order of presentation, across subjects (Spiteri et al., 2000).

3.3.2 OPERANT CONDITIONING

Typically, in operant conditioning, also called instrumental conditioning, an animal learns to perform a specific behavior, i.e., a lever press, in order to obtain a reward or to avoid an aversive consequence. In contrast to classical conditioning in which the animal learns associations about predictable signals (the tone predicts the shock), in operant conditioning the subject learns to associate a behavior with its consequences (perform the behavior and get the reward). The relationship or contingency between the behavioral response and its consequence can be manipulated on four basic schedules. The reward or reinforcement can be provided on a fixed ratio, wherein for every specified number of responses one reward is given, or on a variable ratio schedule, wherein the number of responses necessary for reinforcement is varied around a set average. Two other schedules provide reinforcement after a fixed or a variable length of time performing the response.

Most scheduled forms of operant conditioning in rodents take place in a specially designed operant chamber, which was originally developed by Skinner (1966). The animal is placed in an insulated chamber with controlled light, sound, and temperature and the mechanism on which to perform the behavior, i.e., the lever, and the mechanism to deliver the reward. Modern equipment is controlled by computer software that also records and analyzes the characteristics of the animal's behavior. Various treatments will influence the response profile to a particular schedule of reinforcement (Weiss and O'Donoghue, 1994).

The active avoidance task is another form of operant conditioning and it requires a different style of apparatus. In active avoidance, the animal has to perform a motor behavior to avoid an aversive stimulus. Basically, the animal is placed in a chamber with a floor wired to deliver a brief shock and two compartments separated by a low wall. The animal learns to avoid the shock by jumping to the other side of the chamber. The passive avoidance task uses a divided chamber with one brightly lit and one dark compartment with a doorway allowing open access between the two sides (Alleva and Sorace, 2000). The rat is placed in the lighted side and the latency for it to enter the dark compartment is measured. (Rats naturally prefer dark spaces.) Once the animal enters the dark compartment, it receives a brief foot shock and is removed from the apparatus. After a period of time, the animal is returned to the bright compartment and latency to enter the dark side should be increased if the animal has learned the association between moving into the dark compartment and the consequent shock. Treatments that improve learning should enhance performance in both avoidance tasks.

3.3.3 Motor Learning

Motor learning is a category of operant conditioning in which the animal learns a new behavioral response. Motor learning is not simply the occurrence of a new voluntary movement or a repeated pattern of movement; it involves improvements in the execution of the movement as a result of experience. Rats can learn a wide range of motor behaviors; however, one task in particular, the runway rod walking task, has proved to be a useful and reliable measure of motor learning in a laboratory setting (Lalonde and Botez, 1990; Watson and McElligott, 1984).

The runway rod walking task consists of a straight runway with a platform at either end. Rods extend across the runway and provide obstacles to the subject's progress. Initially, the rods are covered by a plexiglass plate and food- or water-deprived rats learn to move from one end of the runway to the other to receive a reward. Once the behavior is established, the cover plate is removed and the animals must negotiate over the evenly spaced obstacles to reach the goal location. To test the effects of a treatment, the rods are placed in a new irregular pattern, which increases the difficulty of the task. Running time is calculated for a set number of trials over several days of testing. This measure of motor learning has already proved sensitive to nutritional manipulation (Bickford et al., 1999).

3.3.4 Simple Maze Learning

In maze learning, the animal moves from a start position to a goal location by learning information about spatial locations. Performance is motivated either by a food or water reward in animals that have been partially deprived to increase their appetite or by escape from an aversive stimulus such as shock or water. Simple maze learning tasks present a limited number of choices; most frequently used is a Y- or T-shaped runway maze in which the animal faces two choices. The demands of the task can be increased by using delays between trials or by manipulating the pattern of reinforcement (Sarter, 1987). Rats in a Y- or T-maze will spontaneously alternate their left and right arm choices. However, increasing the interval of time between trials will reduce spontaneous alternation in a predictable fashion, enabling the testing of treatments that might be expected to improve performance (Ragozzino et al., 1994). Other versions of the simple two-choice mazes begin by blocking one of the arms on the first trial and allowing the rat to receive the reward at the end of the unblocked arm. On the next trial, both arms are open and the reward can be placed in the same arm, asking the subject to match its performance to the previous trial. This design is called *matching to sample* as opposed to having the blocked arm from the sample trial baited, which is called *nonmatching to sample*. By adjusting the intertrial intervals, the difficulty of both designs can be manipulated to a level at which it is possible to measure positive and negative treatment effects (Ingram et al., 1994). Because rats have well-developed spatial abilities, it is important to pilot behavioral tests to avoid ceiling effects that might prevent accurate assessment of nutritional treatments (Galli et al., 2000; Ragozzino et al., 1994).

3.3.5 COMPLEX MAZE LEARNING

Damage to the hippocampal regions of the brain results in impairments in the ability to form new memories in humans. Disrupting the activity of the hippocampus in rats creates deficits in complex spatial learning and memory tasks but leaves simpler forms of learning intact. Since this discovery, a large variety of tests have been designed to assess the navigational skills of rodents (Alleva and Sorace, 2000; McLay et al., 1999; Olton and Samuelson, 1976). Nowadays, the most commonly used tasks are the radial arm maze and the Morris water maze.

The radial arm maze consists of a circular start area from which radiate eight arms (Olton and Samuelson, 1976; Olton, 1987). At the end of each arm is a food cup that may be filled with a food or liquid reward. Not all of the arms are baited, and the rat is placed in the central start position and allowed to explore the arms for a set period of time. After repeated trials over 10 to 15 days, the animal learns the position of the rewards and rarely enters unbaited arms or arms that have already been visited on that trial. Because there are no visual cues within the maze, rats must rely on distal cues for accurate navigation in the maze.

A benefit of the radial arm maze is that it allows for the distinction between reference and working memory errors. Reference memory is a measure of how well the animal has learned the general rules of the task. Reference memory errors are a count of the number of times that the animal visits an unbaited arm. Working memory is assessed by the number of reentries into an arm on a single trial. Working memory relies on information gained within the trial and is short term in nature, whereas reference memory requires long-term storage of information across trials and days.

The effect of a treatment can be measured by the rate of acquisition of the task and the number and type of errors committed compared with control conditions. Alternatively, once animals meet a set performance criterion, e.g., no more than two errors on five consecutive trials, nutritional manipulations may be introduced to assess their effect on retention. Other versions of the radial arm maze differ by the period of time between trials, the number of arms in the apparatus, or the cues available in the environment (Ingram et al., 1994; McLay et al., 1999).

An important consideration for nutritional neuroscience studies involves the use of food or liquid rewards to motivate performance in many complex spatial learning and memory tasks. Appetite-motivated tests require some level of deprivation to the animals before and during testing. One standard regimen is to provide subjects with 80% of their normal food intake; another is to limit access to water to 1 h/day. Also, subjects within a group may get an unequal number of rewards, typically food pellets, fruit loops, or sweet water, based on their individual performance. Depending on the experimental manipulation under investigation, this may be a concern for studies focused on nutrition. To avoid potential problems related to the use of consumable rewards, many researchers employ water mazes to test spatial learning and memory.

3.3.6 MORRIS WATER MAZE

The Morris water maze is a circular pool, approximately 1 m in diameter, filled with room-temperature water (Morris, 1984). A small platform is hidden just below the surface of the water in the center of one of the imaginary quadrants of the pool. Rats are unable to see the platform and must use distal cues to navigate and locate the hidden platform. Rats swim well and are motivated to escape from the water onto the platform. The escape latency (time to locate and escape onto the platform) and path searched are monitored through a video camera connected to image tracking software. A learning curve for performance on each day as well as multiple measures of retention can be evaluated.

Testing in the water maze begins when the rat is placed gently into the pool facing the rim. The trial ends when the rat finds the platform and climbs onto it. If at the end of a set period of time, usually 1 min, the animal has not found the platform, it is guided to it by the experimenter. The rat remains on the platform for 10 to 20 sec. This time allows the rat to observe its position relative to cues around the room — there are no intramaze cues. After six to ten trials in the maze, a healthy young adult rat can swim to the platform in less than 10 sec from any point on the circumference of the pool. Normal rats perform well on spatial learning tasks and it can sometimes be difficult to detect beneficial effects of treatments. Therefore, performance may be artificially impaired with a low dose of scopolamine, a cholinergic antagonist (Galli et al., 2000; Ragozzino et al., 1994; Wongwitdecha and Marsden, 1996).

There are many versions of the Morris water maze task. [For reviews see Brandeis et al. (1989) and D'Hooge and De Deyn (2001).] A typical training regimen consists of 10 trials a day for 2 to 4 days, with a constant platform location and varying start positions. This allows for the application of treatments before, during, and between sessions. The number of trials per day and the platform position, start location, and intertrial interval can be varied to manipulate the difficulty of the task. A probe trial can be introduced; the platform is removed and the areas searched and the number of crossings of the former platform location evaluated to assess search strategies. Reversal learning can be measured by moving the platform to another quadrant of the pool. A working memory procedure trains the subjects for two trials per day with a short intertrial interval for 4 days (Brandeis et al., 1989; Shukitt-Hale, 1999). The platform location is changed each day. Performance in this version of the swim task primarily examines short-term working memory, based on scores on the second of each pair of trials.

Some investigators measure the angle of the swim path as the rat leaves the start position and the amount of time the animal spends within a set area around the platform location (Brandeis et al., 1989; Zyzak et al., 1995). These qualitative measures are not confounded by the motor abilities of the subjects. A visible platform trial can be completed at the end of testing to determine whether there are any differences between groups in the noncognitive abilities needed to successfully find and mount the platform.

Use of the Morris water maze does not require appetitive rewards, and therefore satiety is not a factor in motivation. This permits a greater number of trials to be run per day and shortens the length of time necessary to complete testing (D'Hooge and De Deyn, 2001). Periods of inactivity do not occur in the water maze, thus avoiding the failure to respond that can happen in the radial arm maze. Also, any possible effects of odor cues that may be present in the radial arm maze are removed in water mazes.

3.3.7 RADIAL ARM WATER MAZE

The radial arm water maze combines many of the advantages of both the radial arm and the Morris water mazes (Hodges, 1996; Shukitt-Hale, 2004). The apparatus consists of a round pool fitted with walls to create eight arms radiating out from a central circular area. The arms are uniform in appearance and there are no intramaze cues. The platform goal is hidden just below the surface of the water at the end of one of the arms. Similar to the traditional radial arm maze, this task allows the simultaneous measurement of reference and working memory but avoids the problems presented by the use of consumable rewards. As in the Morris water maze, the task promotes efficient learning and requires fewer days to complete testing as compared with the radial arm maze.

Testing in the radial arm water maze is generally conducted over 3 to 5 days (Hodges, 1996; Shukitt-Hale, 2004). On each trial, the rat is placed gently into the maze at the end of one of the arms that does not contain the platform. The trial ends when the animal finds the platform or at the end of 2 min. If the platform is not found by the end of the trial, the rat is guided to it. Once on the platform, the rat remains there for 15 sec before being removed from the maze. The time on the platform enables the rat to observe its location relative to distal cues in the environment. A video camera connected to an image tracking software collects information on latency to find the platform, swim speed, order of arm entry, and distance traveled per trial. Performance is assessed by calculating reference and working memory errors and differences in latencies between groups. As memory errors are considered to be independent of motor skills, these variables can be used in cases in which experimental conditions cause a difference in swim speed (Bursova et al., 1985; van der Staay and de Jonge, 1993).

The difficulty of the radial arm water maze can be adjusted by varying the training regimen. Increasing the intertrial interval or decreasing the number of trials per day makes the task more challenging. Once subjects learn the location of the platform, a reversal learning paradigm can be introduced. The platform is moved to a new location, and performance, as the animals relearn the task, is monitored as a measure of how well the original position was learned and how well the subjects are able to shift set and learn the new position. It is not possible to use multiple goal platforms on a single trial as rats will not leave a platform to continue searching for a second or third target. Despite its advantages, the radial arm water maze is used by a relatively small number of laboratories, perhaps in part because the apparatus is not commercially available and must be constructed on site.

3.4   ANIMAL MODELS OF DISORDERS

In the behavioral neurosciences, a wide range of animal models of human disorders has been developed. The most relevant models possess face validity, construct validity, and predictive validity. [For reviews see van der Staay (2000) and Willner (1984).] Face validity refers to how well the behavioral symptoms, underlying pathology, and treatment effects seen in the model resemble those present in humans. Construct validity is related to how well the theoretical underpinnings of the model correspond to those of the human condition. Generally, an animal model is considered to have predictive validity if it predicts how well a therapeutic treatment will work in humans. Given the limitations imposed by using nonhuman animals and by our limited understanding of the neurophysiology of most disorders, it is not surprising that few animal models meet the criteria for all three concepts of validity (Redei et al., 2001; Willner, 1984). However, many models have proved useful for assessing treatment strategies and have provided insight into underlying mechanisms relevant to human disease. It is beyond the scope of this chapter to review the many animal models currently in use. However, three examples of rodent models are addressed in some detail: the forced swim test of depression, the elevated plus maze measure of anxiety, and a genetic model of Alzheimer's disease.

3.4.1   FORCED SWIM TEST OF DEPRESSION

The Porsolt forced swim test is the most extensively used assessment of treatments postulated to have an antidepressant effect (Lucki, 1997; Porsolt et al., 1977). In this task, each rat is individually placed into a bucket-shaped container filled with water to a depth that forces the animal to swim and that also prevents any chance of escape over the walls. Initially the animal seeks to escape, swimming and attempting to climb, then swims in place, and eventually becomes immobile, floating on the surface of the water with only occasional movements. When the rat is returned to the water a day later, the length of time until it assumes an immobile posture is shorter than on the first trial. Antidepressant treatments that are effective in humans lengthen the latency to immobility and increase swimming and climbing on the second trial relative to control conditions. The high reliability of this test has resulted in it being used to assess the efficacy and potency of potential treatments (Lucki, 1997; Redei et al., 2001; Willner, 1984).

Although the predictive validity of the forced swim test is high, it does not score as well on the other measures of validity. In regards to construct validity, the predominant theoretical interpretation of the immobility measure of depressed state corresponds to the learned helplessness theory of depression (Maier, 1984; Redei et al., 2001). In animals, a state of learned helplessness is induced by exposure to an inescapable aversive stimulus (Seligman and Maier, 1967). Once animals are in a conditioned state of learned helplessness, they display motivational, emotional, and cognitive deficits. These deficits are postulated to correspond to depression in humans. However, in humans, a state of learned helplessness is thought to rely on attribution style, the way in which a person explains life events (Seligman, 1991). The underlying mechanism by which long-standing thought patterns lead to the emergence of learned helplessness in people is unlikely to be replicated by the process induced by exposure to the forced swim test in rats. Also, the time course of the effect of antidepressant drugs, in which 2 to 4 weeks of medication is typically required for therapeutic activity in people, is not mirrored in the forced swim animal model. Although these discrepancies weaken the construct and face validity of the task, its strong predictive validity and ease of use support its widespread use as an animal model of depression (Maier, 1984; Redei et al., 2001).

3.4.2 ELEVATED PLUS MAZE MEASURE OF ANXIETY

Anxiety in animals is frequently assessed in the elevated plus maze. [For reviews see Hogg (1996) and Wall and Messier (2001).] In fact, pharmaceutical companies routinely use it as a reliable first screen in testing potential antianxiety drugs. The plus maze apparatus is elevated from the floor and consists of four arms radiating out from a central platform at right angles to each other. Two of the arms are enclosed with high walls and two of the arms are open like gangplanks. Rats freely explore the maze, and passage onto the open arms elicits physiological and behavioral signs of anxiety, i.e., increases in stress hormone levels, production of fecal boli, freezing postures, and flight. Anxiogenic manipulations increase the tendency of subjects to remain in the closed arms, whereas anxiolytic treatments promote exploration in the open arms.

Behavioral assessment in the elevated plus maze begins with placement of the rat on the center start platform. The number of open arm entries and the amount of time spent on the open arms are recorded and expressed as percents of total arm entries and total time in the maze, respectively. The maze is thought to evoke both a drive to explore and a state of anxiety and fear, thus creating an approach-avoidance conflict. Open arm entries and time spent on these arms are considered to be measures of anxiety.

Given the sensitivity of the elevated plus maze, it is important to attend to procedural factors that may influence anxiety levels. Housing, transport, and handling, as well as qualities of the apparatus and the testing environment, can affect results (Hogg, 1996). Raised lips around the edges of open arms, sometimes added to decrease the likelihood of falling, have been shown to increase baseline open arm entries and reduce the chance of measuring an anxiolytic effect. Conversely, testing in an unfamiliar or brightly lit area can decrease baseline entries into open arms. This can introduce a floor effect when assessing anxiogenic manipulations. The definition of what constitutes an arm entry can also affect the outcome of studies. Scoring criteria vary between laboratories as to whether two paws or all four paws must cross the threshold to constitute an arm entry. Automated photo beam systems do not allow for this distinction in paw placement and therefore are not recommended for use with the plus maze.

The elevated plus maze has high predictive validity for testing experimentally induced differences in state anxiety in rats. The sensitivity of the measure to drug manipulations of neurotransmitter levels corresponds to pharmacological effects seen in humans. Despite extensive testing of factors relevant to the face and construct validity of the model, debate continues over the interpretation of results (Wall and Messier, 2001). The relations between novelty, fear, exploration (activity), and motivation and their contributions to anxiety as measured in animals and humans remain ambiguous. Nevertheless, the elevated plus maze remains the most widely used animal model of anxiety.

3.4.3 THE APP + PS1 GENETIC MODEL OF ALZHEIMER'S DISEASE

Once the genetic basis of a disorder is discovered, it enables the use of transgenic engineering techniques to create animal models (Taratino and Bucan, 2000). Changes in two genes have been identified as contributing to the inheritable form of early onset Alzheimer's disease (AD). Mutations in the amyloid precursor protein (APP) and presenilin-1 (PS1) genes are linked with the distinctive changes seen in the brain and behavior of individuals with familial AD. [For reviews see Hardy (1997) and St. George-Hyslop (2000).] Based on these findings, an APP + PS1 double transgenic mouse was developed as an animal model of AD (Arendash et al., 2001; Price and Sisodia, 1998).

To make a transgenic mouse, first a copy of the genetic sequence of interest is isolated or synthesized in the laboratory (Tsien, 2000). An *in vitro* technique is used to insert the genetic material into a fertilized egg. The egg is implanted into the mother and, in time, offspring are tested to determine whether they carry one copy of the inserted genetic sequence, the transgene. Mice containing a single copy are crossbred to produce animals with two copies of the transgene. For double transgenic mice, another step of crossbreeding is required. For example, in the case of the APP + PS1 mouse line, a transgenic mouse with mutated copies of the APP gene is bred with a transgenic mouse with mutated copies of the PS1 gene. The resulting offspring are tested to assess their genetic makeup and further evaluated on behavioral and neuroanatomical parameters.

The APP + PS1 double transgenic mouse displays brain pathology characteristic of AD (Citron et al., 1992; Price and Sisodia, 1998). Behavioral evaluation determined that compared with control subjects, the APP+PS1 mice were impaired on the open field task, the balance beam motor test, on spontaneous alternation behavior in the Y-maze, on acquisition of a Morris water maze task, and on measures of working memory in the radial arm water maze (Arendash et al., 2001; Holcomb et al., 1999). There were no differences between groups on the elevated plus maze or on retention in the Morris water maze. The deficits in balance, learning, and memory correspond to human AD symptomatology. Furthermore, several of the behavioral changes were progressive and corresponded to brain neuropathology. The face and construct validity of the APP+PS1 mouse model are supported by the genetic mutations and resulting brain and behavior alterations. The predictive validity of this model is under investigation; however, one recent study found it to be sensitive to a nutritional treatment targeting neuronal and behavioral functioning (Joseph et al., 2003).

3.5 CONCLUSION

This chapter has only skimmed the surface of a number of animal models of normal and abnormal human behavioral processes. The effects of diet on behavior can be subtle; therefore, behavioral measures must be finely tuned for each study, and choosing accurate, reliable, and valid tests is essential. Even when it is possible to study a nutritional manipulation in humans, it can be advantageous to choose laboratory animals for the amount of control possible over diet, individual history, genetics, and environmental conditions. In humans, it can be difficult to attribute any changes in a behavioral response solely to the dietary manipulation, due in part to lack of control and high variability and in part to nonnutritional factors such as the fact that eating is associated with cultural beliefs regarding foods (Kanarek, 1986). Nutritional neuroscience is a particularly challenging field of study given the complex and dynamic interactions among behavior, environment, nutrition, and neural activity. Although all animal models have their shortcomings, they provide a scientific approach to the study of nutrition–brain–behavior relationships. Caution must be used when extrapolating results; however, animal models often represent the best approximations available for studying nutritional manipulations relevant to humans.

REFERENCES

Alleva, E., Sorace, A. (2000) Important hints in behavioural teratology of rodents. *Current Pharmaceutical Design,* 6, 99–126.

Arendash, G., King, D., Gordon, M., Morgan, D., Hatcher, J., Hope, C. et al. (2001) Progressive age-related behavioral impairments in transgenic mice carrying both mutant amyloid precursor protein and presenilin-1 transgenes. *Brain Research,* 891, 42–53.

Bickford, P., Shukitt-Hale, B., Joseph, J. (1999) Effects of aging on cerebellar noradrenergic function and motor learning: nutritional interventions. *Mechanisms of Ageing and Development,* 111, 141–154.

Brandeis, R., Brandys, Y., Yehuda, S. (1989) The use of the Morris water maze in the study of learning and memory. *International Journal of Neuroscience,* 48, 29–69.

Bursova, O., Bures, J., Oitzel, M., Zahalka, A. (1985) Radial maze in the water tank: an aversively motivated spatial working memory task. *Physiology and Behavior,* 34, 1003–1005.

Casadesus, G., Shukitt-Hale, B., Joseph, J. (2001) Automated measurement of age-related changes in the locomotor response to environmental novelty and home-cage activity. *Mechanisms of Ageing and Development,* 122, 1887–1897.

Citron, M., Oltersdorf, T., Haas, C., McConlogue, L., Hung, A., Seubert, P. et al. (1992) Mutation of the beta-amyloid precursor protein in familial Alzheimer's disease increases beta-protein production. *Nature,* 360, 672–674.

Clark, J., Rager, D., Calpin, J. (1997) Animal well-being: an overview of assessment. *Laboratory Animal Science,* 47(6), 580–585.

D'Hooge, R., De Deyn, P. (2001) Applications of the Morris water maze in the study of learning and memory. *Brain Research Reviews,* 36(1), 60–90.

Galli, R., Fine, R., Thorpe, B., Shukitt-Hale, B., Lieberman, H. (2000) Antisense oligonucleotide sequences targeting the muscarinic type 2 acetylcholine receptor enhance performance in the Morris water maze. *International Journal of Neuroscience,* 103, 53–68.

Galli, R., Shukitt-Hale, B., Youdim, K., Joseph, J. (2002) Fruit polyphenolics and brain aging: nutritional interventions targeting age-related neuronal and behavioral deficits. *Annals of the New York Academy of Sciences,* 959, 128–132.

Greenough, W., Bailey, C. (1988) The anatomy of memory: convergence of results across a diversity of tests. *Trends in Neuroscience,* 11, 142–147.

Hardy, J. (1997) Amyloid, the presenilins and Alzheimer's disease. *Trends in Neuroscience,* 20, 154–160.

Hodges, H. (1996) Maze procedures: the radial-arm and water maze compared. *Cognitive Brain Research,* 3, 167–181.

Hogg, S. (1996) A review of the validity and variability of the elevated plus-maze as an animal model of anxiety. *Pharmacology, Biochemistry and Behavior,* 54(1), 21–30.

Holcomb, L., Gordon, M., Jantzen, P., Hsiao, K., Duff, K., Morgan, D. (1999) Behavioral changes in transgenic mice expressing both amyloid precursor protein and presenilin-1 mutations: lack of association with amyloid deposits. *Behavioral Genetics,* 29, 177–185.

Ingram, D. (1983) Toward the behavioral assessment of biological aging in the laboratory mouse: concepts, terminology, and objectives. *Experimental Aging Research,* 9(4), 225–238.

Ingram, D., Jucker, M., Spangler, E. (1994) Behavioral manifestations of aging. In *Pathobiology of the Aging Rat,* Vol. 2. International Life Sciences Institute Press, Washington, D.C., pp. 149–170.

Joseph, J., Denisova, N., Arendash, G., Gordon, M., Diamond, D., Shukitt-Hale, B., Morgan, D. (2003) Blueberry supplementation enhances signaling and prevents behavioral deficits in an Alzheimer's disease model. *Nutritional Neuroscience,* 6(3), 153–162.

Joseph, J., Shukitt-Hale, B., Denisova, N., Bielinski, D., Martin, A., McEwen, J., Bickford, P. (1999) Reversals of age-related declines in neuronal signal transduction, cognitive, and motor behavioral deficits with blueberry, spinach, or strawberry dietary supplementation. *Journal of Neuroscience,* 19(18), 8114–8121.

Kacew, S., Dixit, R., Ruben, Z. (1998) Diet and rat strain as factors in nervous system function and influence of confounders. *Biomedical and Environmental Sciences,* 11, 203–217.

Kanarek, R., Orthen-Gambill, N. (1986) Complex interactions affecting nutrition-behavior research. *Nutrition Reviews,* 44 (Suppl.), 172–175.

Kelley, A., Cador, M., Stinus, L. (1989) Exploration and its measurement: a psychopharmacological perspective. In *Neuromethods, Vol. 13: Psychopharmacology,* A. Boulton, G. Baker, A. Greenshaw (Eds.). Humana Press, Clifton, NJ, pp. 95–144.

Lalonde, R., Botez, M. (1990) The cerebellum and learning processes in animals. *Brain Research Reviews,* 15, 325–332.

Lucki, I. (1997) The forced swimming test as a model for core and component behavioral effects of antidepressant drugs. *Behavioral Pharmacology,* 8, 523–532.

Maier, S. (1984) Learned helplessness and animal models of depression. *Progress in Neuro-psychopharmacology & Biological Psychiatry,* 8(3), 435–446.

McLay, R., Freeman, S., Harlan, R., Kastin, A., Zadina, J. (1999) Tests used to assess the cognitive abilities of aged rats: their relation to each other and to hippocampal morphology and neurotrophin expression. *Gerontology,* 45, 143–155.

Morris, R. (1984) Developments of a water-maze procedure for studying spatial learning in the rat. *Journal of Neuroscience Methods,* 11, 47–60.

Olton, D., Samuelson, R. (1976) Remembrance of places passed: spatial memory in rats. *Journal of Experimental Psychology: Animal Behavior,* 2, 97–116.

Olton, D. (1987) The radial arm maze as a tool in behavioral pharmacology. *Physiology and Behavior,* 40, 793–797.

Pare, W. (1969) Interaction of age and shock intensity on acquisition of a discriminated conditioned emotional response. *Journal of Comparative and Physiological Psychology,* 68(3), 364–369.

Porsolt, R., LePinchon, M., Jaffre, M. (1977) Depression: a new animal model sensitive to antidepressant treatments. *Nature,* 266, 730–732.

Price, D., Sisodia, S. (1998) Mutant genes in familial Alzheimer's disease and transgenic models. *Annual Reviews in Neuroscience,* 21, 479–505.

Ragozzino, M., Arankowsky-Sandoval, G., Gold, P. (1994) Glucose attenuates the effect of combined muscarinic-nicotinic receptor blockade on spontaneous alternation. *European Journal of Pharmacology,* 256(1), 31–36.

Redei, E.E., Ahmadiyeh, N., Baum, A.E., Sasso, D.A., Slone, J.L., Solberg, L.C. et al. (2001) Novel animal models of affective disorders. *Seminars in Clinical Neuropsychiatry,* 6, 43–67.

St. George-Hyslop, P. (2000) Piecing together Alzheimer's. *Scientific American,* December, 76–83.

Sarter, M. (1987) Measurement of cognitive abilities in senescent animals. *Journal of Neuroscience,* 32, 765–774.

Schmitt, U., Hiemke, C. (1998) Strain differences in open-field and elevated plus-maze behavior of rats without and with pretest handling. *Pharmacology, Biochemistry and Behavior,* 59(4), 807–811.

Seligman, M., Maier, S. (1967) Failure to escape traumatic shock. *Journal of Experimental Psychology,* 74, 1–9.

Seligman, M. (1991) *Learned Optimism.* Norton, New York.

Shukitt-Hale, B. (1999) The effects of aging and oxidative stress on psychomotor and cognitive behavior. *Age,* 22, 9–17.

Shukitt-Hale, B., McEwen, J., Szprengiel, A., Joseph, J. (2004) Effect of age on the radial arm water maze — a test of spatial learning and memory. *Neurobiology of Aging,* 25(2), 223–229.

Skinner, B.F. (1966) What is the experimental analysis of behavior? *Journal of Experimental Analysis of Behavior,* 9, 213–218.

Spiteri, T., LePape, G., Agmo, A. (2000) What is learned during place preference conditioning? A comparison of food- and morphine-induced reward. *Psychobiology,* 28(3), 367–382.

Stohr, T., Wermeling, D., Weiner, I., Feldon, J. (1998) Rat strain differences in open-field behavior and the locomotor stimulating and rewarding effects of amphetamine. *Pharmacology, Biochemistry and Behavior,* 59(4), 813–818.

Taratino, L., Bucan, M. (2000) Dissection of behavior and psychiatric disorders using the mouse as a model. *Human Molecular Genetics,* 9(6), 953–965.

Tsien, J. (2000) Building a brainier mouse. *Scientific American,* April, 62–68.

van der Staay, F., de Jonge, M. (1993) Effects of age on water escape behavior and on repeated acquisition in rats. *Behavioral and Neural Biology,* 60, 33–41.

van der Staay, F. (2000) *The Study of Behavioral Dysfunctions: An Evaluation of Selected Animal Models.* BCN Press, University Groningen, The Netherlands, pp. 5–52.

Vigas, M. (1989) Neuroendocrine responses to psychosocial and somatic stress in rats and humans. In *Stress: Neurochemical and Humoral Mechanisms*, G.R. Van Loon, R. Kvetnansky, R. McCarty, and J. Axelrod (Eds.). Gordon & Breach, New York, pp. 15–28.

Waddell, C.A., Desai, I.D. (1981) The use of laboratory animals in nutrition research. *World Review of Nutrition and Dietetics,* 36, 206–222.

Wall, P.M., Messier, C. (2001) Methodological and conceptual issues in the use of the elevated plus-maze as a psychological measurement instrument of animal anxiety-like behaviour. *Neuroscience and Biobehavioral Reviews,* 25, 275–286.

Watson, M., McElligott, J. (1984) Cerebellar norepinephrine depletion and impaired acquisition of specific locomotor tasks in rats. *Brain Research,* 296, 129–136.

Weindruch, R., Masoro, E. (1991) Concerns about rodent models for aging research. *Journal of Gerontology: Biological Sciences,* 46(3), B87–B88.

Weiss, B., O'Donoghue, J.L. (Eds.) (1994) *Neurobehavioral Toxicity.* Raven Press, New York.

Willner, P. (1984) The validity of animal models of depression. *Psychopharmacology (Berlin),* 83(1), 1–16.

Wongwitdecha, N., Marsden, C.A. (1996) Effects of social isolation rearing on learning in the Morris water maze. *Brain Research,* 715, 119–124.

Yu, B., Masoro, E., McMahan, C. (1985) *Journal of Gerontology,* 40(6), 657–670.

Zyzak, D., Otto, T., Eichenbaum, H., Gallagher, M. (1995) Cognitive decline associated with normal aging in rats: a neurophysiological approach. *Learning and Memory,* 2, 1–16.

4 Electrophysiological Methods: Application in Nutritional Neuroscience

Rubem Carlos Araújo Guedes

CONTENTS

4.1 INTRODUCTION

This chapter presents the main electrophysiological methods currently used to study brain function. Some data obtained with such methods are reported and discussed. Techniques such as the recording of spontaneous activity [electroencephalogram (EEG)] and of sensory-evoked potentials are described. The technical principles of each electrophysiological method are presented in a lucid manner, allowing nonspecialists to read the text easily, using it for a better understanding of how the brain develops and functions under a variety of nutritional conditions.

Particular attention is devoted to the electrophysiological recording of the phenomenon known as spreading depression (SD) of brain electrical activity. SD is an interesting response presented by the brain tissue as a consequence of electrical, mechanical, or chemical stimulation of one point of its surface. Changes in SD electrical features associated with nutritional conditions have been studied by my laboratory over the last two decades and some interesting data are presented and discussed to illustrate how electrophysiological methods can be applied in nutritional neuroscience.

First, a brief account is presented on the relationship between nutritional alterations and brain function. Also, the general principles for the use of electrophysiological techniques in studies involving brain function are briefly described. Then the use of EEG and evoked potential recordings in nutritional neuroscience studies is discussed and, finally, my experience on the recording the phenomenon of cortical SD is detailed.

4.1.1 NUTRITIONAL ALTERATIONS AND BRAIN FUNCTION

The deficiency of one or more nutrients in the diet can undoubtedly disrupt the biochemical and morphological organization of the brain, and this is usually followed by repercussions on its function. Depending on the intensity and duration of the nutritional disturbances, the effects may result in deleterious consequences, which in some cases can be more or less devastating for the whole organism (Morgane et al., 1978, 1993). Basic neural functions such as processing of sensory information and perception of the corresponding sensation, as well as execution of motor tasks, can be affected to more or less extent by nutritional deficiency. This is also the case of the more elaborated functions such as those involving cognition, consciousness, emotion, learning, and memory. Disturbances of such processes can also lead to important pathological conditions (Almeida et al., 2002).

In several parts of the world malnutrition still affects an impressive number of children, and this has influenced several research groups in their decisions to investigate in laboratory animals, as well as in humans, the effects of early malnutrition on the adult central nervous system, thereby generating an extensive body of data on this subject. Similar to what has been extensively documented in animals, malnutrition in children can also have grave consequences on their development, depending on the period of occurrence and the intensity of the nutritional deficiency (Grantham-McGregor, 1995). On the other hand, although less investigated, it is currently accepted that excessive food intake can also interfere with brain development and function (Almeida et al., 2002).

Different approaches have been used to understand to what extent such nutritional disorders affect neuroanatomical, biochemical, and electrophysiological aspects of the brain. It is now clear that such changes are much more severe when malnutrition coincides with the so-called brain growth spurt, which corresponds to the highest speed of neurogenesis, gliogenesis, and cell migration, in the neural tissue (Morgane et al., 1993). Under these conditions, the electrophysiological activity can also be considerably affected in animals, both in the peripheral (Silva et al., 1987) and the central nervous system (Morgane et al., 1978, 1993). Also, reports are available describing increased susceptibility of malnourished rats to processes related to neural excitability, such as higher reactivity to aversive stimuli (Rocinholi et al., 1997) and to experimentally induced seizures (Palencia et al., 1996; Stern et al., 1974). All the evidence documenting nutritionally related changes in neural electrophysiological phenomena and in nervous excitability prompted us to investigate changes in brain electrical activity by using, as a model, the phenomenon known as SD of brain electrical activity, which is addressed in Section 4.4.

4.1.2 GENERAL PRINCIPLES IN NEURAL ELECTROPHYSIOLOGY

In all moments of its history, the living brain has continuously produced electrical signals, and this constitutes what we call the spontaneous electrical activity of the brain. "Spontaneous," in this context, should not be understood as an activity generated without a causal stimulus. It actually means that this activity is generated without the intentional stimulation of the researcher or observer; that is, the causal stimulus is endogenous. Different techniques can be used for recording this cerebral activity in laboratory animals and humans. Valuable information can be obtained with the use of techniques that record the activity produced simultaneously by a group of many thousands of neurons located in close anatomical relationship in the brain. The recording obtained in this manner, called EEG, is obtained as electric potential differences between several pairs of electrodes placed at distinct brain regions. These macroelectrodes (usually fine wire having at the tip metal discs a few millimeters in diameter) are fixed on the scalp by special pastes with high electric conducting properties. As a noninvasive technique, EEG recording is extensively used in human patients to help diagnose certain neurological diseases. In laboratory animals, the electrodes (usually of smaller dimensions, mainly fine wires without any disc at the tip) are preferably fixed (under anesthesia) throughout small holes drilled in the skull, with the electrode tips either just touching the dura mater [recording the activity of the cerebral cortex, what we call electrocorticogram (ECoG)] or deeply inserted into the brain tissue in order to record the activity of a specific subcortical structure. After electrode implantation, the animal can be allowed to recover from anesthesia, so that the brain electrical activity can be recorded on several occasions during periods of normal awaking or sleeping states.

The EEG presents features that are quite reproducible, provided the recording conditions remain invariable. In this situation, the electrical activity produced by the cerebral cortex of a healthy adult organism displays a pattern of continuous oscillations, generating waves of electrical potential with certain frequencies and amplitudes. This pattern is currently very well known and can change depending, for example, on the brain region in which the recording is performed (e.g., frontal vs. occipital) or on the state of consciousness of the individual (whether he or she is awake or sleeping). In the awake condition, the electrical activity can change depending on whether the individual has eyes open or closed or whether he or she is mentally performing a simple arithmetic operation (e.g., multiplying two small numbers). In a sleeping person, the electrical activity changes its pattern when a superficial stage of sleep turns into a deeper one. Figure 4.1 shows examples of EEG recordings under several physiological and pathological conditions. Changes in these patterns

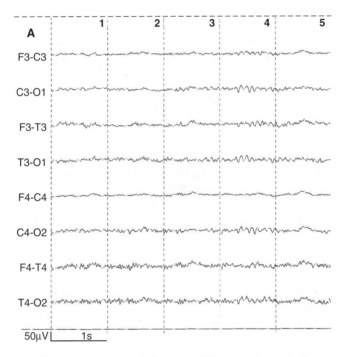

FIGURE 4.1 Examples of human electroencephalograms (EEGs) showing typical patterns in different physiological and pathological states. (A) Normal subject, at resting; illustrates the higher amplitudes of EEG waves at the posterior cerebral regions (alpha rhythm) as compared with the beta rhythm predominant at the anterior areas (compare the posterior traces T3-O1 and T4-O2 with the anterior ones F3-C3 and F4-C4, respectively). (B) Normal subject, at resting; illustrates EEG changes (reduction in amplitude and increase in frequency) induced by opening the eyes (during the fifth and the sixth seconds of the record, marked by the black bar). The big deflections at the upper trace are artifacts provoked by eyelid movements. (C) Normal subject, sleeping, to show the appearance of the so-called fuses of sleep (more easily identifiable at the traces marked by black bars). Note also the general pattern of waves with low frequency and high amplitudes, typical of the sleeping state. (D) Abnormal EEG, showing irritative activity (high-amplitude waves) at the temporal and frontal regions of the left hemisphere (marked by black bars). According to international conventions, even and odd numbers (at the left side of the traces) refer to the right and left hemispheres, respectively; F, P, T, and O refer to the frontal, parietal, temporal, and occipital regions of the brain, respectively; Cz, central electrode (at the vertex of the skull); Rf, reference electrode (in this case at the ears). Numbers at the superior part of the traces indicate time in seconds.

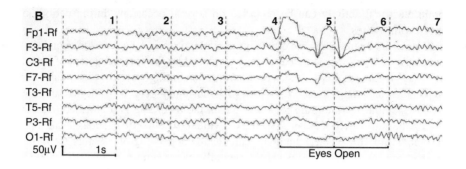

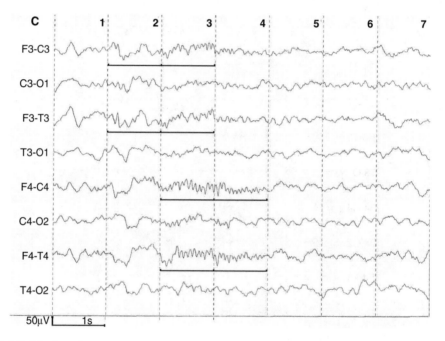

FIGURE 4.1 (Continued)

associated with nutritional alterations can be identified and studied by using the EEG technique. In laboratory animals, it is also possible, by using very fine microelectrodes (glass pipettes with tip diameters ranging from one to a few micrometers), to record the activity of a single neuron, called unit activity. Events occurring in a single neuron can be recorded intracellularly or extracellularly. Depending on the tip dimensions and geometry (as well as on the electrical impedance) of the electrode, extracellular recordings can document the activity of a single cell or of a group of few neurons (multiunit activity; Figure 4.2). These techniques can be very useful to study how nerve impulse-mediated transmission of information between neurons can be influenced by nutritional variables.

In addition, the brain can react to exogenous (sensorial or electrical) stimuli applied intentionally by the researcher. The electrical activity produced in response to an exogenous stimulus is called evoked (or provoked) potential and can be recorded with either macroelectrodes (as extracellular field potentials) or microelectrodes (as intra- or extracellular unit activities). Evoked activity can be distinguished from the spontaneous one, because it is temporally associated with the exogenous stimulus and usually presents very characteristic waveforms, which differ from the spontaneous

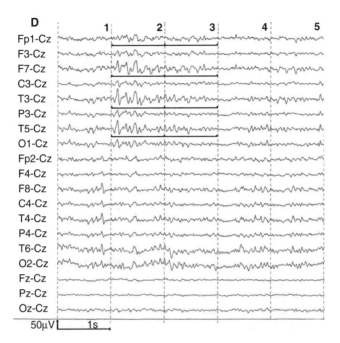

FIGURE 4.1 (Continued)

EEG rhythmic oscillations. The adequate recording of these evoked responses can also represent a helpful tool to investigate the ability of the brain in reacting to incoming information reaching the brain through a specific sensory pathway. Several physiological processes in the brain, such as the wakefulness–sleep cycle, arousal, and dreaming, can be studied by EEG and evoked potential recordings. Similarly, these techniques can help diagnose some pathological states such as epilepsy, brain tumors, coma, and clinical brain death.

A more detailed analysis on some technical and theoretical aspects of such methods is beyond the scope of this chapter. The reader can find this information in several reference sources (see, e.g., Bures et al., 1976).

4.2 THE ELECTROENCEPHALOGRAM (EEG) IN NUTRITIONAL NEUROSCIENCE

Pioneering studies in malnourished children have described abnormal EEG patterns, mostly characterized by reduction in amplitude and frequency of the brain waves as compared with the EEGs of control children and usually associated with other clinical alterations (Engel, 1956). In an EEG study on 46 South African malnourished children, Nelson and Dean (1959) found abnormal focal discharges in 36% of the cases, suggesting an enhancement of excitability, as found in certain pathologies such as epilepsy. Such findings could lead to the hypothesis that malnourished humans would present a higher incidence of epilepsy as compared with well-nourished controls. In humans, however, the relationship between nutritional deficiency and seizure susceptibility has not been much investigated in the last three decades. More recently, follow-up studies have shown a higher prevalence of developmental delay and neurological disabilities in low-birth-weight preterm human newborns (see Almeida et al., 2002), and a tendency of malnourished children to develop epilepsy has also been reported (Nunes et al., 1999; Hackett and Iype, 2001). Although suggestive, these studies did not represent the definitive confirmation of the causal relationship between malnutrition and epilepsy in humans. The acceptance of this hypothesis still needs systematic and robust investigations, similar to those already available in laboratory animals.

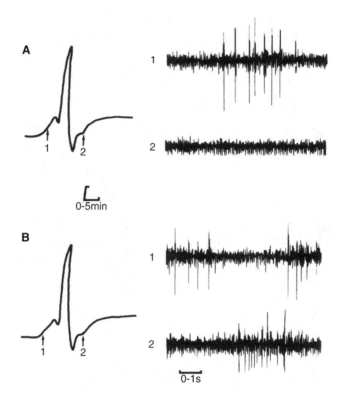

FIGURE 4.2 Multiunit spike responses (right traces) recorded extracellularly with a multibarrel glass micropipette during periods of iontophoretically ejected pulses of the excitatory amino acid glutamate (10 nA; 30 sec) through the same pipette assembly. Left traces are DC recordings of the slow potential change of spreading depression (SD; see Section 4.4). SD was elicited by topical application of KCl (a cotton ball 1 to 2mm in diameter soaked in a 2% KCl solution for 1 min) to a point of the cortical surface of an anesthetized adult rat. Recordings 1 and 2 at the right were taken at time points marked by numbers 1 and 2 at the left tracings (respectively before and during SD). (A) Control condition, showing the glutamate-elicited spike activity (A1), which disappeared during SD (A2). (B) Same as in (A), except that during the entire sequence of SD elicitation and DC recording (showed at left) naloxone hydrochloride (50 nA) was ejected iontophoretically while the neuronal responses to glutamate (shown at right) were recorded through the same pipette assembly. Naloxone ejection during SD (tracing B2) prevented the disappearance of neuronal spike activity (compare with tracing A2 taken during SD in the absence of naloxone), suggesting an antagonistic action of naloxone on SD (see also Table 4.1). (From Guedes, R.C.A. et al. (1987b) *Experimental Brain Research* 39, 113–118. With permission.)

The earlier EEG studies in humans were essentially based on the visual analysis of the EEG by employing experienced observers trained in doing visual EEG scoring and in recognizing in it certain defined electrographic patterns. Although useful for qualitative analysis, this technique cannot avoid certain subjectivity. Quantification of parameters such as amplitude, frequency, and rhythmicity cannot be satisfactorily achieved with the visual analysis technique (Morgane et al., 1978). The employment of spectral analysis and fast Fourier transform algorithms to quantify EEG features (see Morgane et al., 1978) allowed studying the ontogeny of EEG patterns both in children (Schulte and Bell, 1973) and in laboratory animals. In the latter subjects, several studies have provided important information on EEG patterns under nutritional deficiency, reporting alterations in frequencies and amplitudes of the EEG waves (Salas and Cintra, 1975). A delay in the maturation of the EEG pattern in malnourished rats has also been observed (Gramsbergen, 1976). Also, in the 1970s, Peter J. Morgane and coworkers developed a series of EEG studies in the rat, forming an

EEG power spectral atlas describing EEG ontogeny (see Morgane et al., 1978). EEG recordings and power spectral plots from several brain regions at different ages and in distinct stages during the sleep–wakefulness cycle provided a good description of the developmental EEG features in the albino rat. By comparing animals fed a control diet (25% protein) with those fed a protein-deficient one (8% protein), it was found that the main EEG disturbances were present in the hippocampus, the protein-deprived rats having more power in the frequency range of 3.5 to 10 Hz (which includes the theta waves) both during REM sleep and waking (Morgane et al., 1978). A recent EEG study of the rat circadian sleep–wake cycle has confirmed the higher EEG power in the theta range of frequencies in the malnourished group, as compared with the controls; the authors were also able to describe significant homeostatic and circadian alterations in the sleep–wake cycle (Cintra et al., 2002). Interestingly, a lower threshold for inducing seizures by the kindling technique was found in the hippocampus of malnourished rats (Bronzino et al., 1986). Other experiments performed with distinct paradigms of nutritional deficiency and different seizure models also indicate lower threshold for seizures in nutritionally deficient rats (Stern et al., 1974; Palencia et al., 1996). In contrast to this idea, two sets of experiments performed by the Morgane group (Morgane et al., 1978) showed that (1) penthylenetetrazole (PTZ)-induced seizures occurred in malnourished rats after latencies comparable to those of well-nourished animals and (2) motor seizures induced by repeated stimulation of the amygdala (kindling model of epilepsy) revealed higher thresholds in malnourished rats as compared with control rats. Taken together, the results are compatible with the idea that the facilitatory effect of malnutrition on seizure susceptibility in the rat depends on the method used to induce seizures and the brain structure involved in each case. Whether this is also the case for the human brain remains to be determined.

4.3 EVOKED POTENTIALS AND NUTRITION

During an EEG recording session, the spontaneous activity pattern can be modified by the appearance of evoked responses consequent to sensory stimulation. At the peripheral level, it is possible to elicit evoked responses by stimulating sensory receptors with the adequate forms of energy. Under such a situation, an evoked response will appear at the functionally corresponding region of the brain. For example, application of light stimuli to the photoreceptors of the retina will elicit evoked potentials in the occipital cortex; on the other hand, such evoked responses will be recorded in the temporal cortical region when one applies a sound stimulus to the auditory receptors of the cochlea at the inner ear.

Electrical current pulses, instead of sensory stimuli, can also be used to produce evoked responses. In this case, electrical stimulation can be applied either at a peripheral afferent structure (such as the sciatic nerve) or centrally (at the brain tissue). These responses can be recorded either peripherally (at another point on the afferent or efferent pathway that is under stimulation) or centrally (by recording the responses at brain level). In summary, when using electrical stimulation, one can activate a peripheral nerve and record either peripherally, at a remote point of the same nerve, or centrally, at the brain level; responses elicited by central (brain) stimulation can also be recorded either centrally, at another brain area of the same or the opposite hemisphere, or peripherally, at an efferent pathway. In all cases, it is necessary that both stimulated and recorded areas be functionally connected.

Brain-evoked potentials may provide valuable clues on how the nervous system transmits and processes information, both at the peripheral and at the central levels. Disturbances in these processes may be revealed by recording and analyzing evoked responses. Thus, it can be very useful to compare evoked potential features in organisms submitted to distinct nutritional conditions in order to detect the effects of such situations on transmitting and processing peripheral sensory information. Increases in the latencies of visually and transcallosal electrically evoked responses have been found at an early age (14 to 20 days) but not later in adult life (95 to 100 days and 60

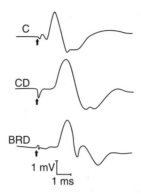

FIGURE 4.3 Examples of compound action potentials recorded at the proximal part of the sciatic nerve in response to electrical stimulation of a distal point of the same nerve in three rats submitted respectively to a control diet (C; with 22% casein), a chow diet (CD; with 14% protein), and the basic regional diet of human populations in northeast Brazil (BRD; with 8% protein). Note the different latencies, indicating distinct conduction velocities (in these examples, respectively 57.1, 40.0, and 26.1 m/sec). (From Silva, A. et al. (1987) *Brazilian Journal of Medical and Biological Research* 20, 383–392. With permission.)

to 65 days, respectively) in early protein-deprived rats as compared with controls (Morgane et al., 1978). Auditory-evoked potentials (Rocinholi et al., 2001) and electrically evoked action potentials in sensory nerves (Segura et al., 2001) of early malnourished rats are also delayed. In anesthetized and paralyzed chronically malnourished rats, sciatic nerve action potentials provoked by electrical stimulation presented nearly a 50% reduction in the conduction velocities as compared with well-nourished animals (Silva et al., 1987; Figure 4.3). Responses recorded in the corticospinal tract following surface stimulation of the motor cortex were reduced about 15.5% in early malnourished rats as compared with well-nourished, age-matched controls (Quirck et al., 1995). In malnourished children, motor and sensory nerve conductions are shown to be significantly impaired both in peripheral and in central pathways (Chopra, 1991; Tamer et al., 1997). In some cases, histological investigation has indicated a decrease in diameter of myelinated fibers as well as signs of delayed myelination and axonal degeneration. All the clinical data suggest altered functional status, which could be associated with clinical signs such as muscle weakness, hypotonia, and hyporeflexia, usually found in malnourished children. Such signs could be involved in generating learning deficits and impairment of hand dexterity and motor coordination, among other functional deficiencies linked to malnutrition (Chopra, 1991).

4.4 NUTRITION AND CORTICAL SPREADING DEPRESSION (SD)

Cortical SD was first described by Leão (1944) as a reversible and propagated wave of reduction (depression) of the spontaneous and evoked electrical activity of the cerebral cortex, with a simultaneous slow potential change (also called DC potential change) of the tissue, in response to the electrical, chemical, or mechanical stimulation of one point of the cortical surface. The term *depression*, in this context, does not refer to the psychiatric disease, named also by that term, but rather has an electrophysiological meaning: it signifies that the amplitude of the EEG activity at a certain cortical region becomes temporarily depressed, i.e., the potential difference between two recording points in the depressed area tends to zero. In some cases, the EEG trace actually becomes isoelectric (the amplitudes of the oscillating EEG waves equal to zero). Simultaneously to the EEG depression, the DC potential of that cortical point (measured with a DC amplifier against a remote point having a fixed potential, as, for example, the nasal bones) starts to change, becoming progressively negative. This slow potential change can attain maximum values ranging from –5 to –30 mV after 1 to 2 min of onset and is fully reverted after a few minutes. The always present

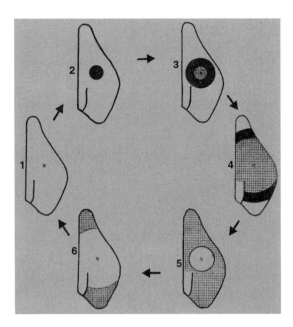

FIGURE 4.4 Scheme illustrating the cycle of events occurring during an episode of SD. Stimulation of a point (x) of the normal cortical surface (1) can elicit SD (dark area in 2). Once elicited, SD usually propagates concentrically to remote regions (dark areas in 3 and 4), while the initially depressed area slowly recovers its normal pre-SD activity (central light area in 5 and 6). Finally, the whole cortical tissue returns back to the control predepression situation, as in 1.

negative DC potential can be eventually preceded, and more frequently followed, by positive deflections, usually of smaller amplitudes than those of the negative one. The depression of EEG activity (as well as the DC potential changes) spreads concentrically from the stimulated point, reaching gradually more and more remote cortical regions, while the originally depressed area starts to recover (Figure 4.4). As a rule, the complete recovery of a depressed region is achieved after about 5 to 10 min, as evaluated by the restoration of the predepression EEG pattern and DC level. In contrast to the EEG, the slow potential change accompanying SD has all-or-none characteristics, being very useful to calculate the velocity of propagation of the phenomenon. Surprisingly, in all vertebrate species so far studied, from fishes to mammals, the velocity of SD propagation was found to be remarkably low (2 to 5 mm/min; Leão, 1944) when compared with the much higher conduction velocity of neuronal action potentials, which in mammals is in the range of meters per second. This peculiar SD velocity has led the authors to postulate a humoral mechanism of propagation based on the release of one or more chemical factors from the neural cells. According to this idea, as these compounds diffuse through the extracellular space, they "contaminate" the neighbor cells, which then become depressed, releasing the SD-eliciting chemical factors, which contaminate other cells, and so on, giving rise to a genuine autoregenerative propagation maintained by that positive feedback loop.

During these six decades since SD description, a very extensive body of information on SD phenomenology has accumulated. However, the clarification of its final mechanisms has not yet been fully achieved, but putative links with three relevant human pathologies — epilepsy, migraine, and brain ischemia — deserve some comments. During SD, while the spontaneous activity is depressed, epileptiform waves, similar to those found in the EEG of epileptic patients, eventually appear (Leão, 1944; Guedes and Do Carmo, 1980; Figure 4.5). This led to the idea that perhaps SD and epilepsy mechanisms have some common features. The description of brain vascular changes during SD, similar to those seen in the classical migraine (migraine with aura), also leads to the association between the two phenomena in terms of common mechanisms (Hadjikhani et

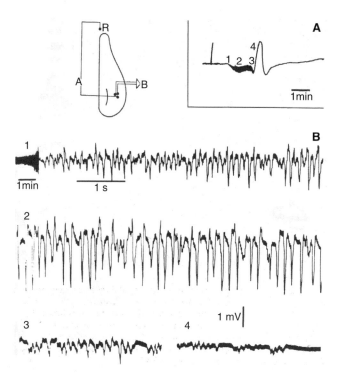

FIGURE 4.5 Recording of a typical DC change of spreading depression (A) and of the EEG epileptiform activity immediately preceding it (B) in an anesthetized rabbit submitted to systemic reduction of the extracellular chloride concentration by means of gastric washing during the 4-h period preceding SD recording. The inset at the upper left shows the positions of the cortical electrodes for recordings (A) and (B) as well as the extracortical position of the common reference electrode (R) at the nasal bones. Records B1 to B4 were taken during the periods indicated by 1 to 4 in (A). Epileptiform activity was fully developed at time point 2 and became totally depressed at time point 4. The calibration pulse in (A) (vertical bar at left) corresponds to –10 mV. (From Guedes, R.C.A. and Do Carmo, R.J. (1980) *Experimental Brain Research* 39, 341–349. With permission.)

al., 2001; Lehmenkühler et al., 1993). Finally, the same logic led some authors to postulate an important role for SD in the physiopathology of ischemia (Takano et al., 1996). In all cases, current discussions often mention the possible involvement of certain ions (Guedes and Do Carmo, 1980; Siesjö and Bengtsson, 1989), free radicals produced in the nervous tissue (El-Bachá et al., 1998; Guedes et al., 1996), or neurotransmitter activity. Different neurotransmitters have been shown to be associated with distinct effects on SD; in some cases, a facilitatory effect has been reported (Guedes et al., 1992) and in others an antagonistic effect has been found (Gorelova et al., 1987; Guedes et al., 1987b, 1988, 2002). An overview of the current knowledge about SD features and research trends can be found in Lehmenkühler et al. (1993) and Gorgi (2001), among others.

As regards malnutrition, electrophysiological studies have documented a facilitatory effect on brain ability of early malnourished rats to propagate SD (De Luca et al., 1977). Data from our laboratory have shown that malnutrition early in life facilitates SD propagation in adulthood, as judged by the SD velocities, higher in the early malnourished animals than in controls (Guedes et al., 1987a). The protein supplementation of a diet deficient in both quantity and quality of its protein led to distinct results on SD propagation, depending on the quality of the protein used in the supplementation: by supplementing the diet with a low quality (vegetable) protein, the effects on SD were not reverted. Reversion of SD alterations was achieved only when the protein used to supplement the diet was of high quality (casein; Andrade et al., 1990). Indeed, even short periods of malnutrition during lactation can long-lastingly facilitate SD propagation (Rocha-de-Melo and

Guedes, 1997). In addition, the SD responses of the brain to substances such as diazepam (Guedes et al., 1992) and glucose (Costa-Cruz and Guedes, 2001) are reduced in early malnourished rats in comparison with the responses in controls.

Besides malnutrition, several other clinically relevant conditions, including nutritional and metabolic variables, associated or not with environmental, hormonal, or pharmacological ones, have been investigated. Such conditions, which are known to affect brain development and functioning, can modify considerably the cortical tissue ability to present and to propagate SD. For example, increased brain susceptibility to SD has been observed in conditions such as (1) systemic reduction of extracellular chloride levels (Guedes and Do Carmo, 1980); (2) deprivation of REM sleep (Amorim et al., 1988); (3) hypoglycemia (Costa-Cruz and Guedes, 2001); (4) systemic increase of the GABAergic activity by diazepam (Guedes et al., 1992); (5) ingestion of ethanol (Guedes and Frade, 1993); (6) hyperthyroidism early in life (Santos, 2000); and (7) dietary deprivation of antioxidant vitamins (El-Bachá et al., 1998; Guedes et al., 1996). In contrast, other experimental situations have been shown to reduce brain susceptibility to SD. These conditions include (1) dietary treatment with lithium (Guedes et al., 1989); (2) hyperglycemia (Costa-Cruz and Guedes, 2001); (3) anesthesia (Guedes and Barreto, 1992); (4) early hypothyroidism (Guedes and Pereira-da-Silva, 1993); (5) aging (Guedes et al., 1996); (6) environmental stimulation (Santos-Monteiro et al., 2000); (7) systemic, topical, and microiontophoretic treatments with the opioid antagonist naloxone (Guedes et al., 1987b); (8) topical cortical application of excitatory amino acid antagonists (Guedes et al., 1988); and (9) pharmacological treatment with drugs that increase brain serotonin activity (Guedes et al., 2002). A summary of studies on the conditions that influence cortical susceptibility to SD and the main effects associated with them are presented in Table 4.1.

SD recording at the cortical surface of the rat brain is performed mostly with the animal under anesthesia, but it can also be performed in the nonanesthetized animal. In this latter case, the previous implantation of the electrodes under anesthesia is necessary; after the electrodes have been fixed on the cortical surface, throughout the skull bones, the animal is allowed to recover from anesthesia and then the recordings can be done in the awake condition (Figure 4.6). By using this technique, we were able to study in rats the effects of two types of anesthesia: (1) that obtained with thionembutal and (2) that produced by a mixture of urethane plus chloralose. In that study, SD features were compared in the same animals during anesthesia and in the awake state (Guedes and Barreto, 1992; see also Table 4.1).

It can be concluded that the electrophysiological recording of the SD phenomenon is a very interesting and valuable tool for studies on both nutrition and brain function and on nutrition and other clinically important conditions that influence the functioning of the nervous system.

4.5 CONCLUDING REMARKS

The generation of electrical activity is the main physiological feature of the nervous tissue. Through this activity, the brain is able to execute the immense repertoire of its actions, from the simplest to the highly complex ones. Therefore, the methods that can be used to record and to analyze the brain electrical activity can provide very important information on the understanding of how it functions under physiological and pathological conditions.

As shown in this chapter, data from electrophysiological investigations, both in laboratory animals and in humans, point to the importance and usefulness of this technique in studying the nervous system under normal conditions and also in understanding how its functioning may be changed by nutritional alterations. For example, attempts to establish the degree of comparability between rat and human sensory evoked potentials, considered as measures of sensory function, have been performed in studies on neurotoxicology in order to examine the extrapolation of data from one species (rat) to the other (humans; Boyes, 1994). It is thereby highly desirable that such techniques become increasingly used in studies in the field of nutritional neuroscience as a

TABLE 4.1
Conditions of Clinical Interest that Facilitate or Antagonize Spreading Depression (SD)

	Animal	Effect	Ref.
Conditions that Facilitate SD			
Reduction of extracellular chloride (by gastric washing)	Rabbit	Increased SD velocity of propagation and EEG epileptiform activity; appearance of a second SD in the opposite hemisphere	Guedes and Do Carmo, 1980
Malnutrition early in life		SD velocities higher than in the controls	Andrade et al., 1990
Deprivation of REM sleep (water-tank technique)	Rat	Increased SD velocities; apomorphine did not change them	Amorim et al., 1988
Hypoglycemia [by (1) insulin or (2) food restriction + insulin]	Rat	SD velocities higher than in normoglycemic rats	Costa-Cruz and Guedes, 2001
Increase of the GABAergic activity by diazepam	Rat	Higher SD velocities after diazepam as compared with predrug values	Guedes et al., 1992
Ingestion of ethanol	Rat	Ethanol ingestion for 7 days increased SD velocities as compared with controls	Guedes and Frade, 1993
Hyperthyroidism early in life	Rat	T4-injected rats displayed higher SD velocities	Santos, 2000
Dietary deprivation of antioxidant vitamins	Rat	Higher SD velocities; increased SD-eliciting action of photoactivated riboflavin	El-Bachá et al., 1998; Guedes et al., 1996
Conditions that Impair SD			
Dietary treatment with lithium	Rat	Lower SD velocities compared with controls	Guedes et al., 1989
Hyperglycemia	Rat	Lower SD velocities compared with controls	Costa-Cruz and Guedes, 2001
Anesthesia	Rat	Lower velocities in anesthetized than in nonanesthetized state in the same rat	Guedes and Barreto, 1992
Early hypothyroidism (by PTU)	Rat	Reduced SD velocities in PTU-treated rats compared with saline-treated controls	Guedes and Pereira-da-Silva, 1993
Aging	Rat and gerbil	Inverse correlation between age and SD velocity; reduced by dietary antioxidant vitamin deficiency	Guedes et al., 1996
Early environmental stimulation	Rat	Reduced SD velocities compared with controls	Santos-Monteiro et al., 2000
Treatment with the opioid antagonist naloxone	Rat and gerbil	Naloxone antagonizes SD incidence and propagation	Guedes et al., 1987b
Topical cortical application of excitatory amino acid antagonists	Rat	MK-801 antagonizes SD propagation. NMDA and kainic acid facilitate or block it, depending on the dose	Guedes et al., 1988
Treatment with drugs that increase brain serotonin activity	Rat	Both D-fenfluramine and citalopram antagonize SD propagation	Guedes et al., 2002

complementary means to understand the relationship among diet, nutrition, and neural development and function.

In our laboratory, we have dedicated much effort and time to this goal by employing the phenomenon of SD as an interesting electrophysiological model. At the beginning (late 1970s), we became very enthusiastic with the publication of a short paper by De Luca et al. (1977) on SD in malnourished rats, because we foresaw the possibility of having, in a short period, a proliferation of research groups interchanging ideas and data about that subject. Unfortunately, with the exception of that pioneer work, to the best of our knowledge no other research groups have decided, till date, to share our efforts on this interface (nutrition and SD) — we consider that a pity. However, we are optimistic about the future of this field of research in view of the recently growing number of

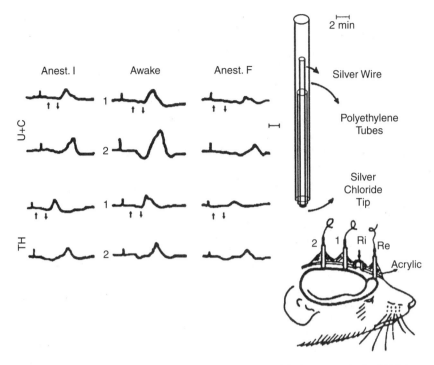

FIGURE 4.6 Representative recordings of the DC slow potential change typical of SD in two rats with implanted Ag-AgCl electrodes (shown in the upper right inset). The animals were anesthetized either with a mixture of urethane plus chloralose (U + C) or with thionembutal (TH). The recording electrodes 1 and 2 were inserted through small holes (about 1 mm in diameter) in the skull until their tips touched the dura mater and were then fixed with dental acrylic. The reference electrode (Re) had its tip implanted inside the nasal bones. Between the reference and the first recording electrodes, a larger hole (2 to 3 mm in diameter) was drilled to expose a portion of the frontal cortical surface and a small plastic ring (Ri) was fixed around it with dental acrylic. Through this hole, the SD-eliciting stimulus (a small cotton ball of to 2 mm in diameter soaked in 2 to 5% KCl solution for 1 min) was applied during successive recording sessions. In the 24- to 48-h period between two consecutive SD recording sessions, this plastic ring was kept filled with a larger cotton ball (2 to 3 mm in diameter) embedded in mineral oil to prevent air drying of the exposed cortical region. In each animal, an initial recording session was held immediately after electrode implantation, with the animal still under anesthesia (recordings ANEST. I; traces at left column). On the following 3 to 7 days, at least two additional recording sessions were performed in the same animals in the awake state (AWAKE, middle column). Finally, the last session was held 4 to 8 days after electrode surgical implantation. This last session consisted of a 2-h baseline recording, after which the animals were anesthetized again and the recordings continued for an additional 2- to 3-h period (ANEST. F; recordings at right column). The horizontal calibration bar equals 1 min. A –5 mV signal is shown at the beginning of each recording. (From Guedes, R.C.A. and Barreto, J.M. (1992) *Brazilian Journal of Medical and Biological Research* 25, 393–397. With permission.)

researchers interested in SD, both in the basic and in the clinical area (see Lehmenkühler et al., 1993; Gorgi, 2001).

Two main reasons motivated us to use the SD model: (1) SD provides an interesting and easy way of studying developmental and nutritional aspects of brain electrophysiology, and (2) we strongly believe that the complete understanding of SD mechanisms could be extremely important in helping develop better knowledge and treatment of human neuropathologies such as epilepsy, migraine, and brain ischemia. In this context, and as a final remark, I cite the impressive words of Professor Charles Nicholson on the importance of studying SD mechanisms. In the preface of the book by Lehmenkühler et al. (1993), on the possible role of SD in human migraine, he wrote:

Spreading depression remains seductive — and important. It is important not only because of the compelling evidence that it is the underlying phenomenon of migraine aura but also because it represents a great challenge to the completeness of our knowledge of the brain. No matter how many channel proteins we sequence, how many neuromodulators we identify and how many neural networks we construct, if we cannot explain spreading depression, we do not understand how the brain works.

Exaggerated? Overestimated? Highly biased opinion? I do not think so, but maybe some readers do. For those readers, here is the challenge: get some more information on the SD phenomenon. This can easily be achieved by consulting the pertinent literature cited in this chapter, as well as other seminal works, which can be found with the help of the Internet. Perhaps after that these readers will change their mind. And if they do not — well, they will find, for sure, an excellent way of learning and practicing electrophysiological techniques currently used to study brain development and function.

REFERENCES

Almeida, S.S., Duntas, L.H., Dye, L., Nunes, M.L., Prasad, C., Rocha, J.B.T., Wainwright, P., Zaia, C.T.B.V. and Guedes, R.C.A. (2002) Nutrition and brain function: a multidisciplinary virtual symposium. *Nutritional Neuroscience* 5, 311–320.

Amorim, L.F., Guedes, R.C.A., Medeiros, M.C., Silva, A.T. and Cabral-Filho, J.E. (1988) Apomorphine does not mimic the effects of REM-sleep deprivation on cortical spreading depression. *Brazilian Journal of Medical and Biological Research* 21, 611–614.

Andrade, A.F.D., Guedes, R.C.A. and Teodósio, N.R. (1990) Enhanced rate of cortical spreading depression due to malnutrition: prevention by dietary protein supplementation. *Brazilian Journal of Medical and Biological Research* 23, 889–893.

Boyes, W.K. (1994) Rat and human sensory evoked potentials and the predictability of human neurotoxicity from rat data. *Neurotoxicology* 15, 569–578.

Bronzino, J.D., Austin-Lafrance, R.J., Siok, C.J. and Morgane, P.J. (1986) Effect of protein malnutrition on hippocampal kindling: electrographic and behavioral measures. *Brain Research* 384, 348–354.

Bures, J., Buresova, O. and Huston, J. (1976) *Techniques and Basic Experiments for the Study of Brain and Behavior*. Elsevier, Amsterdam.

Chopra, J.S. (1991) Neurological consequences of protein and protein-calorie undernutrition. *Critical Reviews in Neurobiology* 6, 99–117.

Cintra, L., Durán, P., Angel-Guevara, M., Aguilar, A. and Castañón-Cervantes, O. (2002) Pre- and postnatal protein malnutrition alters the effect of rapid eye movements sleep-deprivation by the platform-technique upon the electrocorticogram of the circadian sleep-wake cycle and its frequency bands in the rat. *Nutritional Neuroscience* 5, 91–101.

Costa-Cruz, R.R.G. and Guedes, R.C.A. (2001) Cortical spreading depression during streptozotocin-induced hyperglycemia in nutritionally normal and early-malnourished rats. *Neuroscience Letters* 303, 177–180.

De Luca, B., Cioffi, L.A. and Bures, J. (1977) Cortical and caudate spreading depression as an indicator of neural changes induced by early malnutrition in rats. *Activitas Nervosa Superior* 19, 130–131.

El-Bachá, R.S., Lima-Filho, J.L. and Guedes, R.C.A. (1998) Dietary antioxidant deficiency facilitates cortical spreading depression induced by photo-activated riboflavin. *Nutritional Neuroscience* 1, 205–212.

Engel, R. (1956) Abnormal brain wave patterns in kwashiorkor. *EEG & Clinical Neurophysiology* 8, 489–500.

Gorelova, N.A., Koroleva, V.I., Amemori, T., Pavlik, V. and Bures, J. (1987) Ketamine blockade of cortical spreading depression in rats. *Electroencephalography and Clinical Neurophysiology* 66, 440–447.

Gorgi, A. (2001) Spreading depression: a review of the clinical relevance. *Brain Research Reviews* 38, 33–60.

Gramsbergen, A. (1976) EEG development in normal and undernourished rats. *Brain Research* 105, 287–308.

Grantham-McGregor, S. (1995) A review of studies of the effect of severe malnutrition on mental development. *Journal of Nutrition* 125, 2233S–2238S.

Guedes, R.C.A., Amâncio-dos-Santos, A., Manhães-de-Castro, R. and Costa-Cruz, R.R.G. (2002) Citalopram has an antagonistic action on cortical spreading depression in well-nourished and early-malnourished adult rats. *Nutritional Neuroscience* 5, 115–123.

Guedes, R.C.A., Amorim, L.F., Medeiros, M.C., Silva, A.T. and Teodósio, N.R. (1989) Effect of dietary lithium on cortical spreading depression. *Brazilian Journal of Medical and Biological Research* 22, 923–925.

Guedes, R.C.A., Amorim, L.F. and Teodósio, N.R. (1996) Effect of aging on cortical spreading depression. *Brazilian Journal of Medical and Biological Research* 29, 1407–1412.

Guedes, R.C.A., Andrade, A.F.D. and Cabral-Filho, J.E. (1987a) Propagation of cortical spreading depression in malnourished rats: facilitatory effects of dietary protein deficiency. *Brazilian Journal of Medical and Biological Research* 20, 639–642.

Guedes, R.C.A., Andrade, A.F.D. and Cavalheiro, E.A. (1988) Excitatory amino acids and cortical spreading depression. In Cavalheiro, E.A., Lehman, J. and Turski, L. (Eds.), *Frontiers in Excitatory Amino Acid Research*. Alan R. Liss, New York, pp. 667–670.

Guedes, R.C.A. and Barreto, J.M. (1992) Effect of anesthesia on the propagation of cortical spreading depression *Brazilian Journal of Medical and Biological Research* 25, 393–397.

Guedes, R.C.A., Cabral-Filho, J.E. and Teodósio, N.R. (1992) GABAergic mechanisms involved in cortical spreading depression in normal and early malnourished rats. In do Carmo, R.J. (Ed.), *Experimental Brain Research Series, Vol. 23: Spreading Depression*. Springer, Berlin, pp. 17–26.

Guedes, R.C.A., de Azeredo, F.A.M., Hicks, T.P., Clarke, R.J. and Tashiro, T. (1987b) Opioid mechanisms involved in the slow potential change and neuronal refractoriness during cortical spreading depression. *Experimental Brain Research* 69, 113–118.

Guedes, R.C.A. and Do Carmo, R.J. (1980) Influence of ionic alterations produced by gastric washing on cortical spreading depression. *Experimental Brain Research* 39, 341–349.

Guedes, R.C.A. and Frade, S.F. (1993) Effect of ethanol on cortical spreading depression. *Brazilian Journal of Medical and Biological Research* 26, 1241–1244.

Guedes, R.C.A. and Pereira-da-Silva, M.S. (1993) Effect of pre- and postnatal propylthiouracil administration on the propagation of cortical spreading depression of adult rats. *Brazilian Journal of Medical and Biological Research* 26, 1123–1128.

Hackett, R. and Iype, T. (2001) Malnutrition and childhood epilepsy in developing countries. *Seizure — European Journal of Epilepsy* 10, 554–558.

Hadjikhani, N., Del Rio, M.S., Wu, O., Schwartz, D., Bakker, D., Fischl, B., Wong, K.K., Cutrer, F.M., Rosen, B.R., Tootell, R.B.H., Sorensen, A.G. and Moskowitz, M.A. (2001) Mechanisms of migraine aura revealed by functional MRI in human visual cortex. *Proceedings of the National Academy of Sciences (USA)* 98, 4687–4692.

Leão, A.A.P. (1944) Spreading depression of activity in the cerebral cortex. *Journal of Neurophysiology* 7, 359–390.

Lehmenkühler, A., Grotemeyer, K.H. and Tegtmeier, F. (1993) *Migraine: Basic Mechanisms and Treatment*. Urban and Schwarzenberg, Munich.

Morgane, P.J., Austin-La France, R., Bronzino, J., Tonkiss, J., Díaz-Cintra, S., Cintra, L., Kemper, T. and Galler, J.R. (1993) Prenatal malnutrition and development of the brain. *Neuroscience and Biobehavioral Reviews* 17, 91–128.

Morgane, P.J., Miller, M., Kemper, T., Stern, W., Forbes, W., Hall, R., Bronzino, J., Kissane, J. Hawrylewicz, E. and Resnick, O. (1978) The effects of protein malnutrition on the developing central nervous system in the rat. *Neuroscience and Biobehavioral Reviews* 2, 137–230.

Nelson, G.K. and Dean, R.F.A. (1959) The electroencephalogram in African children; effects of kwashiorkor and a note on the newborn. *Bulletin of the World Health Organization* 21, 779–782.

Nunes, M.L., Teixeira, G.C., Fabris, I. and Gonçalves, R.A. (1999) Evaluation of the nutritional status in institutionalized children and its relationship to the development of epilepsy. *Nutritional Neuroscience* 2, 139–145.

Palencia, G., Calvillo, M. and Sotelo, J. (1996) Chronic malnutrition caused by a corn-based diet lowers the threshold for pentylenetetrazol-induced seizures in rats. *Epilepsia* 37, 583–586.

Quirck, G.J., Mejia, W.R., Hesse, H. and Su, H. (1995) Early malnutrition followed by nutritional restoration lowers the conduction velocity and excitability of the corticospinal tract. *Brain Research* 670, 277–282.

Rocha-de-Melo, A.P. and Guedes, R.C.A. (1997) Spreading depression is facilitated in adult rats previously submitted to short episodes of malnutrition during the lactation period. *Brazilian Journal of Medical and Biological Research* 30, 663–669.

Rocinholi, L.F., Almeida, S.S. and De-Oliveira, L.M. (1997) Response threshold to aversive stimuli in stimulated early protein-malnourished rats. *Brazilian Journal of Medical and Biological Research* 30, 407–413.

Rocinholi, L.F., De-Oliveira, L.M. and Colafemina, J.F. (2001) Malnutrition and environmental stimulation in rats: wave latencies of the brainstem auditory evoked potentials. *Nutritional Neuroscience* 4, 199–212.

Salas, M. and Cintra, L. (1975) Development of the electroencephalogram during starvation in the rat. *Physiology and Behavior* 14, 589–593.

Santos, R.S. (2000) Nutrição, hipertiroidismo precoce e desenvolvimento cerebral: estudo em ratos recém-desmamados. Master's Thesis, Recife, Universidade Federal de Pernambuco, 62 pp.

Santos-Monteiro, J., Teodósio, N.R. and Guedes, R.C.A. (2000) Long-lasting effects of early environmental stimulation on cortical spreading depression in normal and early malnourished adult rats. *Nutritional Neuroscience* 3, 29–40.

Schulte, F.J. and Bell, E.F. (1973) An atlas of EEG power spectra in infants and young children. *Neuropädiatrie* 4, 30–45.

Segura, B., Guadarrama, J.C., Gutierrez, A.L., Merchant, H., Cintra, L. and Jiménez, I. (2001) Effect of perinatal food deficiencies on the compound action potential evoked in sensory nerves of developing rats. *Nutritional Neuroscience* 4, 475–488.

Siesjö, B.K. and Bengtsson, F. (1989) Calcium fluxes, calcium antagonists and calcium-related pathology in brain ischemia, hypoglycemia and spreading depression: a unifying hypothesis. *Journal of Cerebral Blood Flow and Metabolism* 9, 127–140.

Silva, A., Costa, F.B.R., Costa, J.A., Teodósio, N.R., Cabral-Filho, J.E. and Guedes, R.C.A. (1987) Sciatic nerve conduction velocity of malnourished rats fed the human "basic regional diet" of the Northeast of Brazil. *Brazilian Journal of Medical and Biological Research* 20, 383–392.

Stern, W.C., Resnick, O. and Morgane, P.J. (1974) Seizure susceptibility and brain amine levels following protein malnutrition during development in the rat brain. *Brain Research* 79, 375–384.

Takano, K., Latour, L.L., Formato, J.E., Carano, R.A.D., Helmer, K.G., Hasegawa, Y., Sotak, C.H. and Fisher, M. (1996) The role of spreading depression in focal ischemia evaluated by diffusion mapping. *Annals of Neurology* 39, 308–318.

Tamer, S.K., Misra, S. and Jaiswal, S. (1997) Central motor conduction time in malnourished children. *Archives of Disease in Childhood* 77, 323–325.

Part II

Macronutrients, Brain Function, and Behavior

5 Diet, Cerebral Energy Metabolism, and Psychological Functioning

David Benton

CONTENTS

Although the brain is only 2% of the adult body weight, it accounts for ca. 20% of the resting metabolic rate. In contrast to other organs, the brain has a limited ability to store energy and is therefore dependent on a continuous supply of its major energy source glucose crossing the blood-brain barrier. The brain cannot regulate its uptake of nutrients or decrease the need for energy when the supply of fuel is limited. Thus, it is widely believed that the evolution of the sophisticated means of maintaining a relatively constant level of blood glucose, ca. 70 to 90 mg/dl, reflects the need to maintain a constant supply of glucose to the brain.

Cerebral stores of glucose are limited, sufficient to satisfy demands for ca. 10 min (Marks and Rose, 1981). Yet, unlike most other organs, the brain does not use free fatty acids for energy. Swanson (1992) showed that glycogen is found almost exclusively in glia, and glycogen metabolism is mobilized to meet the metabolic demands of glia rather than neurons. The small amounts of glycogen in the brain would support functioning for ca. 3 min. Thus, under normal circumstances, the brain depends on a continuous supply of glucose from the blood. The brain oxidizes ca. 120 g of glucose a day, which requires a correspondingly high rate of blood supply, ca. 50 ml/min/100 g tissue. This rate compares with 5 ml/100 g/min for skeletal muscle at rest and 50 ml/100 g/min during vigorous exercise. The use of glucose by the brain is higher in children than in adults (Kalhan and Kilic, 1999). In general, insulin does not influence the overall rate of glucose utilization by the brain, although there are insulin receptors in the brain and there is evidence of the local stimulation of glucose metabolism.

Only ca. 30% of glucose is required for direct energy production. Much of the remainder is used for the synthesis of amino acids, peptides, lipids, and nucleic acids; for example, a source of glucose is essential for the synthesis of physiologically active molecules, including a range of neurotransmitters.

5.1   FROM MEAL TO STARVATION: THE HOMEOSTATIC CONTROL OF BLOOD GLUCOSE

Glucose homeostasis is essential to provide the brain with a constant source of its basic fuel. The level of blood glucose is tightly controlled, reflecting that the central nervous system relies almost entirely on glucose as its source of energy. If blood glucose levels fall to very low levels, the brain's functioning is disrupted quickly and damage and death follow rapidly. The intake of food is intermittent yet the brain's need for energy is continuous. It follows that the excess fuel after a meal must be stored and later released.

Key roles in the storage and release of energy are played by the pancreatic hormones insulin and glucagon. Insulin is secreted in response to a number of stimuli, including an increase in blood glucose levels. The concentrations of glucose and insulin in the blood correspond closely in healthy individuals and both typically rise following a meal. Most tissues depend on insulin for the uptake of glucose from the blood. In its absence, glucose does not readily enter the cell and is hence not available for metabolic activities. The exceptions are the brain and the liver, where cells are permeable to glucose in the absence of insulin, although this does not mean that insulin does not influence neural functioning (see later).

Insulin has anabolic effects: it stimulates the synthesis of glycogen and decreases the activity of several enzymes involved in the gluconeogenic pathway (formation of glucose from noncarbohydrate precursors). Insulin also promotes the use of amino acids in peripheral tissues, thus decreasing the supply of amino acids that are normally a major substrate for gluconeogenesis in the liver. Insulin also influences lipid metabolism: it promotes the storage of excess fuel as triglycerides. In adipose tissue, insulin stimulates the uptake of glucose, where it is used either for fatty acid synthesis or to form glycerol phosphate, the backbone of triglycerides. The overall effect of insulin is to promote triglyceride synthesis while inhibiting its breakdown. Insulin also profoundly affects the metabolism of protein. It stimulates the transport of amino acids into muscle cells, stimulates protein synthesis, and decreases protein breakdown.

Glucagon affects many of the same metabolic pathways as insulin, but its influence is opposite. As blood glucose levels fall, the release of glucagon from the pancreas is stimulated. The principal target of glucagon is the liver, where it stimulates the conversion of glycogen to glucose (glycogenolysis). High levels of amino acids stimulate and fatty acids inhibit the release of glucagon. Gluconeogenesis, the conversion of nonglucose substrates such as amino acids into glucose, is a second mechanism by which blood glucose levels are raised. Glucagon affects lipid metabolism, encouraging the production of ketone bodies, which are used as a source of energy by muscle, sparing glucose for the brain.

Because of the opposing actions of insulin and glucagon, it has been suggested that the ratio between these hormones is important. After a high-carbohydrate meal, the molar ratio of insulin to glucagon can be as high as 30. Before breakfast, the ratio is ca. 2 and after a fast of 1 or 2 days the ratio maybe less than 0.5. From the perspective of ensuring a supply of energy to the brain, a series of stages, from eating to prolonged starvation, can be distinguished:

1. In the postabsorptive stage that follows a meal, glucose is released into the bloodstream following digestion. Glucose stimulates the release of insulin from the pancreas, which

stimulates the liver to synthesize glycogen. Glycogen is a readily mobilized storage form of glucose, a polymer of glucose residues. A man of average weight would have ca. 40 kcal of energy in the form of glucose whereas he would have more than 600 kcal in the form of glycogen even after an overnight fast, although more glycogen would be found in muscle than the liver (Frayn, 1996).

2. A major function of the liver is to take up glucose after a meal when it is plentiful and to release it later as required. Glycogen provides a store of glucose that is made available between meals or during exercise. As blood glucose is used up and levels fall, more glucose is released from the liver following the breakdown of glycogen (glycogenolysis). The enzyme glycogen phosphorylase in the liver initiates the breakdown of glycogen at a rate of ca. 100 mg/min. Glycogen in muscle can be used locally but is not released as glucose into the bloodstream.

3. When fasting exceeds ca. 8 h, gluconeogenesis begins to replace glycogenolysis as the means of supplying glucose to the blood and hence the brain and other glucose-consuming tissues such as erythrocytes. This change preserves glycogen stores, and after a 10-h fast glycogenolysis accounts for 70% of the total production of glucose (Tironte and Brunicardi, 2001). After ca. 24 h, the reserves of glycogen are exhausted. Gluconeogenesis is the synthesis of glucose from other precursors; substrates in increasing order of importance include lactate, alanine, and glycerol. Generally, gluconeogenesis is stimulated by glucagon and inhibited by insulin. In addition, glucagon increases glucose production by also stimulating glycogenolysis. Various counterregulatory hormones also stimulate glucose production: adrenaline promotes glycogenolysis and cortisol promotes gluconeogenesis.

4. Gluconeogenesis depends to a large extent on the breakdown of muscle protein. As ca. 1.75 g of muscle is required to produce 1 g of glucose, ca. 150 g of protein a day will be used to supply the brain with glucose (Frayn, 1996). To spare protein stores during prolonged starvation, other adaptations develop. Fatty acids can be converted by the liver into ketone bodies (ketogenesis), which are used as a source of energy by peripheral tissues and after 1 or 2 days of starvation by the brain. The significant use of ketone bodies by the brain decreases the need for glucose synthesis. Glucagon stimulates the hormone-sensitive lipase (lipolysis), increasing glycerol production, which becomes increasingly important for gluconeogenesis. In this way protein is spared. With prolonged fasting, as much as 50% of the energy requirement of the brain can be obtained from ketone bodies.

The traditional view has been that the existence of these homeostatic mechanisms ensures that the supply of glucose to the brain does not, under most normal conditions, limit functioning. Hence, from this perspective, the nature of the diet and its effect on blood glucose levels should not influence psychological functioning. However, as discussed next, this view is beginning to be questioned.

5.2 GLUCOSE TRANSPORT MOLECULES

Except in the proximal tubules of the kidney and the lumen of the small intestines, the transport of glucose across cell membranes occurs by the process of facilitated (carrier-mediated) diffusion (Pessin and Bell, 1992). Facilitated diffusion differs from active transport in that it is not energy dependent and cannot move glucose against the concentration gradient. Facilitated diffusion involves specific carrier proteins that shuttle glucose across the membrane at a higher rate than would result from diffusion alone. A family of transporter molecules has been identified: GLUT 1 (erythrocyte); GLUT 2 (liver); GLUT 3 (brain); GLUT 4 (muscle/fat); GLUT 5 (small intestine); GLUT 7 (liver microsomes); and GLUT 8, also called GLUTx1 (testis).

GLUT 1, GLUT 3, and GLUT 4 are present in high levels in the brain, although other transporters have been found in small amounts in discrete areas. GLUT 1 is responsible for glucose transport at the blood-brain barrier (capillary walls) and although it has a higher K_m than that of GLUT 3, it is not thought that it normally limits the rate at which glucose enters the brain. Under normal conditions, the rate of glucose transport across the barrier is two to three times higher than the rate of glucose utilization, ensuring that glucose metabolism is not normally limited by its transport into the brain (Pardridge, 1983). Once through the blood-brain barrier, glucose must diffuse through extracellular space and then cross neuronal membranes. GLUT 3 transports glucose into neurons. At relatively low glucose concentrations, this transport molecule is saturated; therefore, it is again thought that there is little variation in the rate of glucose uptake with variation in extracellular glucose levels (Pessin and Bell, 1992).

Glucose transport across the blood-brain barrier is an equilibrating mechanism and as such reflects blood glucose concentrations (Lund-Anderson, 1979). As glucose can move in both directions, the net flux will be in the direction of the concentration gradient, normally from the blood into the brain. It has been calculated that it takes up to 15 min for equilibrium to form after a change in the level of blood glucose (Lund-Anderson, 1979). However, at high glucose concentrations, carrier-mediated transport is downregulated (Gjedde and Crone, 1981). Brain extracellular glucose levels follow the levels of glucose in the blood and are typically ca. 25% of the blood concentration.

In cells that are insulin dependent, such as muscle, GLUT 4 promotes the movement of transporter molecules from intracellular sites into the cell membrane, where they become available for the transport of glucose. Although insulin increases the entry of glucose into muscle, the majority of evidence is that this hormone does not increase the uptake of glucose by the brain. However, this does not mean that insulin does not cross the blood-brain barrier and that insulin receptors are found within the brain (Schwartz et al., 1992). For example, in the hippocampus, glucose metabolism can be stimulated by the action of exogenous insulin acting through GLUT 4 insulin receptors (Hoyer et al., 1996).

5.3 POSITRON EMISSION TOMOGRAPHY

A positron emission tomography (PET) scan provides images of brain activity. A common approach involves the patient receiving an injection of mildly radioactive 2-deoxyglucose, a sugar that enters cells with glucose, although it is not metabolized. The head is scanned, and the emission of positrons that are emitted as the sugar decays is monitored. Thus, a picture is obtained of the level of radioactivity in various areas of the brain; higher levels of radioactivity are associated with greater metabolic activity. The first observation is that the radioactivity is not delivered uniformly throughout the brain; it is concentrated in specific metabolically active areas that differ with the task being performed. The second general observation is that within a few minutes, the radioactive sugar injected into the carotid artery can be detected in metabolically active areas of the brain. That is, metabolically active sites within the brain rely to a large extent on a rapid, continuous transport of glucose from the blood rather than on local stores of energy.

Thus, the use of PET scans gives an indication of the way the brain uses glucose. When faced with cognitive demands, alterations in the rate of glucose utilization in particular brain regions occur rapidly, with utilization increasing selectively in those areas associated with the processing of particular information. For example, the viewing of a complex visual scene increases glucose utilization in the occipital lobe that processes vision. In contrast, listening to music increases the use of glucose in the temporal lobes. PET studies have shown a decreased rate of cerebral metabolism during slow-wave sleep, although similar rates of glucose metabolism are found during rapid eye movement sleep as when awake. The changes in blood flow result from capillary recruitment with the associated provision of glucose. It is well established that regional glucose utilization and regional blood flow are closely related and more metabolically active nuclei contain more capillaries than those that are less active.

5.4 DOES THE SUPPLY OF GLUCOSE TO THE BRAIN MATCH DEMAND?

The important factor in the functioning of the brain is not the concentration of glucose in the blood after a meal but the levels found inside brain cells. As the blood-brain barrier does not limit the entry of glucose, it has often been suggested that the provision of glucose does not limit functioning. However, the important question is not whether glucose crosses the capillary wall quickly enough but whether capillary recruitment is always sufficient to direct sufficient blood to metabolically active areas. Pardridge (1983) speculated that the ability to transport glucose within the brain was not always sufficient to prevent intracellular glucose concentrations from declining. A related question is whether higher levels of blood glucose result in increased stores of glucose, which in the absence of an adequate blood supply maintain neuronal activity during times of higher metabolic demands.

Under normal conditions, it appears that the availability of glucose to the brain plays no part in the regulation of glycolysis. Rather, brain glucose utilization is limited by hexokinase, a high-affinity enzyme that is saturated by very low levels of glucose (Pardridge, 1983). Thus, it is probable that cerebral glycolysis is limited by hexokinase activity and is independent of the levels of blood glucose. The exception to this is a hypoglycemic condition, when the sugar supply is too low to maintain normal functioning. It has been shown that there is a close coupling between glucose phosphorylation and the transport of glucose into the brain. Traditionally, this coupling has been viewed as suggesting that there should be little change in the concentration of glucose when the brain is metabolically active unless the levels of plasma glucose are extremely low.

The traditional model of the supply of glucose to the brain relies on two basic assumptions: (1) the level of glucose is the same throughout the brain; that is, the brain can be treated as one compartment and (2) the level of glucose does not vary; that is, the capacity to transport glucose exceeds the demands placed on the brain. These assumptions create considerable problems when trying to understand reports that the administration of glucose, both peripherally and directly into the brain, enhances cognition (see later). However, both of these assumptions have been questioned.

The concentration of extracellular glucose represents a balance between its use by the cells and its supply by the blood. The assumption that the supply of glucose to the brain and its metabolism are closely coupled has begun to be challenged. For many years it was assumed that there is a single compartment for glucose in the brain with a concentration that was stable, and hence under normal circumstances it did not limit neural activity (Lund-Anderson, 1979). It was assumed that the brain was freely permeable to glucose, and hence the concentration was thought to be similar throughout the brain. If this model were correct, then providing glucose to the brain should not influence its functioning. This view has been challenged by reports that the concentration of brain glucose varies with brain area and with time within a particular area of the brain. It is becoming clear that in resting animals the levels of extracellular glucose vary with the area of the brain considered. Second, behavioral testing has been found to cause fluctuations in extracellular glucose levels, which appear to be localized to areas specifically involved in processing a particular task. Third, short-term variations in the level of extracellular glucose in the brain are not closely linked to the level of blood glucose.

McNay et al. (2000) measured the levels of extracellular glucose in both the hippocampus and striatum while rats were learning a maze. At rest, the level of glucose in the striatum was 70% of that in the hippocampus. When learning a maze, the level of glucose declined by 30% in the hippocampus but it increased by 9% in the striatum. The fall in hippocampal glucose reflected cognitive load. The level of extracellular glucose in the hippocampus declined by 32% when rats learned a four-arm maze but only by 11% when learning a simpler three-arm maze. These findings are consistent with the decline in brain glucose, being a reflection of the metabolic demands placed on the brain. It was particularly interesting that a peripheral injection of glucose both prevented a fall in extracellular glucose in the hippocampus and enhanced maze learning. However, administration of glucose did not change the extracellular concentrations of hippocampal glucose when the animal was at rest. It seems that during learning the demand for glucose by the hippocampus did not match its supply.

These recent findings suggest that providing glucose to the brain can no longer be viewed as occurring in a single compartment. The level of extracellular glucose is differentially controlled in various areas of the brain. The findings are also consistent with the view that one means by which the administration of glucose enhances memory is by increasing the supply of glucose from the blood to those metabolically active brain areas that are involved in memory. McNay found that providing blood glucose improved learning, although at the time of testing the level of blood glucose had returned to previous values. This observation suggested that pretreatment with glucose allows a reserve to be created in the brain. That pretreatment with glucose prevented the depletion of extracellular glucose in the hippocampus strongly supported this explanation. Lund-Anderson (1979) had summarized the evidence that over a period of 10 to 15 min, equilibrium develops between the level of glucose in the blood and that in the brain.

Analogous findings have been reported in humans. Parker and Benton (1995) argued that as the left side of the human brain is more metabolically active when processing verbal information, if the supply of glucose was inadequate, this side of the brain should be more susceptible to the taking of a glucose drink. They reported that the memory for a word list entering the right ear benefited more from a glucose drink than for lists directed to the left ear. Information entering the right ear is directed to the metabolically active left side of the brain.

5.5 HYPOGLYCEMIA

What is the implication of blood glucose levels outside the range that allows the brain to function normally? Development of sophisticated mechanisms to maintain blood glucose within a limited range suggests a functional importance. What happens if blood glucose levels fall to low values? There is no doubt that if blood glucose levels fall to very low levels, the functioning of the brain is quickly disrupted.

A common definition of hypoglycemia is that it occurs when the level of venous blood is less than 40 mg/dl within 6 hours of ingesting a hyperglycemic meal. Such a definition in no way implies that such a level of blood glucose is necessarily abnormal or pathological and mentions no symptoms. A distinction should be made between reactive hypoglycemia so defined and essential reactive hypoglycemia, in which these glucose values are associated with spontaneous symptoms. Repeated demonstrations of blood glucose levels less than 40 mg/dl, at the time when spontaneous symptoms occur, are necessary for the latter diagnosis.

Acute hypoglycemia is associated with changes in electroencephalographic measures; for example, reduced amplitudes in studies of sensory-evoked potentials have been reported. At blood glucose levels below ca. 55 mg/dl, the functioning of the brain begins to be disrupted. The effects may not be immediately obvious to the individual, but can be demonstrated by the use of tests of reaction times or sustained attention. Complex speed-dependent and attention-demanding tasks are mostly impaired, although accuracy may be maintained at the expense of speed (Deary, 1992). It is interesting that cognitive functioning may not fully recover for 1.5 h after normal blood glucose levels are restored. This observation raises questions about the ability of the brain to rapidly maintain optimal levels of glucose under demanding conditions.

Although there can be major individual differences, two types of reaction to very low blood glucose levels can be distinguished: autonomic and neuroglycopenic responses. Neuroglycopenia refers to the signs and symptoms that develop when the supply of carbohydrates to the neuron is inadequate for normal functioning. If the provision of glucose to neurons is compromised, then brain functioning is disrupted and confusion, amnesia, blurred vision, weakness, tiredness, and bizarre behavior result.

Both the sympathetic and parasympathetic autonomic symptoms are activated by centers in the hypothalamus, and symptoms such as sweating, tachycardia, and tremor may result. The stimulation of the release of adrenaline by the sympathetic nervous system increases the intensity of the autonomic responses, resulting in symptoms such as a pounding heart. It is easy to see how such symptoms might be perceived as those of anxiety. In young adults, secretion of counterregulatory hormones begins at a blood glucose level of ca. 70 mg/dl, which is just below the normal blood glucose range (Schwartz et al., 1987) and higher than the level at which psychologically adverse effects of low blood glucose are observed. A blood glucose level of approximately 40 mg/dl is about the level at which marked symptoms may occur, although there are individual differences. The performance of some individuals will be normal below 40 mg/dl, whereas in others adverse reactions occur with values as high as 70 mg/dl. In nondiabetic subjects, Snorgaard et al. (1991) found that those whose hypoglycemic symptoms were relieved by food tended to demonstrate cognitive dysfunction at higher blood glucose levels.

Many in the general population believe that although the intake of carbohydrate may give a short-term boost of energy, several hours later there can be a hypoglycemic response and an associated lowering of mood. The suggestion is that there is a carbohydrate-induced release of insulin, which, over time, causes low levels of blood glucose, resulting in irritability or aggression.

Harris (1924) introduced the term *spontaneous hypoglycemia* when he described nondiabetic patients who in response to their normal diet presented symptoms similar to those that result from insulin-induced hypoglycemia. The treatment simply involved the consumption of frequent small, low-carbohydrate and high-protein meals.

How many people become hypoglycemic? The ability to control blood glucose levels can be assessed by a glucose tolerance test (GTT). After fasting overnight, 50 g of glucose is consumed and the initial rise and subsequent fall in blood glucose are monitored. Lev-Ran and Anderson (1981) found a range of glucose nadirs when the test was administered to 650 normal subjects. The median nadir was 65 mg/dl and the 2.5th percentile was 39 mg/dl. However, when 135 patients suspected of suffering from reactive hypoglycemia were examined, only 4 had blood glucose levels sufficiently low to attract the diagnosis.

There is surprisingly little evidence on the change in blood glucose concentrations in healthy subjects during their everyday lives. Alberti et al. (1975) monitored normally fed individuals and reported very little variation in glucose concentration during the day. They commented that the changes are quite different to those obtained during a GTT. The picture that emerges in normally fed healthy subjects is of apparent blood glucose stability. Values rise following a meal and then fall, but typically not to levels that clinicians require to diagnose hypoglycemia (Benton, 2002). In normal individuals fed in the normal manner, a hypoglycemic response is uncommon. The degree of the elevation of blood glucose depends on the nature, texture, and form of food ingested, but, except in diabetics, to a surprisingly small extent on the actual amount.

Food-stimulated hypoglycemia is an uncommon condition, only rarely the cause of human ailments. Hogan et al. (1983) contrasted blood glucose levels in patients who had been diagnosed as hypoglycemic, who received the traditional GTT and on a second occasion an equivalent amount of glucose in a mixed meal. When 33 patients were examined, symptoms associated with low blood levels were observed in 19 during the GTT. In contrast, when administered with a meal, a similar glucose load did not result in blood glucose levels of less than 75 mg/dl.

5.6 GLUCOSE SENSORS AND HUNGER

The hypothalamus plays an important role in regulating metabolism. It integrates information from other parts of the brain and also contains sensors that monitor the concentration of glucose in the blood. The hypothalamus plays important roles in maintaining the level of blood glucose, including stimulating the release of adrenaline from the adrenal medulla when blood glucose levels are low.

In both rats and humans, feeding is preceded by a small decline in blood glucose levels (Campfield, 1997). It has been suggested that a subset of neurons are sensitive to insulin and these are involved in the control of feeding; for example, there are insulin receptors in the hypothalamus. In both the hypothalamus and liver, there are neurons sensitive to the local level of glucose or some aspect of its metabolism (Langhans and Scharrer, 1992). It is assumed that in the absence of insulin, glucose does not enter these neurons, decreasing the likelihood that action potentials are generated. In contrast, in the presence of insulin, glucose enters the neuron, leading to the generation of action potentials. 2-Deoxy-D-glucose (2-DG) competes with glucose to cross the membrane and when in the cell inhibits its metabolism. In animals, injections of 2-DG stimulate feeding. Feeding is triggered when nutrient levels in the liver are low and an injection of 2-DG into the hepatic portal vein stimulates feeding, whereas the injection of glucose induces satiety. It can be assumed that the decision to feed is made by the brain, which integrates information from the hypothalamus, the liver, and other factors such as the sensory quality of the food.

In the 1950s, researchers assumed that eating was regulated by a system that attempted to maintain blood glucose at a set point. The glucostatic theory of the control of food intake proposed that glucose utilization must be sensed in the brain or information must be relayed to the brain from the periphery (Mayer, 1953). According to this view, we become hungry when our blood glucose levels drop below a set point. In the 1950s, the level of glucose was said to initiate and end meals, whereas lipostatic mechanisms were thought to account for the long-term regulation of food intake. Set-point theories seem at present rather dated in that they fail to recognize the importance of taste, learning, and social factors. In evolutionary terms, it would have made little sense for our ancestors to stop eating as soon as the blood glucose levels increased. Meals would be intermittent and there was a need to build up energy reserves. Attempts to reduce meal size by preconsuming a high-calorie drink have been largely unsuccessful.

More recently, the positive-incentive theory has predominated, which suggests that we eat because we have evolved to enjoy food. Many factors influence the expectation that we will enjoy a meal, including the flavor of the food, the people present during the meal, and the previous experience of that food. Although the level of blood glucose is not of predominant importance, there is evidence that it plays a role. In addition, it is known that glucoreceptors within the brain increase the adrenal medullary secretion of adrenaline as part of the counterregulatory mechanisms when blood glucose levels fall to low values. The ventrolateral and dorsomedial hypothalamus are areas of the brain that respond to low levels of glucose by stimulating feeding and controlling the release of adrenaline. An insulin-sensitive glucose receptor is required to monitor glucose levels. GLUT 4 has been found in the hypothalamus and has been suggested to be part of a metabolic signaling system (Livingstone et al., 1995). High glucose concentrations are presumed to activate hypothalamic neurons, but only in the presence of insulin as would occur after a meal.

5.7 INSULIN AND THE BRAIN

Inevitably, a dietary-induced increase in blood glucose will be associated with the release of insulin, raising the question of whether the hormone may influence neural functioning. Traditionally, it has been said that the brain is not sensitive to the influence of insulin; rather, its peripheral action has attracted virtually all attention. The possibility that insulin has a role in the functioning of the brain is now being considered increasingly (Park, 2001). Many peripheral tissues are described as being insulin sensitive in that the hormone controls glucose uptake by its action on the insulin-sensitive glucose transport molecule GLUT 4. Because early studies found that GLUT 1 at the blood-brain barrier and GLUT 3 at the neuron were not influenced by insulin, the brain was described as insulin insensitive, a description that is now known to be misleading.

There are two main types of insulin receptors in the CNS, one found on neurons and the other peripheral type found on glia cells. Most insulin receptor immunoreactivity is, however, found on neurons. Insulin receptors are not uniformly distributed within the brain and occur in highest concentrations in the olfactory bulbs, hypothalamus, hippocampus, cerebellum, and cerebral cortex (Park, 2001). The insulin-sensitive GLUT 4 has been found in various areas of the brain, including the hippocampus (Vannucci et al., 1998).

There are two possible origins of insulin found in the brain. It may be synthesized by neural tissue or after a meal it may cross the blood-brain barrier after being released from the pancreas in response to increased levels of blood glucose. Park (2001), after reviewing the area, concluded: "the majority of evidence suggests that insulin is not synthesized in the adult brain although this may be dependent on species." Systemic insulin enters the brain through circumventricular organs lacking the blood-brain barrier and by a receptor-mediated transport system in the endothelial cells of micro-blood vessels. The transport of insulin into the cerebrospinal fluid from plasma has been reported in various species, and insulin is transported actively across the blood-brain barrier by a mechanism thought to be specific for insulin. As the transport mechanism is saturated at levels of insulin associated with optimal glucose levels (Banks et al., 1997), it appears that the mechanisms controlling peripheral blood glucose differ from those controlling insulin uptake by the brain.

Insulin influences the activity of enzymes involved in various aspects of carbohydrate metabolism, glycolysis, glycogen synthesis, and gluconeogenesis (Dimitriadis et al., 2000). Many of the enyzmes involved in the mitochrondrial tricarboxylic acid cycle are modulated by insulin, independently of the stimulation of glucose transport (Bessman and Mohan, 1997). The firing rate of neurons in the hippocampus and glucose metabolism in this area is sensitive to insulin (Hoyer et al., 1996). In summary, any description of the brain as insulin insensitive is inaccurate because of the following:

- Insulin is found in high levels in the brain and enters the brain through the blood-brain barrier.
- Insulin receptors and insulin secondary messenger systems are present in neurons.
- Changes in insulin level and insulin sensitivity are associated with cognitive deficits.
- Insulin may have an effect on cognition independently of any glucoregulatory role.
- Various possible mechanisms of action exist. By stimulating the insulin-sensitive GLUT 4 transporter in the hippocampus, glucose uptake may be stimulated.

5.8 GLUCOSE AND MEMORY

The hippocampus is important in modulating memory. The location of insulin receptors in the hippocampus, the ability of insulin to influence the uptake of glucose, and the rate of neuronal firing in this area of the brain suggest a possible role in diet-induced improvements in memory.

Interest in the influence of blood glucose levels on cognitive functioning was stimulated by attempts to develop drugs that facilitate memory and attention. Four types of related evidence suggest the general hypothesis that cognitive-enhancing drugs produce their effect by increasing the availability and uptake of glucose. Some cognitive-enhancing drugs do not cross the blood-brain barrier, others are effective when injected peripherally but not when injected directly into the brain, many cognitive-enhancing drugs are not effective following adrenalectomy, and cognitive functioning correlates with glucose regulation in aged animals and humans. These observations can be accounted for by the theory that many memory-enhancing drugs release adrenaline from the adrenal glands, which in turn releases glucose from the liver. The glucose enters the brain, where it facilitates memory. For example, Hall and Gold (1990), in a study of mice with adrenalectomy-induced memory deficits, found that injecting glucose restored memory as compared with sham-operated animals.

There is growing evidence of a similar phenomenon in healthy humans. For example, Benton and Owens (1993) found in young adults that higher blood glucose levels were associated with better memory. There is, however, no doubt that the memory of elderly rather than younger subjects is more responsive to a glucose drink. In a sample of the elderly, a story was more easily remembered following moderate increases in blood glucose, although a similar but smaller effect was observed in young adults (Hall et al., 1989; Manning et al., 1990). Consumption of a glucose-containing drink improved the memory of a group of healthy elderly people if administered either shortly before or immediately after hearing a story, suggesting a role in the consolidation of memory and not only in the act of learning. Memory improvement outlasted the transient increase in blood glucose and was better 24 h later (Manning et al., 1992). When the influence of glucose on a list of words was considered, its effect was predominantly on words toward the beginning rather than the end of the list, the so-called primacy effect (Messier et al., 1998). The selective effect of glucose on words from one part of a list demonstrates a specific rather than general action on the memory process, although the details are unclear.

Psychologists do not treat memory as a single entity; they divide it into various types. A major distinction is between explicit and implicit memory, also called declarative and nondeclarative memory. A key factor is whether the memory can be brought into conscious thought. Conscious memories that can be declared, that is, stated in words, are classified as declarative memory. In contrast, other memories cannot be verbally reported but are implied from particular tests. They are implicit, and examples include classical conditioning and motor skills such as riding a bicycle.

In the elderly, the ability to recall a story has been repeatedly enhanced by giving a glucose drink; that is, declarative memory is facilitated. Manning et al. (1997) found that glucose benefited declarative but not nondeclarative tasks in the healthy elderly. It has been suggested that glucose has a particular impact on declarative memory, although other functions are also influenced, to a lesser extent. It is widely accepted that the hippocampus and related brain areas mediate declarative memory in humans. It has been proposed that the hippocampus plays a critical role in encoding personal experience and relating it to previous experience—it is essential to the organization of memory. Available data indicate that glucose profoundly influences tasks involving hippocampal functioning. In animals, glucose influences tasks that depend on hippocampal functioning while having little influence on tasks that depend on other brain areas (Winocur and Gagnon, 1998).

5.9 DEMANDS OF THE TASK

Although there is increasing evidence that memory is particularly susceptible to the provision of glucose, there are reports that other activities also benefit, such as reaction times, the Stroop test, face recognition, kinaesthetic memory, and vigilance. However, some tests are susceptible to the level of blood glucose whereas others are not. Why are only some tests sensitive? Donohoe and Benton (1999a, 1999b) suggested that the cognitive demand induced by the task was critical. Although it is difficult to quantify cognitive demand, the duration and complexity of tasks can be considered. The provision of blood glucose has been found to influence difficult rather than easy forms of tests. When the duration of tasks is considered, raising blood glucose tends to be more influential toward the end of test sessions.

5.10 MEALS AND MEMORY

When the influence of breakfast on cognitive functioning has been considered, memory has been reported to be adversely influenced by fasting. In contrast, tests of attention and other aspects of cognition tend not to be affected. For example, Benton and Parker (1998) reported that taking a glucose drink can reverse some of the adverse effects on memory of missing breakfast. Benton and

Sargent (1992) found that in people who had eaten breakfast, memory correlated with the level of blood glucose.

5.11 MECHANISMS UNDERLYING MEMORY IMPROVEMENT

The ability of an injection of glucose into the ventricles to improve the memory of rats clearly demonstrates a central action (Lee et al., 1988). The disruption of memory that results from the peripheral injection of the anticholinergic drug scopolamine can be reversed by injecting glucose into the ventricles (Parsons and Gold, 1992). It seems that glucose can have a central effect, but what is the mechanism by which an improvement in memory is stimulated? Two broad possibilities exist. First, that the increase in glucose provision selectively benefits particular areas of the brain or particular metabolic pathways. By using this approach the impact of glucose on cholinergic mechanisms has attracted the most attention. Alternatively, glucose may act in a more general manner, acting as a metabolic substrate, with the additional energy being used to enhance the performance of energy-dependent cellular mechanisms such as ion pumps or reuptake mechanisms.

The association between acetylcholine-mediated neurotransmission and memory is well accepted. Glucose is the main source of the acetyl groups used to form acetyl-CoA, which when combined with choline forms acetylcholine. Messier et al. (1990) reviewed the topic and concluded that in resting conditions, increased glucose availability has little effect on acetylcholine levels in continuously fed animals. However, when there is a high demand for acetylcholine, a high availability of glucose increases the rate of synthesis of the transmitter. One situation associated with a high demand for acetylcholine is learning. Durkin et al. (1992), by measuring the increased release of acetylcholine from the rat hippocampus during neuronal activity, produced direct evidence that raised glucose levels facilitate acetylcholine synthesis. Taken together, these animal studies are consistent with the view that, under periods of neuronal activity, raising the glucose supply is associated with an increased synthesis of acetylcholine, which benefits memory.

A range of other mechanisms has also been considered. For example, the injection of morphine into the amygdala impaired memory, an effect that was reversed by injecting glucose into the same area of the brain (Gold, 1994). A peripheral injection of glucose decreased the firing of nigrostriatal dopamine-containing neurons (Saller and Chiodo, 1980). Galanin, a peptide found in the brains of mammals, including humans, is found associated with acetylcholine in a subpopulation of neurons that project to the hippocampus (Melander et al., 1986). The adverse effect of galanin on memory can be attenuated by administering glucose. The mechanism by which galanin and glucose affect the brain is unknown, but Stefani and Rokaeus (1998) speculated that the ATP-sensitive K^+ channel may be involved.

Given the importance of the liver in the control of blood glucose levels, the possibility has been considered that the memory-enhancing action of glucose is mediated by its action on this organ. Most of the autonomic nervous system messages from the liver to the brain pass through the coeliac ganglion. Lesions of this ganglion decrease the memory-enhancing effect of large doses of glucose (White, 1991). Fructose does not cross the blood-brain barrier and does not significantly increase the level of blood glucose, yet it improves memory in a manner similar to that of glucose. Two optimal doses of glucose enhance memory. Both fructose and glucose had the same effect at a dose of 2 to 3 g/kg, but only glucose improved memory at the lower optimal dose of 50 to 250 mg/kg (White, 1991). These data suggested that glucose acts by two mechanisms: a peripheral action that is shared with fructose and a direct action on the brain that is produced by glucose only.

Although the improvement in memory is a robust phenomenon, particularly in the elderly, the mechanism is unclear. Rather than a single mechanism, it is more likely that there are several or there is a sequence of effects. There is good evidence that glucose influences both peripheral and central mechanisms.

5.12   CEREBRAL METABOLISM AND NEURODEGENERATION

There is considerable evidence that abnormalities of cerebral metabolism occur in Alzheimer's disease and other neurodegenerative diseases. Gibson et al. (1998) reviewed the evidence that the functionings of mitochondrial enzymes are implicated, including the pyruvate dehydrogenase complex that links glycolysis to Kreb's cycle, the α-ketoglutarate dehydrogenase complex that links Kreb's cycle to glutamate metabolism, and cytochrome oxidase that links Kreb's cycle to oxygen utilization. However, there is need to consider whether this series of metabolic changes is secondary to other aspects of the disorder rather than its origin. There is extensive evidence that the rate of cerebral metabolism is reduced in Alzheimer's disease, although inevitably in part this reflects neurodegeneration. However, decreased cerebral metabolism occurs at the earliest signs of Alzheimer's disease and even precedes neuropsychological and imaging evidence of the disease. The occurrence of changes in metabolism in nonneural tissue suggests that not all these changes are secondary to neurodegeneration.

It has been suggested that an alleviation of the impact of a decline in the rate of cerebral metabolism is a probable mechanism by which glucose drinks benefit the memory of Alzheimer's patients. A major cause of reduced acetylcholine synthesis in this disease is likely to result from an inadequate supply of acetyl-CoA because of reduced pyruvate dehydrogenase activity. It should, however, be remembered that Alzheimer's disease is a heterogeneous disorder with several etiologies and risk factors. Hoyer (1994) proposed that reduced insulin receptor responses in the brain decrease the initiation of glycolysis, with a resulting decrease in acetylcholine production.

The possibility that there is impaired insulin activity in Alzheimer's disease has been considered (Hoyer, 1998). The disease is associated with abnormal levels of insulin and insulin receptors. Systemic insulin sensitivity is impaired in Alzheimer's disease, and this correlates with cognitive performance (Craft et al., 1993). It has been suggested that the insulin receptor may be associated with age-related problems of memory. In the brains of Alzheimer's patients, insulin concentration and the number of insulin receptors are reduced. In addition, administration of insulin enhanced the memory of Alzheimer's patients whose plasma glucose levels were kept constant (Craft et al., 1996). Hoyer (1998) proposed that sporadic late-onset Alzheimer's disease is caused by non-insulin-dependent diabetes mellitus that is confined to the brain.

There are fewer data in healthy subjects. Insulin administered through the nose has been reported to change auditory-evoked potentials, although there it had no influence on the levels of glucose and insulin in the blood (Kern et al., 1999).

5.13   SUMMARY

The brain is the most metabolically active organ in the body yet it has only limited reserves of energy. Homeostatic mechanisms have developed to ensure a continuous supply of glucose as the source of energy for the brain. Although glucose is normally the exclusive source of fuel for the brain, after a few days of fasting it adapts to use ketone bodies as a source of energy. The use of ketone bodies by the infant brain protects it while the supply of blood glucose is marginal. In both animals and humans, an improvement in memory after an increase in blood glucose levels is a robust phenomenon, particularly in the elderly. The evidence that glucose-induced improvements in memory, in part at least, act by central mechanisms conflicts with the traditional assumption that under almost all circumstances the brain is well supplied with glucose. The ability of diet-induced increases in blood glucose to improve memory is, however, consistent with findings that in rats maze learning is associated with a localized decrease in extracellular glucose that can be prevented by an exogenous source of glucose.

REFERENCES

Alberti, K.G.M., Dornhorst, A. and Rowe, A.S. (1975) Metabolic rhythms in normal and diabetic man. *Israel Journal of Medical Sciences,* 11, 571–580.

Banks, W.A., Jasper, J.B. and Kastin, A.J. (1997) Selective physiological transport of insulin across the blood-brain barrier: novel demonstration by species-specific radioimmunnoassays. *Peptides,* 28, 1257–1262.

Benton, D. (2002) Carbohydrate ingestion blood glucose and mood. *Neuroscience and Biobehavioral Reviews,* 26, 293–308.

Benton, D. and Owens, D. (1993) Blood glucose and human memory. *Psychopharmacology,* 113, 83–88.

Benton, D. and Parker, P.Y. (1998) Breakfast blood glucose and cognition. *American Journal of Clinical Nutrition,* 67, 772S–778S.

Benton, D. and Sargent, J. (1992) Breakfast blood glucose and memory. *Biological Psychology,* 33, 207–210.

Bessman, S.P. and Mohan, C. (1997) Insulin as a probe of mitochondrial metabolism in situ. *Molecular and Cellular Biochemistry,* 174, 91–96.

Campfield, L.A. (1997) Metabolic and hormonal controls of food intake: highlights of the last 25 years—1972–1997. *Appetite,* 29, 135–152.

Craft, S., Dagogo-Jack, S.E., Wiethop, B.V., Murphy, C., Nevins, R.T., Fleischman, S., Rice, V., Newcomer, J.W. and Cryer, P.E. (1993) The effects of hyperglycemia on memory and hormone levels in dementia of the Alzheimer type: a longitudinal study. *Behavioral Neuroscience,* 107, 926–940.

Craft, S., Newcomer, J., Kanne, S., Dagogo-Jack, S., Cryer, P., Sheline, Y., Luby, J., Dagago-Jack, A. and Alderson, A. (1996) Memory improvement following induced hyperinsulinemia in Alzheimer's disease. *Neurobiology of Aging,* 17, 123–130.

Deary, I.J. (1992) Diabetes, hypoglycaemia and cognitive performance. In *Handbook of Human Performance,* Vol. 2 (A.P. Smith and D.M. Jones, Eds.), pp. 243–278. Academic Press, London.

Dimitriadis, G.D., Raptis, S.A. and Newsholme, E.A. (2000) Integration of some biochemical and physiologic effects of insulin that may play a role in the control of blood glucose. In *Diabetes Mellitus: A Fundamental and Clinical Text* (D. LeRoith, S.I. Taylor and J.M. Olefsky, Eds.), pp. 161–176. Williams and Wilkins, Philadelphia.

Donohoe, R.T. and Benton, D. (1999a) Cognitive functioning is susceptible to the level of blood glucose. *Psychopharmacology,* 145, 3778–3785.

Donohoe, R.T. and Benton, D. (1999b) Declining blood glucose levels after a cognitively demanding task predict subsequent memory. *Nutritional Neuroscience,* 2, 413–424.

Durkin, T.P., Messier, C., de Boer, P. and Westerink, B.H.C. (1992) Raised glucose levels enhance scopolamine-induced acetylcholine overflow from the hippocampus: an *in vivo* microdialysis study in the rat. *Behavioral Brain Research,* 49, 181–188.

Frayn, K.N. (1996) *Metabolic Regulation.* Portland Press, London.

Gibson, G.E., Sheu, K.F.R. and Blass, J.P. (1998) Abnormalities of mitochrondrial enzymes in Alzheimer's disease. *Journal of Neural Transmission,* 105, 855–870.

Gjedde, A and Crone, C. (1981) Blood-brain barrier transfer: repression in chronic hyperglycaemic. *Science,* 214, 456–457.

Gold, P.E. (1994) Modulation of emotional and non-emotional memories: same pharmacological systems, different neuroanatomical systems. In *Brain and Memory: Modulation and Mediation of Neuroplasticity* (J.L. McGaugh, N.M. Weinberger and G.S. Lynch, Eds.), Oxford University Press, New York.

Hall, J.L. and Gold, P.E. (1990) Adrenalectomy-induced memory deficits: role of plasma glucose levels. *Physiology and Behavior,* 47, 27–33.

Hall, J.L., Gonder-Frederick, L.A., Chewning, W.W., Silveira, J. and Gold, P.E. (1989) Glucose enhancement of performance on memory tests in young and aged humans. *Neuropsychologia,* 27, 1129–1138.

Harris, S. (1924) Hyperinsulinism and dysinsulinism. *Journal of the American Medical Association,* 83, 729.

Hogan, M.J., Service, F.J., Sharbrough, F.W. and Genich, J.E. (1983) Oral glucose tolerance test compared with a mixed meal in the diagnosis of reactive hypoglycemia. *Mayo Clinic Proceedings,* 58, 491–496.

Hoyer, S. (1994) Neurodegeneration, Alzheimer's disease and beta-amyloid toxicity. *Life Sciences,* 55, 1977–1983.

Hoyer, S. (1998) Is sporadic Alzheimer's disease the brain type of non-insulin dependent diabetes mellitus? A challenging hypothesis. *Journal of Neural Transmission,* 105, 415–422.

Hoyer, S., Henneberg, N., Knapp, S., Lannert, H. and Martin, M.L. (1996) Brain glucose metabolism is controlled by amplification and desensitization of the neuronal insulin receptor. *Annals of the New York Academy of Sciences,* 777, 374–379.

Kalhan, S.C. and Kilic, I. (1999) Carbohydrate as nutrient in the infant and child: range of acceptable intake. *European Journal of Clinical Nutrition,* 53 (Suppl. 1), S94–S100.

Kern, W., Born, J., Schreiber, H. and Fehm, H.L. (1999) Central nervous system effects of intranasally administered insulin during euglycemia in men. *Diabetes,* 48, 557–563.

Langhans, W. and Scharrer, E. (1992) Metabolic control of eating. *World Review of Nutrition and Dietetics,* 70, 1–67.

Lee, M.K., Graham, S.N. and Gold, P.E. (1988) Memory enhancement with posttraining intraventricular glucose injections in rats. *Behavioral Neuroscience,* 102, 591–595.

Lev-Ran, A. and Anderson, R.W. (1981) The diagnosis of postprandial hypoglycemia. *Diabetes,* 30, 996–999.

Livingstone, C., Lyall, H. and Gould, G.W. (1995) Hypothalamic GLUT4 expression: a glucose- and insulin-sensing mechanism? *Molecular and Cellular Endocrinology,* 107, 67–70.

Lund-Anderson, H. (1979) Transport of glucose from blood to brain. *Physiological Reviews,* 59, 305–352.

Manning, C.A., Hall, J.L. and Gold, P.E. (1990) Glucose effects on memory and other neuropsychological tests in elderly humans. *Psychological Science,* 1, 307–311.

Manning, C.A., Parsons, M.W., Cotter, E.M. and Gold, P.E. (1997) Glucose effects on declarative and nondeclarative memory in healthly elderly and young adults. *Psychobiology,* 25, 103–108.

Manning, C.A., Parsons, M.W. and Gold, P.E. (1992) Antergrade and retrograde enhancement of 24-h memory by glucose in elderly humans. *Behavioral and Neural Biology,* 58, 125–130.

Marks, V. and Rose, F.G. (1981) *Hypoglycaemia,* 2nd ed. Blackwell Scientific, Oxford.

Mayer, J. (1953) Glucostatic mechanism of regulation of food intake. *New England Journal of Medicine,* 249, 13–16.

McNay, E.C., Fries, T.M. and Gold, P.E. (2000) Decreases in rat extracellular hippocampal glucose concentration associated with cognitive demand during a spatial task. *Proceedings of the National Academy of Sciences of the United States of America,* 97, 2881–2885.

Melander, T., Staines, W.A. and Rokaeus, A. (1986) Galanin-like immuno-reactivity in hippocampal afferents in the rat with special reference to cholinergic and noradrenergic inputs. *Neuroscience,* 19, 223–240.

Messier, C., Durkin, T., Mrabet, O. and Destrade, C. (1990) Memory-improving action of glucose: indirect evidence for a facilitation of hippocampal acetylcholine synthesis. *Behavioral Brain Research,* 39, 135–143.

Messier, C., Pierre, J., Desrochers, A. and Gravel, M. (1998) Dose-dependent action of glucose on memory processes in women: effect on serial position and recall priority. *Brain Research: Cognitive Brain Research,* 7, 221–233.

Pardridge, W.M. (1983) Brain metabolism: a perspective from the blood-brain barrier. *Physiological Reviews,* 63, 1481–1535.

Park, C.R. (2001) Cognitive effects of insulin in the central nervous system. *Neuroscience and Biobehavioral Reviews,* 25, 311–323.

Parker, P.Y. and Benton, D. (1995) Blood glucose levels selectively influence memory for word lists dichotically presented to the right ear. *Neuropsychologia,* 33, 843–854.

Parsons, M.W. and Gold, P.E. (1992) Scopolamine-induced deficits in spontaneous alteration performance: attenutation with lateral ventricle injections of glucose. *Behavioral Neural Biology,* 57, 90–92.

Pessin, J.E. and Bell, G.I. (1992) Mammalian facilitative glucose transporter family: structure and molecular regulation. *Annual Review of Physiology,* 54, 911–930.

Saller, C.F. and Chiodo, L.A. (1980) Glucose suppresses basal firing and haloperidol-induced increases in the firing rate of central dopaminergic neurons. *Science,* 210, 1269–1271.

Schwartz, M.W., Figlewicz, D.P., Baskin, D.G., Woods, S.C. and Porte, D.J. (1992) Insulin in the brain: a hormonal regulator of energy balance. *Endocrinological Reviews,* 13, 387–414.

Schwartz, N.S., Clutter, W.E., Shah, S.D. and Cryer, P.E. (1987) Glycemic thresholds for activation of glucose counterregulatory symptoms are higher than the threshold for symptoms. *Journal of Clinical Investigation,* 79, 777–781.

Snorgaard, O., Lassen, L.H., Rosenfalck, A.M. and Binder, C. (1991) Glycaemic thresholds for hypoglycaemic symptoms, impairment of cognitive function and release of counter-regulatory hormones in subjects with functional hypoglycemia. *Journal of Internal Medicine,* 229, 343–350.

Stefani, M.R. and Gold, P.E. (1998) Intra-septal injections of glucose and glibenclamide attenuate galanin-induced spontaneous alternation performance deficits in the rat. *Brain Research,* 813, 50–56.

Swanson, R.A. (1992) Physiologic coupling of glial glycogen metabolism to neuronal activity in brain. *Canadian Journal of Physiology and Pharmacology,* 70 (Suppl.), S138–S144.

Tironte, T.A. and Brunicardi, M.D. (2001) Overview of glucose regulation. *World Journal of Surgery,* 25, 461–467.

Vannucci, S.J., Koehler-Stec, E.M., Li, K., Reynolds, T.H., Clark, R. and Simpson, I.A. (1998) GLUT-4 glucose transporter expression in rodent brain: effect of diabetes. *Brain Research,* 797, 1–11.

White, N.M. (1991) Peripheral and central memory-enhancing actions of glucose. In *Peripheral Signaling of the Brain* (R.C.A. Frederickson, J.L. McGaugh and D.L. Felton, Eds.), pp. 421–441. Hogrefe and Huber, Toronto.

Winocur, G. and Gagnon, S. (1998) Glucose treatment attenuates spatial learning and memory deficits of aged rats on tests of hippocampal function. *Aging,* 19, 233–241.

6 The Acute Effects of Meals on Cognitive Performance

Caroline R. Mahoney, Holly A. Taylor, and Robin B. Kanarek

CONTENTS

It is well established that poor nutrition, particularly early in life, can have lasting effects on brain functioning and cognitive performance. In contrast, much less is known about the short-term effects of meals on cognitive behavior in well-nourished individuals. Interest in this area of research has primarily stemmed from the desire to improve cognitive performance in either the workplace or classroom, and, consequently, some meals have received more attention than others. However, the results of studies on the effects of meals on cognitive behavior can also increase our understanding of the basic manner in which nutrients affect brain functioning.

This chapter considers how meals may influence cognitive processing in a broad sense. It presents the issues confronted when investigating the effects on meals and cognitive performance and provides a review of experiments that have addressed the consequences of intake of breakfast, lunch, dinner, and snacks on the cognitive performance of individuals across the lifespan.

6.1 WHY IS IT IMPORTANT TO STUDY THE EFFECTS OF MEALS ON COGNITIVE BEHAVIOR?

Studying meals is important because it represents the way that people actually eat. Although studies examining the effects of single nutrients on mental performance may be easier to interpret and produce larger effects than studies using mixed meals, results of studies using single nutrients are difficult to frame in the context of normal eating behavior. People do not consume individual nutrients; rather, they eat meals or snacks containing varying amounts of the three energy-containing nutrients — protein, fat, and carbohydrate. However, although it is important to investigate the effects of actual foods on cognitive behavior, studying meals is complicated. First, it is often difficult to create a comparison group or placebo for a meal, short of eating vs. not eating. The lack of a suitable placebo may affect the experience of the participant on several levels as well as create methodological problems. For example, without a placebo, it is impossible to conduct double-blind experiments, leading to the possibility of subject and experimenter bias.

Results of research on the effects of meals on cognitive behavior differ significantly among studies; for example, some experimenters have found effects of meals on behavior whereas others have not. Several explanations exist for these inconsistencies. Test meals, in terms of calories or macronutrient content, cognitive tasks used, and times between meal consumption and testing, differ greatly across experiments. With little consistency in meals or methodology in experimental studies, the variation in results is not surprising.

6.2 VARIABLES TO CONSIDER WHEN INVESTIGATING THE EFFECTS OF MEALS ON COGNITIVE BEHAVIOR

6.2.1 CHARACTERISTICS OF THE MEAL

Both the size and macronutrient content of a meal contribute to the way in which the meal influences mental performance. With respect to meal size, children displayed greater improvements on measures of creativity, physical endurance, and mathematical ability when they received a high-energy breakfast as compared to a low-energy breakfast (Wyon et al., 1997). In contrast, other studies have suggested that a large lunch is associated with a greater decline in cognitive performance in the afternoon than is a small meal. Participants' normal meal habits also must be considered when evaluating the effects of meal size on cognitive behavior. For example, individuals who regularly consume a large lunch are less affected by the consumption of a large meal at lunchtime than those who normally eat a small meal (Craig and Richardson, 1989). The specific macronutrient content of the meal is also important in determining the cognitive effects of the meal. Studies examining the effects of lunch on subsequent behavior have demonstrated that high-protein meals typically lead to increased distractibility, whereas high-carbohydrate meals are associated with slowing of reaction times (Smith et al., 1988).

6.2.2 TIME

The ways in which a meal affects cognitive performance vary depending on the time of day at which it is consumed. Identical meals eaten in the morning, midday, or evening can have different effects on cognitive behavior. Endogenous biological rhythms may be one explanation for the differential effects of meals on performance. In general, individuals tend to be alert and perform well on cognitive tasks during the early part of the day, where as in the afternoon many individuals suffer from a decrease in alertness and efficiency (Smith et al., 1988). Real-life measures of attention, such as the frequency of nodding off while driving a car or the potential for a train engineer to miss a warning signal, illustrate that performance typically declines around 2:00 p.m. (Folkard and Monk, 1985). Laboratory investigations examining fluctuations in cognitive behavior across the day have also shown that performance efficiency improves until midday, at which time there are reductions in performance on tasks measuring simple reaction times, sustained attention, and ability to perform mathematical problems (Blake, 1967). Later in the day, performance on these tasks again improves. Whether the midday decrement in performance is simply due to endogenous rhythms or due to the consumption of lunch is still not clear. For example, performance on a serial reaction time task was slower in the afternoon than in the morning for subjects, whether they had or had not eaten lunch. In contrast, measures of perceptual discrimination (Craig et al., 1981; Smith and Miles, 1986a) declined during midafternoon in participants who had consumed lunch but not in those who abstained from eating. It is possible that the midday decrement in performance observed on some performance measures may be due to both endogenous rhythms and the consumption of lunch (Smith and Miles, 1986b).

The interval between when a meal is consumed and the time of testing may also affect performance measures. For example, observations made 15 or 30 min after consuming a meal may differ from those obtained 3 h later. In addition, the interval between the test meal and the last meal consumed must be considered. Performance may be quite different following a meal if participants have been previously fasting than if they have recently consumed another meal. These effects may be due to differences in gastric contents, blood glucose levels, or, simply, feelings of satiety.

6.2.3 PERSONAL CHARACTERISTICS

Age can influence the effects of a meal on performance in a number of ways. Younger children are generally more affected by breakfast consumption than are older children (Mahoney et al., unpublished data). In addition, the negative effects of lunch on cognitive performance increase with age (Spring et al., 1982). Personality also influences the way in which meals affect performance. It has been reported, for example, that following a lunchtime meal, subjects who were less anxious tended to show greater performance impairments than those who were more anxious (Craig et al., 1981).

Finally, nutritional status influences the effects a meal has on performance. It has been repeatedly observed that children who are poorly nourished are more adversely affected by missing breakfast than those who are considered well nourished. For example, breakfast consumption significantly improved performance on measures of verbal fluency, memory, arithmetic performance, and perceptual abilities in nutritionally at-risk children, but had minimal effects in children who were adequately nourished (Pollitt, 1995).

6.3 HOW MIGHT MEALS AFFECT COGNITIVE PROCESSES?

One way that meals are thought to affect cognitive function is through the changes in blood glucose concentrations that result from food ingestion (see Chapter 5). Glucose is an essential component for brain metabolism and provides the brain with almost all of its energy (Benton et al., 1996). If the concentration of glucose in the blood becomes too low, the result could range from impairment of mental function to coma or death (Harper, 1957). The synthesis of neurotransmitters essential for fine motor skills and cognitive processes, such as noradrenaline, acetylcholine, and serotonin, is also dependent on glucose metabolism.

There is considerable evidence that even small increases in circulating blood glucose concentration can lead to the enhancement of learning and memory across the lifespan (Gold, 1986, 1995). This enhancement is seen in the form of an inverted-U dose–response curve, with optimal levels of cognitive enhancement typically reported after a glucose load of 25 g (Parsons and Gold, 1992). Glucose-induced improvements in memory do not occur only in those whose blood glucose levels are initially low; rather, they occur irrespective of initial blood glucose level (Benton and Owens, 1993).

Glucose ingestion has been shown to enhance memory for learned material regardless of whether it is administered before the acquisition of new material or recall of the material (Hall et al., 1989; Manning et al., 1990, 1992, 1998; Craft et al., 1992). In addition, glucose improves memory for learned material even when recall is tested 24 h after acquisition (Manning et al., 1992).

It is important to recognize that glucose ingestion does not always lead to improvements in memory. Gonder-Frederick et al. (1987) found no significant differences between individuals given glucose or a placebo on simple memory tests. It appears that glucose enhancement of memory is type specific. For example, Gold (1995) found that glucose ingestion enhanced verbal declarative memory but not implicit memory. He proposed that memory stores are differentially sensitive to blood glucose concentrations, which could account for the selective increase in performance on some but not all memory tasks (Gold, 1995).

Another possible explanation for the effects of food on cognitive performance is that the individual macronutrients consumed within a meal differentially affect cognitive performance (Kaplan et al., 2001; Fischer et al., 2002). A study done by Kaplan et al. (2001) examined this issue. Participants were given a pure protein drink, a pure carbohydrate drink, a pure fat drink, or a nonenergy placebo drink on four separate mornings. Pure dietary protein, carbohydrate, and fat enhanced memory performance 15 min after ingestion, independently of elevations of blood glucose (Kaplan et al., 2001). However, 60 min after ingestion, only the glucose drink was associated with improved memory. In addition, only protein was associated with a decrease in forgetting on a paragraph recall test and only fat and glucose improved performance on the Trails test. In a study examining the effects of the carbohydrate-to-protein ratio in a morning meal on cognitive performance in younger adults, Fischer et al. (2002) found that a protein-rich or a balanced meal resulted in better cognitive performance than a carbohydrate-rich meal. Thus, though energy ingestion may influence some areas of cognitive performance, each macronutrient may have independent effects on cognitive processes. These differences may be the result of varying neurotransmitter synthesis.

The ingestion of a meal can influence the synthesis of brain neurotransmitters (Wurtman, 1982). For example, a meal rich in carbohydrates increases the amount of brain tryptophan, which results in an increase in the synthesis of serotonin (Lieberman et al., 1986). The increase in brain tryptophan is a result of insulin secretion following ingestion of carbohydrates. Insulin causes most of the other amino acids to leave the bloodstream and be taken up into the muscle, whereas tryptophan remains in the blood (Spring, 1986). Consequently, the flow of tryptophan into the brain is increased as the competition with other amino acids for transport across the blood-brain barrier is decreased. The increase of tryptophan to the brain causes an increase in the synthesis and release of serotonin. This effect can easily be reversed, however, if a meal contains even a small amount of protein. For example, when tested in rats, a meal containing only 6% protein reduced brain tryptophan levels (Glaeser et al., 1983).

Consumption of a protein-rich meal leads to an increase in the level of brain tyrosine, which results in an increased synthesis of neurotransmitters such as dopamine and norepinephrine (Lieberman et al., 1986). Because a protein-rich meal contains both tryptophan and tyrosine, it seems likely that this would lead to an increase of both brain tryptophan and tyrosine levels. However, because tryptophan is the least abundant amino acid in a protein-rich meal and all amino acids compete for the same transport molecules to cross the blood-brain barrier, the ingestion of a high-protein meal decreases the amount of tryptophan available to the brain (Wurtman, 1986). On the other hand, the plasma ratio of tyrosine increases after a protein-containing meal, and therefore more tyrosine is available for uptake into the brain. Thus, the synthesis of dopamine and norepinephrine increases after a protein-rich meal, whereas the synthesis of serotonin decreases.

The altered levels of neurotransmitters produced by food consumption may in turn exact changes in mood and cognitive performance. For example, Lieberman et al. (1986) examined the effects of tryptophan and tyrosine on mood, reaction time, and motor performance. They found that when participants consumed a tryptophan pill, they rated themselves as being less alert, having less vigor, and being more fatigued compared to when they consumed a tyrosine pill or a placebo. In addition, participants had slower reaction times after tryptophan than after tyrosine (Lieberman et al. 1986). Several studies looking at the effects of carbohydrate-rich meals compared to protein-rich meals have found similar results: a carbohydrate-rich meal increased drowsiness and calmness compared to a protein-rich meal (Spring et al., 1987).

6.4 BREAKFAST

Breakfast is often described as the most important meal of the day. Because breakfast follows the longest period of fasting, skipping breakfast could result in a decrease in the amount of nutrient availability to the brain and ultimately lead to a decline in cognitive performance (Pollitt, 1995). Despite this, several studies have found that nearly 25 to 60% of children in the U.S. are sent to school each day with nothing to eat between their evening meal at approximately 6:30 p.m. and lunch the next day (Siega-Riz, 1998; Mahoney et al., unpublished data). These findings have led researchers to address the question of how a prolonged fast affects a child's ability to perform in school.

6.4.1 CHILDREN

Results of numerous experiments suggest that missing breakfast can have detrimental effects on cognitive performance (Pollitt, 1995; Pollitt et al., 1983; Wesnes et al., 2003). In a study assessing the effects of skipping breakfast on problem-solving ability in young well-nourished boys, Pollitt et al. (1983) reported that the boys did worse in late-morning tests of problem-solving abilities when they had skipped breakfast than when they had consumed it. Skipping breakfast has also been linked to impairments in attention and episodic memory, which increase in severity over the morning (Wesnes et al., 2003).

The interest in the effects of breakfast on cognitive performance led to several investigations of the National School Breakfast Program. An early study in Los Angeles, CA, looked at children in Grades 3 to 6 in two schools (Lieberman et al., 1976). One school was offered free breakfast and the other was not. Performance measures included school attendance and arithmetic and reading scores. There was no difference in the performance of children who did and did not participate in the school breakfast program. However, the children were not randomly assigned to breakfast or control groups, so that children who did not participate in the school breakfast program were not suffering from the lack of nutrients and were consuming diets as adequate as those in the breakfast program.

More recent studies have indicated that consumption of a school breakfast can have beneficial effects on academic performance. For example, in Lawrence, MA, children in Grades 3 to 6 who participated in a school breakfast program on at least 60% of the school days performed better on a standardized test and had lower school tardiness rates than children who did not partake of the breakfast program. Similarly, research conducted in Philadelphia and Baltimore found that across the school year, children who regularly ate a school-supplied breakfast displayed significantly greater improvements in math scores and significantly greater decreases in rates of school absences and tardiness than those who rarely ate breakfast (Murphy et al., 1998).

Studies in other countries as well as the U.S. provide evidence that breakfast consumption can have positive outcomes on academic performance. In Israel, Vaisman et al. (1996) reported that children who ate a school breakfast that contained 30 g of sugared corn flakes and 200 ml of 3% milk performed better on memory and learning tasks than children who did not eat breakfast or who ate at home. Similarly, studies from Spain demonstrated that children who regularly consumed an adequate breakfast (i.e., more than 20% of total daily caloric intake) scored higher on a test of logical reasoning than did children who consumed a less adequate meal (Lopez-Sobaler et al., 2003). Research in Jamaica demonstrated that consumption of a standard government breakfast positively affected school attendance and arithmetic performance in undernourished children. Other studies, also done in Jamaica, showed that children's nutritional status must be considered when investigating the effect of breakfast on academic skills. For example, Simeon and Grantham-McGregor (1989) found that breakfast intake improved measures of verbal fluency, memory, arithmetic performance, and perceptual abilities in children who were considered nutritionally at risk. In contrast, breakfast did not affect cognitive performance in well-nourished children.

6.4.2 ADULTS

Consumption of breakfast has positive effects on cognitive behavior in adults as well as children. In studies conducted more than 50 years ago, Tuttle et al. (1952) compared effects of a heavy (800 kcal), light (400 kcal) or no breakfast or just coffee on simple and choice reaction times. Results indicated that subjects in the no-breakfast group took longer to respond than subjects in the other three groups. However, the no-breakfast group was the only group that did not receive coffee. Thus, it is possible that the observed effects were due to the caffeine in the coffee rather than the meal vs. no meal manipulation. Caffeine itself can enhance performance on vigilance tasks, increase alertness, and improve complex behavior such as that required for driving a car or flying a plane (Brice and Smith, 2001). Most consistently, caffeine seems to improve performance on tasks that are boring or require sustained attention, especially under conditions of sleep deprivation (Committee on Military Nutrition Research, 2001). Thus, the inclusion of caffeine in a test meal could confound results of measures of cognitive performance if the caffeine is not present in the control condition also.

More recent studies controlling for the presence of caffeine have also demonstrated the beneficial effects of breakfast consumption in adults. For example, Benton and Sargent (1992) found that university students (mean age 21 years) took significantly less time to recall a list of words and to finish a spatial memory task when they had eaten breakfast than when they had not. Similarly, Smith et al. (1994a) reported that subjects recalled more words in a test of free recall and made fewer errors in a test of recognition memory when they had eaten breakfast cereal than when they had not had breakfast.

6.4.3 ELDERLY

The elderly, like the young, may be particularly sensitive to the effects of nutrition on cognitive behavior. Elderly individuals, as a result of disease, a decreasing ability or desire to prepare meals, or problems associated with a reduction in income, are more likely to suffer nutritional deficiencies than their younger counterparts. Thus, breakfast intake may have greater effects on cognitive performance in older than in younger individuals. Studies done in the early 1950s, however, do not support this hypothesis (Tuttle et al., 1952, 1953). Cognitive performance, as measured by a choice reaction task, was similar in elderly adults, aged 60 to 83 years, given no breakfast or one of two isocaloric breakfasts that contained either 25 g protein, 37 g fat, and 80 g carbohydrate or 25 g protein, 28 g fat, and 100 g carbohydrate (Tuttle et al., 1952). In a follow-up to the preceding study, Tuttle et al. (1953) compared the effect of a heavier breakfast (998 kcal) vs. two more basic breakfasts (744 to 750 kcal) on reaction times in adults aged 60 to 84 years. Each participant again participated in all three conditions. In this study, subjects had slower reaction times when they had consumed the heavier breakfast than when they had consumed either of the two less-caloric breakfasts.

Although the early studies cited previously suggest that breakfast intake does not play an important role in mediating cognitive behavior in the elderly, more recent studies counter this suggestion. Elderly adults, aged 60 to 79 years, who ate breakfast cereal every day performed better on a national adult reading test than those who either ate breakfast every day but not always cereal or infrequently consumed breakfast. These results, although indicating that breakfast consumption improves intellectual functioning, can also be interpreted as demonstrating that healthier, more cognitively aware elderly are more likely to consume breakfast than their less fortunate counterparts.

6.4.4 BREAKFAST TYPE

As detailed in the previous sections, breakfast intake is not consistently associated with improvements in cognitive performance (Table 6.1). Differences in the quantity and quality of foods consumed at breakfast may account, in part, for these mixed results. With respect to the quantity of food consumed at breakfast, Wyon et al. (1997) reported that school-aged children did better on tests of creativity, physical endurance, and mathematical ability when they had consumed a high-energy breakfast (bread rolls, light margarine, soft processed cheese, a boiled ham slice, milk, cornflakes, an apple, and orange juice) than when they had consumed a low-energy breakfast (bread rolls, light margarine, diet raspberry jam, and a diet orange cordial). Similarly, Michaud et al. (1991) showed that a high-energy breakfast, as a compared to a low-energy breakfast, improves short-term memory in young adults. Although these studies suggest that consumption of a large breakfast may have beneficial effects on cognitive behavior relative to a small breakfast, they do not take into account the role of the participants' normal patterns of food intake on the dependent variables. With respect to this issue, studies have found that individuals did worse on cognitive tests when the experimental breakfast was either significantly larger or smaller than the breakfast they normally consumed.

The macronutrient content of a morning meal must also be considered when evaluating the consequences of the meal on cognitive behavior. For example, intake of pure protein, fat, or carbohydrate following an overnight fast improved performance on a short-term memory task in healthy elderly subjects; intake of each of the macronutrients affected cognitive behavior in a unique manner. More specifically, carbohydrate intake was associated with more sustained improvement on the memory task than were protein or fat intakes, whereas protein, but not carbohydrate or fat intake, decreased the rate of forgetting on a paragraph recall test (Kaplan et al., 2001). Further support for the importance of the nutrient quality of breakfast in determining subsequent behavior comes from a study by Lloyd et al. (1996), who reported that mood, but not cognitive performance, was improved when subjects consumed a low-fat, high-carbohydrate breakfast relative to either a medium-fat, medium-carbohydrate breakfast, a high-fat, low-carbohydrate breakfast, or no breakfast (Lloyd et al., 1994).

In two recent studies, Mahoney and coworkers (unpublished data) examined the effects of two common breakfast foods, instant oatmeal and ready-to-eat cereal, vs. no breakfast on cognitive performance in elementary-school children. The two cereals were similar in calories but differed in their macronutrient composition, processing characteristics, effects on digestion and metabolism, and glycemic index. More specifically, each cereal contained the same amount and type of sugar, but the oatmeal contained approximately three times the protein and fiber of the ready-to-eat cereal. Each child participated in all conditions separated by a week. In the first experiment, 9- to 11-year-old boys and girls did significantly better on a spatial memory task presented 1 h after eating instant oatmeal than after having no breakfast. In addition, after consuming either breakfast cereal, children more accurately copied a complex visual display than when they had not consumed breakfast. Short-term memory was influenced by breakfast intervention and sex: consumption of the instant oatmeal improved performance for girls compared to when they consumed the ready-to-eat cereal breakfast or no breakfast, but for boys performance did not differ based on breakfast intervention. In the second experiment, 6- to 8-year-old children performed better on a spatial memory task when they had oatmeal breakfast than when they did not. There were no differences in performance between the ready-to-eat cereal condition and the no breakfast condition. Again, as with the older children, analysis of the short-term memory task revealed that girls recalled more when they had oatmeal than in the other two conditions. In addition, performance on an auditory attention task was best after consuming oatmeal, followed by no-breakfast, and finally ready-to-eat cereal.

TABLE 6.1
Effects of Breakfast on Cognitive Performance

Author	Participants	Manipulation	Meal	Cognitive Tasks	Results
Chandler et al. (1995)	Children, Grades 3 and 4	Breakfast vs. no breakfast for undernourished and adequately nourished	Bread, cheese, chocolate milk 2174 kJ, 21.3 g protein Placebo: One fourth of an orange	Visual search Digit span Verbal fluency Information processing	Verbal fluency: undernourished children's performance improved with breakfast
Cromer et al. (1990)	Children, Grade 9	Government breakfast vs. a low calorie breakfast	Government: doughnut, orange juice, chocolate milk 424 kcal, 11.5 g protein, 63.9 g carbohydrate, 14.1 g fat Low-calorie meal: 8 oz sugar-free powdered drink mix and half cup sugar-free gelatin 12 kcal, 1.6 g protein, 1 g carbohydrate, 0 g fat	Continuous performance task (CPT) Matching familiar figures Rey auditory-verbal	No differences
Duam et al. (1955)	Children, 12 to 14 years	Size and content of breakfast	745 kcal, 24.5 g protein, 28.6 g fat, 97.9 g carbohydrate 744 kcal, 24.8 g protein, 36.8 g fat, 97.9 g carbohydrate 797 kcal, 26.6 g protein, 44.9 g fat, 72.6 g carbohydrate 1215 kcal, 40.9 g protein, 53.6 g fat, 143 g carbohydrate	Choice reaction time	No differences between breakfast content or breakfast size
Lieberman et al. (1976)	Children, Grades 3 to 6	School breakfast vs. no breakfast	?	School attendance Reading scores Arithmetic scores	No difference
Lopez et al. (1993)	Children, Grades 4 to 6	Breakfast vs. no breakfast for normal, wasted, and stunted children	394 kcal, 6 g protein No breakfast	Memory test Domino test — problem solving Attention test	No differences
Politt et al. (1981)	Children 9 to 11 years	Breakfast vs. no breakfast	Waffles, syrup, margarine, orange juice, and milk 535 kcal, 15 g protein, 20 g fat, 75 g carbohydrate Or no breakfast	Matching familiar figures test (MFFT) Continuous performance task Hagen central-incidental task (HCI)	Problem solving better with breakfast Fasting better in short-term memory

Study	Population	Comparison	Breakfast composition	Tests	Results
Politt et al. (1996)	Children, Grades 4 and 5	Breakfast vs. no breakfast for at-risk and no-risk children	80 g Small cake 50 g Almilac (similar to milk) Or no breakfast	Number discrimination; Peabody picture vocabulary test; Raven progressive matrices; Reaction time; Stimulus discrimination; Sternberg memory search	No effect; No effect; No effect; No effect; RT of no-risk kids slower with breakfast; RT slower without breakfast for at-risk kids
Simeon and Grantham-McGregor (1989)	Children, 9 to 10.5 years	Breakfast vs. no breakfast for stunted, nonstunted, and previously severely malnourished	Nutribun, milk, cheese, 590 kcal, 29 g protein, 12 g fat, 91 g carbohydrate Or tea sweetened with aspartame	Arithmetic; Digit span; Coding; Fluency; Listening comprehension; MFFT; HCI	Groups 1 and 3 had lower scores in fluency and coding without breakfast; Group 2 had higher arithmetic score with no breakfast; Wasted children performed worse on backwards digit span and MFFT without breakfast; Normal kids performed better on MFFT without breakfast
Vaisman et al. (1996)	Children, 11 to 13 years	School breakfast vs. home breakfast vs. no breakfast; time of breakfast	30 g Sugared corn flakes and 200 ml of 3% fat milk	Rey auditory verbal learning test; Logical memory; Benton visual retention test	Children who ate school breakfast performed better on memory and learning tasks than children who did not eat breakfast or ate at home
Benton and Parker (1998), Study 2	Adults	Breakfast vs. no breakfast; with or without glucose drink	(1) Normal breakfast (2) Normal breakfast with 50 g glucose drink (3) No breakfast (4) No breakfast and 50 g glucose drink	Word list; Wechsler story; Abstract reasoning	Group 4 recalled more than 3; Group 1 recalled more than 3; Breakfast better than none — drink did not influence; No effect
Benton and Sargent (1992)	Adults	Breakfast vs. no breakfast	Nestle Build Up 1370 kJ energy, 18.5 g protein, 37.7 g carbohydrate, 12.2 fat	Spatial memory; Word list	Time to complete both tasks shorter in the breakfast group; No difference in number of errors for either task
Smith et al. (1994), Study 1	Adults	Breakfast vs. no breakfast and caffeine vs. no caffeine	Cooked breakfast: two eggs, two pieces bacon, one slice whole-wheat toast, 10 g margarine; Cereal/toast breakfast: <25g corn flakes, 150 ml semiskimmed milk, one slice whole-wheat toast, 10 g margarine; No breakfast	Simple reaction time; Five-choice serial response task; Repeated digits vigilance task	No effect of breakfast; Caffeine improved performance; No effect of breakfast; No effect of breakfast; Caffeine improved performance

(continued)

TABLE 6.1 (CONTINUED)
Effects of Breakfast on Cognitive Performance

Author	Participants	Manipulation	Meal	Cognitive Tasks	Results
Smith et al. (1994), Study 2	Adults	Breakfast vs. no breakfast and caffeine vs. no caffeine	Cooked breakfast: two eggs, two pieces bacon, one slice whole wheat toast, 10 g margarine No breakfast	Free recall Recognition memory Logical reasoning Semantic processing	More words recalled with breakfast Fewer false alarms with breakfast More accuracy without breakfast No effect of breakfast
Tuttle et al. (1949)	Adults, 22 to 27 years	Breakfast vs. no breakfast	Heavy breakfast: 800 kcal, fruit, cereal and cream, egg, one slice bacon, two slices toast, jam, milk, and coffee Light breakfast: 400 kcal, fruit, one slice toast, butter, milk, coffee Coffee only No breakfast	Simple reaction time Choice reaction time	Reaction times greater for no breakfast conditions
Tuttle et al. (1950)	Adults, 21 to 28 years	Breakfast vs. no breakfast	Breakfast: 749 kcal, two slices white toast, butter, jelly, cereal, whole milk, sugar, fruit No breakfast	Choice reaction time	No breakfast period slower
Smith (1998)	Elderly, 60 to 79 years	Normal breakfast habits	(1) Breakfast cereal everyday (2) Breakfast everyday but not always cereal (3) Irregular breakfast eaters	National adult reading test	Group 1 scored higher than group 3
Tuttle et al. (1952)	Elderly, 60 to 83 years	Breakfast vs. no breakfast	Bacon, egg, milk, toast, and fruit: 750 kcal, 25 g protein, 37 g fat, 80 g carbohydrate Cereal, milk, toast, fruit: 750 kcal, 25 g protein, 28 g fat, 100 g carbohydrate No breakfast	Choice reaction time	No effect of breakfast
Tuttle et al. (1953)	Elderly, 60 to 84 years	Breakfast vs. no breakfast	Bacon, egg, milk, toast, and fruit: 744 kcal, 25 g protein, 37 g fat, 78 g carbohydrate Cereal, milk, toast, and fruit: 750 kcal, 25 g protein, 28 g fat, 100 g carbohydrate Fruit, cereal, cream, sugar, white bread, butter, whole milk, eggs, and bacon: 998 kcal, 40g protein, 48 g fat, 105 g carbohydrate	Choice reaction time	Slower reaction times following the heavier breakfast than the two basic breakfasts

6.4.5 CONCLUSION

Taken together, the results of the preceding studies provide strong evidence for the beneficial effects of breakfast on cognitive behavior across the lifespan. Although there are some differences in exactly what tasks were affected, in almost all the studies breakfast intake led to improvements in cognitive performance. These results have practical and political implications. A substantial number of children in the U.S. go to school every day without eating breakfast. Over time, the small impairments in cognitive performance that could result from the lack of a morning meal could have substantial consequences on academic achievement and later success in life. Thus, it is important for individuals to regularly consume breakfast and for the government to maintain school breakfast programs.

6.5 AFTERNOON MEAL

In contrast to breakfast, consumption of lunch has been most often reported to impair mental performance and negatively alter mood state. In real-life situations, individuals generally perform more poorly on mental tasks after lunch than in the morning or late afternoon hours. For example, falling asleep while driving, lapses of attention by locomotive engineers, and errors by shift workers are most pronounced in the midafternoon hours (Craig, 1986). Similarly, studies conducted in the laboratory have found that a drop in performance of mental tasks often is reported approximately 1 h after consuming lunch and may take several hours to return to the prelunch state. The drop in performance is usually found in measures of speed, accuracy, and sustained attention (Smith and Miles, 1986a,1986b). Mood has also been shown to differ after consuming lunch (Smith et al., 1988, 1991), with self-ratings of more feeble, bored, lethargic, mentally slow, dreamy, and clumsy after consuming lunch compared to before. The decrease in mental ability and mood that occurs following intake of a midday meal has been termed the *postlunch dip*.

It has been suggested that the effects of lunch intake on cognitive behavior may be primarily the results of circadian rhythms in mental alertness (Hildebrant et al., 1974). In support of this suggestion, midafternoon decrements in performance have been observed on some tasks whether or not lunch was consumed. However, for other tasks, mental abilities at midday are impaired to a greater degree in individuals who have consumed lunch than in those who have not, indicating that food intake does contribute to the postlunch dip (Craig, 1981; Smith and Miles, 1986a, 1986b; Wells et al., 1998).

6.5.1 CHILDREN

Studies examining the effect of lunch on cognitive performance in children have been primarily limited to those evaluating the merit of the school lunch program. Although many of these studies suggest that consumption of a school-supplied meal improves nutritional status, weight for age, height for age (Rewal, 1981), and the behavior of children in the classroom compared to when they are hungry (Read, 1973), the general consensus is that there is little effect on school achievement (Gietzen and Vermeerch, 1980). Unfortunately, because of the nature of these studies, control groups are often not possible or inappropriately matched and measures generally consist of course grades or performance on standardized tests. These measures are in essence achievement tests and thus not appropriate for evaluating the effects of lunch vs. no lunch on cognitive performance. In addition, many of these studies are confounded in that they do not control for socioeconomic status, parental education, number of siblings, number of parents present in the household, parental participation in the child's education, nutritional status, and regular meal habits, as all these variables may influence performance either independently or as a function of the meal consumed.

6.5.2 Adults

Studies examining the effects of the noontime meal on performance in adults have been conducted both in the laboratory (Smith and Miles, 1986a,1986b) and work settings (Hildebrant et al., 1974). In general, the research indicates that the consumption of lunch leads to a decline in performance. These findings may suggest potential costs of consuming a midday meal with regard to efficiency and safety in work, academic, or military environments.

Craig et al. (1981) examined whether lunch affected the efficiency of perceptual discrimination. This measure was chosen because of the significance to many work-related tasks such as radar monitoring, product inspection, and x-ray examination, as well as the role it may play in safety when warning signals are presented. Participants performed this task before and after either eating a three-course lunch and or having no lunch. Performance of the participants who consumed lunch declined, whereas performance of those who abstained from lunch did not. The authors conclude that consumption of lunch is an important precursor for the postlunch dip in performance. In addition, they state that the magnitude of the effect found in the present study is comparable to performance after the deprivation of a complete night of sleep (Craig et al., 1981).

Smith and Miles (1986a) also examined whether a postlunch change in performance was observed only in participants who consumed lunch or whether it was observed irrespective of lunch consumption. Participants were either assigned to a lunch condition, which consisted of a three-course meal at the dining hall or no lunch. However, participants in the lunch condition were allowed to eat what they wanted and therefore the size and content of the meals varied. In addition, individuals in the lunch condition were allowed to drink coffee and tea and smoke if they desired following the meal. Participants were tested both during the late morning and early afternoon on a five-choice serial reaction time task and the Stroop task. Results of the study revealed that participants who ate lunch had slower reaction times on the reaction time task, but measures on the Stroop task were not affected by consuming lunch. This study was problematic for several reasons, however. First, the lunch subjects were allowed to consume caffeine and nicotine if they desired before testing and these substances may have affected performance. In addition, nearly all the participants reported that the size of the lunch they ate was larger than their normal lunch. Thus, as previous research has shown, decrements in performance may have occurred due to the deviation from normal meal habits.

Smith and Miles (1986b) also conducted a study to examine the effects of a midafternoon meal on a vigilance task. Again, participants either had a three-course meal at the cafeteria or abstained from lunch until after testing. The content and the size of the meal for the lunch group were not controlled and participants were all allowed to have coffee, tea, or cigarettes before testing. Participants completed a dual task, which consisted of a detection of repeated numbers task and an estimation task. In the detection task, subjects were required to watch sequences of three-digit numbers, each differing from the previous number by one digit, and identify when a number appeared that was identical to the number preceding it. At the same time this task was being completed, participants were also scanning a continuous stream of single letters and digits and making estimations of the percentage of letters presented at specified time intervals. Results from this study revealed that subjects who ate lunch detected fewer targets than those who did not consume lunch. However, participants who had lunch did not show impairment in performance on the estimation task. Results also showed that participants who did not consume lunch performed worse in the afternoon than in the morning session. Thus, the consumption of lunch may affect some aspects of cognitive performance, but there also appears to be an endogenous component.

6.5.3 Lunch Type

Another important issue is the effect of the composition of the meal on performance and mood. In many of the studies reviewed, the participants were required to eat a meal that differed from their normal meal habit. It is possible that some of the effects of meals on performance were due to the novelty of the test meals or the departure from their normal meal habits. To examine this notion, Craig and Richardson (1989) looked at the interaction between normal lunch habit and experimental lunch size. Participants were either classified as light eaters (less than 300 kcal) or heavy eaters (more than 1200 kcal) and were tested in both a light test meal condition (sandwich consisting of 260 kcal) and a heavy test meal condition (three-course meal consisting of 1300 kcal) whose protein to carbohydrate ratios were 2:3. They hypothesized that there were four possible outcomes of the experiment: (1) there would be no difference in performance; (2) performance would differ by time of day but not by lunch type, and thus the differences would be endogenous; (3) changes depend on the experimental meal but not the normal meal habit, and thus the differences would be purely exogenous; and (4) changes depend on both the experimental meal and habitual meal size, and thus a combination of endogenous and exogenous processes occur (Craig and Richardson, 1989). Results of this study showed that the number of errors on a letter cancellation task increased after a big lunch and decreased after a small lunch. In addition, the extent of change was greater if the meal was different from that normally consumed; for example, light eaters who consumed a big lunch performed the worst whereas heavy eaters consuming a small lunch performed the best.

As with many laboratory studies on the effects of meals and performance, participants are often asked to consume a meal that is different from what they would normally consume. Thus, it is possible that the effects observed in many studies are influenced by a combination of the normal eating habits of the participants and the novelty of the test meals. Several studies found that the performance of participants is affected by the interaction of the normal eating habits of the participants and the test meals provided at both breakfast and lunch (Craig et al., 1981; Mahoney et al., unpublished data).

6.6 SNACKING

Snacking contributes to a significant amount of our daily food intake. This is especially true for young children and adolescents, for whom snacks may account for up to one third of the daily calories. There is a popular belief that snacking is a negative habit, especially when sugary snacks are consumed. However, limited research on the effects of snacking on cognitive performance suggests that between-meal food intake may actually be beneficial for learning and memory.

6.6.1 LATE-MORNING SNACK

Research to date suggests that consuming a late-morning snack can improve subsequent cognitive performance. In the first study, performance on two tasks measuring concentration abilities was compared when office workers either had or had not consumed a late-morning snack (two ham sandwiches and hot tea; Hutchinson, 1954). Cognitive tasks consisted of a digit-symbol matching test and a task that required participants to cross out six-figure numbers, of which there were several columns, containing certain pairs of digits. When the office workers had a late-morning snack, performance on the digit-symbol matching test was significantly better than when they did not. There were no differences in performance in the second measure of concentration. There were, however, several confounding variables in this study. First, the caffeine in the tea could have affected some aspects of performance. In addition, there were no controls for what the participants ate for breakfast the morning of testing or when they ate. Finally, the duration of cognitive testing was short. It is possible that the second measure of concentration may also have shown significant differences between snack conditions if the testing session was longer.

A follow-up study was conducted, which alleviated some of the previous confounds (Hutchinson, 1954). The procedures remained the same with two exceptions. Tea was replaced with a fruit drink and a longer testing session was used. Results showed that when the participants had a late-morning snack, performance was significantly better than when they did not. The authors concluded that the ability to concentrate, when the stomach can reasonably be assumed to be empty, is improved after consuming a small amount of food (Hutchinson, 1954). However, in addition to the concerns mentioned previously, one more concern exists. The morning snack was given at 11:50 a.m. and the size of the snack could be consistent with a normal lunch. Although it was noted that in this population lunch often was consumed after 1:00 p.m., both the size of the snack and the time of ingestion raise concerns on whether these studies really examined the effects of a late-morning snack or a light afternoon meal.

In a more recent study, Benton et al. (2001) addressed the question of whether a late-morning snack would interact with breakfast intake in determining psychological functioning. Young women were given breakfasts containing either 10 or 50 g of carbohydrate. After 1.5 h, half the women were given a further 25 g of carbohydrate in the form of corn flakes (snack) and the other half no snack. Results indicated that performance on a verbal memory task was not influenced by breakfast consumption but was positively affected by snack intake.

6.6.2 Afternoon Snacks

By using a paradigm similar to that described for late-morning snacks, early research investigating the effects of a late-afternoon snack on measures of concentration found no differences in performance as a function of food intake. Workers either consumed a snack from 3:50 to 4:00 p.m. or did not and then completed two short tests of concentration at 4:10 p.m. The snack consisted of a plate of two ham sandwiches, cut into eight quarters, with a total weight of 5 oz, and a 9-oz cup of hot tea. Participants were allowed to eat as many of the eight sandwich quarters as they desired. Cognitive tasks consisted of a digit-symbol matching test and a task that required participants to cross out six-figure numbers, of which there were several columns, containing certain pairs of digits. Results showed that an afternoon snack had no effect on performance during either task. However, this study was problematic for several reasons. First, there were no controls for what the participants ate for lunch on the day of testing or when they ate lunch. It is possible that a considerable number of subjects may have still had food in their stomachs as a result of their afternoon meal. In addition, testing took place 10 min after snack consumption and lasted only 3 min. It is possible that the experimenters did not leave enough time in between the snack and testing to allow the levels of blood glucose in the snack condition to rise. Furthermore, the tests measured a relatively small window of concentration time (1.5 min/test). It is possible that this window was not large enough to detect significant differences in performance. Finally, the snack condition was confounded with the consumption of caffeine from the tea, as previous research shows that caffeine itself can affect some aspects of performance (Smith et al., 1994a, 1994b; Committee on Military Nutrition Research, 2001).

More recent studies looking at the effect of an afternoon snack lack the confounds mentioned in the previous study. Kanarek and Swinney (1990) looked at the effect of a late-afternoon snack on the cognitive performance of college-aged men.

Results of this study suggest that ingestion of a calorie-rich snack enhances performance. The snack consisted of an 8-oz fruit-flavored yogurt or a noncaloric diet soft drink without caffeine. College students were fed breakfast and lunch in the laboratory and participated in each of four conditions separated by a week. These conditions were (1) breakfast, lunch, and yogurt snack; (2) breakfast, no lunch, and yogurt snack; (3) breakfast, lunch, and diet soda (noncaloric snack with no caffeine); and (4) breakfast, no lunch, and diet soda. Fifteen minutes after consuming the snack, participants completed tests of memory (digit span), arithmetic reasoning, reading, and attention. There was a practice session for the tasks before the start of the experiment. Results showed that participants recalled more digits during the backward digit span task and responded more quickly when making correct responses during the attention task after the calorie-rich snack than the noncaloric snack. In addition, participants correctly solved more problems in the arithmetic reasoning task and had significantly faster reading times after having the yogurt snack compared to the soft drink.

6.6.3 CONFECTIONERY SNACKS

There is a popular cultural belief that consumption of sugary snacks causes hyperactive behavior in children. More specifically, people believe that as a result of ingesting a sugar snack, children will suffer from a decreased ability to maintain attention, which hinders abilities to carry out cognitive tasks. Hyperactivity includes such components as failure to follow through on tasks, inability to sustain activities for an appropriate period of time, and difficulties in organizing and completing work (Kanarek, 1996). King (1996) suggests that sugar does not affect the behavior or cognitive performance in children. Instead, it may be because activities surrounding ingestion of sugar, such as trick-or-treating, produce excited states in children. Parents then associate the excited state with sugar intake. Thus, parental expectancies may have some impact on how they perceive behavior.

Hoover and Milich (1994) tested the hypothesis that the commonly observed hyperactivity in children after sugar consumption may actually be due to the mother's expectancies rather than the sugar itself. Mothers of thirty-five 5- to 7-year-old boys were told either that their child received a large dose of sugar or that their child received a sugar-free placebo. In actuality, all the children received the sugar-free placebo. The children in the "sugar" condition were rated significantly more hyperactive by their mothers than the children in the "nonsugar" condition (Hoover and Milich, 1994). This study shows that despite the lack of evidence to support the belief that sugar causes hyperactivity in children, the expectancies held by the parents may actually color the way they interpret their own child's behavior. It appears, because there is no evidence that sugar alone can turn a normal child into a hyperactive child, that any adverse effect of sugar is by no means as severe or as prevalent as uncontrolled observation and opinion would suggest (Kinsbourne, 1994).

Wolraich et al. (1994) conducted several studies that looked at the effects of sugar on behavior and cognitive tasks in preschool and school-age children. Children were provided with a special diet by the experimenters for 3 weeks containing varying amounts of sugar and artificial sweeteners. The manipulation of the children's sugar intake did not affect behavior or cognitive performance.

Recent studies looking at the effect of an afternoon confectionery snack indicate that ingestion of a confectionery snack may actually improve several aspects of cognitive performance. Kanarek and Swinney (1990) looked at the effect of a late-afternoon snack on the cognitive performance of college-aged men. College students were all fed breakfast in the laboratory and then either lunch or no lunch and a confectionary snack or noncaloric snack. Fifteen minutes after consuming the snack, participants completed tests of memory, arithmetic reasoning, reading, and attention. Results showed that participants performed better on a memory task and responded more quickly when making correct responses during the attention task after the confectionery snack than the noncaloric snack (Kanarek and Swinney, 1990). No significant differences were found in measures of reading time or arithmetic reasoning.

Three studies conducted by Busch et al. (2002) examined the effects of a confectionery snack on cognitive performance. In the first experiment, 38 male undergraduates ate breakfast and lunch in the laboratory and then came back in the late afternoon for a snack. Fifteen minutes after consuming the snack, they participated in a dual learning task. The snack consisted of either 50 g of a confectionery product or one cup of an artificially sweetened drink. Each student participated in both conditions, separated by a week. Results showed that when participants consumed the confectionery product, they correctly placed more country names on a map, as well as left fewer countries blank, during long-term recall on the primary spatial learning task. However, with regard to the secondary attention task, when consuming the placebo participants had a higher hit rate as well as a lower miss rate (unpublished data).

A second experiment looked at the performance of thirty-eight boys, age 9–11 yrs, on a dual task in the afternoon after ingestion of either 25 g of a confectionery product or half a cup of an artificially sweetened beverage. Unlike the first experiment, children received either the confectionery product or the artificially sweetened beverage. In addition to the dual task, children also completed a vigilance attention task. Results showed that the children who had consumed the confectionery snack placed more items correctly on a map as well as left fewer countries blank in short-term and long-term memory recall of the spatial learning task. In contrast with the first experiment, the children who consumed the confectionery snack had a higher hit rate and a lower miss rate for a secondary attention task than children who consumed the placebo. However, when tested on a separate vigilance attention task, the children who consumed the placebo performed better (unpublished data).

6.7   EVENING MEAL

To date, there is very limited data examining the effect of an evening meal on cognitive performance. This may be because the evening meal is often consumed at home and does not impede on performance at work or in the classroom. However, this is not altogether true because both children and young adults often study for exams or complete homework after the evening meal. Thus, there is a need for additional research.

In studies on evening meals, Smith et al. (1994) examined the effects of evening meals on cognitive performance in college students. Forty-eight male and female undergraduates either ate a 1200- to 1500-kcal meal between 6:00 and 6:45 p.m. or did not eat. Within those conditions, participants had 150 ml of decaffeinated coffee either with or without a caffeine tablet mixed in. Subjects performed a battery of cognitive tasks an hour before the meal and 90 and 180 min after the start of the meal. The cognitive tasks consisted of free recall, vigilance attention, logical reasoning, simple reaction time, a five-choice serial response task, and delayed recognition memory. Results show that participants who had an evening meal completed more sentences during the logical reasoning task than those who did not have an evening meal. In addition, they reported feeling stronger, more interested, and more proficient than participants who did not consume a meal. However, there were no effects of meal on any of the other tasks. The authors conclude that when a meal of similar quality is consumed in the evening, it has very different effects on performance than when it is consumed as an afternoon meal (Smith et al., 1994b). Although time of day may mediate the effects of a meal on some areas of cognitive performance, note that the authors did not control for what the subjects had consumed earlier that day. Subjects were asked not to eat or drink anything an hour before testing, but if they had a late lunch or a late-afternoon snack, it is possible that some participants may have still had food in their stomach even if they were in the no-meal group.

6.8   SUMMARY

The consumption of a meal produces acute effects on some types of cognitive processing in well-nourished children and adults. These effects depend on many variables, such as age, sex, time of testing, size of meal, energy and macronutrient content of meal, normal meal habits of participants, and types of tasks used.

In general, consuming breakfast improves performance on several types of cognitive measures. The types of tasks that tend to be most affected are those assessing short-term memory, vigilance attention, arithmetic performance, and problem solving. The type of breakfast consumed may also affect cognition: a high-energy breakfast improves performance compared to a low-energy breakfast on measures of creativity, physical endurance, concentration, short-term memory, and mathematical addition. Finally, foods that differ in the amount of protein, fiber, or processing characteristics may affect cognitive processing differently.

The consumption of lunch is usually followed by a decrease in performance on measures of sustained attention and alertness. However, there is no clear consensus on whether this reported dip is because of consuming a meal or simply the time of day. In addition, the extent of the dip depends on characteristics of the person eating the meal, size of meal, and types of task tested. In terms of personality, subjects who are more anxious tend to show lesser performance impairments after consuming lunch compared with those who are less anxious (Craig et al., 1981). With regard to the size of the meal, a larger meal tends to produce a larger decline in performance compared to a smaller meal, and participants' normal meal habits also interact with the type of meal consumed — participants who regularly consume a large meal do not experience decrements to the same extent as those who normally eat a small meal (Craig et al., 1981). In addition, the effects on performance observed after lunch may increase with age (Spring et al., 1982). Finally, the macro-nutrient composition of the meal also influences performance: high-protein meals lead to increased distractibility and high-carbohydrate meals lead to slowing of reaction times (Smith et al., 1988).

While the postlunch dip in performance tends to peak about an hour after meal consumption and persist for several hours thereafter, the consumption of an afternoon snack alleviates performance decrements produced by the consumption of lunch. In addition, the consumption of caffeine, either with a meal or alone, produces improvements in performance on tasks such as sustained attention.

Finally, the limited research looking at the consumption of an evening meal suggests that when an evening meal is consumed, performance on a logical reasoning task is improved. In addition, subject feelings of strength, proficiency, and interest are greater after consuming an evening meal.

Although there is not a clear consensus of how mixed meals affect performance, several mechanisms have been considered, including increases in the level of blood glucose and individual effects of macronutrients on neurotransmitter synthesis. Although it is hard to determine how macronutrients affect performance when studied in mixed meals rather than tests of single nutrients, mixed-meal studies are valuable in that represent the way people normally eat.

REFERENCES

Benton, D. and D. S. Owens (1993). Blood glucose and human memory. *Psychopharmacology* 113: 83–88.

Benton, D., P. Parker and R. T. Donahoe (1996). The supply of glucose to the brain and cognitive functioning. *The Journal of Biosocial Science* 28: 463–469.

Benton, D., M.-P. Ruffin, T. Lassel, S. Nabb, M. Messaoudi, S. Vinoy, D. Desor and V. Lang (2002). The delivery rate of dietary carbohydrates affects cognitive performance in both rats and humans. *Psychopharmacology* 166: 86–90.

Benton, D. and J. Sargent (1992). Breakfast, blood glucose and memory. *Biological Psychology* 33: 207–210.

Benton, D., O. Slater and R. T. Donohoe (2001). The influence of breakfast and a snack on psychological functioning. *Physiology and Behavior* 74: 559–571.

Blake, M. J. F. (1967). Time of day effects on performance in a range of tasks. *Psychonomic Science* 9(6): 349–350.

Brice, C. and A. Smith (2001). The effects of caffeine on simulated driving, subjective alertness and sustained attention. *Human Psychopharmacology* 16(7): 523–531.

Busch, C. R., H. A. Taylor, R. B. Kanarek and P. J. Holcomb (2002). The effects of a confectionery snack on attention in young boys. *Physiology and Behavior* 77: 333–340.

Committee on Military Nutrition Research, Food and Nutrition Board, Institute of Medicine (2001). *Caffeine for the Sustainment of Mental Task Performance: Formulations for Military Operations.* Washington, D.C., National Academies Press.

Craft, S., G. Zallen and L. D. Baker (1992). Glucose and memory in mild senile dementia of the Alzheimer's type. *Journal of Clinical and Experimental Neuropsychology* 14: 253–267.

Craig, A., K. Baer and A. Diekmann (1981). The effects of lunch on sensory-perceptual functioning in man. *Internal Archives of Occupational and Environmental Health* 49: 105–114.

Craig, A. and E. Richardson (1989). Effects of experimental and habitual lunch size on performance, arousal, hunger and mood. *Internal Archives of Occupational and Environmental Health* 61: 313–319.

Fischer, K., P. C. Colombani, W. Langhans and C. Wenk (2002). Carbohydrate to protein ratio in food and cognitive performance in the morning. *Physiology and Behavior* 75: 411–423.

Folkard, S. and T. H. Monk (1985). *Hours of Work: Temperal Factors in Work Scheduling*. Chichester, Wiley.

Gietzen, D. and J. A. Vermeerch (1980). Health status and school achievement of children from Head Start and Free School Lunch programs. *Public Health Reports* 95(4): 362–368.

Gonder-Frederick, L., J. L. Hall, J. Vogt, D. J. Cox, J. Green and P. E. Gold (1987). Memory enhancement in elderly humans: effects of glucose ingestion. *Physiology and Behavior* 41: 503–504.

Glaeser, B. S., T. J. Mahar and R. J. Wurtman (1983). Changes in brain levels of acidic, basic and neutral amino acids after consumption of single meals containing various proportions of protein. *Journal of Neurochemistry* 41: 1016–1021.

Gold, P. E. (1986). Glucose modulation of memory storage processing. *Behavioral and Neural Biology* 45: 342–249.

Gold, P. E. (1995). The role of glucose in regulating the brain and cognition. *American Journal of Clinical Nutrition* 61: 981–995.

Hall, J. L., L. Gonder-Frederick, W. W. Chewning, J. Silveira and P. E. Gold (1989). Glucose enhancement on performance on memory tests of young and aged humans. *Neuropsychologia* 27: 1129–1138.

Harper, H. A. (1957). *Review of Physiological Chemistry*. Los Altos, CA, Lange Medical Publications.

Hildebrant, G., W. Rohmert and J. Rutenfranz (1974). Twelve and twenty-four hour rhythms in error frequency of locomotive drivers and the influence of tiredness. *International Journal of Chronobiology* 2: 175–180.

Hoover, D. W. and R. Milich (1994). Effects of sugar ingestion expectancies on mother-child interactions. *Journal of Abnormal Child Psychology* 22: 501–515.

Hutchinson, R. C. (1954). Effects of gastric contents on mental concentration and production rate. *J. Appl. Physiol.* 7: 143–147.

Kanarek, R. B. and D. Swinney (1990). Effects of food snacks on cognitive performance in male college students. *Appetite* 14: 15–27.

Kaplan, R. J., C. E. Greenwood, G. Winocur and T. M. S. Wolever (2001). Dietary protein, carbohydrate and fat enhance memory performance in the healthy elderly. *American Journal of Clinical Nutrition* 74: 687–693.

King, E. A. (1996). Effects of sugar on children. *Journal of Family Practice* 42: 344–345.

Kinsbourne, M. (1994). Sugar and the hyperactive child. *The New England Journal of Medicine* 330: 355–356.

Lieberman, H. M., I. F. Hunt, A. H. Coulson, V. A. Clark, M. E. Swendseid and L. Ho (1976). Evaluation of a ghetto school breakfast program. *Journal of the American Dietetic Association* 68: 132–138.

Lieberman, H. R., B. J. Spring and G. S. Garfield (1986). The behavior effects of food constituents: strategies used in studies of amino acids, protein, carbohydrate and caffeine. *Nutrition Reviews* 44 Suppl: 61–69.

Lloyd, H. M., M. W. Green and P. J. Rogers (1994). Mood and cognitive performance effects of isocaloric lunches differing in fat and carbohydrate content. *Physiology and Behavior* 56(1): 51–57.

Lopez-Sobaler, A. M., R. M. Ortega, M. E. Quintas, B. Navia and A. M. Requejo (2003) Relationship between habitual breakfast and intellectual performance (logical reasoning) in well-nourished schoolchildren of Madrid (Spain). *European Journal of Clinical Nutrition* 57: S49–S53.

Manning, C. A., J. L. Hall and P. E. Gold (1990). Glucose effects on memory and other neuropsychological tests in elderly humans. *Psychological Science* 1: 307–311.

Manning, C. A., M. W. Parsons and P. E. Gold (1992). Antrograde and retrograde enhancement of 24-hour memory by glucose in elderly humans. *Behavioral and Neural Biology* 58: 125–130.

Manning, C. A., W. S. Stone, D. L. Korol and P. E. Gold (1998). Glucose enhancement of twenty-four hour memory retrieval in healthy elderly humans. *Behavioral Brain Research* 93: 71–76.

Michaud, C. A., N. Musse, J. P. Nicolas and L. Mejean (1991). Effect of breakfast size on short term memory, concentration, mood, and blood glucose. *Journal of Adolescent Health* 12: 53–57.

Murphy, J. M., M. E. Pagano, J. Nachmani, P. Sperling, S. Kane and R. E. Kleinman (1998). The relationship of school breakfast and psychosocial and academic functioning. *Archives of Pediatrics and Adolescent Health* 152: 899–907.

Parsons, M. W. and P. E. Gold (1992). Glucose enhancement of memory in elderly humans: an inverted-U dose-response curve. *Neurobiology of Aging* 13: 401–404.

Pollitt, E. (1995). Does breakfast make a difference in school? *Journal of the American Dietetic Association* 95(10): 1134–1139.

Pollitt, E., N. L. Lewis, C. Garza and R. J. Shulman (1983). Fasting and cognitive function. *Journal of Psychiatric Research* 17: 169–174.

Read, M. S. (1973). Malnutrition, hunger and behavior. II. Hunger, school feeding programs and behavior. *Journal of the American Dietetic Association* 63: 386–391.

Rewal, S. (1981). Results of a school lunch program in India. *Food and Nutrition Bulletin* 3(4): 42–47.

Seiga-Riz, A. M., B. M. Popkin and T. Carson (1998). Trends in breakfast consumption for children in the United States from 1965–1991. *American Journal of Clinical Nutrition* 67: 748S–756S.

Simeon, D. T. and S. Grantham-McGregor (1989). Effects of missing breakfast on cognitive functions of schoolchildren of differing nutritional status. *American Journal of Clinical Nutrition* 49: 646–653.

Smith, A., A. M. Kendrick, A. L. Maben and J. Salmon (1994a). Effects of breakfast and caffeine on cognitive performance, mood, and cardiovascular functioning. *Appetite* 22: 39–55.

Smith, A., S. Leekam, A. Ralph and G. McNeill (1988). The influence of meal composition on post-lunch changes in performance efficiency and mood. *Appetite* 10: 195–203.

Smith, A., A. L. Maben and P. Brockman (1994b). Effects of evening meals and caffeine on cogntiive performance, mood, and cardiovascular functioning. *Appetite* 22: 57–65.

Smith, A., A. Ralph and G. McNeill (1991). Influences of meal size in post-lunch changes in performance efficiency, mood, and cardiovascular functioning. *Appetite* 16: 85–91.

Smith, A. P., A. M. Kendrick and A. L. Maben (1992). Effects of breakfast and caffeine on performance and mood in the late morning and after lunch. *Neuropsychobiology* 26: 198–204.

Smith, A. P. and C. Miles (1986a). Effects of lunch on selective and sustained attention. *Neuropsychobiology* 16: 117–120.

Smith, A. P. and C. Miles (1986b). The effects of lunch on cognitive vigilance tasks. *Ergonomics* 29(10): 1251–1261.

Spring, B. J. (1986). Effects of foods and nutrients on the behavior of normal individuals. In *Nutrition and the Brain* (R. J. W. Wurtman, and J.J. Wurtman, Eds.), pp. 1–47. New York, Raven Press.

Spring, B. J., J. Chiodo and D. J. Owen (1987). Carbohydrates, tryptophan and behavior: a methodological review. *Psychological Bulletin* 102: 234–256.

Spring, B. J., O. Maller, R. J. Wurtman, L. Digman and L. Cozolin (1982). Effects of protein and carbohydrate meals on mood and performance: interactions with sex and age. *Journal of Psychiatric Research* 17: 155–167.

Tuttle, W., K. Daum, B. Randall and M. T. Schumacher (1952). Effect of omitting breakfast on the physiologic response of the aged. *Journal of the American Dietetic Association* 28: 117–123.

Tuttle, W. (1953). Physiologic response to size and content of breakfast by men over 60. *Journal of Abnormal Child Psychology* 29: 34–40.

Vaisman, N., H. Voet, A. Akivis and E. Vakil (1996). The effects of breakfast timing on the cognitive function of elementary school students. *Archives of Pediatrics and Adolescent Medicine* 150: 1089–1092.

Wells, A. S., N. W. Read, C. Idzikowski and J. Jones (1998). Effects of meals on objective and subjective measures of daytime sleepiness. *Journal of Applied Physiology* 84: 507–515.

Wesnes, K. A., C. Pincock, D. Richardson, G. Helm and S. Hails (2003). Breakfast reduces declines in attention memory over the morning in schoolchildren. *Appetite* 41: 329–331.

Wolraich, M. L., S. D. Lindgren, P. J. Stumbo, L. D. Steglink, M. I. Appelbaum and M. C. Kiristy (1994). Effects of diets high in sucrose or apsartame on the behavior and cognitive performance of children. *The New England Journal of Medicine* 330(5): 301–307.

Wurtman, R. J. (1982). Nutrients that modify brain function. *Scientific American* 246: 50–59.

Wurtman, R. J. (1986). Ways that foods can affect the brain. *Nutrition Reviews* 44 Suppl: 2–5.

Wyon, D. P., L. Abrahamsson, M. Jartelius and R. J. Fletcher (1997). An experimental study on the effects of energy intake at breakfast on the test performance of 10-year-old children in school. *International Journal of Food Sciences and Nutrition* 48: 5–12.

7 Regulation of Macronutrient Preference — Component of Food Selection

Jason C.G. Halford and John E. Blundell

CONTENTS

7.1 WHAT IS MACRONUTRIENT PREFERENCE?

Because nutrients are continually used (as well as stored) by the body, there is a constant demand for the intake of nutrients. At any particular moment, this demand does not apply equally to all macronutrients. This requirement involves a control over behavioral processes (of food intake) by physiological systems in the body. The scientific approach therefore seeks to disclose how the apparent selective intake of nutrients is related to the requirement of the nutrients by the body. The process is not completely determined by physiology and also involves the development of adaptive links between physiology and the environment.

The term *regulation* implies the operation of a control system. In the area of macronutrient intake, this implies a balance between the drive arising for a biological requirement (nutrient) and the intermittent (or episodic) termination of that drive by the stimulation of some receptor or system of receptors. The notion of regulation of a macronutrient requires the existence of a detector that signals deviation from some optimal (or preferred) value together with a mechanism that organizes the behavioral correction. In simple terms, these processes represent the operation of a homeostatic system.

7.2   REGULATION VS. HEDONICS

It can be claimed that preference is an expressed willingness to consume (humans) or a directional form of behavior involving a choice among alternatives (humans and animals). When we say that animals or humans prefer a particular type of food that contains a substantial amount of a macronutrient (or consists solely of that nutrient), what evidence is there that the preference is caused by the actual macronutrient content rather than by other contextual features of the food such as taste, texture, and mode of presentation, any of which could give value to the consumption? This issue gives rise to one of the most enduring methodological problems underlying research on the control of macronutrients in animals or humans. The problem is how to allow the subject to display a preference (either in a choice situation or with a single commodity) that is not biased by collateral features of the nutrient, which creates attraction or aversion and therefore lead to spurious conclusions about preferences. It is often difficult to demonstrate unequivocally the behavioral operation of such a mechanism uncontaminated by biases. It is possible to demonstrate the preference for a macronutrient per se when the nutrient is contained in a food whose contextual qualities provide pleasure or positive reward value.

A good example is the preference for fat and sugar. It is widely believed that the preferences of most humans (and many animals) for foods that have properties of fattiness or sweetness constitute a genetic trait (or biobehavioral disposition). Consumption of such materials can be presumed to be linked to an experience of pleasure or to the stimulation of reward pathways. Such a mechanism would be biologically useful because the properties of sweetness and fattiness normally signal the existence of valuable energy-yielding commodities. In other words, the consumption (and possible overconsumption) of fat and sugar depends not on the need for the nutrients themselves but on the hedonic properties. Recognizing this argument (about fat and sugar) leads to certain circularity, but it still leaves the question of whether consumption of fat and sugar represents the operation of some regulatory process or the potency of hedonic properties. In this particular area of research, the principle of regulation comes into conflict with the principle of hedonics. It may not be possible to untangle these two principles, but the investigation of macronutrient preferences must recognize both systems.

7.3   BACKGROUND

Regulation of nutrient supply to the brain depends in large part on the selection of nutrients in the diet and therefore demands a scientific interest in the control of food consumption. Eating behavior is the means by which both the body and the brain receive their supply of nutrients, and in turn the nutrient content of food is a clear factor in determining the pattern of food consumption. Over the past 30 years, the study of the brain and appetite control has been in part driven by the need to understand the etiology of and developed therapies for human obesity, now a critical health issue in both the developed and the developing world. Global changes in both diet and lifestyle have conspired to produce an obesogenic environment by encouraging a positive energy balance (in which energy intake becomes greater than expenditure), which in turn promotes adiposity. Changes in the patterns of food consumption and the composition of foods consumed have been linked with the rise in obesity. Specifically, over the past 50 years there has been an increase in the consumption of dietary fat. This is at least in part due to an increase in the consumption of between-meal snacks, which usually consist of processed foods high in fats and sugars and are very energy dense. This increase in dietary fat is positively associated with the prevalence of overweightedness and obesity in people. There is also mounting evidence for a link between sugar consumption and weight gain.

Therefore, it is not just the overconsumption of calories per se but also the excessive intake of such palatable energy-dense high-fat foods that appears to be the critical dietary factor. In this context, the regulation of nutrient supply assumes particular significance. A particular theme in this area is therefore related to control of intake of fat and sugar, or sometimes fat plus sugar.

7.3.1 NATURE OF APPETITE

Appetite regulation can be conceptually viewed as a homeostatic process by which current energy intake is altered to meet the energy expenditure, thereby maintaining energy balance. However, both personal experience of appetite regulation and the current obesity pandemic demonstrate that the body defends against energy deficits far better than it does against excesses. Overconsumption, which leads to the accumulation of body fat, does not appear to generate a biological drive to undereat, and, at least for the obese, excess body fat does not appear to feedback to reduce further consumption. Therefore, two principles of appetite regulation become clear. First, biological processes exert a strong defense against undereating, which serves to protect the body from energy deficit (a debilitating and life-threatening state). Those who have or are trying to diet to lose weight can attest to this. At this stage, the epiphenomenon of dieting observed over the past 50 years in the West has not halted the rising levels of obesity. Second, the biological defenses against overconsumption are weak and for the many obese appear completely inadequate. From a historical perspective, it is likely that nonbiological factors such as a more physically demanding environment and a relative scarcity of nutritionally adequate food were more than sufficient to keep adiposity in check. Thus, appetite regulation appears to be an asymmetric phenomenon rather than a tightly controlled homeostatic process.

The key purpose of eating behavior is also to ensure that a sufficient amount of each essential macro- and micronutrient required by the body is ingested and absorbed. In addition to small amounts of vitamins and minerals, the body requires protein, carbohydrate, and fats, the major nutritional constituents of the foods we eat. Appetite regulation can achieve this by promoting general dietary diversity and by employing more specific mechanisms of adjustment, such as nutrient-deficit-driven appetites. Omnivorous diets consist (by definition) of a diverse range of foods. This adaptation firstly reflects the omnivore's ability to digest a variety of foods, necessitated by their opportunistic food acquisition strategy. This enables the omnivore to literally eat what comes to hand. However, omnivores are less able to satisfy their metabolic needs from the limited diets consumed by more specialized eaters and therefore must consume a variety of food to provide all the essential nutrients (those that cannot be synthesized) they require. Although diversity prompts a balanced diet, it cannot correct a deficit in any particular nutrient. This would require some form of feedback to reorient or adjust food-seeking behavior toward the required nutrient.

With regards to adjustment, foods of varying nutritional composition engage differently with the sensory, cognitive, postingestive, and postabsorptive mechanisms critical to appetite regulation. These differences, detected within the central nervous system (CNS), will in turn revise food choice to promote the subsequent consumption of foods that differ markedly in composition. This change would not be expressed as a craving or preference for the pure macronutrient in deficit as we tend to eat whole foods rather than their constituents. However, the changes in food preferences would be based on the sensory characteristics (such as sweetness, taste intensity, and mouth feel) reflecting the chosen food's nutritional content. Consequently, appetite regulation is concerned with meeting a variety of key metabolic needs with differing substrates rather than just matching calories consumed with those expended.

7.3.2 THE APPETITE CASCADE

To coordinate the intake of such a number of nutrients, appetite regulation must be underpinned by a multiplicity of nutrient-sensitive systems. Globally, their collective actions initiate and terminate episodes of eating behavior and alter the subjective experiences of hunger and satiety. Separately, they must alter food choice to satisfy their various demands. Both these objectives can be conceptualized in the appetite cascade (Figure 7.1). The appetite cascade represents a progression of events that determine the pattern of eating common to most omnivores and consists of discrete episodes of feeding behavior that can be termed as *meals* (to which humans have added the "snack"). The cascade makes us consider the following:

1. Preprandial events that stimulate eating and motivate organisms to seek food.
2. Behavioral actions that form the topography of meal and antagonistic processes underlying them in this prandial phase.
3. Postprandial processes that follow meal termination.

Internal factors determining meal and nonmeal periods are the psychological experiences of hunger and satiety and the transition between the two. Hunger can be defined as the motivation to seek and consume food that initiates feeding behavior. The preprandial stimulation of hunger by the

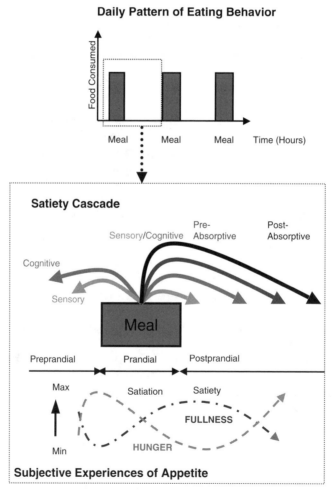

FIGURE 7.1 The appetite cascade: sensory, cognitive, preabsorptive, and postabsorptive signals, generally prior to, during, and after consumption, which control the pattern of episodic food intake.

sight, smell, and even thought of food is termed the *cephalic phase*. These stimuli continue to arouse hunger throughout the meal, although their effects diminish as satiation develops. The within-meal process that brings the period of eating to an end is termed *satiation*. Satiation leads to the postprandial state of satiety, during which hunger and further eating behavior are suppressed. Critical to the stimulatory and inhibitory factors contained within the preprandial, prandial, and postprandial phases are sensory, cognitive, postingestive, and postabsorptive signals. These are derived from the food consumed and are at least in part based on the food's macronutrient composition.

In addition to acting collectively, systems may act differentially to promote or inhibit the intake of specific nutrients by altering the preprandial (before consumption) cravings and preference for certain foods and the prandial (during consumption) palatability and pleasure derived from them. Within a meal or at the next meal, the differential modulation of food choice by sets of nutrient-sensitive systems promotes the varied diet required to guarantee the adequate intake of a variety of nutrients.

7.3.3 NEURAL CONTROL OF APPETITE

Food intake and choice are behavioral events dependent on the neural control of appetite. The CNS must integrate information generated by the sensory experience of eating and from the periphery indicating the ingestion, absorption, metabolism, and storage of different macronutrients. In addi-tional to the sensory modalities of sight, taste, and smell (constituting the cephalic phase), the CNS detects macronutrients more directly by three main routes:

1. Peripheral nutrient-sensitive chemoreceptors in the gut and nutrient-induced metabolic changes in the liver generate afferent signals relayed through the vagus to the nucleus of the solitary tract (NTS)/area postrema (AP) complex in the brain stem.
2. Receptors within the CNS, particularly in the brain stem (such as glucose-sensing neurons), detect circulating nutrients, their metabolites, or increases in metabolic hor-mones produced in response to macronutrient detection, absorption, metabolism, and storage.
3. Macronutrients (such as carbohydrate) can facilitate neurotransmitter precursor crossing of the blood-brain barrier. In addition, dietary protein consists of many amino acids that are themselves neurotransmitter precursors. These directly alter CNS neurochemical activity, particularly in the key hypothalamic nuclei.

The precise integration of all the key loci involved in appetite regulation is beyond the scope of this chapter. However, sites in the brain stem and the limbic system, specifically the hypothalamus, appear critical in regulating energy balance and the expression of appetite. These sites include the arcuate nucleus (ARc), amygdala (AMYG), lateral hypothalamus (LH), nucleus accumbens (NAc), paraventricular nucleus (PVN), and the ventromedial hypothalamus (VMH).

7.4 NUTRIENT-SENSITIVE SYSTEMS

7.4.1 SENSORY AND COGNITIVE ESTIMATIONS

The first indication of food intake is the sight and smell and then the taste of food, which generate anticipatory changes. Such sensory stimulation promotes food intake and the physiological response that prepares the body for digesting and absorbing the nutrients. This response is the cephalic (head) phase and includes salivation; increased insulin, pancreatic polypeptide (PP), glucagon, and cate-cholamines; and an increase in energy expenditure. These changes prepare the body for the digestion, absorption, and metabolism of nutrients and continue during the initial phase of food consumption. The magnitude of the physiological responses observed may be directly related to the sensory stimulation or palatability of the food being consumed (Teff, 2000; Le Blanc, 2000). Chemosensory detection of food by smell and taste appears to be critical to the cephalic response to foods. This orosensory stimulation (by sites in the brain stem) elicits both parasympathetic and sympathetic nervous system activity (Teff, 2000; Le Blanc, 2000; Powley, 2000).

Such sensory factors can also be inhibitory and contribute to the initiation of the processes of satiation and satiety. Nutrients can be sensed in the oral cavity by their taste (the classic modalities such as sweetness) and their mouth feel (oiliness and viscosity), which allows the first estimation of the food's energy content and its composition. This may serve to confirm cognitive estimates (based on matching visual cues and past experience) of how filling or satisfying a meal will be. Prolonged chewing of a sweet food suppresses subsequent intake (Lavin et al., 2002). If during the meal consumption changes to different foods, these estimates will be adjusted and appetite restimulated. Initial postingestive detection of nutrients in the gut, produced after the swallowing the first mouthfuls of the meal, is fed back to the CNS. Such feedback results in a change in the sensory pleasure derived from that food. These signals appear to inhibit the continued consumption of that particular food (nutrient-specific feedback) and promote the consumption of foods containing different nutrients. This change in sensation is termed *negative alliesthesia* (Cabanac, 1999).

7.4.2 PREABSORPTIVE PEPTIDE SIGNALS

A second class of nutrient-sensitive signals consists of a variety of gut peptides released into the blood stream in response to the presence of food within the gut. These include cholecystokinin (CCK), gastric-releasing peptide (GRP), glucagon-like peptide-1 (GLP-1), enterostatin, apolipoprotein A-IV (apo A-IV), and somatostatin. At least three of these signals are fat sensitive, explaining why unabsorbed fat within the gut produces a strong satiety response (Greenberg et al., 1990).

CCK is a hormone that has been demonstrated to be critical in mediating satiation and early phase satiety (Kissileff et al., 1981; Moran and Schwartz, 1994; Degan et al., 2001). The consumption of both fat and protein, detected by chemoreceptors within the gut wall, stimulates the release of endogenous CCK. This CCK activates CCK-A receptors in the pyloric region of the gut, and the signals generated are then transmitted within the CNS by vagal afferents to the nucleus tractus solitarius (NTS) within the brainstem, where they are relayed onto various hypothalamic nuclei, including the PVN and VMH (Moran and Schwartz, 1994; Lieverse et al., 1994; Gutzwiller et al., 2000). Data suggest that within these loci, the satiety effect of CCK is mediated by serotonin 5-HT_{2C} receptors. Serotonin (5-HT) has been closely implicated in the processes of satiation and the end state of satiety (Blundell, 1977). The role of serotonin in macronutrient preference will be considered later, but note that the CCK–5-HT link provides a mechanism by which the ingestion of dietary fat alters CNS neurochemistry (Blundell and Halford, 1998).

Other preabsorptive signals also indicate the consumption of dietary fat. The conversion of pancreatic procolipase to colipase, a process required for fat digestion, also produces enterostatin, a five-amino-acid peptide (Erlanson-Albertsson et al., 1991; Okada et al., 1991). Endogenous enterostatin levels in the periphery rise in response to a high-fat meal. This indicates that enterostatin selectively regulates the intake of fat. Interestingly, administration of exogenous enterostatin reduces food intake per se when the diet contains a significant level of fat and selectively reduces the intake of high-fat diets when rats are offered a choice. Peripheral enterostatin activates neurons in the NTS, the PVN, and the VMH through afferent vagal mechanisms. Centrally administered enterostatin directly acts on the PVN and the AMYG to reduce intake. The hypophagic effect of enterostatin appears to be moderated centrally both by 5-HT and by the opioidergic system. This suggests that endogenous enterostatin released by fat intake, like CCK, enhances satiety by a central hypothalamic 5-HT mechanism (Erlanson-Albertsson and York, 1997; Lin and York, 1997). A third possible indicator of dietary fat is the intestinal glycoprotein apo A-IV. Endogenous apo A-IV is produced in the human small intestine and released into the intestinal lymph in response to dietary lipids. In rats, administration of exogenous apo A-IV reduces both intake and meal size in a manner consistent with satiety. However, whether apo A-IV acts peripherally by vagal stimulation or directly by a central mechanism is yet to be determined, as have the central mechanisms (5-HT, NPY, etc.) that moderate apo A-IV-induced hypophagia. Although recent data suggest that endogenous apo A-IV is secreted within the hypothalamus, this secretion is dependent on fat ingestion (Liu et al., 2001).

Not all preabsorptive signals are fat sensitive. The incretin peptide GLP-1 is produced in response to duodenal carbohydrate. The rise in endogenous GLP-1 levels within the bloodstream is closely associated with the self-reported suppression of appetite and reduced energy intake (Lavin et al., 1998). Exogenous administration of GLP-1 into the CNS produces a marked decrease in food intake in rodents. GLP-1 infused into the periphery produces a similar reduction in food intake along with increased feelings of fullness and decreased hunger in humans (Flint et al., 1998; Näslund et al., 1999). GLP-1 administration into the CNS activates neurons in the PVN and the AMYG. Moreover, the brain stem (AP/NTS) and the hypothalamus sites, such as the ARc, AMYG, and the PVN, have been implicated in GLP-1-induced hypophagia (Turton et al., 1996). However, the interaction of GLP-1 with the established CNS mechanism associated with hunger and satiety, such as by neuropeptide Y (NPY) or 5-HT, respectively, remains to be established. The continued release of factors initially triggered by the cephalic stage, such as insulin, glucagons, and amylin, may be sustained by the preabsorptive detection of nutrients. These hormones when directly infused into the CNS reduce food intake (Komenami et al., 1996; Rushing et al., 2000; Geary, 1999).

7.4.3 POSTABSORPTIVE NUTRIENT SIGNALS

Almost immediately as food enters the gut, some nutrients can directly cross into the body without the need for digestion. Dietary glucose, the basic brain fuel, is quickly absorbed into the bloodstream and carried to the brain and other organs shortly after the food is swallowed. Certain neurons within the hypothalamic nuclei (and in the NTS) associated with feeding appear sensitive to plasma levels of glucose (Levin et al., 1999). Neurons that increase firing in response to glucose are termed *glucose responsive (GR)*. These appear to be located in hypothalamic areas associated with metabolic control, their firing inhibiting the metabolic drive to consume. They could modulate the function of a variety of neurotransmitters and peptides associated with appetite regulation, particularly NPY. Neurons that decrease their firing in response to glucose are termed *glucose sensitive (GS)* and may be one signal that initiates feeding behavior when glucose levels fall (Levin et al., 1999; Levin, 2001).

The brain is not the only glucose-sensitive organ capable of generating inhibitory signals. Infusions of glucose into the hepatic portal vein decrease food intake and reduce meal size, suggesting a key role for the liver in generating within-meal satiety (Tordoff and Friedman, 1986; Surina-Baumgartner et al., 1996; Langhans et al., 2001). The liver is critical in the metabolism of different fuels (glucose and fatty acids). Hepatic energy metabolism appears to generate signals that are carried from the liver to the CNS by vagal afferents. This appears to be directly related to the liver's energy status, as represented by its ATP content (Hong et al., 2000).

As stated previously, cephalic stimulation and preabsorptive stimulation produce hormones such as insulin, glucagon, and amylin, which are critical in control of postabsorptive plasma glucose levels and metabolism. Although the release of these hormones is predominantly pre- rather than postabsorptive, they respond to postabsorptive glucose levels and like postabsorptive nutrients these hormones reduce meal size and food intake through the liver (LeSauter and Geary, 1991). They also appear to directly act in the CNS (Geary, 1999). Infusion of exogenous glucagon into the CNS appears to selectively reduce the nocturnal intake of carbohydrate in rats (Komenami et al., 1996).

Nutrients in the bloodstream crossing the blood-brain barrier alter brain neurochemistry. Amino acids such as tryptophan, tyrosine, and histidine are precursors of the classical neurotransmitters serotonin, dopamine, noradrenaline, and histamine, all of which have been implicated in appetite control. Moreover, the presence of other nutrients can aid or hinder the passage of these neurotransmitter precursors across the blood-brain barrier. 5-HT function has been linked with satiation and satiety for 25 years (Blundell, 1977; Blundell and Halford, 1998). Thirty years ago, it was also proposed that 5-HT neurons functioned as plasma amino acid ratio sensors (Fernstrom and Wurtman, 1973). The authors noted a relationship between the ratio of tryptrophan to large neutral amino acids (T:LNAAs) in the plasma and the relative proportion of dietary carbohydrate and protein consumed. Increased carbohydrate consumption raised the T:LNAAs, preferentially increasing tryptophan entry into the brain (Teff et al., 1989; Fernstrom, 1987). This in turn increased CNS 5-HT synthesis and release. Therefore, given that CNS 5-HT levels reduce food intake and enhance satiety and that 5-HT function is directly related to the status of specific nutrients within the body, 5-HT has been considered a nutrient-specific satiety system. Alteration in the T:LNAAs ratio, reflected in changes in neuronal 5-HT, could act as a diet composition signal to the brain. However, the T:LNAAs ratio is only one way by which specific nutrients influence CNS 5-HT systems. CNS 5-HT also integrates fat-derived signals mediated by CCK and enterostatin release.

7.4.4 SIGNALS FROM ENERGY STORAGE

Ultimately, nutrients are metabolized, used, secreted, or stored. Factors derived from the nutrient-specific metabolite process and nutrient storage may provide potent nutrient-specific satiety signals. Of all macronutrients, fat is preferentially stored in the body as adipose tissue. Until 8 years ago, the fat-storing cells that comprise adipose tissue were not of much interest to the study of appetite regulation. However, then it was discovered that these cells secreted an adiposity-indicating hormone, leptin (ob-protein; Zhang et al., 1994; Pelleymounter et al., 1995; Halaas et al., 1995). Leptin production in the adipose tissue directly reflected the amount of energy in storage, and the detection of circulating leptin by the CNS inhibited food intake. Subsequent research demonstrated that leptin has a potent effect on a number of neuropeptides in the key hypothalamic sites associated with appetite regulation (such as NPY, melanocortins, and orexin). Thus, excess fat intake, resulting in increased fat storage, inhibits further food intake through leptin secretion. Leptin is most prominent in a new class of storage-derived signals and has led us to consider adipose tissue as an active endocrine organ rather than a mere depository for excess energy (Mohamed-Ali et al., 1998; Halford and Blundell, 2000). Other such substances may include satietin, adipsin, or cytokine signals such as interleukin-6 (IL-6) and tumor necrosis factors (such as TNFα). Insulin is another adiposity indicator whose release is not dependent solely on intake-induced stimulation. Insulin, like leptin, is also secreted in direct proportion to adiposity, directly entering the brain and interacting with hypothalamic neurons (Woods and Seeley, 2001). However, there is considerable interest in the selective modulation of leptin by dietary carbohydrate and fat (Havel, 1999, 2000). This raises the issue of whether leptin could form part of a circuit involved in macronutrient regulation or whether changes in leptin constitute a passive response to the diet.

7.5 CENTRAL MECHANISMS OF MACRONUTRIENT INTAKE STIMULATION

Nutrients generate differing signals through sensory or cognitive preabsorptive hormone release, postabsorptive nutrient detection, and factors derived from long-term energy storage. But how effective are these signals in altering subsequent food intake both within the current meal and at the next and subsequent meals? Do deficits in or the absence of carbohydrate-induced signals such as sweet taste, GLP-1 release, or glucose-responsive neurons firing individually or collectively stimulate the selective consumption of dietary carbohydrate? Do deficits in dietary fat and adiposity-derived signals such as CCK, enterostatin, and leptin selectively stimulate the intake of fats?

A review of a large number of animal studies provides overall evidence that the exclusive intake of, or a deficit in, the intake of a given macronutrient leads to appropriate readjustment of intake at the next eating opportunity. In humans, the picture is less clear. Human participants given preloads predominately of fat or of carbohydrates do not always adjust their intake at the subsequent meal when they are offered a wide variety of foods. In such experimental situations, they do not compensate for the earlier fat or carbohydrate load by consuming less fat or carbohydrate at the next meal. However, if the subsequent intake is monitored over 2 or more days, such adjustments can be observed (De Castro, 1998). This would support the hypothesis that nutrient-derived signals selectively inhibit further intake of the given nutrient by altering food choice, as conversely would their absence. If so, what CNS mechanisms stimulate or inhibit the subsequent intake of macro-nutrients? How are these CNS mechanisms modulated by nutrient-specific signals derived for the previous intake?

7.5.1 METHODOLOGICAL ISSUES

Before reviewing a selection of animal studies examining the role of CNS mechanisms in the control of food preference and choice, a note of caution should be sounded. Diet selection studies are fraught with methodological problems, which have in turn led to a lack of definitive, and sometimes contradictory, results within the field. In the earliest studies, rodents were often offered a choice between two manipulated diets (a two-choice paradigm). In these diets, two of the three macronutrients were altered, for example, to produce a high-carbohydrate (low-protein) diet and a high-protein (low-carbohydrate) diet (Wurtman and Wurtman, 1979). Often, the levels of other nutrients in the diet were not held constant, producing extraneous variables that could account for the observed results (such as differences in the fat content, energy density, or palatability between diets).

In later, better-controlled two-diet choice paradigm studies, the level of the third macronutrient was held constant. However, the design still produced inherent problems. Such procedures precluded the observation of any effect of the manipulation on the third macronutrient. Second, how were the appropriate low and high levels for each macronutrient within the test diets to be defined? Finally, when the animals expressed a preference, was this a drive to consume the predominant macronutrient in the preferred diet or avoid the predominant macronutrient in the other? In an effort to deal with some of these problems, researchers employed a three-choice paradigm. Rodents were offered a choice among three diets, each consisting of predominantly one macronutrient or a pure source of the macronutrient (with essential micronutrients added). This procedure allowed research-ers to better ascertain whether manipulation-induced changes in food selection were due to pref-erence or avoidance. However, without prior habituation to these diets, researchers were placing animals in a novel situation. Unfamiliarity on exposure and lack of experience of the postingestive consequences of consuming these diets (satisfaction or malaise) would have had a dramatic effect on the animals' initial behavioral response to the diets. Moreover, animals like humans tend to consume foods rather than pure macronutrients. In the choice paradigms, a variety of contextual variables, such as palatability, quality of the test diets, form or type of macronutrient offered, prior exposure to the diet, and the metabolic state of the animal (fasted or satiated), are among the many factors that can have a marked effect on any observed result.

Recent studies have employed dietary-induced obesity models and the cafeteria diet paradigm to study macronutrient preferences. Rats provided highly palatable diets in addition to their labo-ratory chow become obese over time. Often, these additional obesity-inducing diets vary markedly in their macronutrient content, enabling such paradigms to be used to assess the role of various systems in food preference. In such paradigms, animals are offered a choice between a standard familiar laboratory chow and other foods provided in addition. These can be palatable versions of the same chow mixed together with additional macronutrients (such as pure forms of fat, carbo-hydrate, or protein). Alternatively, these additional palatable foods could be human snack foods such as peanuts, potato chips, chocolate, cookies, and high-sugar sweets (all highly acceptable to the laboratory rat!). In such long-term paradigms, the food choice is less forced. The animals always have the familiar laboratory chow, and they are allowed time to become acquainted with the novel diets and thus in time are able to express a more representative appetite. These studies have also revealed a fascinating and usable phenomenon. Within these paradigms, rats of the same strain, parentage, and kept under the same constant laboratory conditions develop remarkably distinct food preferences. In groups of rats exposed to the cafeteria diet, some animals develop a strong preference for high-carbohydrate foods whereas others develop preferences for the high-fat items. Interestingly, humans can also be defined as high- or low-fat "preferers" or consumers.

7.5.2 CLASSICAL NEUROTRANSMITTERS

The distinctive roles of monoamine transmitter systems in appetite have been studied for nearly 35 years. Blundell (1977) first proposed that CNS 5-HT was critical to the development of within-meal satiation and postmeal satiety. Studies had shown than when 5-HT or its precursors (tryptophan and 5-HTP) were administered to rodents, not only food intake but also meal size and eating rate were significantly reduced. Drugs that increased synaptic 5-HT also produced similar changes to food intake without disrupting behavior, suggesting that endogenous 5-HT systems control eating behavior. Of the numerous 5-HT receptors, those directly implicated in appetite regulation are the presynaptic 5-HT_{1A} and the postsynaptic 5-HT_{1B} ($5\text{-HT}_{1D}\beta$ in humans) and 5-HT_{2C} receptors. The role of these receptors was first disclosed in animals by agonist–antagonist studies in rodents and then in knockout mice models. Human studies demonstrated that drugs that increase synaptic 5-HT (fenfluramine, D-fenfluramine, fluoxetine, sertraline, and sibutramine) and selective agonists of $5\text{-HT}_{1D}\beta$ in 5-HT_{2C} receptors reduced food intake and meal size and also reduced hunger and strengthened satiety.

The role of 5-HT in macronutrient preference has been exhaustively studied and warrants its own lengthy review. CNS 5-HT appears to modulate CCK and enterostatin signals. Peripheral administration of CCK stimulates neuronal 5-HT release in hypothalamic areas such as the PVN. Antagonists to the CCK-A receptor can block 5-HT-induced reductions in food intake. Conversely, 5-HT antagonists can block CCK-induced hypophagia. As the release of both CCK and enterostatin is triggered by preabsorptive fat signals, CNS 5-HT would be sensitive to the consumption of dietary fat. Endogenous administration of both the octo-peptide CCK-8 and enterostatin selectively decreases fat consumption. Such macronutrient-selective effects could be, at least in part, mediated by 5-HT activation. However, neuronal 5-HT levels are also sensitive to changes in dietary carbo-hydrate. As stated previously, increased carbohydrate increases the amount of the 5-HT precursor tryptophan crossing the blood-brain barrier by increasing the T:LNAAs ratio. Carbohydrate-induced insulin secretion reduces the levels of LNAAs in the bloodstream, enabling more tryptophan to be transported into the brain promoting 5-HT synthesis. [For a review, see Blundell and Halford (1998).]

Hypothalamic areas in and around the PVN have been associated with 5-HT-induced hypo-phagia and mediation of macronutrient preferences. Interestingly, blocking 5-HT synthesis or antagonist 5-HT receptors increases NPY functioning in the PVN. NPY potently stimulates food intake per se and has been implicated in carbohydrate selection (see later). Therefore, without PVN 5-HT, consumption per se and carbohydrate consumption in particular may not be kept in check. 5-HT-stimulatory drugs, including an agonist of the $5\text{-HT}_{1B/2C}$ receptors, appear to inhibit endog-enous NPY levels within the PVN (Dryden et al., 1996), suggesting that 5-HT activation could reduce intake and selectively alter food choice.

Many diet selection studies, predominantly of the two-choice paradigm, employing 5-HT agents suggested that serotonin activation was most potent in reducing the intake of high-carbohydrate diets (Wurtman and Wurtman, 1979). When rodents were offered three diets in some studies, they selectively decreased carbohydrate consumption (Leibowitz et al., 1992). However, in the majority of studies animals administered 5-HT drugs (peripherally or centrally) or 5-HT directly administered into the PVN selectively reduced consumption of fat (Kanarek and Duskin, 1988). This effect was most marked in rats that already displayed a baseline preferred fat (Smith et al., 1999). As with animal studies, human research is affected by a similar range of analogous methodological issues. Initial human studies only examined the effect of 5-HT drugs on differences in the consumption of high-carbohydrate and high-protein diets (fat held constant). In these studies, serotoninergic manipulation reduced the intake of high-carbohydrate foods (Wurtman et al., 1985). However, later studies showed that 5-HT drugs can selectively suppress fat intake in humans and generally suppressed the intake of both high-carbohydrate and high-fat foods (Lawton et al., 1995). Perhaps the most notable finding was that these drugs potently reduce the intake of snack foods that are often high in both fats and sugars. A free-living study in the obese has demonstrated that the serotoninergic D-fenfluramine, given over a 3-month period, most potently suppressed the intake of snacks. This was part of an overall reduction in energy intake over the period of the study, strikingly characterized by a decrease in the percentage of energy from fat (Lafreniere et al., 1993). These data collectively suggest that not only is CNS 5-HT sensitive to carbohydrate and fat, but also its activation may inhibit their consumption, specifically when they occur in sweet, high-fat, energy-dense snack foods.

7.5.3 NEUROPEPTIDES

The role of a few neuropeptides in appetite regulation has been established over the past 15 years, but since the identification of leptin in 1994 numerous novel appetite-regulatory peptides have been identified [orexins, melanocortins, and cocaine-amphetamine related transcript (CART)]. The role of most of these in macronutrient preference has yet to be determined. One of the earliest peptides identified, NPY, is a 36-amino-acid peptide found throughout the CNS. Data show that increased NPY levels within the PVN produce a potent increase in food intake initiating meals, an effect opposed by endogenous 5-HT (Dryden et al., 1996). Hypothalamic NPY is stimulated by food deprivation and is also sensitive to circulating levels of insulin and leptin (indictors of carbohydrate and fat signals, respectively). There is also evidence that NPY inhibition mediates the anorectic effects of amylin. Thus, the hypothalamic NPY system is sensitive to input from both nutrient-specific consumption and to the metabolic status of the animal. Maintaining rats on a diet of high carbohydrate has been shown to augment the synthesis and release of hypothalamic NPY. A rapid increase in hypothalamic NPY can also be produced in response to a single high-carbohydrate meal or by injecting glucose at the start of feeding. This finding demonstrates that endogenous NPY reacts swiftly to the consumption of carbohydrate (Wang et al., 1999). Hypothalamic NPY not only appears to be highly responsive to dietary carbohydrate but also selectively stimulates its consumption. A series of studies that used NPY ligands and selective antagonists indicated the critical role of the NPY receptors Y1 and Y5 in mediating NPY-induced hyperphagia. Antagonizing NPY Y1 receptors has also been shown to selectively block carbohydrate intake (Leibowitz et al., 1992). Conversely, increased NPY function and selective stimulation of the NPY Y1 receptor specifically stimulate carbohydrate consumption in various studies (Stanley et al., 1985).

Another CNS peptide that stimulates feeding behavior and demonstrates an effect of macronutrient choice is galanin. When infused directly into the PVN, exogenous galanin stimulates fat intake (Leibowtiz et al., 1989; Barton et al., 1995). Moreover, rats overexpressing galanin mRNA in the PVN demonstrate a marked preference for a high-fat diet. CNS galanin appears to be inhibited by both enterostatin and fatty acid oxidation. It remains unclear what other nutrient-derived and metabolic signals stimulate or inhibit endogenous CNS galanin to subsequently alter food choice.

The endogenous opioid system, implicated in the sensory pleasure derived from food consumption, has also been shown to produce selective macronutrient consumptive effects (Barton et al., 1995). The endogenous opioid system may comodulate the fat intake suppressing effects of enterostatin. Recently, the role of endogenous opioid systems in the stimulation of fat intake produced by the endogenous melanocortin system antagonist agouti-related peptide (AgRP) has been demonstrated (Hagan et al., 2001). Studies with opioid antagonists in humans have shown that these agents selectively reduce the intake of fats (Yeomans and Gray, 1996), which can be interpreted as supporting the concept that the endogenous opioid system selectively regulates fat intake. However, studies have also demonstrated that opioid antagonists block the consumption of sweet high-carbohydrate foods in rats (Levine et al., 1995) and the preference for and liking of sweet foods in humans (Fantino et al., 1986; Yeomans and Gray, 1996). It would then seem that the endogenous opioid system probably stimulates the consumption of preferred foods, be they sweet carbohydrates or fats. Indeed, opioid antagonists block the intake of the preferred diet of rats (Glass et al., 1996) and block the intake of the high-fat, high-sugar foods in human binge eaters (Drewnowski et al., 1995). Therefore, the endogenous opioid system may not drive nutrient-specific appetites but instead mediate the rewarding aspect of eating palatable foods (Glass et al., 1996).

7.6 FUNCTIONAL PHENOTYPES — CASE STUDY IN HUMAN MACRONUTRIENT SELECTION

It is possible to probe human macronutrient preferences by experimental interventions (similar to those used in animal studies) to provoke an acute change in the direction of food choice. An alternative is based on identifying and selecting individuals according to their natural habitual selection of foods. Such individuals have been termed *behavioral phenotypes*. A phenotype is normally defined as a stable cluster of measurable characteristics that separate one type from another. This type is classically regarded as the consequence of a particular genotype. The approach therefore lends itself to the development of a taxonomy of unambiguously defined types; it is therefore a typology. In principle, a number of phenotypes could be defined based on taste preferences (for foods), patterns of eating, beliefs about foods, or motivational responses. The approach favored here emphasizes behavior, because this can be defined more rigorously and unambiguously than attitudes or subjective perceptions.

The power of this approach resides in the capacity to make clear distinctions between different phenotypes, that is, between groups of individuals with contrasting and habitual patterns of food selection. In turn, the term *selection* is defined rigorously by reference to foods that are actually consumed. These phenotypes can be distinguished according to habitual dietary intakes (of foods that have been selected). Because of the importance of dietary fat to energy balance, body weight regulation, and other aspects of health (Blundell et al., 1996), we have initiated this approach to food selection through the study of high-fat and low-fat (high-carbohydrate) phenotypes.

First, any investigation of actual food selection in natural free-living subjects must face the problem of accuracy of measurement of food intake behavior. A dilemma for nutritional science has developed related to this issue, which has been termed a *misreporting problem,* the most common form of which is underreporting (Blundell, 2000). When subjects are brought into a laboratory or research unit, an experimental control makes it possible to accurately record the amount and type of food consumed. Food intake can be measured directly. The problem is that eating in these rather sterile environments may be quite different to the eating that occurs under natural circumstances. Measurement in the real world requires an indirect rather than a direct approach; that is, it relies on some type of self-report or self-monitoring. Unfortunately, and surprisingly, many people find it very difficult to accurately report the foods they habitually consume on a daily basis (Macdiarmid and Blundell, 1998).

An example of this can be seen in the records of people choosing foods high in sugar or high in fat. In the field of obesity research, there is an active debate on the relative contribution of high-fat or high-sugar foods to the development of obesity. Interestingly, some epidemiological data (Bolton-Smith and Woodward, 1994) have confirmed the existence of the sugar–fat seesaw. This means that obesity is associated with high-fat or low-sugar intakes and leanness with the converse. A similar phenomenon can be seen in the analysis of another large database (Macdiarmid et al., 1998). However, in these analyses, it is important to minimize the effect of misreporting. This can be done by establishing a threshold value for the subjective records of energy intake such that estimates below this value can be regarded as being physiologically implausible. By convention, this threshold is set at $1.2 \times$ RMR (where RMR is normally calculated from equations). When underreporters were excluded from the database, analysis indicated that sugar intakes remained constant in lean, normal-weight, overweight, and obese subjects. However, fat intakes rose as the degree of obesity increased (see Macdiarmid et al., 1998). When the analysis was directed to particular types of foods (sweet and fatty foods) as opposed to the intakes of nutrients (sucrose and fat), the degree of underreporting had a major impact on interpretation of the data. Analysis of the plausible dietary records indicated that the most obese women consumed the highest amounts of sweet or fat foods. However, if the entire group of subjects were analyzed (including the underreporters) the obese women appeared to consume the smallest quantities of sweet or fat foods (Macdiarmid et al., 1998).

This type of outcome means that when food-selection phenotypes are being identified, it is essential to assess the degree of underreporting. Any records derived from food frequency questionnaires (FFQs) or food diaries that show physiologically implausible energy intakes should lead to those subjects being excluded (even though they may show interesting food selection patterns).

7.7 HABITUAL CONSUMERS OF HIGH- AND LOW-FAT FOODS

High-fat (HF) and low-fat (LF) phenotypes are classified according to the type of diet habitually consumed, measured by an FFQ and diary record and with underreporters excluded. The records indicate that these groups habitually consume different types of foods and display different patterns of eating (Macdiarmid et al., 1996). The research strategy is designed to allow individuals characterized in a population by their macronutrient intakes to be tracked so that their particular food choices and consumption of specific foods can be identified (see Figure 7.2).

In a first series of studies, the characteristics of young adult males have been examined. When subjected to energy and macronutrient challenges in order to evaluate the responses of the appetite control system, clear differences between the groups were demonstrated. Initially, HF displayed higher initial hunger levels, with a much sharper decline in hunger in response to meals or nutrient loads (Cooling and Blundell, 1998a). After eating, hunger recovered more rapidly in HF compared with LF. In addition, the size of a test meal consumed was closely related to the suppression of hunger in HF; in contrast, the appetite response system in LF appeared to be somewhat insensitive and damped. This relationship between habitual fat intake and hunger is reminiscent of a previous finding. French et al. (1996) found that during 2 weeks of high-fat overfeeding to normal-weight subjects, which caused a significant gain in weight, subjects displayed a progressive increase in hunger and a decrease in fullness before a test meal. Taken together, these findings indicate that eating a high-fat diet may facilitate feelings of hunger.

A further feature of these behavior studies was that HF and LF differed in the control over meal size when offered an unlimited range of either high-fat or high-carbohydrate foods. HF consumed a similar weight of food on both diets and therefore took in a much higher amount of energy with the high-fat (high-energy-dense) foods. In contrast, LF consumed a much smaller amount of the high-fat foods and consequently took in a similar amount of energy on both diets. These findings suggest that signaling systems for meal termination (satiation) and postmeal inhibition of appetite (satiety) operate with differing strengths in HF and LF. This finding may not be surprising in view of the fact that the gastrointestinal tract has become adapted to dealing with quite different dietary components and this factor would have exerted a priming effect on specific satiety signals.

The existence of distinctive profiles of appetite control in HF and LF indicates different patterns of physiological responses to food ingestion. The possibility of other physiological differences was investigated by indirect calorimetry to measure basal metabolic rate (BMR), respiratory quotient (RQ), and dietary-induced thermogenic responses to specific fat and carbohydrate loads (Cooling and Blundell, 1998b; Blundell et al., 2002).

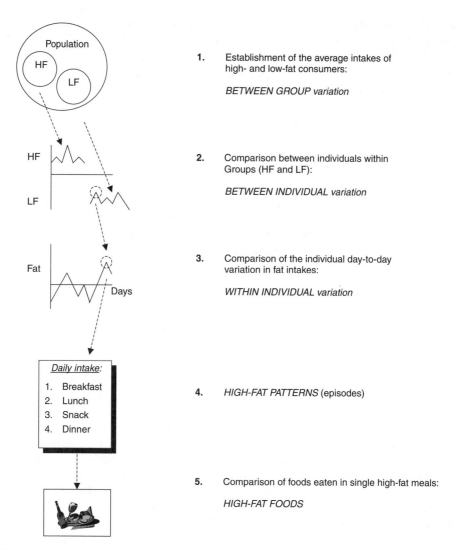

FIGURE 7.2 Sequence of steps for deriving the specific foods eaten in high-fat meals (step 5) from high-fat consumers identified in the population (step 1). HF = high-fat consumers, LF = low-fat consumers.

The results indicated that HF has a lower RQ than does LF; this finding confirmed that fat oxidation is higher in HF, as would be expected because of the habitual high intake of fat-containing foods. However, an unexpected finding was the significantly higher BMR in HF than LF, together with different profiles of thermogenic responses to the high-fat and high-carbohydrate loads (Blundell et al., 2002). A further important finding was that HF had higher plasma leptin levels than LF did (Cooling et al., 1998), despite having similar levels of body fat.

This case study has indicated one approach to the study of human macronutrient preferences that links particular selection patterns to important variables such as energy balance and body weight control. The measurable differences in the types of foods habitually selected pose the question of whether the choices are biologically driven (by particular tissue needs, physiological requirements, or neurosensory characteristics) or incidentally picked up from the environment. In either case, it will be necessary for the physiological system to adapt to the ingestion of large amounts of specific macronutrients. This therefore provides an opportunity to identify biological markers of habitual food selection even if they are not the actual causes.

REFERENCES

Barton, C., Lin, L., York, D.A., and Bray, D.A. (1996) Differential effect of enterostatin, galanin and opioids on high fat diet consumption. *Brain Res.,* 702, 55–60.

Blundell, J.E. (1977) Is there a role for serotonin (5-hydroxy-tryptamine) in feeding? *Int. J. Obes.,* 1, 15–42.

Blundell, J.E. (2000) What foods do people habitually eat? A dilemma for nutrition, an enigma for psychology. *Am. J. Clin. Nutr.,* 71, 3–5.

Blundell, J.E., and Halford, J.C.G. (1998) Serotonin and appetite regulation: implications for the pharmacological treatment of obesity. *CNS Drugs,* 9, 473–495.

Blundell, J.E., Lawton, C.L., Cotton, J.R., and Macdiarmid, J.I. (1996) Control of human appetite: implications for the intake of dietary fat. *Annu. Rev. Nutr.,* 16, 285–319.

Bolton-Smith, C., and Woodward, M. (1994) Dietary composition and fat to sugar ratios in relation to obesity. *Int. J. Obes.,* 18, 820–828.

Cabanac, M. (1999) Sensory pleasure and alliesthesia. In *Elsevier's Encyclopedia of Neuroscience,* 2nd ed. (G. Adleman and B.H. Smith, Eds.), pp. 1836–1837. Elsevier Science, Amsterdam.

Cooling, J., and Blundell, J.E. (1998a) Are high-fat and low-fat consumers distinct phenotypes? Differences in the subjective and behavioural response to energy and nutrient challenges. *Eur. J. Clin. Nutr.,* 52, 193–201.

Cooling, J., and Blundell, J.E. (1998b) Differences in energy expenditure and substrate oxidation between habitual high fat and low fat consumers (phenotypes). *Int. J. Obes.,* 22, 612–618.

Cooling, J., Barth, J., and Blundell, J.E. (1998) The high-fat phenotype: is leptin involved in the adaptive response to a high fat (high energy) diet? *Int. J. Obes.,* 22, 1132–1135.

De Castro, J.M. (1998) Prior day's intake has macronutrient-specific delayed negative feedback effect on spontaneous food intake of free-living humans. *J. Nutr.,* 128, 61.

Degan, L., Matzinger, D., Drewe, J., and Beglinger, C. (2001) The effect of cholecystokinin in controlling appetite and food intake in humans. *Peptides,* 22, 1265–1269.

Drewnowski, A., Krahn, D.D., Demitrack, M.A., Nairn, K., and Gosnell, B.A. (1995) Naloxone, an opiate blocker, reduces the consumption of sweet high-fat foods in obese and lean female binge eaters. *Am. J. Clin. Nutr.,* 61, 1206–1212.

Dryden, S., Wang, Q., Frankish, H.M., and Williams, G. (1996) Differential effects if the 5-HT1B/2C agonist mCPP and the 5-HT1A agonist flesinoxan on hypothalamic NPY in the rat. *Peptides,* 6, 943.

Erlanson-Albertsson, C., Mei, J., Okada, S., York, D.A., and Bray, G.A. (1991) Pancreatic procolipase propeptides, enterostatin, specifically inhibits fat intake. *Physiol. Behav.,* 49, 1191–1194.

Erlanson-Albertsson, C., and York, D.A. (1997) Enterostatin: a peptide regulating fat intake. *Obes. Res.,* 5, 360–372.

Fantino, M., Hosotte, J., and Apfelbaum, M. (1986) An opioid antagonist, naltrexone, reduces preference for sucrose in humans. *Am. J. Physiol. Regul. Integr. Comp. Physiol.,* 251, R91–R96.

Fernstrom, J.D. (1987) Food-induced changes in brain serotonin synthesis: is there a relationship to appetite for specific macronutrients? *Appetite,* 8, 163.

Fernstorm, J.D., and Wurtman, R.J. (1973) Control of brain 5-HT content by dietary carbohydrates. In *Serotonin and Behaviour* (J. Barchas and E. Usdin, Eds.), pp.121–128. Academic Press, New York.

Flint, A., Raben, A., Astrup, A., and Holst, J.J. (1998) Glucagon-like peptide 1 promotes satiety and suppresses energy intake in humans. *J. Clin. Invest.,* 101, 515–520.

French, S.J., Murray, B., Rumsay, R.D.E., Fadzlin, R., and Read, N.W. (1995) Adaptation to high-fat diets: effects on eating behaviour and plasma cholecystokinin. *Br. J. Nutr.,* 73, 179–189.

Geary, N. (1999) Effect of glucagons, insulin, amylin and CGRP on feeding. *Neuropeptides,* 33, 400–405.

Glass, M.J., Grace, M., Cleary, J.P., Billington, C.J., and Levine, A.S. (1996) Potency of naloxone's anorectic effect in rats is dependent on diet preference. *Am. J. Physiol. Regul. Integr. Comp. Physiol.,* 271, R217–R221.

Greenberg, D., Smith, G.P., and Gibbs, J. (1990). Intraduodenal infusions of fats elicit satiety in sham-feeding fats. *Am. J. Physiol. Regul. Integr. Comp. Physiol.,* 259, R110–R118.

Gutzwiller, J.-P., Drewe, J., Ketterer, S., Hildebrand, A.K., and Beglinger, C. (2000) Interaction between CCK and a preload on reduction of food intake is mediated by CCK-A receptors in humans. *Am. J. Physiol. Regul. Integr. Comp. Physiol.,* 279, R189–R195.

Hagan, M.M., Rushing, P.A., Benoit, S.C., Woods, S.C., and Seeley, R.J. (2001) Opioid receptor involvement in the effect of AgRP (83-132) on food intake and food selection. *Am. J. Physiol. Regul. Integr. Comp. Physiol.,* 280, R814–R821.

Halaas, J.L., Gajiwala, K.S., Maffei, M., Cohen, S.L., Chati, B.T., Rabiowitx, D., Lallone, R.L., Burley, S.K., and Friedman, J.M. (1995) Weight reducing effects of the plasma protein encoded by the obese gene. *Science,* 269, 543–546.

Halford, J.C.G., and Blundell, J.E. (2000) Separate systems for serotonin and leptin in appetite control. *Ann. Med.,* 32, 222–232.

Havel, P.J. (1999) Mechanisms regulating leptin production: implications for control of energy balance. *Am. J. Clin. Nutr.,* 70(3): 305–306.

Havel, P.J. (2000) Role of adipose tissue in body–weight regulation: mechanisms regulating leptin production and energy balance. *Proc. Nutr. Soc.,* 59(3): 359–371.

Hong, J.I., Graczyk-Milbrandt, G., and Friedman, M.I. (2000) Metabolic inhibitors synergistically decrease hepatic energy status and increase food intake. *Am. J. Physiol. Regul. Integr. Comp. Physiol.,* 278, R1579–R1582.

Kanarek, R.B., and Duskin, H. (1988) Serotonin administration selectively reduces fat intake in rats. *Pharmacol. Biochem. Behav.,* 13, 122–133.

Kissileff, H.R., Pi-Sunyer, X., Thornton, J., and Smith, G.P. (1981) C-terminal octapeptide of cholecystokinin decreases food intake in man. *Am. J. Clin. Nutr.,* 34, 154–160.

Komenami, N., Su, F-H., and Thibault, L. (1996) Effect of central glucagon infusion on macronutrient selection in the rat. *Physiol. Behav.* 59, 383–388.

Lafreniere, F., Lambert, J., and Rasio, E. (1993) Effects of dex fenfluramine treatment on body weight and post-prandial thermogenesis in obese subjects: a double-blind, placebo-controlled study. *Int. J. Obes.,* 17, 25–30.

Langhans, W., Grossman, F., and Geary N. (2001) Intrameal hepatic-portal infusion pf glucose reduced spontaneous meal size in rats. *Physiol. Behav.,* 73, 499–507.

Lavin, J.H., French, S.J. Ruxton, C.H.S., and Read, N.W. (2002) An investigation of the role of oro-sensory stimulation in sugar satiety. *Int. J. Obes.,* 26, 384–388.

Lavin, J.H., Wittert, G.A., Andrews, J., Wishart, J.M., Morris, H.A., Morley, J.E., Horowitz, M., and Read, N.W. (1998). Interation of insulin, glucagon-like peptide 1, gastric inhibitory polypeptide, and appetite in response to intraduodenal carbohydrate. *Am. J. Clin. Nutr.,* 68, 591–598.

Lawton, C.L., Wales, J.K., Hill, A.J., and Blundell, J.E. (1995) Serotoninergic manipulation, meal induced satiety and eating pattern: effects of fluoxetine in obese female subjects. *Obes. Res.,* 3, 345–356.

Le Blanc, J. (2000) Nutritional implications of cephalic phase thermogenic response. *Appetite,* 34, 214–216.

Leibowitz, S.F. (1989) Hypothalamic neuropeptide Y, galanin and amines: concepts of coexistence in relation to feeding behaviour. *Ann. N.Y. Acad. Sci.,* 575, 221–235.

Leibowitz, S.F., Xuereb, M., and Kim, T. (1992) Blockade of natural and neruopeptide Y induced carbohydrate feeding by a receptor antagonist PYX-2. *NeuroReport,* 3, 1023–1026.

Le Sauter, J., and Geary, N. (1991) Hepatic portal glucagons infusion decreases spontaneous meal size in rats. *Am. J. Physiol. Regul. Integr. Comp. Physiol.,* 261, 154–161.

Levin, B.E. (2001) Glucosensing neurons do more than just sense glucose. *Int. J. Obes.,* 25 (Suppl. 5), S68–S72.

Levin, B.E., Dunn-Meynell, A.A., and Routh, V.H. (1999) Brain glucose sensing and body energy homeostatis: role in obesity and diabetes. *Am. J. Physiol. Regul. Integr. Comp. Physiol.,* 276, 1223–1231.

Levine, A.S., Weldon, D.T., Grace, M., Cleary, J.P., and Billington, C.J. (1995) Naloxone blocks that portion of feeding driven by sweet taste in food-restricted rats. *Am. J. Physiol. Regul. Integr. Comp. Physiol.,* 268, R248–R252.

Lieverse, R.J., Jansens, J.B.M.J., Masclee, A.A.M., Rovati, L.C., and Lamers, C.B.H.W. (1994) Effect of a low dose of intraduodenal fat on satiety in humans: studies using the type A cholecystokinin receptor antagonist loxiglumide. *Gut,* 35, 501–505.

Lin, L., and York, D.A. (1997) Enterostatin actions in the amygdala and PVN to suppress feeding in the rat. *Peptides,* 18, 1341–1347.

Liu, M., Doi, T., Shen, L., Woods, S.W., Seeley, R.J., Zheng, S., Jackman, A., and Tso, P. (2001) Intestinal satiety protein apolipoprotein AIV is synthesised and regulated in the rat hypothalamus. *Am. J. Physiol. Regul. Integr. Comp. Physiol.,* 280, R1382–R1387.

Macdiarmid, J.I., and Blundell, J.E. (1998) Assessing dietary intake: who, what and why of under-reporting. *Nutr. Res. Rev.,* 11, 1–24.

Macdiarmid, J.I., Cade, J.E., and Blundell, J.E. (1996) High and low fat consumers, their macronutrient intake and body mass index: further analysis of the National Diet and Nutrition Survey of British Adults. *Eur. J. Clin. Nutr.,* 50, 505–512.

Macdiarmid, J.I., Cade, J.E., and Blundell, J.E. (1998) The sugar-fat relationship revisited: differences in consumption between men and women of varying BMI. *Int. J. Obes.,* 22, 1053–1061.

Mohamed-Ali, V., Pinkney, J.H., and Coppack, S.W. (1998) Adipose tissue as an endocrine and paracrine organ. *Int. J. Obes.,* 22, 1145–1158.

Moran, T.H., and Schwartz, G.J. (1994) Neurobiology of cholecystokinin. *Crit. Rev. Neurobiol.,* 9, 1–28.

Näslund, E., Barkeling, B., King, N., Gutnaik, M., Blundell, J.E., Holst, J.J., Rössner, S., and Hellström, P.M. (1999). Energy intake and appetite are suppressed by glucagons-like peptide-1 (GLP-1) in obese men. *Int. J. Obes.,* 23, 304–311.

Okada, S., York, D.A., Bray, G.A., and Erlanson-Albertsson, C. (1991) Enterostatin (Val-Pro-Asp-Pro-Arg), the activation peptide of procolipase, selectively reduced fat intake. *Physiol. Behav.,* 49, 1185–1189.

Pelleymounter, M.A., Cillen, M.J., Baker, M.B., Hecht, R., Winters, D., Boone T., and Collins, F. (1995). Effects of the obese gene product on the body weight regulation in the ob/ob mice. *Science,* 269, 540–543.

Powley, T.L. (2000) Vagal circuitry mediating cephalic-phase response to food. *Appetite,* 34, 184–188.

Rushing, P.A., Hagan, M.M., Seeley R.J., Lutz, T.A., and Woods, S.C. (2000) Amylin: a novel action in the brain to reduce body weight. *Endocrinology,* 414, 850–853.

Smith, B.K., York, D.A., and Bray, G.A. (1999) Activation of hypothalamic serotonin receptors reduces intake of dietary fat and protein but not carbohydrate. *Am. J. Physiol.,* 277, R802–R811.

Stanley, B.G., Anderson, K.C., Grayson, H.M., and Leibowitz, S.F. (1985) Paraventricular nucleus injection of peptide YY and neuropeptide Y preferentially enhance carbohydrate ingestion. *Peptides,* 6, 1205–1211.

Surina-Baumgartner, D.M., Arnold, M., Moses, A., and Langhans, W. (1996) Metabolic effect of a fat- and carbohydrate-rich meal in rats. *Physiol. Behav.,* 59, 973–981.

Teff, K. (2000) Nutritional implications of the cephalic-phase reflexes: endocrine responses. *Appetite,* 34, 206–213.

Teff, K.L., Young, S.N., and Blundell, J.E. (1989) The effect of protein or carbohydrate breakfasts on subsequent plasma amino acid levels, satiety and nutrient selection in normal males. *Pharmacol. Biochem. Behav.,* 34, 410–417.

Tordoff, M.G., and Friedman, M.I. (1986). Hepatic portal glucose infusion decrease food increased food preference. *Am. J. Physiol. Regul. Integr. Comp. Physiol.,* 251, R192–R196.

Turton, M.D., O' Shea, D., Gunn, I., Beck, S.A., Edwards, C.M.B., Meeran, K., Choi, S.J., Taylor, G.M., Heath, M.M., Lambert, P.D., Wilding, J.P., Smith, D.M., Ghatei, M.A., Herbert, J., and Bloom, S.R. (1996) A role for glucagon-like peptide-1 in the central regulation of feeding. *Nature,* 379, 69–72.

Wang, J., Dourmashkin, J.T., Yun, R., and Leibowtiz, S.F. (1999) Rapid changes in hypothalamic neuropeptide Y produced by carbohydrate-rich meals that enhance corticosterone and glucose levels. *Brain Res.,* 484, 124–136.

Woods, S.C., and Seeley, R.J. (2001) Insulin as an adiposity signal. *Int. J. Obes.,* 25 (Suppl. 5), s35–s38.

Wurtman, J.J., and Wurtman, R.J. (1979) Drugs that enhance central serotonimergic transmission diminish elective carbohydrate consumption in rats. *Life Sci.,* 24, 895–890.

Wurtman, J.J., Wurtman, R.J., and Mark, M. (1985) D-Fenfluramine selectively suppresses carbohydrate snacking in obese people. *Int. J. Eat. Disord.,* 4. 89–99.

Yeomans, M.R., and Gray, R.W. (1996) Selective effects of naltrexone on food pleasantness and intake. *Physiol. Behav.,* 60, 439–446.

Zhang, Y., Proenca, R., Maffei, M., Barone, M., and Friedman, J.M. (1994) Positional cloning of the mouse obese gene and its human homologue. *Nature,* 372, 425–432.

8 Central Regulation of Feeding: Interplay between Neuroregulators

Allen S. Levine, Pawel K. Olszewski, Charles J. Billington, and Catherine M. Kotz

CONTENTS

Eating is a complex behavior and requires the interplay of a variety of regulatory processes, ranging from neurohormonal changes to cognitive functions, to maintain an organism in homeostatic balance. The appetitive behaviors needed to maintain feeding-related homeostasis are produced by the interaction between the brainstem, which plays the key role in integrating visceral information, and forebrain structures, such as the hypothalamus, striatum, and amygdale, which are mainly responsible for equilibrium of hormonal and autonomic responses (Levine and Billington, 1997).

The complexity of brain networks involved in food intake regulation is reflected in behavior. Studies with transgenic and knockout animals indicate that elimination of a single, apparently crucial, neuroregulator does not alter feeding behavior in the predicted manner. Even lesions of entire brain regions involved in feeding may affect consumption only under specific conditions, such as when selected foods are paired with certain environmental cues. In addition, humans and animals alter their eating patterns for reasons other than energy needs. Consumption may be initiated, maintained, or terminated because of stress, boredom, taste, food availability, time of day, and social stimuli.

Various neurochemical systems play significant roles in the central control of feeding. Many neurochemicals have been implicated in the control of food intake, including biogenic amines, endogenous cannabinoids, and excitatory and inhibitory amino acids and peptides. Some of these compounds have been defined as satiety factors, for example, corticotropin releasing hormone (CRH), cholecystokinin (CCK), α-melanocyte stimulating hormone (α-MSH), oxytocin (OT), and glucagon-like peptide-1 (GLP1). On the other hand, neuropeptide Y (NPY), galanin, opioids, and agouti-related protein (AgRP), among others, stimulate consummatory behavior (Woods et al., 1998). Within the brain, neurons containing these substances form pathways and wider networks. By the organization of this circuitry, these neuropeptides often interact with one another in the process of regulating feeding, for example, by affecting each other's release or by reaching the same target cells. Thus, the regulation of consummatory behavior can be attributed to a dynamic balance involving peptides within brain networks. When studying the impact of a single neuropeptide on food intake, it is essential to define this substance's function as part of a larger feeding-related central circuitry.

In this chapter, we focus on representative neuropeptides, each involved in different and multiple aspects of feeding regulation. We show that these substances affect food intake by being part of a central network and, consequently, that changes in consummatory behavior occur as a result of the interplay between these peptides.

8.1 OPIOIDS: MAINTENANCE AND REWARD ASPECTS OF FEEDING

The endogenous opioid peptides represent one family of peptides that is part of a larger central network affecting food intake. Results of some research suggest that opioids selectively affect intake of dietary fat, whereas results of other experiments demonstrate a strong relationship between opioids and sweet tasting foods and solutions or indicate that diet preference is the major factor in opioid-related feeding (Glass et al., 1999).

Many investigators argue that opioids are involved in macronutrient selection, whereas others interpret the data to indicate that these peptides reinforce intake of preferred foods. Some of the earliest studies indicated that morphine and other opioid agonists increased fat intake more effectively than carbohydrate intake when the two macronutrients were offered concurrently, either in complete diets or as individual macronutrients (Marks-Kaufman, 1982). However, others found that opioids increased intake of sweet solutions and that opioid blockade was particularly effective in decreasing intake of sweet solutions (Levine et al., 1985). This led to the concept that opioids affect intake of preferred diets. In support of this concept, Gosnell et al. (1990) found that morphine did not specifically increase intake of a high-fat diet, but rather increased intake of a preferred diet. Rats offered a high-fat diet and a high-carbohydrate diet were segregated into tertiles, depending on their preference for fat or carbohydrate diets. When both diets were offered concurrently, rats injected with morphine ingested more of those diets that were preferred at baseline. Along similar lines, Weldon et al. (1996) reported that naloxone, an opioid receptor antagonist, was much more effective in decreasing intake of a high-sucrose diet than of a high-cornstarch or high-polycose diet in food-restricted rats. Others have also shown that naloxone has little or no anorectic effect on nonpreferred diets, whereas it is an extremely potent anorectic agent when rats are given a highly preferred diet (Glass et al., 1999).

A variety of reports further indicate that opioids are involved in the rewarding aspects of feeding (Levine and Billington, 1997). Opioid receptor-deficient mice (CXBK) do not prefer saccharin as much as wild-type animals do. Naloxone reduces intake of a 10% sucrose solution in food-deprived and non-food-deprived sham drinking rats, indicating that naloxone's effects on fluid intake are not due to postabsorptive signals. Several investigators have noted that opioid blockade can decrease the development of a sucrose preference in developing and adult rats. Naltrexone, another opioid antagonist, reduces the positive hedonic properties of sucrose in rats and humans.

One can easily imagine that under certain conditions, initiation of food intake is associated with energy needs, whereas final stages of a meal may reflect rewarding aspects of eating. Operant studies indicate that opioids are involved maintaining rather than initiating feeding. In one model, rats were trained to press a lever 80 times to receive their first food pellet and then press the lever 3 times to obtain each additional pellet (Rudski et al., 1994b). Naloxone did not decrease intake of the first pellet, but did impact total food intake; thus, antagonism of opioid receptors appears to interfere with meal maintenance. Opioid receptor agonists such as butorphanol and buprenorphine increase operant responding for food in rats (Rudski et al., 1994a, 1995). These agonists do not affect the tasks associated with meal initiation, but increase the total amount of time spent on continuing the meal. In another intriguing model, naloxone was given to grasshopper mice and their food intake was evaluated. When grasshoppers are placed in a cage containing grasshopper mice, the mice kill their prey and then eat them. In an unpublished study, we noted that naloxone decreased intake of the grasshoppers, but did not alter the number of grasshoppers killed. Such data support the hypothesis that opioids affect reward- and palatability-related feeding rather than energy-deficit-induced feeding.

Although opioids affect ingestion of rewarding substances, ingestion of rewarding solutions also alters peptide levels and gene expression of opioid peptides. Dum and Herz (1984) reported that ingestion of chocolate milk or candy decreases β-endorphin levels in the hypothalamus of rats, perhaps due an increase in β-endorphin release. β-endorphin-like immunoreactivity was elevated in the pituitary of rats made obese by prolonged feeding of palatable foods (Levine and Billington, 1997). Prodynorphin gene expression in the paraventricular nucleus of the hypothalamus and the supraoptic nucleus correlated positively with consumption of a high-fat diet (Levine and Billington, 1997). Our group noted that dynorphin gene expression was elevated in rats exposed to a diet containing fat and sucrose (Welch et al., 1996). Further evidence that palatable diets can impact opioid release or opioid receptors, or both, comes from studies of opioids and pain. For example, chronic ingestion of a glucose–saccharin solution decreased morphine-induced analgesia (Bergmann et al., 1985). Oral infusions of sucrose to rat pups increased paw-lift latencies in the hotplate analgesia test and decreased distress vocalization, and both the changes were reversed by naloxone administration (Blass et al., 1987). Kanarek et al. (2001) reported that rats drinking a 32% sucrose solution displayed significantly higher levels of antinociception induced by μ- and κ-opioid receptor agonists injected into the periaqueductal gray area than did rats drinking water.

There appears to be a relationship between diet or taste preferences and self-administration of opioids and other drugs that affect the reward circuitry. Carroll and Boe (1982) found that rats self-administered less of the opiate etonitazene when a glucose–saccharin solution was present than when it was absent. Carroll et al. (2002) also examined heroin self-administration in rats during food satiation and food restriction, noting that food restriction increased heroin infusions by an average of 96% in both females and males. Similarly, Gosnell et al. (1995) noted that rats selected for high saccharin intake self-administered more morphine than rats selected for low-saccharin intake. Data demonstrating that methadone or heroin addicts report an increased intake of sugars and report a greater craving for sweets before than after drug taking provide further evidence for a relationship between intake of palatable food and drug-taking behavior

Other data support the relationship among opioids, palatable ingestants, and drug dependence. Colantuoni et al. (2001, 2002) presented rats with a 25% glucose solution together with laboratory chow for a 12-h period, followed by a 12-h period with no food or glucose. Rats increased their glucose consumption during a 10-day period, particularly early in the access period. Binding to μ-1 opioid receptor increased in several brain nuclei known to be involved in both reward and drug abuse. Also, behaviors associated with opiate withdrawal were observed after naloxone was injected in rats that had consumed glucose for 30 days. Pomonis et al. (2000) found that naloxone elevated levels of the early gene transcription factor cFos in the central nucleus of the amygdala in rats drinking a 10% sucrose solution for 10 days. The central nucleus of the amygdala is involved in emotional aspects of behavior, in addition to receiving signals from energy centers in the brain and relaying signals to reward centers.

Despite this evidence, it is probably an oversimplification to impute the function of reward to all opioids at all brain sites. For example, Glass et al. (2000) found that naltrexone injection into the paraventricular nucleus in the hypothalamus decreased ingestion of fat and carbohydrate diets, whereas infusion of this substance into the amygdala decreased intake of preferred diets only. Carr's laboratory conducted a series of studies evaluating opioid binding in food-restricted rats and noted that after food restriction, opioid receptor binding was altered in 5 of 50 regions studied and dynorphin levels increased in a variety of regions, including the dorsomedial, ventromedial, and medial preoptic hypothalamic areas (Carr, 1996). Further, selected opioid receptor antagonists decreased feeding induced by food restriction or deprivation, indicating that the opioid antagonism can block more than reward-based eating. Finally, Egawa et al. (1993) found that β-endorphin decreased sympathetic nerve activity in brown adipose tissue, suggesting a role for opioids in thermogenesis. Intravenous morphine increased oxygen consumption and carbon dioxide production in pigs. It is possible that a brain site confers the function of opioids, with some sites more involved in reward-based feeding and others in energy-based feeding. It is also possible that reward- and energy-dependent brain mechanisms share information, processing, and common neural output pathways.

Recent years have brought the exciting discovery of a feeding-related peptide that appears to be closely related to the opioid family. Meunier et al. (1995) isolated a heptadecapeptide that, similarly to opioid receptor agonists, had antinociceptive effects. Because of its ability to lower the perception threshold for painful stimuli, the peptide was named nociceptin. The substance was described simultaneously by Reinscheid et al. (1995), who referred to it as orphanin FQ, an endogenous ligand of the "orphan" G_i/G_o-coupled opioid receptor-like 1 receptor (ORL1R).

Interestingly, nociceptin/orphanin FQ (N/OFQ) and the ORL1R display a high level of homology with opioid peptides (dynorphin A in particular) and receptors, respectively. However, despite being structurally analogous, N/OFQ does not bind to opioid receptors and opioid receptor agonists do not serve as ligands of the ORL1R. Not only do the structures of N/OFQ and the ORL1R resemble those of members of the opioid family, but even the N/OFQ precursor, preproN/OFQ, as well as the gene encoding this precursor molecule, share similarities with their classical opioid counterparts. [For a review, see Harrison and Grandy (2000).]

Resemblance between N/OFQ and opioid peptides prompted the question of whether N/OFQ, similar to opioids, plays a significant role in regulating ingestive behavior. Organization of the N/OFQ system (i.e., distribution of N/OFQ-containing neuronal elements and of the ORL1 receptor) in the brain provides a solid neuroanatomical foundation supporting the notion that N/OFQ regulates food intake through central mechanisms. Based on the presence of N/OFQ-positive fiber terminals, N/OFQ is synthesized and released in several brain areas involved in the regulation of ingestive behavior, including the striatum, hypothalamus (lateral hypothalamus, and paraventricular, supraoptic, arcuate, dorsomedial, and ventromedial nuclei), and brainstem (nucleus of the solitary tract and parabrachial nucleus) (Neal et al., 1999b). In addition, these feeding-related structures contain the ORL1 receptor (Neal et al., 1999a).

Results of initial experiments have suggested that central administration of N/OFQ stimulates food intake in a fashion similar to that of opioid peptides. In the first study addressing this issue, Pomonis et al. (1996) found that N/OFQ injected intracerebroventricularly (ICV) caused moderate increases in food intake in *ad libitum* fed, satiated rats. Hyperphagia has been observed following injection of N/OFQ into specific sites of the brain where the ORL1R is present, namely, the nucleus accumbens, hypothalamic ventromedial, paraventricular, and arcuate nuclei, and the nucleus of the solitary tract (Polidori et al., 2000). The feeding response to ICV N/OFQ occurs relatively soon after injection; animals begin eating ca. 8 to 19 min after the treatment, and hyperphagic effects are observed almost exclusively during the first hour postinjection (Pomonis et al., 1996). The latency of the feeding response to N/OFQ is relatively comparable to that observed following opioid injections and agrees with data showing that orexigenic doses of this peptide evoke mild hypolo-comotion (Polidori et al., 2000).

In addition to generating food intake in sated animals, ICV administration of N/OFQ in overnight-deprived rats before refeeding increases the amount of food ingested. N/OFQ treatment before the onset of the dark phase, the time of vigorous feeding activity in rats, also increases consumption. These two findings suggest that N/OFQ, similar to opioid peptides, may participate in maintaining food intake.

Blockade of opioid receptors decreases intake of palatable foods and solutions (especially those high in sucrose or fat) more readily than ingestion of neutral or nonpreferred flavors. Unlike morphine and other opioid ligands, the injection of N/OFQ in rats that had long-term *ad libitum* access to both high-carbohydrate (sweet) food and high-fat food does not increase consumption of a favored diet; instead, it elevates intake of both diets in fat-preferring rats only (Olszewski et al., 2002). Also, ICV-administered N/OFQ reduced intake of ethanol in alcohol-preferring rats, an effect opposite to that evoked by opioid receptor agonists. Thus, N/OFQ's action on consummatory behavior appears to be wide ranging, from opioid-like to antiopioid effects (Polidori et al., 2000).

Although the majority of behavioral and physiological effects (e.g., nociception and control of water–electrolyte balance) induced by N/OFQ are not reversed by administering opioid receptor antagonists, N/OFQ-stimulated ingestive behavior appears to be sensitive to blockade of opioid receptors. Pretreatment with subcutaneous naloxone or ICV naltrexone decreases the hyperphagic response to N/OFQ (Pomonis et al., 1996; Leventhal et al., 1998). That N/OFQ and ligands of classical opioid receptors do not compete for the same receptors allows us to infer that N/OFQ's orexigenic action is mediated through an opioid-containing central circuitry. That N/OFQ and opioids seem to play different roles in the hedonic aspects of consumption suggests that N/OFQ's orexigenic effects do not stem solely from the interaction between the N/OFQ and opioid systems.

8.2 OXYTOCIN (OT) AND VASOPRESSIN (VP): FROM SATIETY TO AVERSION

As described in the preceding sections, laboratory animals and humans eat for a variety of reasons, ranging from energy needs to hedonism. Just as initiation of consumption may be triggered by numerous physiological (e.g., hunger) and environmental or external stimuli (e.g., food presentation), termination of ingestive behavior is also determined by a wealth of factors.

Under normal conditions, termination of food intake occurs when the animal is sated, that is, when the nutritional needs are met and no other reason, such as hedonic value, prolongs a meal. In the classical approach to regulating ingestive behavior, the physiological bases of the termination of feeding have been primarily discussed with regard to mechanisms associated with satiety. However, it is important to remember that inhibition of consummatory behavior may be determined by factors other than satiety. In this section, we present a broader perspective on termination of food intake by enriching our deliberations with aspects of feeding termination related to an osmotic challenge and to consumption of toxic (thus taste aversion-inducing) ingestants. We focus on the neuropeptides oxytocin (OT) and vasopressin (VP), hypothesized to lead to the inhibition of food consumption for osmotic and toxic reasons, as well as the central networks of which the two substances are a part.

The nonapeptides OT and VP have been studied for many years. Although initially OT and VP were described as neurohypophyseal hormones, the paracrine and neurotransmitter modes of their action have been well documented (Buijs, 1990). The first described biological effects of OT were stimulation of uterine contractions to induce parturition and contraction of myoepithelial cells in the mammary gland to promote milk ejection. Historically, VP has been implicated in vasoconstriction and regulation of water resorption by the kidney. At present, a much broader range of functions has been assigned to these peptides. VP has been shown to facilitate memory, influence thermoregulation, and coordinate autonomic, metabolic, and behavioral responses to stressful stim-uli. Similarly to VP, OT responds to stress, perturbations in blood pressure, and water balance, as well as osmotic challenge. [For a review, see Buijs (1990).] Importantly, both OT and VP have been proposed to support the termination of ingestive behavior for a particularly broad variety of reasons, including satiety, osmolality, and aversion [see, e.g., Olson et al. (1991) and Langhans et al. (1991)].

The complex structural organization of the OT and VP peptide/receptor systems reflects a vast range of functions, including their involvement in feeding control. In the brain, neurons producing these peptides are localized primarily in two hypothalamic sites: the paraventricular and supraoptic nuclei. VP neurons are present also in the suprachiasmatic nucleus, which serves as a primary component of the mammalian biological clock. Relatively sparse so-called accessory OT and VP cells are scattered in groups in the forebrain, primarily throughout the hypothalamus. OT and VP neuronal populations send projections to various brain areas, ranging from the autonomic centers in the brainstem to limbic structures, pituitary, and neocortex (Buijs, 1990).

The two largest and morphologically diverse populations of OT and VP neurons, those that originate in the paraventricular and supraoptic nuclei, have been most extensively studied at the anatomical level as well as in relation to their role in regulating food intake. The PVN contains two types of OT and VP neurons that have been classically referred to as magno- and parvocellular (terminology based on a large or small size of perikarya, respectively), whereas neurons in the supraoptic nucleus are magnocellular.

Magnocellular OT and VP neurons of both the paraventricular and supraoptic nuclei send projections almost exclusively to the neural lobe of the pituitary, where OT and VP are released into the general circulation. Although the issue of the influence of circulating OT and VP on food consumption is still controversial, elevated levels of these hormones are detected in the blood at the time of meal completion (Verbalis et al., 1986a; Stricker and Verbalis, 1986). The release of OT and VP from the pituitary has been linked to particular physiological events that promote or result in inhibition of consummatory behavior, such as an increase in plasma osmolality (a rise in osmolality is typically observed due to the absorption process following food intake) or stomach distention. Also, OT and VP injected peripherally reduce food intake and gastric motility, but these anorexigenic effects are achieved only with high doses of each of the two peptides (Arletti et al., 1989; Langhans et al., 1991). These findings suggest that circulating OT and VP, even if not directly involved in a satiety-related decrease in consumption, appear to be part of homeostatic mechanisms controlling ingestive behavior and contribute to the biochemical milieu conducive to termination of ingestive behavior.

Some of the parvocellular OT and VP neurons in the PVN project to the portal capillaries of the median eminence, and this projection has been linked to the regulation of adenohypophyseal ACTH release and thus to control of activation of the hypothalamic-pituitary-adrenal (HPA) axis. Taking into account that HPA axial hormone secretion is elevated following food ingestion and that food intake is altered during a stress response (which is typically accompanied by activation of the HPA axis), several authors have suggested that this parvocellular system in the PVN modulates ingestive behavior by affecting activation of the HPA axis (Kiss et al., 1994; Wideman and Murphy, 1991).

Finally, a subpopulation of parvocellular OT and VP cells has been studied extensively in relation to the control of feeding. Several investigators have reported the presence of OT and VP fibers and fiber terminals in areas not directly related to neuroendocrine pathways, leading to the discovery of extensive OT- and VP-ergic circuitry in the brain. Parvocellular OT and VP neurons, especially those encompassed within the posterior portion of the paraventricular nucleus of the hypothalamus, appear to be the source of these central projections (Buijs et al., 1990). OT and VP neuronal projections to the nucleus of the solitary tract and dorsal motor nucleus of the vagus play a particularly important role in the regulation of food intake. For example, inhibition of food intake caused by electrical stimulation of the PVN can be reversed by microinjecting an OT receptor antagonist into the dorsal motor nucleus of the vagus (Flanagan et al., 1992b). Hypothalamic paraventricular-derived input to the dorsal motor nucleus of the vagus and to the nucleus of the solitary tract has been implicated in inhibition of gastric motility and gastric emptying (Olson et al., 1992). In addition, studies have shown that in rats stimulated to eat by overnight deprivation, the level of Fos immunoreactivity of parvocellular OT and VP neurons in the PVN is significantly higher at the end of a meal than at the beginning. This increased activation of the two groups of cells coinciding with a voluntary (therefore, most likely, satiation-related) termination of food intake indicates that OT and VP derived from the parvocellular region (thus reaching central targets) contribute to the process of achieving satiety. This notion is supported by results of injection studies in which intracranially administered OT or VP resulted in an early termination of consummatory behavior.

As mentioned earlier, consummatory behavior—besides its essential value in survival as the way of providing an organism with water and nutrients—poses a certain challenge to homeostatic balance. Because food is a source of not only calories but also osmotic particles, there are mechanisms that cause animals to stop eating when feeding-related salt loading jeopardizes osmotic balance. OT and VP synthesized in both the magno- and parvocellular populations appear to be part of these mechanisms (Huang et al., 2000; Blackburn et al., 1995). Intake of ingestants that are high in osmolar content induces a very powerful response of the OT and VP systems, as reflected by an elevated plasma profile of the two peptides and by a robust activation (Fos immunoreactivity) of the magno- and parvocellular neurons containing these substances (Verbalis et al., 1995; Blackburn et al., 1992). Consequently, salt loading and dehydration decrease food intake (Flynn et al., 1995).

Food may contain toxic material and it is therefore essential that animals ingesting such food cease consumption as soon as the toxins can be detected by chemoreceptors. Learning to avoid this food in the future has a significant adaptive value. In the laboratory setting, inadvertent intake of tainted food can be mimicked by injecting toxins immediately after animals have consumed an ingestant with a novel and characteristic flavor (Nachman, 1963). When a novel taste is associated with a short-term unpleasant gastrointestinal sensation (caused by the injected toxin), animals learn to avoid this flavor, a phenomenon known as a conditioned (or learned) taste aversion (CTA). Various chemical agents induce aversive effects when paired with novel flavors (Nachman, 1963; Verbalis et al., 1986b); lithium chloride (LiCl) is particularly effective in generating CTA and consequently results in a significant reduction in food intake, gastric motility, and gastric emptying (Flanagan et al., 1989, 1992a). Peripheral injection of LiCl results in the onset of complex neural and endocrine mechanisms that underlie the development of anorexic and aversive responses. It produces pronounced c-Fos immunoreactivity (activation) in several brain regions engaged in regulating ingestive behavior, including the nucleus of the solitary tract, paraventricular nucleus, and supraoptic nucleus (Olson et al., 1993). Dose-dependent neurohypophyseal secretion of OT and VP and activation of paraventricular and supraoptic nucleus-encompassed neurons containing these two peptides have been observed following administration of LiCl. OT and (with some exceptions) VP neurons respond to virtually any treatment that causes toxic poisoning and unpleasant gastrointestinal sensation, and therefore OT and VP are thought to be involved in the development of aversive responses (Verbalis et al., 1986b; Olszewski et al., 2000). In summary, the OT and VP play an important role in a vast array of processes associated with inhibition of food intake, ranging from satiety to homeostatic aspects of feeding termination, that is, aversion or osmotic imbalance.

Such a wide scope of consumption-related inhibitory mechanisms in which OT and VP are engaged stems from the fact that OT and VP neurons are part of (in some cases a final component of) pathways mediating termination of ingestive behavior. OT and VP cells appear to integrate diverse neuropeptidergic input; this cross-talk between OT or VP and other brain peptides results in a behavioral output, namely, modification of feeding. It has been found that OT and VP neurons receive innervation that originates in the areas of the brain implicated in food intake regulation, with brainstem and hypothalamic regions being a powerful source of input. These neurons express receptors for numerous orexigenic (e.g., opioids and N/OFQ) and anorexigenic (e.g., GLP-1) neuropeptides. Also, fibers that contain these neuropeptides terminate in the proximity of OT and VP perikarya (Larsen et al., 1997). Taken together, these data suggest that OT and VP neurons are part of a larger feeding-related network.

Unfortunately, our knowledge of anatomical and functional characteristics of this circuitry remains relatively limited. Considering OT and VP neurons integrate (as inferred through the presence of appropriate receptors or fiber terminals) neuropeptidergic signals that cause increases in feeding, as well as those that promote termination of ingestive behavior, it seems likely that a balance or competition between these signals determines a given feeding response. To illustrate this, we briefly delineate feeding-related evidence regarding a functional relationship between the OT/VP systems and orexigenic (opioids and N/OFQ/anorexigenic (GLP1) peptides that affect activation of OT and VP cells.

8.3 OT AND VP AS PART OF A FEEDING-RELATED NEURAL NETWORK

8.3.1 OT/VP–Opioid Interactions

Opioids appear to be involved in maintaining feeding, perhaps by impacting the rewarding characteristics of food. On the other hand, OT and VP are involved in decreasing intake by signaling unpleasant consequences of feeding, such as gut distension, or by ceasing intake because of the presence of novel or toxic substances. Opioids and N/OFQ are part of the network that encompasses OT and VP neurons. The ORL1R and opioid receptors are expressed by OT and VP neurons, and N/OFQ- and opioid-containing fibers have been detected in the areas where OT/VP cells are localized (Neal et al., 1999b; Russell et al., 1995). Neuroanatomical analyses that suggest the presence of N/OFQ- and opioid-OT/VP pathways have been complemented by functional studies. Taste aversion experiments have revealed that N/OFQ or opioid receptor agonists injected before LiCl prevent the development of CTA and, importantly, decrease LiCl-induced Fos IR of OT and VP neurons in the hypothalamic paraventricular and supraoptic nuclei. Similarly, Flanagan et al. (1988) found that naloxone potentiates the effects of LiCl on OT secretion, gastric motility, and hypophagia (Flanagan et al., 1988). Feeding studies have shown that ICV N/OFQ-induced maintenance of a meal is accompanied by a delayed activation of OT and VP neurons (Figure 8.1). On the basis of these observations, we propose that N/OFQ and opioid peptides prevent satiety- and aversion-related termination of ingestive behavior by suppressing activation of OT and VP neurons.

8.3.2 OT/VP–Glucagon-Like Peptide-1 (GLP1) Interactions

OT and VP neurons receive strong input responsible for development of hypophagic and aversive responsiveness. A significant portion of this input is derived from the nucleus of the solitary tract, a sensory relay cell group strategically located in the brainstem, which integrates the majority of information associated with ingestive behavior. Changes in levels of neuronal activation in the nucleus of the solitary tract have been correlated with inhibition of feeding as well as observed on injection of agents that cause anorexic responses or taste aversion. Within this hindbrain structure, GLP1, a posttranslational product of preproglucagon, is one of the key players thought to mediate various aspects of feeding termination. [For an overview, see Tang-Christensen et al. (2001)]. GLP1-positive neurons, which are present exclusively in the nucleus of the solitary tract, exhibit elevated Fos IR at the time of completion of a meal. ICV and site-specific (e.g., intra-paraventricular hypothalamic) injections of this peptide cause inhibition of food intake. Although it is generally accepted that this anorexigenic effect is related to GLP1's presumed role in energy balance, some

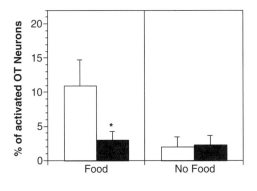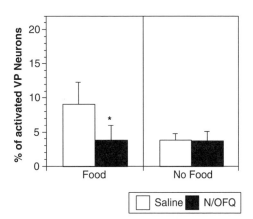

FIGURE 8.1 Nociceptin/orphanin FQ (N/OFQ) decreases feeding-induced activation of oxytocin (OT) and vasopressin (VP) neurons. Percentages of Fos-positive OT (left panel) and VP (right panel) neurons in the hypothalamic paraventricular nucleus in rats that consumed 7 g of chow within 1 h following overnight food deprivation. Animals that had no access to food served as controls. Rats were injected ICV with 10 nmol N/OFQ or vehicle (saline) immediately before the time of food presentation. Perfusions took place 1 h after completion of a meal. Asterisks indicate statistical significance ($P < 0.05$); $N = 4$ to 5/group.

studies have suggested that GLP1 may also mediate aversive responsiveness. Indeed, ICV infusions of this substance (but not injections into the paraventricular nucleus, except for very high doses of GLP1) induce CTA, whereas antagonism of the GLP1 receptor alleviates aversion. Thus, similarly to OT and VP, GLP1 is thought to mediate both satiety and aversion (Tang-Christensen et al., 2001; Kinzig et al., 2002). Intriguingly, projections of GLP1 cells terminate in the paraventricular nucleus in the proximity of OT (Figure 8.2) and VP perikarya, which have been shown to contain the receptor for GLP1 (Tang-Christensen et al., 2001). Thus, neurons expressing these peptides appear to form a pathway. Several studies have provided functional evidence of the involvement of this pathway in feeding control. For example, it has been observed that injections of GLP1 into the hypothalamic paraventricular nucleus and ICV at doses that cause termination of consummatory behavior induce activation of OT and VP calls. Hence, it is likely that an increase in the activity

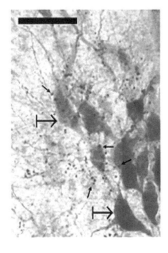

FIGURE 8.2 Glucagon-like peptide-1 (GLP1) fibers terminate in the proximity of OT-containing neurons. A photomicrograph depicting a coronal section through the hypothalamic paraventricular nucleus. Tissue was stained immunohistochemically for the presence of OT and GLP1. Small arrows, GLP1 fibers; large arrows, OT-containing perikarya. Scale bar 0.05 mm.

of this pathway contributes to the inhibition of consumption (Tang-Christensen et al., 2001; Kinzig et al., 2002). Thus, we suggest that GLP1 exerts its anorexigenic action (to some degree) by stimulating OT and VP neurons.

We conclude that OT and VP neurons, which support termination of ingestive behavior, are part of larger circuitry that encompasses neuropeptides responsible for maintaining feeding as well as substances that decrease consumption. A dynamic balance within this circuitry determines a behavioral output, that is, magnitude of feeding response. In this network, peptides that inhibit food intake, such as GLP1, exert their action (to a certain degree) by activating the OT and VP neurons. Conversely, orexigenic properties of neuropeptides that promote ingestive behavior, for example, opioids and N/OFQ, may be linked to their suppressing influence on activation of OT/VP cells.

8.4 PLEASURE–AVERSION CONTINUUM AS A CONSEQUENCE OF ACTIVITY WITHIN A DYNAMIC NETWORK

Based on the previous discussion, we hypothesize that a pleasure–aversion continuum exists and it stems from the interplay between peptides within the central nervous system. For example, ingestion of sweet high-fat foods in relatively small volumes is generally a positive experience, whereas ingestion of extremely large amounts of a palatable diet can cause an unpleasant sensation that leads to the temporary avoidance of the particular food type, and thus the effect resembles aversion. We posit that opioids and OT are important regulators of this pleasure–aversion continuum. Stimuli that contribute to termination of food intake, such as distension of the stomach or gastrointestinal illness induced by injection of LiCl or copper sulfate, activate OT-containing cells in the paraventricular and supraoptic nuclei of the hypothalamus. In contrast, opioids are thought to be involved in rewarding aspects of feeding and to affect meal maintenance. Blockade of opioid receptors results in early satiety, perhaps (in part) due to some aversion-like consequences.

8.5 OREXINS/HYPOCRETINS: FEEDING AND ACTIVITY MODULATORS

Ascribing one function to one molecule is an approach too simplistic in any study of physiology, especially of the brain. All possible behavioral and physiological functions of a peptide are almost never measured concurrently. For example, it is unlikely that a laboratory specializing in feeding behavior would measure effects on blood pressure, and vice versa. Also, several factors influence the measured variable, including site of injection, metabolic and hormonal status, circadian influences, temperature, and other known and unknown factors. Thus, although differences in the measured outcome between control and treated animals indicate that behavioral output is affected by the treatment given, the relationship between this outcome and what occurs in normal physiology is unclear.

This ambiguity of function is exemplified by the discovery and characterization of hypocretins (also known as orexins), which were reported simultaneously by two separate laboratories. The term *hypocretin,* coined by the deLecea group, stems from merging parts of the words *(hypo)thalamus* and *in(cretin)*, referring to the hypothalamic location of hypocretin cell bodies and structural similarity to incretin gut hormones, respectively (de Lecea et al., 1998). The term *orexin,* named for the Greek term *orexis* for appetite, refers to the finding by the Sakurai laboratory that injection of this peptide into the ventricles resulted in a feeding (orexigenic) response (Sakurai et al., 1998). Based on this finding, there began an intensive research effort by several laboratories, including ours, to define the role of orexins in feeding behavior. These investigations revealed that specific brain sites are involved in orexin A-induced feeding; endogenous levels of orexin respond to alterations in energy homeostasis; orexin neurons communicate with leptin, NPY, and AgRP neurons; and the feeding response induced by orexin A is modest and short lived relative to other orexigenic compounds (Sweet et al., 1999, Willie et al., 2001, de Lecea et al., 2002). All these findings suggested a role for orexins in feeding behavior. However, laboratories studying aspects of physiology unrelated to feeding revealed a very different role for orexin: the regulation of sleep and wakefulness (Sakurai, 2002; de Lecea et al., 2002). It is now well documented that orexin enhances arousal. Narcolepsy in dogs is due to an inherited mutation in the gene encoding the orexin 2 receptor, and human narcolepsy has been associated with an absence of orexin-producing neurons within the hypothalamus and low orexin levels in the cerebrospinal fluid (CSF; Taheri et al., 2002). The association of orexins with wakefulness may explain some of the behavioral observations following orexin A and B administration, which include activities such as face washing, grooming, searching, and burrowing behavior (Ida et al., 1999). We have recently confirmed that orexin A increases spontaneous physical activity after injection into the lateral hypothalamus and the hypothalamic paraventricular nucleus (Kotz et al., 2002). Therefore, a critical question is whether orexin A increases feeding based on an interoceptive cue indicating some level of hunger or whether the enhanced feeding after orexin A administration is a consequence of increased arousal. More detailed behavioral studies are needed to answer the question. Regardless of the findings, the outcome of elevated orexin levels is feeding, and because normal feeding behavior occurs for many reasons, it is likely that enhanced arousal is one route by which this feeding occurs. It is also possible that the function of orexin depends on the brain site of the orexin receptor population stimulated, with some sites inferring feeding function and other sites governing arousal.

8.6   DOES FUNCTION DEPEND ON SITE OF ACTION?

A difference in potency of various substances injected in different brain sites is one example of how a brain location influences behavioral outcome. For instance, urocortin, a recently identified peptide with anorectic properties, shows widely different effects on food intake, depending on the site of injection. As shown in Figure 8.3, urocortin injected into the lateral hypothalamus has no influence on deprivation-induced feeding, whereas ventromedial hypothalamic injection of this peptide enhances short-term feeding (although longer-term feeding, 2- to 24-h postinjection, is reduced), and urocortin in the hypothalamic paraventricular nucleus and the intermediate lateral septal area inhibits deprivation-induced feeding by 20 and 50%, respectively. Based on the enhanced anorectic efficacy observed after urocortin injection into the intermediate lateral septal area, one conclusion might be that this region is the primary site of action for urocortin. Several neuroanatomic studies support this conclusion. Urocortin has a dense fiber network in the intermediate portion of the lateral septum, and both urocortin peptide and CRH-R2 (the receptor to which urocortin binds with highest affinity) are present in the intermediate lateral septal area (Vaughan et al., 1995; Bittencourt et al., 1999; Li et al., 2002). Further, urocortin given into the cerebral ventricles results in elevated c-*fos* expression (an indicator of cellular activation) in neurons expressing CRH-R2 in the intermediate lateral septal area (Vaughan et al., 1995). However, another equally valid interpretation is that urocortin has many sites of action, which include the hypothalamic paraventricular nucleus and the intermediate lateral septal area, and potentially other yet untested sites. The function of urocortin within each of these sites may be somewhat different. For instance, the intermediate lateral septal region is considered part of the limbic system, and as such it might be involved in the rewarding aspects of feeding. Lesions of the lateral septal area result in stimulation of feeding behavior and produce increased sensitivity to specific tastants (Fried, 1972; Koppell and Sodetz, 1972; Vasudev et al., 1985). In contrast, the hypothalamic paraventricular nucleus is involved primarily in energy homeostasis and may be associated with feeding related to hunger. Therefore, enhanced anorectic efficacy of urocortin within the intermediate lateral septal region may indicate that urocortin release in this region inhibits a reward aspect of feeding, whereas urocortin within the hypothalamic paraventricular nucleus plays a minor yet significant role in inhibiting hunger-related feeding. Regardless of interpretation, what is apparent by the complexity of feeding behavior is that no one site or neurochemical is a predominant regulator of feeding, be it hunger or reward related, because several brain sites and neurochemicals are regarded as being involved in such processes. Further, neuropeptides and brain sites cannot operate in isolation, as is evidenced by the numerous characterized interactions between neuropeptides within many brain sites.

8.7 NEUROPEPTIDE Y: ENERGY-RELATED FEEDING

NPY is a potent orexigenic agent that appears to be involved in eating driven by energy needs (Clark et al., 1984). NPY is synthesized in the arcuate nucleus and is axonally transported to the paraventricular nucleus of the hypothalamus. Many laboratories have demonstrated that food deprivation and restriction result in an increase in gene expression of NPY in the arcuate nucleus and an increase in NPY peptide levels in the paraventricular nucleus (Levine and Billington, 1997). Other states that result in altered energy metabolism also impact synthesis and release of NPY. Streptozotocin-induced diabetic rats have increased NPY mRNA levels in the arcuate nucleus and increased peptide levels in the PVN. Similarly, in lactating rats, gene expression of NPY in the arcuate nucleus is increased and NPY peptide levels are increased in a variety of hypothalamic sites. Kalra et al. (1991) showed that NPY is released within the paraventricular nucleus of the hypothalamus in schedule-fed rats immediately before initiation of a meal.

Injection of NPY increases the motivation to eat, as measured by operant responding. By using a progressive ratio schedule, we showed that NPY injection in sated rats increased responding on a lever in a fashion similar to that of animals that were food deprived for 36 to 48 h (Jewett et al., 1992). This operant schedule increases the number of presses required to obtain a food pellet in a progressive manner and thereby reflects motivation to eat. The observation that NPY-injected sated rats respond in this operant environment in a manner similar to that of noninjected deprived animals further supports the notion that NPY is involved in energy-induced feeding. In contrast, rats injected with the opiate methadone fail to work for food when placed in an operant environment, but eat food when it is presented freely in the cage.

8.8 NEUROPEPTIDE INTERACTIONS WITHIN A NETWORK: OREXIN–UROCORTIN AND NPY–OPIOID

We have argued that selected peptides might increase or decrease eating because of reward, aversion, or energy needs. However, this is a clear oversimplification. As described for urocortin and opioids, the behaviors observed following administration of these agents are dependent on the site in the brain in which these peptides are injected. Furthermore, peptides supposedly involved in one aspect of feeding may interact with peptides involved in other aspects of feeding.

One example of a neural interaction impacting feeding is that observed between urocortin in the intermediate lateral septal area and orexin A in the lateral hypothalamus. The lateral septal area receives neural input from the lateral hypothalamus and also sends monosynaptic projections to the lateral hypothalamus (Risold and Swanson, 1997b). Electrical stimulation of the mediolateral portion of the lateral septum results in elevated cFos immunoreactivity in the lateral hypothalamus (Varoqueaux and Poulain, 1999). Oliveira et al. (1990) demonstrated that lesions of the lateral septal area enhanced lateral hypothalamus-stimulated feeding behavior, whereas stimulation of the lateral septal region inhibited lateral hypothalamus-stimulated feeding behavior. We have recently found that stimulation of the lateral septal area with a direct infusion of urocortin decreases feeding induced by administration of orexin A in the lateral hypothalamus (Wang and Kotz, 2002). Taken together, these data suggest that urocortin in the lateral septal area and orexin A in the lateral hypothalamus are involved in signaling information important to feeding behavior. The chemical phenoptype of the neurons affected by urocortin in the intermediate lateral septal area is unknown. However, several peptides present within the septal area, including N/OFQ, calbindin, enkephalin, dynorphin, neurotensin, somatostatin, GRH, substance P, and NPY, are involved in feeding and may be influenced by urocortin within the lateral septal area (Risold and Swanson, 1997a).

As shown in Figure 8.3, urocortin within the hypothalamic paraventricular nucleus decreases deprivation-induced feeding; we have also found that urocortin in this region decreases feeding induced by NPY given into the same area (Wang et al., 2001). In contrast, whereas administration of cocaine-amphetamine related transcript (CART) also inhibits feeding induced by NPY in the paraventricular nucleus, it does not decrease deprivation-induced feeding (Wang et al., 2000). This is a surprising result, as high levels of NPY brought about by food deprivation appear to be a driving force behind the robust feeding response following deprivation, and CART is effective in blocking feeding induced by NPY injection. Although CART in this region has not been tested for potential aversive consequences, the lack of effect on deprivation-induced feeding suggests that CART is unlikely to result in malaise. Central CART administration results in alterations in motor activity (Aja et al., 2001), and it is possible that decreases in feeding due to CART may be a consequence of decreased ability to obtain or chew food, although again, that does not fit with the lack of ability of CART to influence deprivation-induced feeing. Thus, two feeding-related compounds, urocortin and CART, given into the same brain site yield very different behavioral outcomes, further exemplifying the complicated interplay between feeding-regulatory neuropeptides.

Another example of communication between different neuropeptides in separate brain sites (thus a presumed existence of a wider network) is that observed between NPY in the hypothalamic paraventricular nucleus and opioids in the hindbrain nucleus of the solitary tract. Several laboratories have shown that NPY in the hypothalamic paraventricular nucleus promotes a robust feeding response (Levine and Billington, 1997). The hindbrain nucleus of the solitary tract receives projections from the hypothalamic paraventricular nucleus (Barraco, 1994), contains opioid receptors (Akil et al., 1984), and administration of opioid agonists into this region increases feeding (Kotz et al., 1997). We found that feeding induced by NPY administration into the paraventricular nucleus was completely inhibited when naltrexone, an opioid receptor antagonist, was concurrently administered into the nucleus of the solitary tract (Kotz et al., 1995). There was a similar outcome when naltrexone was given into the amygdala (Giraudo et al., 1998). Collectively, these data suggest that NPY in the hypothalamic paraventricular nucleus engages opioid pathways in the amygdala and the nucleus of the solitary tract to generate a feeding response.

These are just a few of the many examples of different neuropeptides acting at sites distal to each other to influence the feeding outcome produced by one or the other and demonstrate communication between brain sites in the regulation of feeding. This form of communication is indicative of a network model of feeding regulation.

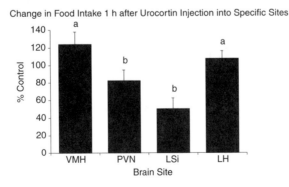

Change in Food Intake 1 h after Urocortin Injection into Specific Sites

FIGURE 8.3 Feeding response to urocortin injected into the ventromedial hypothalamus (VMH), hypothalamic paraventricular nucleus (PVN), intermediate lateral septal area (LSi), or LH lateral hypothalamus. Animals were food deprived for 18 h and then given urocortin (30 pmol) in the VMH, PVN, LSi, or LH. Data are presented as percentage of intake in the control (vehicle-injected) animals. N = 12 to 18/group. (PVN and LSi data derivations of data from Wang, C. et al. (2001) *Am J Physiol Regul Integr Comp Physiol*, 280, R473–R480, and Wang, C. and Kotz, C. M. (2002) *Am J Physiol Regul Integr Comp Physiol*, 283, R358–R367. With permission.)

8.9 LEPTIN AND BODY FAT CONTENT

Leptin is secreted by the adipose tissue in proportion to the total amount of adipose tissue, and when leptin is given in the right circumstance it can reduce food intake and body weight (Porte et al., 2002). The leptin signal must be sensed in the brain, and a high concentration of leptin receptors has been located in the arcuate nucleus of the hypothalamus. Leptin receptors found on particular neurons in the arcuate nucleus include those that produce NPY and AgRP and neurons that produce proopiomelanocortin (POMC), from which is derived the melanocortin agonist α-MSH as well as β-endorphin (Porte et al., 2002). NPY is a known orexigenic agent with significant potency. A potent orexigenic role has been defined for the peptide AgRP, which works by antagonizing α-MSH action at melanocortin receptors and which colocalizes with NPY (Morton and Schwartz, 2001). Leptin can suppress the signaling provided by these orexigenic peptides while increasing signaling dependent on peptides such as MSH, which normally suppress food intake. Input provided by these peptides is generally thought to be received in the paraventricular nucleus of the hypo-thalamus. Taken together, these mechanisms provide a significant source of regulation responsive to body fat stores.

The significance of central melanocortin signaling pathways first became apparent through study of the agouti mouse (Marks et al., 2002). This genetically obese strain is abnormal because of production of the agouti protein, which confers a yellow skin coat but most significantly blocks the action of α-MSH at melanocortin receptors. Probably the most crucial melanocortin receptors are located in the PVN, where a tonic α-MSH signal from the arcuate nucleus is received. α-MSH suppresses food intake, so that presence of the agouti protein blocks this normal suppressive influence and thereby increases food intake and ultimately results in the obesity of this mouse strain.

α-MSH is regulated in part by the reception of the peripheral leptin signal (Wisse and Schwartz, 2001). Leptin receptors are expressed on the neurons in the arcuate nucleus, which make α-MSH from the precursor molecule POMC, and action of leptin at this arcuate receptor leads to increased α-MSH synthesis and subsequent action at the melanocortin receptors in PVN. The full function of this system is, however, complicated by an additional component, the agouti-related protein (AgRP). Neurons in the arcuate nucleus, which produce the orexigenic NPY, also synthesize AgRP, and AgRP functions as a natural antagonist of α-MSH action at the melanocortin receptor. Blocking an anorexic influence of α-MSH means that AgRP functions as an additional orexigenic factor. Further, the neurons in the arcuate nucleus that produce NPY and AgRP are also in part responsive to leptin signaling, with leptin action suppressing the production and function of NPY and AgRP. The cross-talk among NPY, AgRP, and α-MSH within the projections from the arcuate to the paraventricular nucleus amplifies the significance of the leptin signal.

Abnormalities of leptin action have been associated with the expected changes in these systems, and it is therefore apparent that the NPY and melanocortin signaling pathways are crucial compo-nents of body weight regulation that are leptin sensitive (Porte et al., 2002). The significance of the pathways is further underscored by recognition that mutation of components in the melanocortin signaling pathway can result in changed eating behavior and body weight (Wisse and Schwartz, 2001). Besides the agouti mouse, there is an example in humans of altered POMC generation resulting in a syndrome of which obesity is a prominent part. In addition, there is evidence that allelic variation at a site on human chromosome 2 that encodes for the melanocortin receptor is associated with variation in body weight. It therefore seems likely that melanocortin action, inde-pendent of leptin, may have significance with respect to eating behavior and body weight regulation.

8.10 NEUROREGULATION OF FEEDING BEHAVIOR: NETWORK
VS. LINEAR MODEL

An illustration of how interpretation of behavioral outcome due to various stimulations with feeding-regulatory compounds becomes important is the ongoing debate over the mechanism behind what is commonly known as leptin resistance, that is, the lack of a hypophagic response following leptin administration in animals or humans. This concept is based on the view that leptin is the primary mediator and the starting point (regulator) of anorectic responses. This phenomenon of leptin resistance is observed in various obese states, including diet-induced obesity, obesity stemming from age and genes, and also after manipulations such as chronic leptin overexpression through vector delivery. The idea that the lack of an anorectic response to leptin in these states results from some defect in leptin signaling is the prevailing view and one that relies on a linear model of body weight regulation (Ahima and Flier, 2000). In other words, the absence of an anorectic response to leptin requires, in this model, that defective leptin signaling be the root cause of this nonresponse and that this particular defect allows continued dysregulation of body weight. However, as discussed previously, what is profoundly apparent in studies of appetite and body weight regulation is that there are multiple mediators whose actions and functions rely on inputs from each other and from peripheral signals. For instance, the ability of several orexigenic neurotransmitters, including NPY and orexins, to enhance feeding is mitigated by lesions of and stimulation of specific brain areas with various other energy-regulatory neuropeptides. Likewise, the ability of anorectic neuropeptides, including corticotrophin releasing hormone and urocortin, to decrease feeding depends on a specific milieu of signals in other brain areas that provide input to regions responsive to urocortin and corticotrophin releasing hormone. These interactions indicate that no one regulator operates alone or within only one brain area to determine the behavioral response level of food intake. Instead, these neuropeptides appear to form a dynamic neural network encompassing several sites that communicate with each other, and activity within this network determines behavioral output, that is, a feeding response. Further, this model is inconsistent with the idea of a regulator of appetite and body weight and includes the notion of a multitude of mediators of appetite and body weight. Therefore, this is not a static linear model with one initiation point and a unidirectional sequence of activation through the brain, but a multipoint model in which known mechanisms of plasticity, or the brain's ability to modify and form new connections, exist and one in which the dynamic interplay of all mediators of appetite, central and peripheral, is incorporated.

8.11 CONCLUSION

Thus far, many investigations on the role of neuroregulators in the control of food intake have focused on studying only one selected substance of interest. A typical approach has been, for example, to measure food consumption following the central or peripheral injection of one regulator or to determine synthesis or release of this substance due to energy state manipulations. This might have contributed to a relatively prevalent impression that a single substance, as a lone player, can be solely responsible for controlling a particular aspect of consummatory behavior, such as intake of a given macronutrient or a given tastant. In this chapter, we stressed the importance of interplay between neuroregulators in producing a particular feeding response. We chose to illustrate functional relationships between neuropeptides within the brain circuitry by selecting only a few substances known to be involved in regulating ingestive behavior, namely, opioids, NPY, leptin, urocortin, orexins, OT, and VP. One should remember that the list of neural modulators of feeding is much longer and that network-dependent interactions similar to those described appear to be the physiological basis of functioning of any endogenous substance that affects consumption.

We propose that all the neuropeptides discussed appear to participate in a dynamic interplay between multiple mediators of feeding behavior, acting within and between several brain areas, and responding to changes in diet and other environmental cues, therefore suggesting a plastic, multipoint model of feeding behavior.

REFERENCES

Ahima, R.S. and Flier, J.S. (2000). Leptin. *Annu Rev Physiol,* 62, 413–437.

Aja, S., Sahandy, S., Ladenheim, E.E., Schwartz, G.J., and Moran, T.H. (2001). Intracerebroventricular CART peptide reduces food intake and alters motor behavior at a hindbrain site. *Am J Physiol Regul Integr Comp Physiol,* 281(6), R1862–1867.

Akil, H., Watson, S.J., Young, E., Lewis, M.E., Khachaturian, H., and Walker, M.E. (1984). Endogenous opioids: biology and function. *Ann Rev Neurosci,* 7, 223.

Arletti, R., Benelli, A., and Bertolini, A. (1989). Influence of oxytocin on feeding behavior in the rat. *Peptides,* 10(1), 89-93.

Barraco, I.R.A. (1994). *Nucleus of the solitary tract.* Boca Raton: CRC Press Inc.

Bergmann, F., Lieblich, I., Cohen, E., and Ganchrow, J.R. (1985). Influence of intake of sweet solutions on the analgesic effect of a low dose of morphine in randomly bred rats. *Behav Neural Biol,* 44(3), 347–353.

Bittencourt, J.C., Vaughan, J., Arias, C., Rissman, R.A., Vale, W.W., and Sawchenko, P.E. (1999). Urocortin expression in rat brain: evidence against a pervasive relationship of urocortin-containing projections with targets bearing type 2 CRF receptors. *J Comp Neurol,* 415(3), 285–312.

Blackburn, R.E., Demko, A.D., Hoffman, G.E., Stricker, E M., and Verbalis, J.G. (1992). Central oxytocin inhibition of angiotensin-induced salt appetite in rats. *Am J Physiol,* 263(6 Pt 2), R1347–1353.

Blackburn, R.E., Samson, W.K., Fulton, R.J., Stricker, E.M., and Verbalis, J.G. (1995). Central oxytocin and ANP receptors mediate osmotic inhibition of salt appetite in rats. *Am J Physiol,* 269(2 Pt 2), R245–251.

Blass, E., Fitzgerald, E., and Kehoe, P. (1987). Interactions between sucrose, pain and isolation distress. *Pharmacol Biochem Behav,* 26, 483–489.

Buijs, R.M. (1990). Vasopressin and oxytocin localization and putative functions in the brain. *Acta Neurochir Suppl (Wien),* 47, 86–89.

Buijs, R.M., Van der Beek, E.M., Renaud, L.P., Day, T.A., and Jhamandas, J.H. (1990). Oxytocin localization and function in the A1 noradrenergic cell group: ultrastructural and electrophysiological studies. *Neuroscience,* 39(3), 717–725.

Carr, K.D. (1996). Opioid receptor subtypes and stimulation-induced feeding. In S.J. Cooper and P.G. Clifton (Eds.), *Drug receptor subtypes and ingestive behaviour* (pp. 167–192). London: Academic Press.

Carroll, M.E. and Boe, I.N. (1982). Increased intravenous drug self-administration during deprivation of other reinforcers. *Pharmacol Biochem Behav,* 17(3), 563–567.

Carroll, M.E., Morgan, A.D., Lynch, W.J., Campbell, U.C., and Dess, N. K. (2002). Intravenous cocaine and heroin self-administration in rats selectively bred for differential saccharin intake: phenotype and sex differences. *Psychopharmacology (Berl),* 161(3), 304–313.

Clark, J.T., Kalra, P.S., Crowley, W.R., and Kalra, S.P. (1984). Neuropeptide Y and human pancreatic polypeptide stimulate feeding behavior in rats. *Endocrinology,* 115, 427–429.

Colantuoni, C., Rada, P., McCarthy, J., Patten, C., Avena, N.M., Chadeayne, A., and Hoebel, B.G. (2002). Evidence that intermittent, excessive sugar intake causes endogenous opioid dependence. *Obes Res,* 10(6), 478–488.

Colantuoni, C., Schwenker, J., McCarthy, J., Rada, P., Ladenheim, B., Cadet, J.L., Schwartz, G.J., Moran, T.H., and Hoebel, B.G. (2001). Excessive sugar intake alters binding to dopamine and mu-opioid receptors in the brain. *Neuroreport,* 12(16), 3549–3552.

de Lecea, L., Kilduff, T.S., Peyron, C., Gao, X., Foye, P.E., Danielson, P.E., et al. (1998). The hypocretins: hypothalamus-specific peptides with neuroexcitatory activity. *Proc Nat Acad Sci USA,* 95(1), 322–327.

de Lecea, L., Sutcliffe, J.G., and Fabre, V. (2002). Hypocretins/orexins as integrators of physiological information: lessons from mutant animals. *Neuropeptides,* 36(2-3), 85–95.

Dum, J. and Herz, A. (1984). Endorphinergic modulation of neural reward systems indicated by behavioral changes. *Pharmacol Biochem Behav,* 21(2), 259–266.

Egawa, M., Yoshimatsu, H., and Bray, G.A. (1993). Effect of beta-endorphin on sympathetic nerve activity to interscapular brown adipose tissue. *Am J Physiol,* 264(1 Pt 2), R109–115.

Flanagan, L.M., Blackburn, R.E., Verbalis, J.G., and Stricker, E.M. (1992). Hypertonic NaCl inhibits gastric motility and food intake in rats with lesions in the rostral AV3V region. *Am J Physiol,* 263(1 Pt 2), R9–14.

Flanagan, L.M., Olson, B.R., Sved, A.F., Verbalis, J.G., and Stricker, E.M. (1992). Gastric motility in conscious rats given oxytocin and an oxytocin antagonist centrally. *Brain Res,* 578(1-2), 256–260.

Flanagan, L.M., Verbalis, J.G., and Stricker, E.M. (1988). Naloxone potentiation of effects of cholecystokinin and lithium chloride on oxytocin secretion, gastric motility and feeding. *Neuroendocrinology,* 48(6), 668–673.

Flanagan, L.M., Verbalis, J.G., and Stricker, E.M. (1989). Effects of anorexigenic treatments on gastric motility in rats. *Am J Physiol,* 256(4 Pt 2), R955–961.

Flynn, F.W., Curtis, K.S., Verbalis, J.G., and Stricker, E.M. (1995). Dehydration anorexia in decerebrate rats. *Behav Neurosci,* 109(5), 1009–1012.

Fried, P.A. (1972). Septum and behavior: a review. *Psychol Bull,* 78(4), 292–310.

Giraudo, S.Q., Billington, C.J., and Levine, A.S. (1998). Effects of the opioid antagonist naltrexone on feeding induced by DAMGO in the central nucleus of the amygdala and in the paraventricular nucleus in the rat. *Brain Res,* 782(1-2), 18–23.

Glass, M.J., Billington, C.J., and Levine, A.S. (1999). Opioids and food intake: distributed functional neural pathways? *Neuropeptides,* 33(5), 360–368.

Glass, M.J., Billington, C.J., and Levine, A.S. (2000). Naltrexone administered to central nucleus of amygdala or PVN: neural dissociation of diet and energy. *Am J Physiol Regul Integr Comp Physiol,* 279(1), R86–92.

Gosnell, B.A., Krahn, D.D., and Majchrzak, M.J. (1990). The effects of morphine on diet selection are dependent upon baseline diet preferences. *Pharmacol Biochem Behav,* 37(2), 207–212.

Gosnell, B.A., Lane, K.E., Bell, S.M., and Krahn, D.D. (1995). Intravenous morphine self-administration by rats with low versus high saccharin preferences. *Psychopharmacology (Berl),* 117(2), 248–252.

Harrison, L.M. and Grandy, D.K. (2000). Opiate modulating properties of nociceptin/orphanin FQ. *Peptides,* 21(1), 151–172.

Huang, W., Sved, A.F., and Stricker, E.M. (2000). Vasopressin and oxytocin release evoked by NaCl loads are selectively blunted by area postrema lesions. *Am J Physiol Regul Integr Comp Physiol,* 278(3), R732–740.

Ida, T., Nakahara, K., Katayama, T., Murakami, N., and Nakazato, M. (1999). Effect of lateral cerebroventricular injection of the appetite-stimulating neuropeptide, orexin and neuropeptide Y, on the various behavioral activities of rats. *Brain Res,* 821(2), 526–529.

Jewett, D.C., Cleary, J., Levine, A.S., Schaal, D.W., and Thompson, T. (1992). Effects of neuropeptide Y on food-reinforced behavior in satiated rats. *Pharmacol Biochem Behav,* 42, 207–212.

Kalra, S.P., Dube, M.G., Sahu, A., Phelps, C.P., and Kalra, P.S. (1991). Neuropeptide Y secretion increases in the paraventricular nucleus in association with increased appetite for food. *Proceedings of the National Academy of Sciences of the United States of America,* 88, 10931–10935.

Kanarek, R.B., Mandillo, S., and Wiatr, C. (2001). Chronic sucrose intake augments antinociception induced by injections of mu but not kappa opioid receptor agonists into the periaqueductal gray matter in male and female rats. *Brain Res,* 920(1-2), 97–105.

Kinzig, K.P., D'Alessio, D.A., and Seeley, R. J. (2002). The diverse roles of specific GLP-1 receptors in the control of food intake and the response to visceral illness. *J Neurosci,* 22(23), 10470–10476.

Kiss, A., Jezova, D., and Aguilera, G. (1994). Activity of the hypothalamic pituitary adrenal axis and sympathoadrenal system during food and water deprivation in the rat. *Brain Res,* 663(1), 84–92.

Koppell, S. and Sodetz, F.J. (1972). Septal ablation in the rat and bar pressing under appetitive and aversive control. *J Comp Physiol Psychol,* 81(2), 274–280.

Kotz, C.M., Billington, C.J., and Levine, A.S. (1997). Opioids in the nucleus of the solitary tract are involved in feeding in the rat. *Am J Physiol,* 272(4 Pt 2), R1028–1032.

Kotz, C.M., Grace, M.K., Briggs, J., Levine, A.S., and Billington, C.J. (1995). Effects of opioid antagonists naloxone and naltrexone on neuropeptide Y- induced feeding and brown fat thermogenesis in the rat. Neural site of action. *J Clin Invest,* 96(1), 163–170.

Kotz, C.M., Teske, J.A., Levine, J.A., and Wang, C. (2002). Feeding and activity induced by orexin A in the lateral hypothalamus in rats. *Regul Pept,* 104(1-3), 27–32.

Langhans, W., Delprete, E., and Scharrer, E. (1991). Mechanisms of vasopressin's anorectic effect. *Physiol Behav,* 49(1), 169–176.

Larsen, P.J., Tang-Christensen, M., and Jessop, D.S. (1997). Central administration of glucagon-like peptide-1 activates hypothalamic neuroendocrine neurons in the rat. *Endocrinology,* 138(10), 4445–4455.

Leventhal, L., Mathis, J.P., Rossi, G.C., Pasternak, G.W., and Bodnar, R.J. (1998). Orphan opioid receptor antisense probes block orphanin FQ-induced hyperphagia. *Eur J Pharmacol,* 349(1), R1–3.

Levine, A.S. and Billington, C.J. (1997). Why do we eat? A neural systems approach. *Annu Rev Nutr,* 17, 597–619.

Levine, A.S., Morley, J.E., Gosnell, B.A., Billington, C.J., and Bartness, T.J. (1985). Opioids and consummatory behavior. *Brain Res Bull,* 14(6), 663–672.

Li, C., Vaughan, J., Sawchenko, P.E., and Vale, W.W. (2002). Urocortin III-immunoreactive projections in rat brain: partial overlap with sites of type 2 corticotrophin-releasing factor receptor expression. *J Neurosci,* 22(3), 991–1001.

Marks, D.L., Butler, A.A., and Cone, R.D. (2002). Melanocortin pathway: animal models of obesity and disease. *Ann Endocrinol (Paris),* 63(2 Pt 1), 121–124.

Marks-Kaufman, R. (1982). Increased fat consumption induced by morphine administration in rats. *Pharmacol Biochem Behav,* 16, 949–955.

Meunier, J.C., Mollereau, C., Toll, L., Suaudeau, C., Moisand, C., Alvinerie, P., Butour, J.L., Guillemot, J.C., Ferrara, P., Montserrat, B., Mazarguil, H., Vassart, G., Parmentier, M., and Costentin, J. (1995). Isolation and structure of the endogenous agonist of opioid receptor-like ORL1 receptor. *Nature,* 377(6549), 532–535.

Morton, G.J. and Schwartz, M.W. (2001). The NPY/AgRP neuron and energy homeostasis. *Int J Obes Relat Metab Disord, 25 Suppl* 5, S56–62.

Nachman, M. (1963). Learned taste aversion to the taste of lithium chloride and generalization to other salts. *Journal of Comparative Physiological Psychology*(56), 343–349.

Neal, C.R., Jr., Mansour, A., Reinscheid, R., Nothacker, H.P., Civelli, O., Akil, H., and Watson, S.J., Jr. (1999). Opioid receptor-like (ORL1) receptor distribution in the rat central nervous system: comparison of ORL1 receptor mRNA expression with (125)I-[(14)Tyr]-orphanin FQ binding. *J Comp Neurol,* 412(4), 563–605.

Neal, C.R., Jr., Mansour, A., Reinscheid, R., Nothacker, H.P., Civelli, O., and Watson, S.J., Jr. (1999). Localization of orphanin FQ (nociceptin) peptide and messenger RNA in the central nervous system of the rat. *J Comp Neurol,* 406(4), 503–547.

Oliveira, L.A., Gentil, C.G., and Covian, M. R. (1990). Role of the septal area in feeding behavior elicited by electrical stimulation of the lateral hypothalamaus of the rat. *Braz J Med Biol Res,* 23, 49–58.

Olson, B.R., Drutarosky, M.D., Stricker, E.M., and Verbalis, J.G. (1991). Brain oxytocin receptor antagonism blunts the effects of anorexigenic treatments in rats: evidence for central oxytocin inhibition of food intake. *Endocrinology,* 129(2), 785–791.

Olson, B.R., Freilino, M., Hoffman, G.E., Stricker, E.M., Sved, A.F., and Verbalis, J.G. (1993). C-Fos expression in rat brain and brainstem nuclei in response to treatment that alter food intake and gastric motility. *Mol Cell Neurosci,* 4, 93–106.

Olson, B.R., Hoffman, G.E., Sved, A.F., Stricker, E.M., and Verbalis, J.G. (1992). Cholecystokinin induces c-fos expression in hypothalamic oxytocinergic neurons projecting to the dorsal vagal complex. *Brain Res,* 569(2), 238–248.

Olszewski, P.K., Grace, M.K., Sanders, J.B., Billington, C.J., and Levine, A.S. (2002). Effect of nociceptin/orphanin FQ on food intake in rats that differ in diet preference. *Pharmacol Biochem Behav,* 73(3), 529–535.

Olszewski, P.K., Shi, Q., Billington, C.J., and Levine, A.S. (2000). Opioids affect acquisition of LiCl-induced conditioned taste aversion: involvement of OT and VP systems. *Am J Physiol Regul Integr Comp Physiol,* 279(4), R1504–1511.

Polidori, C., de Caro, G., and Massi, M. (2000). The hyperphagic effect of nociceptin/orphanin FQ in rats. *Peptides,* 21(7), 1051–1062.

Pomonis, J.D., Billington, C.J., and Levine, A.S. (1996). Orphanin FQ, agonist of orphan opioid receptor ORL1, stimulates feeding in rats. *Neuroreport,* 8(1), 369–371.

Pomonis, J.D., Jewett, D.C., Kotz, C.M., Briggs, J.E., Billington, C.J., and Levine, A.S. (2000). Sucrose consumption increases naloxone-induced c-Fos immunoreactivity in limbic forebrain. *Am J Physiol Regul Integr Comp Physiol,* 278(3), R712–719.

Porte, D., Jr., Baskin, D.G., and Schwartz, M.W. (2002). Leptin and insulin action in the central nervous system. *Nutr Rev,* 60(10 Pt 2), S20–29; discussion S68–84, 85–27.

Reinscheid, R.K., Nothacker, H.P., Bourson, A., Ardati, A., Henningsen, R.A., Bunzow, J.R., Grandy, D.K., Langen, H., Monsma, F.J., Jr., Civelli, O. (1995). Orphanin FQ: a neuropeptide that activates an opioidlike G protein- coupled receptor. *Science,* 270(5237), 792–794.

Risold, P.Y. and Swanson, L.W. (1997). Chemoarchitecture of the rat lateral septal nucleus. *Brain Res Rev,* 24, 91–113.

Risold, P.Y. and Swanson, L.W. (1997). Connections of the rat lateral septal complex. *Brain Res Rev,* 24(2–3), 115–195.

Rudski, J.M., Billington, C.J., and Levine, A.S. (1994a). Butorphanol increases food-reinforced operant responding in satiated rats. *Pharmacol Biochem Behav,* 49(4), 843–847.

Rudski, J.M., Billington, C.J., and Levine, A.S. (1994b). Naloxone's effects on operant responding depend upon level of deprivation. *Pharmacol Biochem Behav,* 49(2), 377–383.

Rudski, J.M., Thomas, D., Billington, C.J., and Levine, A.S. (1995). Buprenorphine increases intake of freely available and operant- contingent food in satiated rats. *Pharmacol Biochem Behav,* 50(2), 271–276.

Russell, J.A., Leng, G., and Bicknell, R.J. (1995). Opioid tolerance and dependence in the magnocellular oxytocin system: a physiological mechanism? *Exp Physiol,* 80(3), 307–340.

Sakurai, T. (2002). Roles of orexins in regulation of feeding and wakefulness. *Neuroreport,* 13(8), 987–995.

Sakurai, T., Amemiya, A., Ishii, M., Matsuzaki, I., Chemelli, R.M., Tanaka, H., Williams, G.C., Richardson, J.A., Kozlowski, G.P., Wilson, S., Arch, J.R., Buckingham, R.E., Haynes, A.C., Carr, S.A., Annan, R.S., McNulty, D.E., Liu, W.S., Terrett, J.A., Elshourbagy, N.A., Bergsma, D.J., and Yanagisawa, M. (1998). Orexins and orexin receptors: a family of hypothalamic neuropeptides and G protein-coupled receptors that regulate feeding behavior [see comments]. *Cell,* 92(4), 573–585.

Stricker, E.M. and Verbalis, J.G. (1986). Interaction of osmotic and volume stimuli in regulation of neurohypophyseal secretion in rats. *Am J Physiol,* 250(2 Pt 2), R267–275.

Sweet, D.C., Levine, A.S., Billington, C.J., and Kotz, C.M. (1999). Feeding response to central orexins. *Brain Res,* 821(2), 535–538.

Taheri, S., Zeitzer, J.M., and Mignot, E. (2002). The role of hypocretins (orexins) in sleep regulation and narcolepsy. *Annu Rev Neurosci,* 25, 283–313.

Tang-Christensen, M., Vrang, N., and Larsen, P.J. (2001). Glucagon-like peptide containing pathways in the regulation of feeding behaviour. *Int J Obes Relat Metab Disord, 25 Suppl* 5, S42–47.

Varoqueaux, F. and Poulain, P. (1999). Projections of the mediolateral part of the lateral septum to the hypothalamus, revealed by Fos expression and axonal tracing in rats. *Anat Embryol,* 199, 249–263.

Vasudev, R., Gentil, C.G., and Covian, M. R. (1985). Taste preferences in a free-choice situation following electrical stimulation and lesion of septal area in rats. *Physiol Behav,* 34(4), 619–624.

Vaughan, J., Donaldson, C., Bittencourt, J., Perrin, M.H., Lewis, K., Sutton, S., Chan, R., Turnbull, A.V., Lovejoy, D., Rivier, C., Rivier, J., Sawchenko, P.E., and Vale, W. (1995). Urocortin, a mammalian neuropeptide related to fish urotensin I and to corticotropin-releasing factor. *Nature,* 378(6554), 287–292.

Verbalis, J.G., Blackburn, R.E., Hoffman, G.E., and Stricker, E.M. (1995). Establishing behavioral and physiological functions of central oxytocin: insights from studies of oxytocin and ingestive behaviors. *Adv Exp Med Biol,* 395, 209–225.

Verbalis, J.G., McCann, M.J., McHale, C.M., and Stricker, E.M. (1986). Oxytocin secretion in response to cholecystokinin and food: differentiation of nausea from satiety. *Science,* 232(4756), 1417–1419.

Verbalis, J.G., McHale, C.M., Gardiner, T.W., and Stricker, E.M. (1986). Oxytocin and vasopressin secretion in response to stimuli producing learned taste aversions in rats. *Behav Neurosci,* 100, 466–475.

Wang, C. and Kotz, C.M. (2002). Urocortin in the lateral septal area modulates feeding induced by orexin A in the lateral hypothalamus. *Am J Physiol Regul Integr Comp Physiol,* 283(2), R358–367.

Wang, C., Mullet, M.A., Glass, M.J., Billington, C.J., Levine, A.S., and Kotz, C.M. (2001). Feeding inhibition by urocortin in the rat hypothalamic paraventricular nucleus. *Am J Physiol Regul Integr Comp Physiol,* 280(2), R473–480.

Wang, C.F., Billington, C.J., Levine, A.S., and Kotz, C.M. (2000). Effect of CART in the hypothalamic paraventricular nucleus on feeding and uncoupling protein gene expression. *Neuroreport,* 11(14), 3251–3255.

Welch, C.C., Kim, E.M., Grace, M.K., Billington, C.J., and Levine, A.S. (1996). Palatability-induced hyperphagia increases hypothalamic Dynorphin peptide and mRNA levels. *Brain Res,* 721(1–2), 126–131.

Weldon, D.T., O'Hare, E., Cleary, J., Billington, C.J., and Levine, A.S. (1996). Effect of naloxone on intake of cornstarch, sucrose, and polycose diets in restricted and nonrestricted rats. *Am J Physiol,* 270(6 Pt 2), R1183–1188.

Wideman, C.H. and Murphy, H.M. (1991). Effects of vasopressin replacement during food-restriction stress. *Peptides,* 12(2), 285–288.

Willie, J.T., Chemelli, R.M., Sinton, C.M., and Yanagisawa, M. (2001). To eat or to sleep? Orexin in the regulation of feeding and wakefulness. *Annu Rev Neurosci,* 24, 429–458.

Wisse, B.E. and Schwartz, M.W. (2001). Role of melanocortins in control of obesity. *Lancet,* 358(9285), 857–859.

Woods, S.C., Seeley, R.J., Porte, D., Jr., and Schwartz, M.W. (1998). Signals that regulate food intake and energy homeostasis. *Science,* 280(5368), 1378–1383.

9 Amino Acids, Brain Metabolism, Mood, and Behavior

Simon N. Young

CONTENTS

9.1 INTRODUCTION

The regulation of flux down metabolic pathways is controlled in a number of ways. One of the less common methods in mammalian metabolism is through precursor availability; that is, the flux down the pathway depends on the concentration of the substrate of the first and rate-limiting enzyme in the pathway. Interestingly, this method of control occurs in several metabolic pathways involved in converting essential amino acids into psychoactive metabolites in the brain. Essential amino acids are transported into the brain by an active transport system, and the amount transported into the brain depends, in part, on the availability of the amino acids in blood. Therefore, both dietary intake and ingestion of purified amino acids can influence the concentration of metabolic products in the brain. In some circumstances, this can alter aspects of mood and behavior. This chapter (1) reviews some of the biochemical mechanisms involved in converting dietary amino acids into psychoactive metabolites, (2) summarizes what is known about the effects of amino acids on human mood and behavior, and (3) discusses how the sale of amino acids is regulated.

9.2 TRYPTOPHAN AND SEROTONIN

9.2.1 Biochemical Aspects

One minor pathway (quantitatively) of tryptophan metabolism is its conversion to serotonin (5-hydroxytryptamine, 5-HT). Most of this conversion takes place in neurons and mast cells in the intestine, but it also occurs in the brain. Serotonin does not cross the blood-brain barrier and must be synthesized from tryptophan in the brain. Tryptophan is transported across the blood-brain barrier by a system of capillaries that actively transports all large neutral amino acids (LNAAs; tryptophan, phenylalanine, tyrosine, leucine, isoleucine, valine, methionine, threonine, serine, and cysteine). The affinity of various amino acids for the transporter is such that there is competition between individual amino acids (Pardridge, 1986). Therefore, the amount of tryptophan in the brain depends not only on the amount of tryptophan in plasma but also inversely on the levels of the other LNAAs in plasma (Wurtman et al., 1981). Tryptophan is the only amino acid present in plasma partly in free solution, but mostly bound loosely to albumin. Only a small part of the albumin-bound tryptophan is available to the brain, and brain tryptophan levels tend to follow the free (non-albumin-bound) plasma level. Tryptophan can be displaced from its binding site on albumin by free fatty acids, thereby increasing free plasma tryptophan levels. Thus, an increase in plasma fatty acid levels can increase brain tryptophan, and, as a consequence, serotonin levels (Sainio et al., 1996).

In the brain, the conversion of tryptophan to serotonin occurs only in serotoninergic neurons. The first and rate-limiting enzyme in this pathway is tryptophan 5-hydroxylase, which converts tryptophan to 5-hydroxytryptophan (5-HTP). Tryptophan hydroxylase uses molecular oxygen and also tetrahydrobiopterin (BH4) for activity. Tryptophan, oxygen, and BH4 are abnormally present in the brain at concentrations that do not saturate the active site of tryptophan hydroxylase. As a result, alterations in the level of any of these compounds will alter the rate of serotonin synthesis. The concentration of tryptophan in the brain is usually about equal to its K_m for the enzyme. (That is, the concentration of tryptophan is such that the enzyme acts at half its maximum velocity.) Thus, decreases in brain tryptophan will cause a lowering of serotonin synthesis, and increases will increase the rate of serotonin synthesis up to a maximum of twofold (Wurtman et al., 1981).

The conversion of 5-HTP to serotonin occurs by the action of aromatic amino acid decarboxylase, which has a high activity. Therefore, in the brain, the level of 5-HTP is low because 5HTP is rapidly converted into serotonin. 5-HTP, like tryptophan, is an LNAA that when given orally can be transported into the brain and converted into serotonin. 5-HTP is present in the diet in small amounts in some types of beans. However, because aromatic amino acid decarboxylase is present in the liver and kidney as well as in the brain, most 5-HTP ingested orally will be converted to serotonin peripherally and will not alter brain serotonin levels. When used clinically, 5-HTP is usually given with an inhibitor of aromatic amino acid decarboxylase that does not cross the blood-brain barrier, therefore inhibiting its conversion to serotonin in the periphery but not in the brain.

In the brain, serotonin is broken down by a two-step pathway, initiated by monoamine oxidase, to 5-hydroxyindoleacetic acid (5-HIAA). In the pineal, serotonin (formed there) can be converted to melatonin by the action of serotonin-*N*-acetyltransferase and hydroxyindole-*O*-methyltransferase. Although the pineal is located in the center of the brain in humans, neuroanatomically it is part of the peripheral nervous system. Neural connections from the brain to the pineal occur only through the superior cervical ganglion. Furthermore, the pineal does not have anything similar to the blood-brain barrier. However, melatonin released by the pineal can act on the brain and some of the effects of tryptophan may be mediated in part by melatonin. Melatonin shows a marked diurnal rhythm. At night, the activities of serotonin-*N*-acetyltransferase and melatonin synthesis are high, whereas in the day activities of both enzymes are low. However, melatonin synthesis also depends in part on serotonin availability, and providing tryptophan will increase melatonin levels somewhat during both the day and night (Huether et al., 1992).

9.2.2 MOOD AND BEHAVIORAL ASPECTS

9.2.2.1 Rationale for Acute Tryptophan Depletion Studies

Most theories that relate serotonin to psychopathology suggest that low serotonin levels or function is involved [see, e.g., Meltzer and Maes (1995)]. Most of the evidence for these theories is indirect. For example, studies of various measures related to serotonin (e.g., levels of the serotonin metabolite 5-HIAA in cerebrospinal fluid) suggest that lowered serotonin synthesis or function occurs in some depressed patients. However, an association does not necessarily imply cause and effect. Does low serotonin cause depression, does depression cause low serotonin, or are low serotonin levels an epiphenomenon? Furthermore, although some antidepressant drugs potentiate serotonin function, this does not necessarily imply that low serotonin function causes depression. (For example, anticholinergic drugs have a therapeutic effect in Parkinson's disease, although low dopamine, not high acetylcholine, is involved in the etiology of Parkinson's disease.)

Decreasing serotonin synthesis by lowering tryptophan levels has been used to test some of the hypotheses relating low serotonin function to various types of psychopathologies [reviewed by Young and Leyton (2002)]. The acute tryptophan depletion (ATD) technique makes it possible to reduce brain levels of serotonin in humans to see what symptoms may emerge. This can be done under controlled conditions in a laboratory setting, and the subjects' biochemistry can be normalized (by giving tryptophan) before they leave the laboratory. Because the depletion is short term and done under close supervision, risks associated with trying to elicit symptoms similar to those that occur in psychiatric disorders are minimized. To date, no adverse effects of ATD have been reported in more than 100 published studies.

In the ATD technique, subjects ingest an amino acid mixture containing all the essential amino acids except tryptophan (T- mixture). The control is a similar amino acid mixture containing the correct amount of tryptophan for a good protein source (nutritionally balanced or B mixture). Normally, when people ingest a protein meal, the protein is broken down into individual amino acids in the gastrointestinal tract. The amino acids are absorbed into the bloodstream. Then, most of the amino acids are incorporated into labile protein stores in the liver and muscle, which prevents free amino acids levels in plasma from rising too much. The B and T- mixtures also induce protein synthesis. However, as the T-mixture contains no tryptophan, tryptophan that is incorporated into protein comes from free tryptophan stores in the blood and tissues. As tryptophan is incorporated into protein, plasma levels of the amino acid fall dramatically. When a human ingests a 100-g T-mixture (the amount of amino acids in ca. 1 lb of steak) plasma tryptophan falls by 70 to 90% within 5 h. This decline in tryptophan availability is accompanied by a decline in serotonin synthesis, as indicated by a decline in the levels of 5-HIAA in the cerebrospinal fluid (Carpenter et al., 1998).

Positron emission tomography, which measures the rate of serotonin synthesis, showed that the ATD procedure lowered the rate of serotonin synthesis in the human brain by 90% or more (Nishizawa et al., 1997). The decline in serotonin synthesis was greater than that in plasma tryptophan because the amino acid mixtures also contain LNAAs, which tend to inhibit transport of tryptophan into the brain. This is true for both B and T- mixtures; therefore, there will presumably be a small decline in brain serotonin synthesis with the control B mixture.

9.2.2.2 Practical Aspects of Acute Tryptophan Depletion

The procedure in ATD studies varies somewhat, but most commonly 100 g of a tryptophan-deficient amino acid mixture is provided to subjects. The amino acids induce protein synthesis, and as tryptophan is incorporated into protein its level in the plasma and tissues declines. The greatest depletion occurs after 5 to 7 h when plasma tryptophan reaches 10 to 30% of its baseline level. At this point, suitable measurements can be made and then the subjects are repleted with tryptophan by allowing them to eat normally, or better by giving them a tryptophan supplement (Young and Leyton, 2002).

In ATD studies, the control treatment is usually a similar amino acid mixture that contains 2.3 g of tryptophan. The relative amounts of the different amino acids in the control mixture are based on the amino acid content of human milk, which is presumably optimized for human metabolism. The control mixture raises plasma tryptophan, but the plasma ratio of tryptophan to the other LNAAs declines because of the higher level of other LNAAs in the mixture. This implies a small decline in brain tryptophan, and thus the control treatment is a conservative one.

9.2.2.3 Effects of Acute Tryptophan Depletion

The effect of ATD on mood depends on the individuals being studied [detailed reviews in Moore et al. (2000) and Young and Leyton (2002)]. In about half the studies that used healthy volunteers, a modest lowering of mood was reported (relative to the effects of the control amino acid mixture). Women are more likely to show a lowering of mood than men. This may reflect their greater susceptibility to depression, as epidemiological studies show a greater incidence of depression in women than in men. Healthy male volunteers with a family history of depression usually show a greater lowering of mood than those with no family history of the disease, also suggesting that the effects of ATD may depend on a susceptibility to depression. In depressed patients who have recently recovered and are on antidepressant drugs, a dramatic lowering of mood, sometimes described as the reappearance of clinical depression, is often seen with ATD, but this lowering of mood remits when tryptophan levels return to normal. The effect of ATD on mood in depressed patients is usually seen when patients are treated with specific serotonin reuptake inhibitors but not with noradrenaline reuptake inhibitors (Delgado et al., 1999). Therefore, the effect of ATD in newly recovered depressed patients who are still being treated may be more relevant to the mechanism of action of the antidepressant than to the etiology of depression. In formerly depressed patients who have recovered and are off antidepressant drugs, a lowering of mood in the clinical range is seen in a minority of subjects. There is a suggestion that a marked mood response in these patients may predict future relapse (Young and Leyton, 2002). Overall these results suggest that (1) reductions in brain serotonin can contribute to lowered mood, (2) a reduction in serotonin by itself is not enough to lower mood, (3) a reduction in serotonin is probably involved in the etiology of depression in some depressed patients, and (4) reduced serotonin synthesis counteracts the antidepressant action of serotonin reuptake inhibitors.

The effect of ATD on anxiety is different from its effect on mood. ATD by itself does not increase anxiety in either healthy subjects or subjects with a history of anxiety disorders. However, ATD often increases anxiety in people with a history of mood disorders, but this increase in anxiety is seen only in conjunction with a lowering of mood. Although ATD does not alter baseline levels of anxiety, it often increases anxiety in response to an anxiogenic challenge (Anderson and Mortimore, 1999). The anxiogenic challenges include both psychological (e.g., public speaking) and biological challenges. The main biological challenges that have been used are treatments that induce panic attacks in both patients with panic disorder and healthy individuals. These include breathing elevated levels of carbon dioxide, infusing cholecystokinin-4, and administering various anxiogenic drugs. In about half the studies on healthy individuals, ATD increased the anxiety seen in response to an anxiogenic challenge. However, the effect of ATD on anxiety is more reliable in patients with anxiety disorders than in those without these disorders. Therefore, although susceptibility to anxiety may play some role in the effect of ATD, an enhancement of anxiety is usually seen only when there is an anxiogenic challenge. That ATD has no effect, in the absence of an anxiogenic challenge, on patients with anxiety disorders is in sharp contrast to its effect on mood. The data suggest that low serotonin may have a direct effect on mood, but it probably has only has a modulatory effect on anxiety.

A wealth of animal data supports the idea that a low level of serotonin in the brain predisposes an animal to aggression, a finding that is supported by correlational data in humans. [For example, aggressive individuals tend to have low levels of 5-HIAA in their cerebrospinal fluid (Miczek et al., 2002).] In human participants, the effect of ATD has been studied by a variety of laboratory tests of aggression. For example, in one test, participants play a game against an opponent (which, unknown to the participants, is a computer program) in which they can win money or prevent their opponent from winning. The measure of aggression is the number of times the participants prevent their opponent from winning money. Obviously, this type of test suffers from a certain artificiality, but practical and ethical considerations preclude the direct study of ATD on aggressive behavior. In the majority of studies, ATD increased aggressive types of responses, and the effect was often greater or seen only in subjects with preexisting hostile traits. Increases in self-rated hostile or irritable moods have also been seen in patients with conditions often associated with irritability or impulsivity, such as bulimia and premenstrual syndrome (Young and Leyton, 2002). In one study on healthy individuals with a genetic predisposition to alcoholism (often associated with impulsivity or aggression), ATD increased commission errors in a go/no no task; that is, when asked to press a button in response to some stimuli but not others, individuals tended to press the button when they should not. This is taken as an indication of impulsive behavior. Overall, the results suggest that low serotonin can contribute to aggressive behavior in humans and possibly other types of impulsive behavior as well. The effect on aggression is unlikely to be mediated by only an increase in impulsivity, as ATD can also increase hostile mood.

ATD seems to have some subtle cognitive effects, but it is difficult to discern a pattern from the results that have been reported. In a small number of studies, ATD had no effect on hunger or food intake, but in one study ATD was associated with an increase in caloric intake in women with bulimia (Moore et al., 2002). Finally, in one study, ATD had no effect on pain responses in the cold pressor test, but completely abolished the analgesic effect of morphine in that test (Moore et al., 2000). This supports the idea that the analgesic effect of morphine in the cold pressor test is mediated by enhanced release of serotonin, possibly in the spinal cord.

Evidence to date from ATD studies supports the idea that a low level of serotonin in the brain may lower mood and increase irritability or aggression, particularly in those with a susceptibility to depression or aggression. Low serotonin may also increase anxiety in response to stress. Further studies are needed on the other effects of ATD, including its effects on cognition, pain responses, food intake, sexual activity, and any other aspect of central nervous system (CNS) function thought to be mediated by serotonin.

9.2.2.4 Metabolic Considerations in the Clinical Use of Tryptophan

Because of the theories that relate low serotonin function to a variety of different types of psychopathology, tryptophan has been tested for its therapeutic action in a number of conditions [reviewed by Young (1986)]. The daily dietary intake of tryptophan for an adult is ca. 1 g. In clinical use, the dose of tryptophan ranges from 1 to 12 g/day. Tryptophan is metabolized relatively quickly and plasma tryptophan levels return to normal levels within 8 h of a single 3-g dose. Therefore, to keep the rate of serotonin synthesis maximized throughout any 24-h period, tryptophan should be given as a divided dose, three times per day. Clinical trials, which failed to find effects of tryptophan administration on behavior, used single daily dosing, which probably ensured that serotonin synthesis was increased for less than half of each day.

The most convenient way to give tryptophan is with meals, which tends to minimize nausea, one of the side effects sometimes seen. Some articles suggest that tryptophan should be given only with carbohydrate and not with protein. The argument is that other LNAAs in protein will inhibit the transport of tryptophan into the brain. On the other hand, carbohydrate will cause release of insulin. Insulin lowers the level of the branched-chain amino acids leucine, isoleucine, and valine in plasma by stimulating their net uptake into muscle. As the branched-chain amino acids are all LNAAs, this will stimulate tryptophan uptake into the brain. However, this argument does not take into account the dynamic aspects of tryptophan metabolism throughout the day. When a single dose of tryptophan is given, there is a large increase in plasma tryptophan. This probably increases brain tryptophan enough to saturate tryptophan hydroxylase with tryptophan. Therefore, a modest decrease in uptake of tryptophan into the brain will not decrease the rate of serotonin synthesis. If protein had been ingested later, between doses, when tryptophan levels in plasma had fallen from their peak, this might have decreased serotonin synthesis. Thus, ingesting tryptophan with meals is better both because it minimizes side effects and maximizes the rate of serotonin synthesis throughout the day.

9.2.2.5 Side Effects in the Clinical Use of Tryptophan

Side effects associated with tryptophan use are in general infrequent and mild (Young, 1986). Nausea and lightheadedness are the most common. In the largest and longest comparison of tryptophan and placebo, tryptophan (1 g TID) produced no more side effects than placebo (Thomson et al., 1982). However, tryptophan can interact with other drugs that influence serotonin function. For example, when tryptophan is added to monoamine oxidase inhibitors, it potentiates their side effects. In rare cases, the combination of tryptophan and another drug that potentiates serotonin function can cause the serotonin syndrome (Mason et al., 2000). Symptoms of the serotonin syndrome are variable but include (1) cognitive and behavioral, (2) autonomic nervous system, and (3) neuromuscular effects. The cognitive and behavioral changes can include confusion and agitation, as well as anxiety and hypomania, or in severe cases coma. Among the symptoms of autonomic system disturbances, diaphoresis (profuse sweating) and hyperthermia are common. Among the neuromuscular symptoms, myoclonus, hyperreflexia, and muscle rigidity are the most frequently reported. Treatment of the serotonin syndrome is usually symptomatic, but a serotonin receptor antagonist such as methysergide might help. Although deaths have been reported from the serotonin syndrome, these have always been due to drugs other than tryptophan. There are no reports in the literature of fatalities due to tryptophan use (when the pure drug was taken).

In late 1989, an epidemic of a new disorder appeared in the U.S called the eosinophilia myalgia syndrome (EMS; Belongia et al., 1992). EMS is an inflammatory syndrome characterized by eosinophilia, myalgia, periomyositis, fasciitis, and neuropathies. In the U.S., there were more than 1500 cases and 38 deaths. EMS occurred in various European countries and Japan also, but there were not as many cases as in the U.S. EMS was found to be associated with the ingestion of tryptophan from several lots made by a single manufacturer after various changes in the manufacturing process. Various trace contaminants in those batches are suspected as the cause, but there is no definitive evidence as to which compounds caused EMS. Since the initial epidemic, no cases of EMS have been associated with the use of tryptophan.

9.2.2.6 Tryptophan in Mood Disorders

Four placebo-controlled studies conducted in the 1960s demonstrated that tryptophan potentiates the antidepressant effect of monoamine oxidase inhibitors [reviewed by Young (1991)]. However, this combination is rarely used, and when used is only in treatment-resistant patients, because tryptophan also potentiates the side effects of monoamine oxidase inhibitors. This is the only clear conclusion from studies on the use of tryptophan in depressed patients. Unfortunately, the literature contains many studies with small sample sizes and lack of placebo control. The finding that tryptophan is not significantly different from a standard antidepressant in a small trial does not mean that it would have been better than placebo. Because of the variability in response to antidepressants and the relatively small difference between the effects of placebo and standard antidepressants, results of small trials are difficult to interpret. A small number of studies suggest that tryptophan may cause a small augmentation in the effects of tricyclic antidepressants, specifically serotonin reuptake inhibitors and electroconvulsive therapy. However, the majority of studies of this type have not found that tryptophan potentiates the antidepressant effect of these drugs, and in cases in which a statistically significant effect was seen it was not always clinically significant (Young, 1986).

The antidepressant effect of tryptophan when given by itself is also controversial. My own interpretation of the literature is that tryptophan is an antidepressant in mild to moderate depression, but there is no evidence for efficacy in severe depression (Young, 1986). In the largest and best-designed study to date, tryptophan was compared with placebo and the tricyclic antidepressant amitriptyline as given by general practitioners to mildly or moderately depressed outpatients for 12 weeks. Tryptophan was equivalent in its therapeutic effect to amitriptyline and significantly better than placebo (Thomson et al., 1982).

Three small studies indicate that tryptophan has a therapeutic effect in acute mania, although the magnitude of the effect is certainly not sufficient to treat a seriously disturbed manic patient. Various clinical reports also suggest that tryptophan may augment the mood-stabilizing effect of lithium in patients with bipolar mood disorder (Young, 1986).

9.2.2.7 Tryptophan, Impulsivity, Irritability, and Aggression

Two small placebo-controlled studies have reported a therapeutic effect of tryptophan in pathological aggression (Young, 1986). In one study, tryptophan decreased the incidence of uncontrolled behaviors, and in the other tryptophan decreased the amount of neuroleptics needed to control the patients. In a study on patients with premenstrual dysphoric disorder (premenstrual syndrome), tryptophan (given intermittently for a few days before and during the premenstrual phase) decreased irritability and well as improved mood.

A recent study looked at the effect of tryptophan on social interaction in healthy subjects in everyday life (Moskowitz et al., 2001). The method used in this study was a form of ecological momentary assessment, a type of methodology developed by social psychologists. Subjects filled in a one-page questionnaire describing their behaviors after each important social interaction throughout the day. Behavior was assessed on two independent axes, agreeable-quarrelsome and dominant-submissive. These behaviors varied greatly from one interaction to another, but mean values were relatively constant over a sufficient number of interactions, ca. 12 days of measurements. In the study, 100 healthy people received tryptophan or placebo, each for 12 days. Tryptophan decreased quarrelsome behaviors and increased dominance. Thus, although there is a popular but mistaken belief that dominance can be enhanced only through aggression, in this study tryptophan promoted more constructive behavior by increasing dominance while decreasing quarrelsome behaviors. The study is also of interest because some psychoactive drugs, for example, antidepressants, seem to have little effect in healthy people. In the study described previously, tryptophan altered normal behavior in everyday life, suggesting that serotonin function is relevant to more than just psychopathology.

9.2.2.8 Tryptophan and Sleep

Many studies have looked at the effect of tryptophan on sleep [reviewed by Hartmann and Greenwald (1984)]. Overall, these studies indicate that tryptophan, in the dose range of 1 to 5 g taken ca. 45 min before bedtime, can decrease the time taken to fall asleep without altering sleep architecture. Tryptophan is effective in healthy people with long sleep latencies and in patients with mild insomnia. It is not as effective as benzodiazepines in patients with moderate or severe insomnia. However, tryptophan has fewer side effects than benzodiazepines and does not have any detrimental effects the following day. Given that tryptophan can increase melatonin, and that melatonin also has a hypnotic effect, it is unclear to what extent the effect of tryptophan is mediated by an increase of serotonin in the brain or by an increase of melatonin in the pineal gland.

9.2.2.9 Effect of Tryptophan on Pain

The effect of tryptophan has been tested acutely by using experimental pain and in patients suffering from various types of pain (Young, 1986). Some studies found a beneficial effect of tryptophan and others did not. Because of the variety of conditions studied, it is difficult to come to any firm conclusion. However, the situation in which tryptophan has shown a therapeutic effect most frequently is chronic pain associated with deafferentation or neural damage.

9.2.2.10 Tryptophan and Food Intake

The reason why tryptophan availability alters serotonin synthesis in not known, but one theory is that it is part of a mechanism that regulates food intake. However, in a small number of studies tryptophan failed to alter total food intake or macronutrient selection (Young, 1986).

9.3 PHENYLALANINE, TYROSINE, AND CATECHOLAMINES

9.3.1 BIOCHEMICAL ASPECTS

There are many similarities between the roles of tryptophan and tyrosine in regulating the synthesis of their respective neurotransmitter products, but there are also some important differences. Phenylalanine and tyrosine are precursors of the catecholamines dopamine and noradrenaline. The conversion of phenylalanine to tyrosine occurs mainly in the liver, but tyrosine hydroxylase in the brain can also form tyrosine from phenylalanine. The metabolic pathway from tyrosine to dopamine is similar to that from tryptophan to serotonin. Tyrosine is hydroxylated to form dihydroxyphenylalanine (DOPA), which is decarboxylated to produce dopamine. Tyrosine hydroxylase, like tryptophan hydroxylase, is the rate-limiting enzyme in this pathway and requires molecular oxygen and also BH4 for activity. Tryptophan hydroxylase is ca. 50% saturated with tryptophan, whereas tyrosine hydroxylase is probably ca. 75% saturated with tyrosine. Furthermore, tyrosine hydroxylase (unlike tryptophan hydroxylase) is regulated in part by product inhibition, that is, it is inhibited by its products dopamine and noradrenaline. These two facts led to the idea that catecholamine synthesis would not be influenced by substrate availability. However, sufficient evidence has accumulated to indicate that under some circumstances changes in tyrosine levels can alter catecholamine synthesis. Increases in catecholamine synthesis due to tyrosine administration seem to occur most often when firing of catecholamine neurons is activated and demand for catecholamine synthesis is greatest (Wurtman et al., 1981).

DOPA is decarboxylated to dopamine by aromatic amino acid decarboxylase. DOPA, like 5-HTP, is an LNAA and is transported into the brain. Also, like 5-HTP, it occurs in some foods, mainly a few varieties of beans, but when ingested as part of the diet does not affect the brain because it is decarboxylated in the periphery and catecholamines do not cross the blood-brain barrier.

In dopaminergic neurons, dopamine is converted by pathways involving monoamine oxidase to dihydroxyphenylacetic acid (DOPAC) and homovanillic acid (HVA). In noradrenergic neurons, dopamine is converted to noradrenaline by dopamine-β-hydroxylase. The main catabolic product of noradrenaline in the brain is 3-methoxy-4-hydroxyphenylethylene glycol (MHPG).

Tyrosine administration can increase catecholamine synthesis somewhat, but given that tyrosine hydroxylase is close to saturation with tyrosine, there is greater scope for decreasing catecholamine synthesis by lowering tyrosine levels. Acute phenylalanine/tyrosine depletion (APTD) works in the same way as ATD, except that the amino acid mixture is deficient in phenylalanine and tyrosine instead of tryptophan. In monkeys, APTD decreases levels of HVA and MHPG in the cerebrospinal fluid, suggesting that it decreases synthesis of both catecholamines.

9.3.2 Mood and Behavioral Aspects

9.3.2.1 Effects of Acute Phenylalanine/Tyrosine (APTD) Depletion

Relatively few APTD depletion studies have been carried out to date [reviewed by Booij et al. (2003)], and therefore the conclusions given here must be considered tentative. In healthy individuals, APTD may cause small changes in mood, including a decreased feeling of well-being and alertness and an increase in anxiety but not the increase in irritability seen with ATD (which is thought to be related more specifically to serotonin). However, APTD did not modulate the increase in anxiety seen with pentagastrin, a drug that can cause panic attacks. APTD did cause small declines in various measures of memory. APTD has also been reported to decrease intake of alcohol in healthy subjects undergoing a taste test (in which the real measure was the amount ingested). Furthermore, several studies indicate that APTD decreases the psychostimulant effect of amphetamine (thought to be related to enhanced dopamine function) but not the anorexic effect of amphetamine (thought to be related to noradrenergic function). An intriguing preliminary report indicates that APTD temporarily reverses some of the symptoms of manic patients (McTavish et al., 2002). The results suggest that (1) APTD probably decreases dopaminergic function but not noradrenergic function, (2) lowered dopamine decreases mood and arousal, and (3) decreasing dopamine function may be therapeutic in mania.

9.3.2.2 Effects of Tyrosine on Behavior

Many of the considerations that apply to the clinical use of tryptophan also apply to tyrosine. Doses several times the dietary intake are appropriate and should be given in divided doses throughout the day. Doses that have been used range up to 10 g/day. The small number of studies done have mostly used few participants and been of relatively short duration. Side effects have not been studied systematically but seem to be few, as with tryptophan. Tyrosine seems to have little effect on mood in healthy subjects and in some studies had no therapeutic effect in depressed patients. However, at least three studies have reported some antidepressant effect (e.g., van Praag, 1990).

One important role of catecholamines is to regulate blood pressure. Effects are complex and in animals tyrosine can raise blood pressure in hypotensive animals and lower it in hypertensive animals (Ekholm and Karppanen, 1987). The former effect has been attributed to an acceleration of catecholamine synthesis in sympathoadrenal cells and the latter to potentiation of CNS noradrenergic function. However, in humans, tyrosine supplements failed to affect mild essential hypertension.

In rats, tyrosine supplementation reverses the decline in brain noradrenaline and reverses some of the behavioral deficits seen in stressed animals. Several studies have looked at the effects of tyrosine on performance in healthy humans undergoing stressful activities. Tyrosine decreased lowered mood, performance deficits, and memory deficits in subjects exposed to cold or hypoxia. It improved various aspects of cognition in subjects exposed to a loud noise or while multitasking. It reversed the decline in vigilance and performance associated with a period of extended wakefulness and improved memory and decreased systolic blood pressure in cadets undergoing combat training (Deijen et al., 1999).

9.4 METHIONINE AND S-ADENOSYLMETHIONINE

9.4.1 BIOCHEMICAL ASPECTS

S-Adenosylmethionine (SAMe) is the major methyl donor in tissues, including the brain, and is involved in more than 35 methylation reactions (Bottiglieri and Hyland, 1994). SAMe is formed from methionine by the action of methionine synthase, also known as 5-methyltetrahydrofolate homocysteine methyltransferase, on homocysteine. Methionine synthase is a vitamin B12-dependent enzyme that obtains methyl groups from 5-methyltetrahydrofolate. The enzyme is normally saturated with its cofactors, but not with methionine. Administration of methionine increases the level of SAMe in the rat brain (Rubin et al., 1974), but the implications of this are not well understood.

9.4.2 MOOD AND BEHAVIORAL ASPECTS

Views of the human psychopharmacology of methionine are colored by the knowledge that in 10 studies carried out several decades ago, methionine exacerbated symptoms of schizophrenic patients (Cohen et al., 1974). The doses used, from 5 to 20 g/day, were high, and in most studies methionine was given with other psychotropic agents, including monoamine oxidase inhibitors. The mechanism for this adverse effect is unknown, but it has essentially prevented consideration of the use of methionine for other purposes. When methionine is given to healthy subjects, it is usually innocuous. In metabolic studies it has been given at doses up to 5 g/day for periods of weeks with no reports of adverse effects. Furthermore, methionine has been used at a dose of 1 g/infant/day as a urinary acidifier (for treating diaper rashes).

   The fact that methionine can increase brain SAMe raises the question of whether it has antidepressant properties. SAMe was better than placebo in treating depression in about a dozen studies (Bottiglieri and Hyland, 1994). Furthermore, it had few side effects. The dose of SAMe used is usually ca. 1 g/day, suggesting that if methionine is also an antidepressant it should be effective at doses lower than those that had an adverse effect in patients with schizophrenia. To date, there have been no reports on the putative antidepressant effects of methionine.

9.5 HISTIDINE AND HISTAMINE

9.5.1 BIOCHEMICAL ASPECTS

Histidine is converted to histamine in the brain by the action of histidine decarboxylase. Histidine loading can increase brain histamine levels by somewhat more than twofold in rodents, suggesting that histamine is regulated more than catecholamines or serotonin are by precursor availability. However, in the rat, histidine increases brain histamine but it does not increase, or only increases very slightly, the level of the main histamine metabolites tele-methylhistamine and tele-methylimidazoleacetic acid. Furthermore, when brain histamine is depleted with an inhibitor of histidine decarboxylase, histidine still raises brain histamine levels (Prell et al., 1996). This suggests that histidine loads may cause nonspecific decarboxylation of histidine (probably by aromatic amino acid decarboxylase) and that the histamine formed may not all be present in histaminergic neurons. Therefore, histidine loading will not necessarily enhance brain histamine function, even if brain histamine levels are elevated.

9.5.2 MOOD AND BEHAVIORAL ASPECTS

Histamine-releasing neurons project from the hypothalamus to most areas of the brain. In keeping with this broad anatomical distribution, histamine has been implicated in a variety of CNS functions, including arousal, anxiety, pain perception, and control of food intake. A few studies have looked at the behavioral effects of histidine in rodents. A variety of effects have been noted, but no consistent pattern has been reported. There is certainly need for studies looking at the effect of histidine loading in humans.

9.6 THREONINE AND GLYCINE

9.6.1 METABOLIC ASPECTS

Glycine is a nonessential amino acid that is also a neurotransmitter in the spinal cord. In the CNS, glycine is synthesized from glucose, serine, or threonine. Glycine is formed by the action of serine hydroxymethyltransferase (SHMT) on serine. This enzyme employs folic acid. The one-carbon unit removed from serine during the formation of glycine becomes one of the major sources of methyl groups in mammalian metabolism and is used in the formation of SAMe. SHMT also acts on threonine to form glycine (and is sometimes referred to as threonine aldolase). The enzyme is fully saturated with serine, but it is not fully saturated with threonine. Also, unlike serine, threonine is an essential amino acid that is transported into the central nervous system by the transport system that is active toward all large neutral amino acids. As might be expected in these circumstances, giving a rat threonine will increase brain and spinal cord glycine levels (Maher and Wurtman, 1980).

9.6.2 MOOD AND BEHAVIORAL ASPECTS

In the spinal cord, glycine is an inhibitory transmitter controlling muscle tone. This has led to a few small open and placebo-controlled trials of threonine (4 to 7.5 g/day) in treating spasticity (e.g., Growdon et al., 1991) in some studies associated with other conditions such as multiple sclerosis or amyotrophic lateral sclerosis. In some studies, a modest but definite therapeutic effect was found, but it was not always considered clinically significant. Fortunately, the modest therapeutic effect was accompanied by minimal side effects. In the brain, glycine acts to modulate the activity of a particular class of glutamate receptors called the NMDA receptors. The lack of side effects may reflect the finding from animal studies that threonine increases glycine levels more in the spinal cord than in the brain.

9.7 EFFECTS OF PROTEIN AND CARBOHYDRATE MEALS ON BRAIN NEUROTRANSMITTER PRECURSOR LEVELS

All the precursor amino acids discussed are LNAAs. Therefore, their concentration in the brain will depend on the plasma ratio of the individual amino acid levels to that of the sum of the other LNAAs. Administering the amino acids in purified form does not alter the level of the other LNAAs and the brain level of the relevant amino acid rises. After ingestion of protein, the situation is more complex and depends in part on the relative amount of the individual amino acids in the protein (Young, 1996). Tryptophan is among the least abundant amino acids in most proteins, and after ingestion of a protein meal the plasma ratio of tryptophan to the other LNAAs usually declines. Tyrosine is more abundant in most proteins, as is its precursor phenylalanine. Therefore, a protein meal raises the plasma ratio of tyrosine to the other LNAAs. Threonine is also a relatively abundant amino acid and its plasma ratio also rises. Methionine and histidine are less abundant. However, there is one report that the plasma ratios for methionine, brain methionine, and brain SAMe rise after rats ingest a protein meal (Rubin et al., 1974). The reason for this is not clear, but it is a topic that needs further investigation.

Protein meals have a variable effect on the brain levels of different LNAAs, but the effect of a carbohydrate meal is the same for all the amino acids discussed (Young, 1996). Carbohydrate causes the release of insulin, which stimulates the net uptake of the branched-chain amino acids leucine, isoleucine, and valine into the muscle. All branched-chain amino acids are LNAAs, and therefore the plasma ratio for tryptophan, tyrosine, methionine, histidine, and threonine tends to rise after a carbohydrate meal.

One unanswered question is why protein and carbohydrate meals should have effects on neurotransmitter precursor levels and therefore potentially on neurotransmitter function. One suggestion related this phenomenon to the alteration that occurs in serotonin function and the regulation of macronutrient intake (Wurtman and Wurtman, 1988). Pharmacological enhancement of serotonin function in rodents decreases food intake, but the greatest effect seems to be on selection of carbohydrate. Thus, the theory suggests that a carbohydrate meal will raise brain serotonin, which will decrease carbohydrate intake at the next meal. A protein meal will lower serotonin, which will increase intake of carbohydrate at the next meal. In this way, intake of protein and carbohydrate will be regulated within certain limits. Unfortunately, the weight of evidence is against the idea that this mechanism operates in humans. First, real meals do not contain pure protein or carbohydrate. Therefore, a relevant question is what happens to the plasma tryptophan ratio with mixed macronutrient meals. Data show that adding as little as 4% protein to a carbohydrate meal prevents any increase in the plasma tryptophan ratio (Teff et al., 1989a). Even foods popularly regarded as carbohydrate have a protein content this much or more. For example, in orange juice 4% of the calories come from protein, and in bread the figure is more than 10%. Second, meals can certainly alter serotonin levels in rodents, but the extent to which this occurs in humans is uncertain. The amount of food ingested every day, as a percentage of body weight, is much higher for rats than for humans. As a result, the changes in plasma amino acid ratios after ingesting food are greater in rats than humans. In humans, a carbohydrate meal might raise the plasma tryptophan ratio by 20%. This might result in an increase in brain tryptophan of 10% and an increase of brain serotonin synthesis of 5%. A study of human cerebrospinal fluid tryptophan and 5-HIAA after the subjects ingested protein or carbohydrate meals found that the meals did not cause any significant changes in the levels of the two compounds (Teff et al., 1989b). This suggests that any effect of meals on brain serotonin is less than the normal variability in the neurotransmitter. The large changes in tryptophan availability associated with ATD caused no important effects on food intake, and it is highly unlikely that the much smaller change associated with ingesting a meal would have any important action. Third, there is no evidence that humans can regulate their macronutrient intake in the same way that rodents can.

Meals certainly can have effects on mood and behavior in humans, and these sometimes differ depending on the macronutrient content of the meals. For example, carbohydrate meals tend to have a sedative effect, which may be experienced as calmness or tiredness, depending on the individual (Spring et al., 1983). Tryptophan can also have a sedative effect, but this does not necessarily imply that the effect of the carbohydrate meal is mediated by an increase in brain serotonin. As argued previously, the effect of the carbohydrate meal on brain serotonin, if any, would be very much smaller than that of tryptophan. Furthermore, carbohydrate meals are known to have effects unrelated to serotonin. For example, glucose enhances memory, an effect thought to be related to an increase in the release of acetylcholine (Gold, 1995). There may also be other unknown effects. Thus, the fact that the synthesis of a variety of psychoactive substances is determined in part by availability of their amino acid precursors, which in turn is altered by dietary intake, seems to have no known physiological significance.

9.8 ARE AMINO ACIDS FOODS OR DRUGS?

The question of whether amino acids are foods or drugs is one in which opinion and politics play a role as great as science. Different countries regulate amino acids differently. For example, in the

U.S. they are treated as dietary supplements whereas in many European countries and Canada they are regulated as drugs. Part of the problem is that there is no clear delineation between a food and a drug; there is a continuum and governments need to put a dividing line somewhere along this continuum. I consider some of the factors involved.

Traditionally, isolated compounds most often have been regulated as drugs. For example, a cup of coffee is never regulated as a drug. However, caffeine can be extracted from coffee and taken as a tablet. A 200-mg tablet of caffeine is in many countries regulated as a drug, but a cup of coffee containing 200 mg of caffeine is not. By using precedents like this, amino acids would be treated as drugs.

The safety of anything taken in tablet form depends in part on the manufacturing process. For example, the epidemic of eosinophilia-myalgia syndrome (EMS) was associated with a trace contamination of tryptophan produced by bacteria and then purified (but apparently not sufficiently). Safety is certainly a factor in food production, but the issues associated with production of purified amino acids are closer to those of compounds that are unambiguously drugs than those associated with food production.

One important consideration in assessing the status of purified amino acids is the motivation people have for taking them. They are taken usually to overcome some unwanted aspect of mental functioning (e.g., the use of tryptophan to treat depression) or to enhance some aspect of normal functioning (e.g., the use of tyrosine to enhance functioning in stressful situations). In this respect, they fit the definition of a drug as something used to treat or prevent a disease or disorder. When amino acids are regulated as drugs, they can only be sold when regulatory agencies have been persuaded of their efficacy in treating a specific disorder and of their safety. When they are regulated as dietary components, the manufacturer is not usually able to make claims about their effects. In these circumstances, sales of amino acids depend on beliefs about their effect, which may or may not be true. These beliefs are taken from publications (print or electronic) that are, of course, not regulated. Although the manufacturer of an amino acid cannot make any claim as to its effect, there is nothing to stop the owner of a store from placing bottles of amino acids next to sources of information that are totally erroneous about their effect. Although a listing of myths concerning the action of amino acids is beyond the scope of this chapter, a brief survey of Web sites produced the "information" that tryptophan raises serotonin the way nature intended and is often needed by aging individuals. It is useful for treating weight reduction, chemical addiction, alcohol withdrawal, Down's syndrome, chronic fatigue syndrome, migraine, anxiety, and senile dementia. This type of misinformation suggests that in countries in which amino acids are not regulated as drugs, they are widely used in treating disorders for which they have no effect (but still may produce side effects).

The belief that amino acids should not be classified as drugs is usually derived from the fact that they are considered natural treatments. The word *natural* implies that the metabolic effects of pure amino acids are similar to those of foods containing protein. This, of course, is incorrect. When protein is ingested, after absorption of the amino acids there is synthesis of labile protein stores, which attenuate the rise of plasma amino acid levels. No such effect occurs after ingesting a pure amino acid, and the increase the plasma level is much greater. Furthermore, for all the amino acids discussed, the increase in the other LNAAs attenuates any increase in their level in the brain after ingesting a protein. This does not occur when a single amino acid is taken. For tryptophan, the tendency for brain tryptophan and serotonin to decline after a meal containing tryptophan (i.e., protein) is in sharp contrast to the large increase in brain tryptophan and serotonin after ingesting tryptophan tablets. Even for tyrosine, the increase in its concentration in the brain after a protein meal is very much lesser than after taking tyrosine tablets. Thus, ingestion of purified amino acids cannot be considered a natural or dietary treatment. Although the issues discussed suggest that purified amino acids are more similar to drugs than foods, the decision on how to regulate them is ultimately a political one.

9.9 CONCLUSIONS

Several amino acids are precursors of psychoactive substances, and giving those amino acids has useful therapeutic effects in some circumstances. Although tryptophan is available as a drug for treating depression or as a hypnotic in some countries, there is currently little research on other indications of tryptophan or on uses of the other amino acids. Research on the psychopharmacological effects of amino acids peaked in the 1970s and 1980s but is no longer popular. This is partly because most uses of amino acids cannot be patented, leading to a lack of interest from pharmaceutical companies, and is partly related to fashion among researchers. This is unfortunate, as there are many promising areas for future investigation. Two examples are the potential use of methionine as an antidepressant and the use of tryptophan in treating aggression.

The psychopharmacological use of amino acids is not a fashionable topic for researchers, but the use of amino acid depletion strategies to investigate the role of neurotransmitters in the etiology of various mental symptoms is an active area of research and promises further insights into the psychopathology of mental disorders.

One area in need of attention is the wide divergence between the results of research and popular beliefs about the effects of amino acids. There is no simple solution to the mass of misinformation available. Books such as this one will help, but they will never, unfortunately, receive the same exposure as misinformation available on the Web.

REFERENCES

Anderson I.M., and Mortimore C. (1999) 5-HT and human anxiety: evidence from studies using acute tryptophan depletion. In *Advances in Experimental Medicine and Biology, Vol. 467: Tryptophan, Serotonin and Melatonin: Basic Aspects and Applications* (Heuther G., Kochen W., Simat T.J., and Steinhart H., Eds.), pp. 43–55. Kluwer, New York.

Belongia E.A., Mayeno A.N., and Osterholm M.T. (1992) The eosinophilia-myalgia syndrome and tryptophan. *Annu. Rev. Nutr.* 12, 235–256.

Booij L., Van der Does A.J., and Riedel W.J. (2003) Monoamine depletion in psychiatric and healthy populations: review. *Mol. Psychiatr.* 8, 951–973.

Bottiglieri T., and Hyland K. (1994) S-Adenosylmethionine levels in psychiatric and neurological disorders: a review. *Acta Neurol. Scand.* 89, 19–26.

Carpenter L.L., Anderson G.M., Pelton G.H., Gudin J.A., Kirwin P.D.S., Price L.H., Heninger G.R., and McDougle C.J. (1998) Tryptophan depletion during continuous CSF sampling in healthy human subjects. *Neuropsychopharmacology* 19, 26–35.

Cohen S.M., Nichols A., Wyatt R., and Pollin W. (1974) The administration of methionine to chronic schizophrenic patients: a review of ten studies. *Biol. Psychiatr.* 8, 209–225.

Deijen J.B., Wientjes C.J., Vullinghs H.F., Cloin P.A., and Langefeld J.J. (1999) Tyrosine improves cognitive performance and reduces blood pressure in cadets after one week of a combat training course. *Brain Res. Bull.* 48, 203–209.

Delgado P.L., Miller H.L., Salomon R.M., Licinio J., Krystal J.H., Moreno F.A., Heninger G.R., and Charney D.S. (1999) Tryptophan-depletion challenge in depressed patients treated with desipramine or fluoxetine: implications for the role of serotonin in the mechanism of antidepressant action. *Biol. Psychiatr.* 46, 212–220.

Ekholm S., and Karppanen H. (1987) Cardiovascular effects of L-tyrosine in normotensive and hypertensive rats. *Eur. J. Pharmacol.* 143, 27–34.

Gold P.E. (1995) Role of glucose in regulating the brain and cognition. *Am. J. Clin. Nutr.* 61, S987–S995.

Growdon J.H., Nader T.M., Schoenfeld J., and Wurtman R.J. (1991) L-Threonine in the treatment of spasticity. *Clin. Neuropharmacol.* 14, 403–412.

Hartmann E., and Greenwald D. (1984) Tryptophan and human sleep: an analysis of 43 studies. In *Progress in Tryptophan and Serotonin Research* (Schlossberger H.G., Kochen W., Linzen B., and Steinhart H., Eds.), pp. 297–304. Walter de Gruyter, Berlin.

Huether G., Hajak G., Reimer A., Poeggeler B., Blomer M., Rodenbeck A., and Ruther E. (1992) The metabolic fate of infused L-tryptophan in men: possible clinical implications of the accumulation of circulating tryptophan and tryptophan metabolites. *Psychopharmacology* 109, 422–432.

Maher T.J., and Wurtman R.J. (1980) L-Threonine administration increases glycine concentrations in the rat central nervous system. *Life Sci.* 26, 1283–1286.

Mason P.J., Morris V.A., and Balcezak T.J. (2000) Serotonin syndrome: presentation of 2 cases and review of the literature. *Medicine* 79, 201–209.

McTavish S.F., McPherson M.H., Harmer C.J., Clark L., Sharp T., Goodwin G.M., and Cowen P.J. (2002) Antidopaminergic effects of dietary tyrosine depletion in healthy subjects and patients with manic illness. *Br. J. Psychiatr.* 179, 356–360.

Meltzer H.Y., and Maes M. (1995) The serotonin hypothesis of major depression. In *Psychopharmacology: The Fourth Generation of Progress* (Bloom F.E., and Kupfer D.J., Eds.), pp. 933–944. Raven Press, New York.

Miczek K.A., Fish E.W., De Bold J.F., and De Almeida R.M. (2002) Social and neural determinants of aggressive behavior: pharmacotherapeutic targets at serotonin, dopamine and gamma-aminobutyric acid systems. *Psychopharmacology* 163, 434–458.

Moore P., Landolt H.P., Seifritz E., Clark C., Bhatti T., Kelsoe J., Rapaport M., and Gillin J.C. (2000) Clinical and physiological consequences of rapid tryptophan depletion. *Neuropsychopharmacology* 23, 601–622.

Moskowitz D.S., Pinard G., Zuroff D.C., Annable L., and Young S.N. (2001) The effect of tryptophan on social interaction in every day life: a placebo-controlled study. *Neuropsychopharmacology* 25, 277–289.

Nishizawa S., Benkelfat C., Young S.N., Leyton M., Mzengeza S., de Montigny C., Blier P., and Diksic M. (1997) Differences between males and females in rates of serotonin synthesis in human brain. *Proc. Natl. Acad. Sci. USA* 94, 5308–5313.

Pardridge W.M. (1986) Blood-brain barrier transport of nutrients. *Nutr. Rev.* 44 (Suppl.), 15–25.

Prell G.D., Hough L.B., Khandelwal J., and Green J.P. (1996) Lack of a precursor-product relationship between histamine and its metabolites in brain after histidine loading. *J. Neurochem.* 67, 1938–1944.

Rubin R.A., Ordonez L.A., and Wurtman R.J. (1974) Physiological dependence of brain methionine and S-adenosylmethionine concentrations on serum amino acid pattern. *J. Neurochem.* 23, 227–231.

Sainio E.L., Pulkki K., and Young S.N. (1996) L-Tryptophan: biochemical, nutritional and pharmacological aspects. *Amino Acids* 10, 21–47.

Spring B.J., Maller O., Wurtman J.J., Digman L., and Cozolino L. (1983) Effects of protein and carbohydrate meals on mood and performance: interactions with sex and age. *J. Psychiatr. Res.* 17, 155–167.

Teff K.L., Young S.N., and Blundell J.E. (1989a) The effect of protein or carbohydrate breakfasts on subsequent plasma amino acid levels, satiety and nutrient selection in normal males. *Pharmacol. Biochem. Behav.* 34, 829–837.

Teff K.L., Young S.N., Marchand L., and Botez M.I. (1989b) Acute effect of protein and carbohydrate breakfasts on human cerebrospinal fluid monoamine precursor and metabolite levels. *J. Neurochem.* 52, 235–241.

Thomson J., Rankin H., Ashcroft G.W., Yates C.M., McQueen J.K., and Cummings S.W. (1982) The treatment of depression in general practice: a comparison of L-tryptophan, amitriptyline, and a combination of L-tryptophan and amitriptyline with placebo. *Psychol. Med.* 12, 741–751.

van Praag H.M. (1990) Catecholamine precursor research in depression: the practical and scientific yield. In *Amino Acids in Psychiatric Disease* (Richardson M.A., Ed.), pp. 77–97. American Psychiatric Press, Washington.

Wurtman R.J., Hefti F., and Melamed E. (1981) Precursor control of neurotransmitter synthesis. *Pharmacol. Rev.* 32, 315–335.

Wurtman R.J., and Wurtman J.J. (1988) Do carbohydrates affect food intake via neurotransmitter activity? *Appetite* 11(Suppl.), 42–47.

Young S.N. (1986) The clinical psychopharmacology of tryptophan. In *Nutrition and the Brain, Vol. 7: Food Constituents Affecting Normal and Abnormal Behaviors* (Wurtman R.J., and Wurtman J.J., Eds.), pp. 49–88. Raven Press, New York.

Young S.N. (1991) Use of tryptophan in combination with other antidepressant treatments: a review. *J. Psychiatr. Neurosci.* 16, 241–246.

Young S.N. (1996) Behavioral effects of dietary neurotransmitter precursors: basic and clinical aspects. *Neurosci. Biobehav. Rev.* 20, 313–323.

Young S.N., and Leyton M. (2002) The role of serotonin in human mood and social interaction: insight from altered tryptophan levels. *Pharmacol. Biochem. Behav.* 71, 857–865.

10 Modulation of Feeding Behavior by Amino Acid-Deficient Diets: Present Findings and Future Directions

Thomas J. Koehnle and Dorothy W. Gietzen

CONTENTS

People have long been fascinated by accounts of the "wisdom of the body" in helping them to make adequate food choices. Although the folk psychological intuitions about current food choices being driven by past choices are largely incorrect, the literature on nutrient selection continues to grow. Here we briefly outline the history of the study of amino acid homeostasis. We provide an up-to-date review on the behaviors and neural circuits mediating responses to amino acid-deficient diets. Finally, we speculate about the mechanisms that might regulate protein metabolism in animals, particularly how behavioral rules might assist in this regulation.

10.1 MECHANISTIC ACCOUNTS OF NUTRIENT SELECTION

Since Richter (1936, 1956) published his classic papers on the regulation of salt intake in rats, public and academic audiences have been fascinated by stories about animals somehow "knowing" what nutrients they need and being able to obtain them in their diets. Although Richter's assertions about the unlearned preference for sodium chloride taste during periods of sodium deficiency proved correct, attempts to demonstrate unlearned preferences for other nutrients have failed. Most notoriously, three decades of research allegedly supporting unlearned preference for B-vitamin-laced foods during periods of B-vitamin deficiency were overturned when Rozin (1967) demonstrated that the innate preference for thiamin-laced foods was really a learned aversion to the deficient foods. Similarly, there are as many studies purporting to show that rats could obtain a balanced diet by self-selecting from among an array of foods as there are studies demonstrating no such ability [reviewed in Galef (1991, 2000)]. Thus, our folk-psychological intuitions about how animals compose their diets are often in error. However, understanding the mechanisms that enable animals to make food choices remains an intense area of research.

All omnivores and nonspecialist herbivores must choose a diet with sufficient levels of all the indispensable amino acids to ensure adequate protein for growth and tissue maintenance. With regard to human health, it is generally agreed that plant proteins are adequate for adult humans in the modern world when they are available in enough quantity to satisfy energy demands [reviewed in Millward (1999)]. When single plant protein sources are not available in sufficient quantity, it may be necessary to obtain complementary foodstuffs to avoid the suppression of food intake typically resulting from ingestion of amino acid-deficient diets (see later). Young infants, for example, might not receive adequate nutrition from plant proteins alone outside the developed world (Millward, 1999). Moreover, we have shown that rats fed amino acid-deficient diets have increased susceptibility to seizures (Gietzen et al., 1996). This finding is supported by a parallel study undertaken by Palencia et al. (1996), showing that seizure susceptibility was increased in rats fed a diet with corn as the sole protein source. They speculated that the increased incidence of epilepsy among developing nations might be related to inadequate protein nutrition. Thus, the need for adequate levels of all indispensable amino acids and the mechanisms that enable animals to make effective dietary choices are of great interest.

In food choice paradigms, rats have a limited ability to compose nutritionally sufficient meals from foods that are individually imbalanced with respect to amino acids (Fromentin and Nicolaidis, 1996). However, like the cafeteria paradigms discussed previously (Galef, 1991, 2000), this ability is imperfect and its relevance outside the laboratory context is unknown. We discuss the role that indispensable amino acids play in the control of food intake, with particular reference to recent studies of the behaviors and neurobiological mechanisms that support adaptive food choices of animals, both in the laboratory and in the wild.

Dietary protein and amino acid levels are known to influence feeding behavior in omnivores. Rats (*Rattus norvegicus*) have provided an ideal experimental model to examine the interactions between alterations in diet and changes in feeding behavior. With respect to protein, rats will increase their intake over their minimal requirements when given a choice between two diets differing in protein content, but they will not maintain their intake within a narrow range (e.g., Harper and Peters, 1989). Also, unlike the specific drive for sodium chloride, rats do not have a specific hunger, or unlearned preference, for protein (Galef, 2000), though they can be conditioned to respond to the postingestive consequences of particular nutrients in a given foodstuff (Booth, 1985).

There is no specific hunger for protein (Galef, 2000), and there is also no specific hunger for particular indispensable amino acids. Markison et al. (1999) showed that rats maintained on either lysine- or threonine-deficient diets do not promptly increase their intake of the appropriate amino acid when choosing from an array of water bottles spiked with different amino acids. Later work (Markison et al., 2000) showed that lysine-deficient rats did not make appropriate choices between lysine- and threonine-spiked water bottles until more than 30 min after presentation of the choice. Thus, although the response of amino acid-deficient rats to the opportunity to obtain the deficient amino acid is rapid, it is not specific to the taste of the missing amino acid, but requires some time for learning to occur. Similar results were obtained in studies of white-crowned sparrows. Murphy and King (1989) found that it took many days for sparrows given individually inadequate foods to learn to combine them in complementary ways. Follow-up research showed that some balancing could be accomplished within 2 h of exposure to the deficient foods (Murphy and Pearcy, 1993), but the efficacy of balancing by sparrows *decreased* with increasing severity of amino acid imbalance, the exact opposite of outcomes predicted by both optimal foraging theory and conventional models of nutrient selection.

10.2 AMINO ACID-DEFICIENT DIETS ALTER FEEDING BEHAVIOR

Harper (1976) defined an amino acid-imbalanced diet by its increased proportion of one or more amino acids relative to an initial low-protein diet, while the growth-limiting amino acid is held constant. In an imbalanced diet, the additional amino acids are not elevated to toxic concentrations. His laboratory and his students have also used amino acid-devoid diets, wherein one or more nonlimiting amino acids are increased and the limiting amino acid is completely absent from the diet (Harper et al., 1970; Harper, 1976).

In this chapter, the generic term *amino acid-deficient diet* refers to both the imbalanced and devoid diet or more recently developed modifications of these diets, as the behavioral effects of the various diets are essentially identical (Koehnle et al., 2003, 2004) and appear to be mediated by the same underlying neuronal circuitry. Where appropriate, the specific nature of each diet in a particular study is cited.

In the older literature, rats were brought into the laboratory and fed a low-protein basal diet for 5 to 21 days before testing on the devoid or imbalanced diet commenced (Harper et al., 1970; Harper, 1976). The basal diet has approximately 50% of the dietary requirement for all the amino acids except the amino acid of interest. In the threonine basal diet, for example, all the amino acids except threonine are held near 50% of their requirement while threonine is kept a bit lower. This low-protein pretreatment makes the animals more sensitive to the imbalanced and devoid diets, but is not necessary for recognition of amino acid deficiency (Koehnle, unpublished dissertation research; Leung et al., 1968b).

Although diets deficient with respect to an indispensable amino acid have been known to impact growth and wound healing since the aftermath of the Napoleonic wars (Magendie, 1816), the effects of amino acid-deficient diets on feeding behavior per se were not discovered until much later. Work completed in the early part of the 20th century, particularly by Osborne, Mendel, and Rose, demonstrated that rats decreased their food intake and growth when fed amino acid-deficient diets [reviewed in Gietzen et al. (1986a) and Gietzen (1993)].

The initially reported decreases in growth resulting from consumption of amino acid-deficient diets are not the cause of the food intake depression. If rats are force-fed an amino acid-imbalanced diet, their growth is normal (Leung et al., 1968a). In both the rat and the chick, reduction in food intake is the cause of the deficits in animal growth. These observations made exploration of the behavioral responses to deficiency a central feature in research on amino acid-deficient diets.

Rats quickly identify and respond to amino acid-deficient diets. Given a choice between the basal diet and the threonine-imbalanced diet, rats develop a clear preference for the basal diet within the first 24 h [reviewed in Gietzen et al. (1986a)]. Gietzen et al. (1986a) reported that ingestion of as little as 0.5 g of a threonine-imbalanced diet was sufficient to establish avoidance of the food cup containing the imbalanced diet. Bar pressing by rats to obtain food was significantly reduced when the diet was devoid of the indispensable amino acid threonine (Gietzen et al., 1986a). Most rats ceased bar pressing by 60 min, with one rat stopping its bar pressing in only 28 min. When fed a threonine-devoid diet after previous exposure to a corrected diet, rats ate the devoid diet as if it were corrected for the first 15 min and then radically decreased their intake over the 16- to 30-min interval (Figure 10.1; Gietzen et al., 1986a). Erecius et al. (1996) found that the delay between the first and second meals was increased by an isoleucine-imbalanced diet. There were few published reports documenting the complete array of behavioral changes in the 0- to 30-min timeframe after 1996, and work on the neural transduction mechanisms continued to focus on the 30- to 180-min timepoint.

The emphasis on later timepoints resulted primarily from the work of Beverly et al. (1990a), who did not show a rapid effect of threonine deficiency on the number of meals eaten, meal duration, or the delay between meals. His results did show an effect of threonine deficiency on meal size

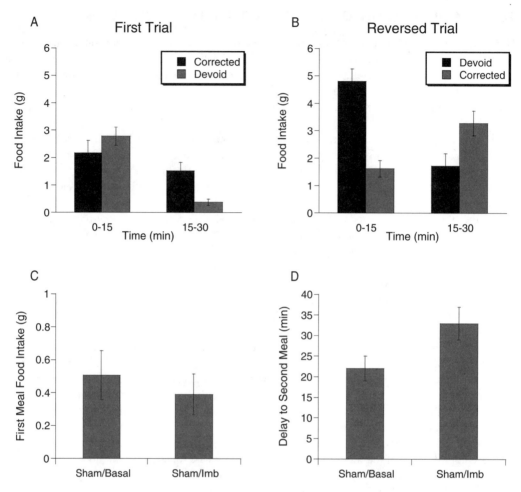

FIGURE 10.1 Early evidence for a rapid response to amino acid deficiency. (A) Rats were maintained for 1 week on a basal diet and on the test day given either a threonine-devoid diet (hatched bars) or a threonine-corrected diet (solid bars). Food intake declined for both groups between 15 and 30 min, but the suppression was higher for the devoid group. (B) The same rats shown in (A) were given a basal diet for another week and then given the opposite diet in a reversal trial. Rats previously fed the devoid diet (hatched bars) failed to recognize the corrected diet by smell or taste within the first 15 min. Rats previously fed the corrected diet also did not detect the devoid diet by smell or taste, but did decrease their intake between 15 and 30 min. (C) Sham-operated rats participating in a vagotomy study showed a slight decrease in food intake during their first meal when fed a threonine-imbalanced diet. (D) The same rats as in (C) demonstrated an increase in the delay between the first and second meal in response to the imbalance. (Data for (B) adapted from Gietzen, D.W. et al. (1986a) In *Interaction of the Chemical Senses with Nutrition* (Kare, M.G., and Brand, J.G., Eds.), pp. 415–463, Academic Press, Orlando, FL, and for (C) adapted from Erecius, L.F. et al. (1996) *J. Nutr.* 126: 1722–1731. With permission.)

(Beverly et al., 1990a), but because he aggregated data over multiple hours, any chance of visualizing rapid changes in feeding behavior was lost. In addition, the traditional focus on taste aversions resulting from amino acid deficiency (e.g., Simson and Booth, 1974) led most workers not only to aggregate data but also to look for effects at later timepoints. Feurte et al. (2002), for example, collected a great deal of data showing that the delay between meals was increased between 3 and 6 h after ingesting a threonine-devoid diet and that meal size and the rate of eating were decreased between 6 and 12 h later. But again, the data were aggregated into 3-h bins and little effort was made to examine individual meals between the 0- and 3-h timepoints.

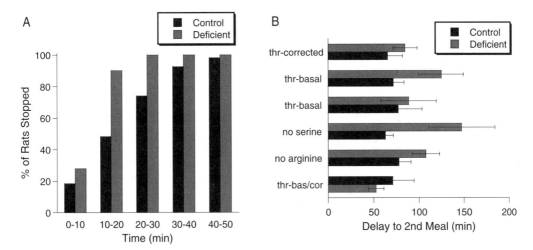

FIGURE 10.2 Summary of recent findings on behavioral responses to amino acid deficiency. (A) The most reliable behavioral finding we have documented is a robust decrease in first meal duration. The Y-axis shows the proportion of rats that has stopped eating at various timepoints on the X-axis. This figure includes data from seven separate experiments. About 85% of the rats in the deficient groups (hatched bars, $n = 50$) stop eating by 20 min, whereas only ~50% of rats in the control groups (solid bars, $n = 54$) have stopped at that time. (B) Consistent with the findings shown in Figure 10.1 (Panel D), rats typically increase the delay between the first and second meal in response to amino acid-deficient diets. Categories on the Y-axis denote the control diets used in each study. All test diets were deficient in threonine, except for the fifth (no arginine) test group, which was deficient in isoleucine. Note that control diets missing the dispensable amino acids serine or arginine did not produce increases in the delay between meals. (Data adapted from Koehnle, T.J. et al. (2003) *J. Nutr.* 133: 2331–2335, Koehnle, T.J. et al. (2004) *Physiol. Behav.* 8: 15–21, and an unpublished study by Koehnle and Gietzen. With permission.)

Why this was neglected, especially after repeated indications that something was happening as early as the first meal (Gietzen, 1986a; Erecius et al., 1996), is unclear. More recently, we have shown that rats are able to recognize diets deficient with respect to the essential amino acids threonine or isoleucine within 20 min of the first presentation, manifested by large decreases in first meal food intake and first meal duration (Figure 10.2; Koehnle et al., 2003, 2004). Because the deficient and complete diets cannot be discriminated on the basis of taste or smell alone [reviewed in Gietzen et al. (1986a) and Koehnle et al. (2003)], these observations suggested that the rapid behavioral recognition of amino acid-deficient diets requires detection by some type of interoceptor system.

In contrast, more information was historically available on the timing of behavioral responses to repletion of the dietary-limiting amino acid after long periods of prior feeding of a deficient diet. Rogers and Leung (1973) found that rats fed isoleucine-imbalanced or -devoid diets for 14 days and then offered a complete, corrected diet rapidly increased their food intake. This response took only 30 min in rats fed the devoid diet and within the first meal when the diet was imbalanced. Markison et al. (2000) showed that rats given a lysine-deficient diet for 4 days increased their intake of water spiked with lysine within 30 min.

Monda et al. (1997) fed rats a threonine-imbalanced diet and then injected threonine bilaterally into the anterior piriform cortex (see later). They found increased food intake in only 30 min in these rats; intake returned to preinjection levels after 1 h. Beverly et al. (1991) found that the half-life of free radiolabeled threonine injected into the APC was approximately 1 h, providing evidence that the timing of the food intake response corresponds to the concentration of free threonine. Beverly et al. (1990a, 1990b) did not observe differences in food intake until nearly 6 h after the threonine injection. However, careful reexamination of their data shows a small effect during the

0- to 6-h interval. It is unclear whether they missed a small, early effect on feeding behavior. In one subsequent study of animals fed a threonine-imbalanced diet and injected with threonine into the APC, we found that the threonine-injected animals had first meals of longer duration and greater food intake within the first meal than saline-injected animals (Russell et al., 2003). These results were consistent with and extended the findings of Monda et al. (1997).

10.3 PHYSIOLOGICAL AND NEUROANATOMICAL CORRELATES OF AMINO ACID DEFICIENCY

Our present working hypothesis holds that after ingestion of an amino acid-deficient diet, amino acid profiles in the plasma are rapidly altered. In our laboratory paradigm, digestion and absorption of amino acids are not likely to be rate limiting in this respect. Feeding humans stable-isotope-labeled amino acids leads to detectable levels of exhaled labeled CO_2 within 30 min (e.g., Mariotti et al., 2000). Because contemporary laboratory animal studies of the responses to deficiency typically use free amino acid mixtures in lieu of whole protein sources, digestion is not necessary. Moreover, infusion of nutrient mixtures into the stomachs of rats at rates similar to those found during normal feeding have shown that gastric emptying proceeds at a high rate throughout a meal and slows down only after the cessation of eating (e.g., Kaplan et al., 1997). These two factors likely facilitate a rapid uptake of amino acids in the duodenum. Therefore, the rapid alterations in behavior documented in response to laboratory diets should be taken as an upper boundary on the rate at which such processes might occur in the wild. Changes in plasma amino acid concentrations outside the laboratory context or using altered prefeeding paradigms could take more time.

For example, feeding a diet with a very high protein content before exposure of rats to an imbalanced diet delays the decrease in food intake (e.g., Leung et al., 1968b). However, rats accustomed to eating a more normal amino acid content show some signs of rapid recognition of an imbalanced diet. In one study, rats fed a threonine-corrected diet for 3 days were presented with a threonine-imbalanced diet. Their first meal durations were decreased relative to their mean intake over the 3 days of corrected diet intake (Koehnle, unpublished dissertation research). More work is needed to clarify to what extent animals must be maintained on marginal diets before investigating rapid changes in behavior in response to dietary amino acid deficiency.

Although amino acid concentrations are normally maintained within narrow limits in the blood and brain independent of protein intake, ingestion of an amino acid-imbalanced diet depletes the plasma of the limiting amino acid over the course of 30 min to several hours (e.g., Feurte et al., 1999). This decrease in the concentration of the limiting amino acid will eventually reduce the rate of protein synthesis in the body (Anthony et al., 2001), though the bolus of amino acids from the intake of the meal first briefly elevates the rate of protein synthesis (Yokogoshi et al., 1992). This initial increase in protein synthesis probably aggravates the amino acid deficiency. Given that free amino acids are not stored (Geiger, 1948), animals must either quickly obtain the deficient amino acid from dietary sources or endure negative nitrogen balance until the next meal (Munro, 1976). Elman (1939) found that amino acid deficiency had to be corrected within a few hours to avoid these consequences.

In addition to its consequences for whole-body protein synthesis, a decrease of the concentration of the limiting amino acid also generates a disadvantage for its transport into the central nervous system (CNS) at the blood-brain barrier (e.g., Tews et al., 1981). This leads to a reduction in the brain concentration of the limiting amino acid within 2 to 3 h (Peng et al., 1972). This decrease in the CNS permits animals to detect amino acid deficiency.

In both rats and cockerels, infusion of small amounts of the most limiting amino acid into the carotid artery prevents the suppression of food intake normally associated with ingestion of amino acid-imbalanced diets [reviewed in Gietzen (1993)]. Similar infusions into the jugular vein failed to modify food intake. There are afferents in the liver that respond to amino acids (Tanaka et al., 1990), but neither dennervation of the liver nor ablation of the olfactory bulbs modifies food intake responses to imbalanced diets [reviewed in Gietzen (1993)]. Although taste receptors on the tongue respond differently to different amino acids (Pritchard and Scott, 1982), diets deficient with respect to the dispensable amino acids arginine and serine do not produce the rapid anorexia typical of diets deficient in an indispensable amino acid (Koehnle et al., 2003). Lesions of the anterior piriform cortex (APC) and amygdala, however, result in increased intake of amino acid-deficient diets (review in Gietzen, 1993).

The amygdala is traditionally associated with the development of learned aversions to imbalanced diets. However, the APC responds rapidly by electrophysiological and biochemical changes to an imbalance. The threonine-devoid diet alters the structure of average evoked potentials recorded in the APC of rats within 3 h and shows some small changes while the rat is still eating (Hasan et al., 1998). The level of phosphorylated mitogen-activated protein (MAP) kinase, a marker for neuronal activation, is increased in the APC just 40 min after ingestion of an imbalanced diet (Sharp et al., 2002). Expression of the immediate early gene c-*fos*, another marker for neuronal activation, is also elevated in the APC following ingestion of an amino acid-imbalanced diet (Wang et al., 1996).

Injecting small doses of the dietary-limiting amino acid into the APC results in increased food intake and delayed recognition of an imbalanced diet (Russell et al., 2003; Beverly et al., 1990a, 1990b; Monda et al., 1997), whereas injecting the limiting amino acid into the amygdala does not impact recognition of the imbalanced diet (Beverly et al., 1990a). Injecting a nonlimiting amino acid such as isoleucine into the APC of rats fed a diet imbalanced with respect to threonine does not prevent recognition of the imbalance either (Beverly et al., 1990b).

Gietzen et al. (1986b, 1998) reported that diets imbalanced with respect to threonine significantly reduce threonine concentration in the APC a few hours after first exposure to the diet. The dietary-limiting amino acid was not decreased in the paraventricular nucleus, lateral hypothalamic nucleus, or ventromedial hypothalamus in response to amino acid-imbalanced diets (Gietzen et al., 1989). However, the threonine-imbalanced diet decreased threonine concentration in the anterior cingulate cortex, dorsomedial hypothalamus, and parabrachial nucleus within 2.5 h of ingesting the test diets (Gietzen et al., 1998). In the same studies, only the APC and dorsomedial hypothalamus showed parallel decreases in isoleucine concentration in response to an isoleucine-imbalanced diet. More recently, we have shown that the level of the limiting amino acid in the APC is decreased within 21 min of the first exposure to a deficient diet (Figure 10.3; Koehnle et al., in press).

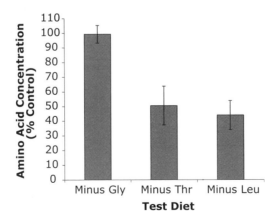

FIGURE 10.3 Diets lacking indispensable amino acids rapidly deplete the most limiting amino acid in the anterior piriform cortex (APC). Basal test diets missing the indispensable amino acids threonine (minus Thr) or leucine (minus Leu) are shown to cause a ca. 50% depletion of the respective amino acid relative to control basal diets 21 min after rats first began to eat them. A diet lacking the dispensable amino acid glycine (minus Gly), however, did not result in a decrease in glycine concentration. (Data adapted from Koehnle, T.J. et al. (2004). *J. Nutr.* 134:2365–2371. With permission.)

Taken together, these data indicate that dietary amino acid deficiency is detected in the APC of rats and that reciprocal changes in the concentration of the limiting amino acid produce reciprocal changes in behavior within a few hours of exposure to the test diet; that is, increases in the concentration of the limiting amino acid increase food intake, whereas decreases in the concentration of the limiting amino acid decrease food intake.

10.4 PUTATIVE MECHANISM FOR RECOGNITION OF DEFICIENCY

Decreases of the dietary-limiting amino acid in the APC could alter neuronal membrane potential or firing patterns by several alternative routes. First, there might be ion channels or metabotropic receptors that are directly gated by indispensable amino acids. At least one calcium channel responds stereospecifically to extracellular L-amino acid concentrations (Conigrave et al., 2000), but not in a manner specific to the indispensables. This possibility thus remains highly speculative.

Second, decreased transport of the limiting amino acid into neuronal cell bodies in the APC could result in a net increase of other nonlimiting amino acids. Blais et al. (2003) exposed primary neuronal cultures from rat APC to media with and without threonine. They documented a 70% depletion of threonine inside the neurons after just 15 min of incubation in the deficient medium. At the same time, they showed small increases in the concentration of serine, glycine, valine, isoleucine, and phenylalanine as a result of increased System A transport activity. The increases are likely the result of decreased competition for access to the transporter, as the System A transporter can facilitate uptake of numerous amino acids (Albers et al., 2001). Curiously, only neurons from the APC and not cerebellar or hippocampal neurons or glia showed increased transport of System A substrates in response to threonine deficiency (Blais et al., 2003). These increases could alter amino acid metabolism in the APC, but recent experiments *in vivo* have not demonstrated that the increases of other amino acids documented *in vitro* are seen in the functional APC (Koehnle et al., in press).

Lastly, it is possible that depletion of the limiting amino acid itself is the primary signal of deficiency. Much recent work in our laboratory and among our collaborators has begun to point in this direction. In this scenario, depletion of an indispensable amino acid from the intracellular pool would lead to a decrease in the number of aminoacylated transfer RNA molecules. This uncharged tRNA is useless for protein synthesis and in yeast leads to increased phosphorylation of the eukaryotic initiation factor 2α (eIF2α; Dong et al., 2000). We have recently shown that within 20 min of ingesting a diet devoid of threonine, eIF2α phosphorylation is increased in the APC of rats (Gietzen et al., 2004).

Together, these data support a mechanism for detection of dietary amino acid deficiency involving the generation of uncharged tRNA as its first step (Figure 10.4). How a change in the level of charged tRNA might alter membrane potential or neuronal firing is not clear at this point, but it might do so by a mechanism analogous to the stringent response in bacteria [reviewed in Chaterji and Ojha (2001)] or by modulating glucose metabolism [reviewed in Rabinovitz (1995)]. Neuronal output from the APC would then impinge on other feeding circuits, outlined next.

10.5 CIRCUITRY MEDIATING BEHAVIORAL RESPONSES TO DEFICIENCY

The anterior piriform cortex is a division of the primary olfactory cortex located in the ventral forebrain [reviewed in Haberly (1990)]. The APC receives most of its input from mitral cells in the olfactory bulb, which synapse with neurons in the APC in layer 1 (Haberly, 1990). It also receives cholinergic inputs from the basal forebrain, noradrenergic, serotonergic, and dopaminergic inputs from the brainstem, and histaminergic inputs from the hypothalamus. The APC sends its output through connections to the insular cortex, amygdaloid complex, thalamus and hypothalamus,

A

B

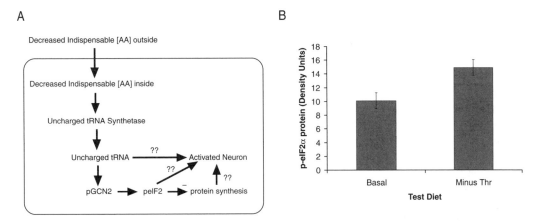

FIGURE 10.4 Putative mechanism for sensing of indispensable amino acid deficiency. (A) Schematic diagram illustrating the pathways activated by amino acid deficiency. The presence of uncharged transfer RNA leads to the phosphorylation of the enzyme GCN2 (pGCN2), which in turn phosphorylates eIF2α (peIF2), making it unavailable for protein synthesis. How this pathway leads to neuronal activation is unclear at present. (B) Rats fed a diet devoid of threonine (minus Thr) show more phosphorylation of eIF2α, a marker for the presence of uncharged tRNA, than rats fed a control diet (basal). Data are adapted from Western blots (Gietzen et al., 2004) taken from the APC 20 min after presentation of each diet.

limbic system, hippocampus, and striatum (Haberly, 1990). The pattern of inputs and outputs from the APC more closely resembles that of the traditional association cortex rather than the architecture of primary sensory cortices (Johnson et al., 2000).

The insular cortex and the adjacent gustatory cortex are well known for their involvement in the development of responses to taste and aversions to taste and odor. The amygdaloid complex, in particular the central nucleus, also plays a role in establishing conditioned taste aversions to amino acid-deficient diets (e.g., Aja et al., 2000). Among its many functions the hippocampus is well known for its involvement in learning and memory, and the connections between the APC and the hippocampus might facilitate memory of cues associated with amino acid-deficient diets, such as location (Fromentin et al., 1998). The APC also has strong connections with the lateral hypothalamus (Monda et al., 1997; Aja, unpublished dissertation research), an area commonly thought to be important in controlling food intake.

Monda et al. (1997) found that injecting threonine into the APC increased intake of a threonine-imbalanced diet within 30 min. They also recorded neuronal output from the lateral hypothalamus at the same time and found increased neuronal firing rates in association with threonine injection by 30 min, which tapered off to baseline levels after 60 min. Cell bodies in the lateral hypothalamus are therefore responsive to the level of the dietary-limiting amino acid in the APC. The simplest hypothesis would then invoke these neurons as the controllers of food intake in response to deficiency. However, electrolytic lesions involving the entire lateral hypothalamus have no impact on the recognition of amino acid-imbalanced diets (Scharrer et al., 1970). More selective lesions of only the excitatory glutamatergic cell bodies in the lateral hypothalamus by ibotenic acid do not disrupt the early recognition of amino acid deficiency (Bellinger et al., 2003), indicating that this brain nucleus is not involved in the recognition of deficiency per se.

Gietzen et al. (1998) proposed that the behavioral changes seen in response to amino acid-deficient diets are mediated by the outputs from the APC to the lateral hypothalamus, while other pathways act through the central nucleus of the amygdala to generate a conditioned taste aversion. The connections between the APC and lateral hypothalamus have been outlined previously. Lesions of the APC disrupt recognition of amino acid-deficient diets [reviewed in Gietzen (1993)] but not the development of conditioned taste aversions. Contrariwise, lesions of the amygdala disrupt the formation of conditioned taste aversions but not the recognition of the amino acid deficiency (Gietzen et al., 1998). All this is in keeping with the known functions of the APC and amygdala, but a role for the lateral hypothalamus as the amino acid sensor is now suspect. Clearly, it has functional connections to the APC (Monda et al., 1997), but the failure of lesions of the lateral hypothalamus to impact the rapid anorexia that typically results from ingestion of amino acid-deficient diets (Bellinger et al., 2003; Scharrer et al., 1970) makes it less likely as a candidate involved in the behavioral changes documented to date.

One possibility is that the lateral hypothalamus is involved in the formation of associations between the deficient state and cues associated with the deficient diet. Ono and coworkers found that neurons in the lateral hypothalamus responded to cues associated with repletion of the limiting amino acid (Tabuchi et al., 1991; Yokawa et al., 1995). Moreover, Torii et al. (1996) found increased activity in the lateral hypothalamus by functional magnetic resonance imaging in response to peripheral infusion of the limiting amino acid. Taken together, these studies (Monda et al., 1997; Torii et al., 1996; Tabuchi et al., 1991; Yokawa et al., 1995) provide some evidence that the lateral hypothalamus is involved in detection of amino acid repletion after periods of prior depletion. Indeed, Sharp et al. (2002) looked for phosphorylated MAP kinase in the lateral hypothalamus 40 min after feeding amino acid-deficient diets and found low levels of this marker of neuronal depolarization, reinforcing the idea that the lateral hypothalamus may not be involved in detection of amino acid deficiency.

By contrast, the paraventricular and dorsomedial hypothalamic nuclei showed very high levels of phosphorylation of MAP kinase, making these areas better candidates in the rapid responses to amino acid deficiency. Another possibility is that outputs from the APC to the reticular thalamus are activated by amino acid deficiency, and the reticular thalamus in turn inhibits feeding by suppressing activity in the lateral hypothalamus (Aja, unpublished dissertation research). A schematic figure representing our current thinking is shown in Figure 10.5.

10.6 AMINO ACIDS AND FOOD CHOICE

Because nutrients are not evenly distributed among food sources, omnivores and generalist herbivores must compose diets from multiple foods as they forage. Although it is tempting to conceive of food intake at one timepoint as being dependent on food choices made at a prior timepoint, studies in free-feeding humans have shown that such simple mechanisms cannot account for more than ~5% of the variance (e.g., de Castro, 1998). Alternatively, animals might be able to avoid the need for tracking nutrient levels on a moment-to-moment (i.e., meal-to-meal or day-to-day) basis by using a simple rule to guide their behavior (Gigerenzer and Selten, 2001), anticipating, rather than reacting to, deficiencies (Collier, 1986). Murphy and Pearcy (1993) tested the ability of white-crowned sparrows, *Zonotrichia leuchophrys gambelii*, to select diets from two food sources with individually incomplete amino acid profiles, but complete if eaten in combination. The sparrows had little difficulty in adjusting to this task, though their selection was less than perfect.

Murphy and Pearcy's work on sparrows provides an ideal link between a heavily studied neural system and the ecological consequences of that system's operation. In addition to birds (e.g., Murphy and King, 1989) and rats (e.g., Gietzen, 2000), the ability to detect and respond to amino acid deficiency has been demonstrated in taxa as diverse as opossums (DeGabriel et al., 2002) and molluscs (Delaney and Gelperin, 1986). That so many species with such phylogenetic diversity possess the ability to detect amino acid deficiency is not evidence that they regularly do so in their daily lives, but it indicates that this ability is evolutionarily conserved or correlated to another trait that is conserved (Gould and Lewontin, 1979). The ability to track changes in dietary amino acid content, however, is not evidence that animals can do so by using optimal food choices. Indeed, as has been found in rats (Fromentin and Nicolaidis, 1996) and birds (Murphy and Pearcy, 1993), foods with complementary amino acid profiles are not selected at the optimal level.

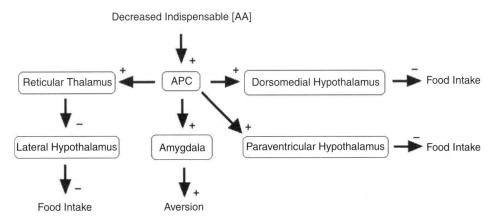

FIGURE 10.5 Schematic diagram of brain circuits underlying behavioral responses to amino acid deficiency. The APC is activated by dietary deficiency. In turn, its projections activate the amygdala, reticular thalamus, and dorsomedial and paraventricular hypothalamus. The connections to the amygdala mediate the development of a conditioned aversion to the taste and location of the diet, whereas modulation of the hypothalamus serves to decrease food intake.

10.7 SIMPLE HEURISTICS IN THE RODENT DIET?

It is well known that rats reject diets deficient with respect to essential amino acids. Detection of the deficiency takes ca. 30 min and is followed by increases in foraging behavior, as assessed by increased locomotor behavior and increased food spillage (Gietzen, 2000). Rats select diets containing the missing amino acid or, in the absence of a choice, dramatically reduce their food intake (Gietzen et al., 1986a).

Threonine-imbalanced diets are perceived as highly aversive by rats, leading to negative hedonic expression and disruption of the normal satiety sequence that follows a balanced meal (Feurte et al., 2000). Maintenance of rats on a lysine-deficient diet for as little as 4 days results in increased anxiety in several behavioral paradigms (Smriga et al., 2002). The aversion that develops is not only to the food source, but also to the taste and location of the food in the home cage. The aversion developed in response to the amino acid deficiency is so strong that rats select a lethal protein-free diet over an amino acid imbalanced diet (Leung et al., 1968b). Rats also increase their ingestion of nonnutritive substances, such as cage bedding, kaolin, and cellulose, in response to amino acid deficiency (J.W. Sharp and D.W. Gietzen, unpublished observations; Barrett, unpublished master's thesis).

Force feeding an amino acid-imbalanced diet allows normal growth (Leung et al., 1968a). Because rats can adapt to some, but not all, amino acid deficiencies (e.g., amino acid-devoid diets) simply by increasing their intake of the deficient food, it makes sense for there to be adaptive rules guiding them to seek out other food sources. Taken together, these data indicate that the rats might have a simple rule they employ in response to amino acid-deficient diets: to stop eating it and seek out something — *anything* — different to eat. Whether analogous rules regarding amino acid nutrition guide food choices in humans is unknown.

In general, the rate of protein turnover in adult humans strongly resists excursions due to alterations in the level of dietary protein intake [reviewed in Millward (1999)]. This apparent lack of responsiveness is a strong indicator that protein synthesis and degradation are tightly regulated variables (Booth, 1987). Given the potential importance of amino acid deficiency to human health (Gietzen et al., 1996; Palencia et al., 1996) and the fact that ~95% of the variance in human food choices cannot be explained by prior food choices (de Castro, 1998), it might be profitable to look for rules that guide human food choices based on protein content, anticipating shortages rather than reacting to them. Stuart, for example, found that the Nahuat Indians of Central America use linguistic rules to obtain sufficient protein intake (Stuart, 1978). We believe that, on the simple heuristics approach (Gigerenzer and Selten, 2001), it may be possible to explain the control of food intake without reference to explicit regulatory mechanisms.

REFERENCES

Aja, S., Sisouvong, S., Barrett, J.A., and Gietzen, D.W. (2000) Basolateral and central amygdaloid lesions leave aversion to dietary amino acid imbalance intact. *Physiol. Behav.* 71: 533–541.

Albers, A., Broer, A., Wagner, C.A., Setiawan, I., Lang, P.A., Kranz, E.U., Lang, F., and Broer, S. (2001) Na⁺ transport by the neural glutamine transporter ATA1. *Pflugers. Arch.* 443: 921–101.

Anthony, T.G., Reiter, A.K., Anthony, J.C., Kimball, S.R., and Jefferson, L.S. (2001) Deficiency of dietary EAA preferentially inhibits mRNA translation of ribosomal proteins in liver of meal-fed rats. *Am. J. Physiol. Endocrinol. Metab.* 281: E430–E439.

Bellinger, L.L., McIntosh, J.E., and Gietzen, D.W. (2003) Hyperphagia and attenuation of the anorexia induced by an imbalanced amino acid diet but no change in body weight following ibotenic acid lesions of the lateral hypothalamus in rats. In *Annual Meeting of the Society of Neuroscience,* New Orleans, LA, November 8–14, 2003.

Beverly, J.L., Gietzen, D.W., and Rogers, Q.R. (1990a) Effect of dietary limiting amino acid in prepyriform cortex on meal patterns. *Am. J. Physiol.* 259: R716–R723.

Beverly, J.L., Gietzen, D.W., and Rogers, Q.R. (1990b) Effect of dietary limiting amino acid in prepyriform cortex on food intake. *Am. J. Physiol.* 259: R709–R715.

Beverly, J.L., Hrupka, B.J, Gietzen, D.W., and Rogers, Q.R. (1991) Distribution of dietary limiting amino acid injected into the prepyriform cortex. *Am J. Physiol.* 260: R525–R532.

Blais, A., Huneau, J.F., Magrum, L.J., Koehnle, T.J., Sharp, J.W., Tomé, D., and Gietzen, D.W. (2003) Threonine deprivation rapidly activates the system A amino acid transporter in primary cultures of rat neurons from the essential amino acid sensor in the anterior piriform cortex. *J. Nutr.* 133: 2156–2164.

Booth, D.A. (1985) Food-conditioned eating preferences and aversions with interoceptive elements: conditioned appetites and satieties. *Ann. N.Y. Acad. Sci.* 443: 22–37.

Booth, D.A. (1987) Central dietary "feedback onto nutrient selection": not even a scientific hypothesis. *Appetite* 8: 195–201.

Chatterji, D., and Ojha, A.K. (2001) Revisiting the stringent response, ppGpp and starvation signaling. *Curr. Opin. Microbiol.* 4: 160–165.

Collier, G. (1986) The dialogue between the house economist and the resident physiologist. *Nutr. Behav.* 3: 9–26.

Conigrave, A.D., Quinn, S.J., and Brown, E.M. (2000) L-amino acid sensing by the extracellular Ca^{2+}-sensing receptor. *Proc. Natl. Acad. Sci. USA* 97: 4814–4819.

de Castro, J.M. (1998) Prior day's intake has macronutrient specific delayed negative feedback effects on the spontaneous food intake of free-living humans. *J. Nutr.* 128: 61–67.

DeGabriel, J., Foley, W.J., and Wallis, I.R. (2002) The effect of excesses and deficiencies in amino acids on the feeding behaviour of the common brushtail possum (*Trichosurus vulpecula*). *J. Comp. Physiol. B* 172: 607–617.

Delaney, K., and Gelperin, A. (1986) Post-ingestive food-aversion learning to amino acid deficient diets by the terrestrial slug *Limax maximus. J. Comp. Physiol.* 159: 281–295.

Dong, J., Qiu, H., Garcia-Barrio, M., Anderson, J., and Hinnebusch, A.G. (2000) Uncharged tRNA activates GCN2 by displacing the protein kinase moiety from a bipartite tRNA-binding domain. *Mol. Cell* 6: 269–279.

Elman, R. (1939) Time factor in retention of nitrogen after intravenous injection of a mixture of amino acids. *Proc. Soc. Exp. Biol. Med.* 40: 484–487.

Erecius, L.F., Dixon, K.D., Jiang, J.C., and Gietzen, D.W. (1996) Meal patterns reveal differential effects of vagotomy and tropisetron on responses to indispensable amino acid deficiency. *J. Nutr.* 126: 1722–1731.

Feurte, S., Nicolaidis, S., and Berridge, K.C. (2000) Conditioned taste aversion in rats for a threonine-deficient diet: demonstration by the taste reactivity test. *Physiol. Behav.* 68: 423–429.

Feurte, S., Nicolaidis, S., Even, P.C., Tome, D., Mahe, S., and Fromentin, G. (1999) Rapid fall in plasma threonine followed by increased intermeal interval in response to first ingestion of a threonine-devoid diet in rats. *Appetite* 33: 329–341.

Feurte, S., Tome, D., Gietzen, D.W., Even, P.C., Nicolaidis, S., and Fromentin, G. (2002) Feeding patterns and meal microstructure during development of a taste aversion to a threonine devoid diet. *Nutr. Neurosci.* 5: 269–278.

Fromentin, G., Feurte, S., and Nicolaidis, S. (1998) Spatial cues are relevant for learned preference/aversion shifts due to amino-acid deficiencies. *Appetite* 30: 224–234.

Fromentin, G., and Nicolaidis, S. (1996) Rebalancing essential amino acids intake by self-selection in the rat. *Br. J. Nutr.* 75: 669–682.

Galef, B.J. (1991) A contrarian view of the wisdom of the body as it relates to dietary self-selection. *Psychol. Rev.* 98: 218–223.

Galef, B. (2000) Is there a specific appetite for protein. In *Neural and Metabolic Control of Macronutrient Intake* (Berthoud, H., and Seeley, R.J., Eds.), pp. 19–28. CRC Press, Boca Raton, FL.

Geiger, E. (1948) The role of the time factor in feeding supplementary proteins. *J. Nutr.* 36: 813–819.

Gietzen, D.W. (1993) Neural mechanisms in the responses to amino acid deficiency. *J. Nutr.* 123: 610–625.

Gietzen, D.W., Dixon, K.D., Truong, B.G., Jones, A.C., Barrett, J.A., and Washburn, D.S. (1996) Indispensable amino acid deficiency and increased seizure susceptibility in rats. *Am. J. Physiol.* 271: R18–R24.

Gietzen, D.W., Erecius, L.F., and Rogers, Q.R. (1998) Neurochemical changes after imbalanced diets suggest a brain circuit mediating anorectic responses to amino acid deficiency in rats. *J. Nutr.* 128: 771–781.

Gietzen, D.W., Leung, P.M.B., Castonguay, T.W., Hartman, W.J., and Rogers, Q.R. (1986a) Time course of food intake and plasma and brain amino acid concentrations in rats fed amino acid-imbalanced or -deficient diets. In *Interaction of the Chemical Senses with Nutrition* (Kare, M.G., and Brand, J.G., Eds.), pp. 415–463. Academic Press, Orlando, FL.

Gietzen, D.W., Leung, P.M.B., and Rogers, Q.R. (1986b) Norepinephrine and amino acids in prepyriform cortex of rats fed imbalanced amino acid diets. *Physiol. Behav.* 36: 1071–1080.

Gietzen, D.W., Leung, P.M., and Rogers, Q.R. (1989) Dietary amino acid imbalance and neurochemical changes in three hypothalamic areas. *Physiol. Behav.* 46: 503–511.

Gietzen, D.W., Ross, C.M., Hao, S., and Sharp, J.W. (2004) Phosphorylation of eIF2alpha is involved in the signaling of indispensable amino acid deficiency in the anterior piriform cortex of the brain in rats. *J. Nutr.* 134: 717–723.

Gigerenzer, G., and Selten, R. (2001) *Bounded Rationality.* MIT Press, Cambridge, MA.

Gould, S.J., and Lewontin, R.C. (1979) The spandrels of San Marco and the Panglossian paradigm: a critique of the adaptationist programme. *Proc. R. Soc. Lond.* 205: 281–288

Haberly, L.B. (1990) Olfactory cortex. In *The Synaptic Organization of the Brain* (Shepard, G.M., Ed.), pp. 317–345. Oxford University Press, New York.

Hao, S., Ross, C.M, Rudell, J.B., and Gietzen, D.W. (2003) The role of unchanged tRNA in the earliest mechanisms of amino acid (AA) recognition in the mammalian anterior piriform cortex (APC). *Soc. Neurosci.* Abst. #929.6.

Harper, A.E. (1976) Protein and amino acids in the regulation of food intake. In *Hunger: Basic Mechanisms and Clinical Implications* (Novin, D., Wyrwicka, W., and Bray, G., Eds.), pp. 103–113. Raven Press, New York.

Harper, A.E., Benevenga, N.J., and Wohlhueter, R.M. (1970) Effects of ingestion of disproportionate amounts of amino acids. *Physiol. Rev.* 50, 428–558.

Harper, A.E., and Peters, J.C. (1989) Protein intake, brain amino acid and serotonin concentrations, and protein self-selection. *J. Nutr.* 119: 677–689.

Hasan, Z.A., Woolley, D.E., and Gietzen, D.W. (1998) Responses to indispensable amino acid deficiency and replenishment recorded in the anterior piriform cortex of the behaving rat. *Nutr. Neurosci.* 1: 373–381.

Johnson, D.M., Illig, K.R., Behan, M., and Haberly, L.B. (2000) New features of connectivity in piriform cortex visualized by intracellular injection of pyramidal cells suggest that "primary" olfactory cortex functions like "association" cortex in other sensory systems. *J. Neurosci.* 20: 6974–6982.

Kaplan, J.M., Siemers, W., and Grill, H.J. (1997) Effect of oral versus gastric delivery on gastric emptying of corn oil emulsions. *Am. J. Physiol.* 273: R1263–R1270.

Koehnle, T.J., Russell, M.C., and Gietzen, D.W. (2003) Rats rapidly reject diets deficient in essential amino acids. *J. Nutr.* 133: 2331–2335.

Koehnle, T.J., Russell, M.C., Morin, A.S., Erecius, L.F., and Gietzen, D.W. (2004) Diets deficient in indispensable amino acids rapidly decrease the concentration of the limiting amino acid in the anterior piriform cortex of rats. *J. Nutr.* 134: 2365–2371.

Koehnle, T.J., Stephens, A.L., and Gietzen, D.W. (2004) Threonine imbalanced diet alters first meal microstructure in rats. *Physiol. Behav.* 8: 15–21.

Leung, P.M.B., Rogers, Q.R., and Harper, A.E. (1968a) Effect of amino acid imbalance in rats fed ad libitum, interval fed, or force-fed. *J. Nutr.* 95: 474–482.

Leung, P.M.B., Rogers, Q.R., and Harper, A.E. (1968b) Effect of amino acid imbalance on dietary choice in the rat. *J. Nutr.* 95: 483–492.

Magendie, M.F. (1816) Sur les proprietes nutritives des substances qui ne contiennent pas d'azote. *Anal. Chim. Physiq.* 3: 66–77.

Mariotti, F., Mahe, S., Luengo, C., Benamouzig, R., and Tome, D. (2000) Postprandial modulation of dietary and whole-body nitrogen utilization by carbohydrates in humans. *Am. J. Clin. Nutr.* 72: 954–962.

Markison, S., Gietzen, D.W., and Spector, A.C. (1999) Essential amino acid deficiency enhances long-term intake but not short-term licking of the required nutrient. *J. Nutr.* 129: 1604–1612.

Markison, S., Thompson, B.L., Smith, J.C., and Spector, A.C. (2000) Time course and pattern of compensatory ingestive behavioral adjustments to lysine deficiency in rats. *J. Nutr.* 130: 1320–1328.

Millward, D.J. (1999) The nutritional value of plant-based diets in relation to human amino acid and protein requirements. *Proc. Nutr. Soc.* 58: 249–260.

Monda, M., Sullo, A., De Luca, V., Pellicano, P., and Viggiano, A. (1997) L-Threonine injection into PPC modifies food intake, lateral hypothalamic activity, and sympathetic discharge. *Am. J. Physiol.* 273: R554–R559.

Munro, H.N. (1976) Eukaryote protein synthesis and its control. In *Protein Metabolism and Nutrition* (Cole, D.J., Boorman, K.N., Buttery, P.J., Lewis, D., Neale, R.J., and Swan, H., Eds.), pp. 3–18. Butterworths, London.

Murphy, M.E., and King, J.R. (1989) Sparrows discriminate between diets differing in valine or lysine concentrations. *Physiol. Behav.* 45: 423–430.

Murphy, M.E., and Pearcy, S.D. (1993) Dietary amino acid complementation as a foraging strategy for wild birds. *Physiol Behav.* 53: 689–698.

Palencia, G., Calvillo, M., and Sotelo, J. (1996) Chronic malnutrition caused by a corn-based diet lowers the threshold for pentylenetetrazol-induced seizures in rats. *Epilepsia* 37: 583–586.

Peng, Y., Tews, J.K., and Harper, A.E. (1972) Amino acid imbalance, protein intake, and changes in rat brain amino acids. *Am. J. Physiol.* 222: 314–321.

Pritchard, T.C., and Scott, T.R. (1982) Amino acids as taste stimuli. I. Neural and behavioral attributes. *Brain Res.* 253: 81–92.

Rabinovitz, M. (1995) The phosphofructokinase-uncharged tRNA interaction in metabolic and cell cycle control: an interpretive review. *Nucleic Acids Symp. Ser.* 33: 182–189.

Richter, C.P. (1936) Increased salt appetite in adrenalectomized rats. *Am. J. Physiol.* 113: 155–161.

Richter, C.P. (1956) Salt appetite of mammals: its dependence on instinct and metabolism. In *L'Instict dans le Comportement des Animaux et de l'Homme* (Autuori, M. and Grassé, P.P., Eds.), pp. 577-632. Masson et Cie, Paris.

Rogers, Q.R., and Leung, P.M.B. (1973) The influence of amino acids on the neuroregulation of food intake. *Fed. Proc.* 32: 1709–1719.

Rozin, P. (1967) Specific aversions as a component of specific hungers. *J. Comp. Physiol. Psychol.* 64: 237–242.

Russell, M.C., Koehnle, T.J., Barrett, J.A., Blevins, J.E., and Gietzen, D.W. (2003) The rapid anorectic response to a threonine imbalanced diet is decreased by injection of threonine into the anterior piriform cortex of rats. *Nutr. Neurosci.* 6: 247–251.

Scharrer, E., Baile, C.A., and Mayer, J. (1970) Effect of amino acids and protein on food intake of hyperphagic and recovered aphagic rats. *Am. J. Physiol.* 218: 400–404.

Sharp, J. Magrum, L., and Gietzen, D. (2002) Role of MAP kinase in signaling indispensable amino acid deficiency in the brain. *Mol. Brain Res.* 105: 11–18.

Simson, P.C., and Booth, D.A. (1973) Effect of CS-US interval on the conditioning of odour preferences by amino acid loads. *Physiol. Behav.* 11: 801–808.

Smriga, M., Kameishi, M., Uneyama, H., and Torii, K. (2002) Dietary L-lysine deficiency increases stress-induced anxiety and fecal excretion in rats. *J. Nutr.* 132: 3744–3746.

Stuart, J.W. (1978) Subsistence ecology of the Isthmus Nahuat Indians of Southern Veracruz, Mexico. Dissertation, University of California, Riverside, CA.

Tabuchi, E., Ono, T., Nishijo, H., and Torii, K. (1991) Amino acid and NaCl appetite, and LHA neuron responses of lysine-deficient rat. *Physiol. Behav.* 49: 951–964.

Tews, J.K., Bradford, M.A., and Harper, A.E. (1981) Induction of lysine imbalance in rats: relation to competition for lysine transport into brain *in vitro. J. Nutr.* 111: 954–967.

Torii, K., Yokawa, T., Tabushi, E., Hawkins, R.L., Mori, M., Knodoh, T., and Ono, T. (1996) Recognition of deficiens nutrient intake in the brain of rat with L-lysine deficiency monitored by functional magnetic resonance imaging, electrophysiologically and behaviorally. *Amino Acids* 10: 73–81.

Wang, Y., Cummings, S.L., and Gietzen, D.W. (1996) Temporal-spatial pattern of c-fos expression in the rat brain in response to indispensable amino acid deficiency I. The initial recognition phase. *Mol. Brain Res.* 40: 27–34.

Yokawa, T., Tabuchi, E., Takezawa, M., Ono, T., and Torii, K. (1995) Recognition and neural plasticity responding to deficient nutrient intake scanned by a functional MRI in the brain of rats with L-lysine deficiency. *Obes. Res.* 3: 685S–688S.

Yokogoshi, H., Hayase, K., and Yoshida, A. (1992) The quality and quantity of dietary protein affect brain protein synthesis in rats. *J. Nutr.* 122: 2210–2217.

11 Role of Dietary Polyunsaturated Fatty Acids in Brain and Cognitive Function: Perspective of a Developmental Psychobiologist

Patricia E. Wainwright and Danica Martin

CONTENTS

11.1 INTRODUCTION

Lipids comprise 50 to 60% of the dry weight of the adult brain, of which approximately 35% are in the form of long-chain polyunsaturated fatty acids (LCPUFAs), mainly arachidonic acid (AA, 20:4 n-6) and docosahexaenoic acid (DHA, 22:6 n-3) (Sastry, 1985). LCPUFAs are derived through biosynthesis from their respective dietary essential fatty acid (EFA) precursors linoleic acid (LA, 18:2 n-6) and α-linolenic acid (ALA, 18:3 n-3) or they can be obtained directly from dietary sources. AA is found at relatively high levels in many tissues, whereas DHA is found at high levels in only a few tissues outside the central nervous system (CNS), such as the testes, but it represents a high proportion of the lipids of the retina and gray matter of the brain (Tinoco, 1982). AA and DHA accrue rapidly in the human brain during the third trimester and the early postnatal period, when the rate of brain growth is maximal (Clandinin et al., 1980a, 1980b). Based on studies that show that n-3 fatty acid (FA) deficiency is associated with changes in some aspects of visual and behavioral function in various species, including human infants, it has been suggested that DHA plays a unique role in the function of excitable membranes and that it may be an essential dietary component during the developmental period when the brain is growing rapidly [reviewed in Innis (2000) and Lauritzen et al. (2001)].

This chapter focuses on the role LCPUFAs play in the functional development of the CNS, as measured by cognitive and behavioral outcomes. The perspective taken is that of developmental psychobiology, in which the development of the brain is seen as a complex epigenetic process in which genes and their environment interact through an intricate series of feedforward and feedback loops (Gottlieb, 1998). Nutritional factors are integral to this process, not only in terms of providing the necessary energy but also in ensuring the supply of specific nutrients that serve either as building blocks, for example, proteins and lipids, or as essential components of biochemical pathways. The chapter is organized into three sections. The first addresses the process of brain development, with an emphasis on the establishment of the neural systems that support behavioral function, including higher-level cognitive processes. This includes a discussion of the various ways in which nutritional variables might influence the developmental outcomes. The next section focuses on the role of lipids in brain function. This includes a brief overview of the biochemistry of various lipids in the brain. The emphasis is on the putative role of LCPUFAs in the brain, including the mechanisms involved, and how brain FA composition may be influenced by diet. The latter is illustrated by work done in our laboratory on various rodent models. The final section addresses the functional effects of the n-3 FAs with respect to cognitive and behavioral outcomes. Here the focus is on work in which we have investigated behavioral effects of dietary supplementation with DHA, based on a hypothesis of changes in the functions of brain dopamine systems.

11.2 BRAIN DEVELOPMENT

11.2.1 DEVELOPMENTAL NEUROBIOLOGY

Adaptive behavioral capabilities depend ultimately on the operations of neural systems established during the process of brain development involving complex patterns of synaptic connections. The mammalian brain develops through a sequence of events that appear to be fairly consistent across species and involve increasing cell differentiation and specialization [reviewed in Bayer (1989) and Nowakowski and Hayes (1999)]. The initial formation of the primitive neural tube is followed by a period of further cell proliferation and diversification into neurons and glia. Neurons arise in the ventricular germinal zone and migrate to their final position, with the timing of origin being related to their final cortical destination. The process of migration also plays a role in defining neuronal identity and functional properties. For example, in the formation of the well-defined layers of the cerebral cortex, migration often occurs on a scaffold of radial glial cells and the pattern of development is inside-out, such that neurons born at the early stages end up in the deepest cortical layers. Neurons then begin to differentiate by extending axons and dendrites. The axon is guided to its target through specialized cell-adhesion molecules that allow for adhesive contacts between the growth cone at its tip and other cells or the extracellular matrix. Further differentiation results in the expression of enzymes and receptors associated with specific neurotransmitters. Cellular interactions play an important role in this process: in making their final connections axons may be attracted (or repelled) by chemicals released by target cells, and neurotrophins such as nerve growth factor are integral to neural survival and development (Huang and Reichardt, 2001). The process of synaptic connectivity is characterized by an initial exuberant increase in the number of synaptic contacts, which is followed by programmed regression in the form of apoptosis, as well as by synaptic and axonal pruning. The latter processes, which continue through the second decade of life in humans, are influenced strongly by afferent stimulation, derived from both intrinsically generated neural activity and external stimulation. This appears to be particularly true of more recently evolved brain structures such as the cerebral cortex, for which input from afferent pathways has a considerable influence on resulting organizational patterns (Pallas, 2001). Myelination by the oligodendroglia is another late-occurring process, and although almost complete by the end of the second year of human life, it takes to nearly the end of the second decade to myelinate completely the intracortical association areas. Through this period of extended plasticity and exposure to input from varying environmental sources, many of the individual differences in the cortical organization of the adult brain come into being. Thus, although the genome plays an integral role in the development of the structural architecture of the brain, the final outcome depends on the orchestration of information from many different sources. This has led to the suggestion that this protracted period of brain development, by allowing the opportunity to adapt to prevailing environmental contingencies, may be the origin of human behavioral flexibility (Saugstad, 1998). Such flexibility does not, however, come without cost; in some instances, alterations in the orchestration of the developmental events that serve to establish functional neural systems may contribute to pathological outcomes. One such example is schizophrenia, a condition that current theories characterize as a neurodevelopmental disorder with both genetic risk and environmental contributions (Lewis and Levitt, 2002). It is thought that the emergence of the characteristic symptoms of this condition during adolescence is related to changes in the normal maturational patterns of the brain, which then interact with other environmental risk factors to precipitate the first episode of the illness (Keshavan and Hogarty, 1999). One intriguing hypothesis relates this devastating condition to alterations in the function of membrane phospholipids (Horrobin, 1999). Another neurodevelopmental dysfunction in which brain lipids may be involved is the attention deficit hyperactivity disorder (ADHD),which is discussed in Section 11.4 in relation to supplementation with LCPUFAs.

11.2.2 DEVELOPMENTAL PSYCHOBIOLOGY

The study of brain development can be seen as integral to the objectives of cognitive neuroscience, which are to identify the neural systems involved in specific cognitive processes. There is ongoing debate about the degree to which these systems are specified as innate modules and the relative contribution of environmental input during development (Quartz and Sejnowski, 1997; Karmiloff-Smith, 1998). A psychological perspective on brain development takes into account that species have evolved to function optimally in a particular ecological niche. It therefore considers not only the physiological, biochemical, and anatomical factors involved but also the ordered series of inputs from the species-typical environment that would be encountered during the normal developmental process, an example being patterns of maternal care (Hall and Oppenheim, 1987; Fleming et al., 1999; Francis and Meaney, 1999). From this perspective, sequential dependencies play an important role not only in developmental processes involving the timing of interactions among various brain regions during development, but also in those involving input from the environment. In mammals, for example, specific types of visual input are necessary during discrete time intervals to develop the functional capacity of the visual system (Shatz, 1992). These time intervals, often referred to sensitive periods, are defined as the period during which the organism is ready (primed) to respond to specific types of signals expected from the environment. If for any reason these signals are not forthcoming, the developmental trajectory may be altered irrevocably, resulting eventually in anomalies of structure and function. Black (1998) characterizes this response to the species-typical environment during specific developmental periods as an experience-expectant process. This is in contrast to an experience-dependent process, which also changes the brain, but in a way that optimizes an individual's adaptation to specific and sometimes unique features of the environment. In rats, for example, environmental enrichment or training on specific tasks is associated with increases in cortical thickness, dendritic branching, number of synapses, and acetylcholine levels (Kolb et al., 1998; Klintsova and Greenough, 1999). That environmental enrichment also enhances subsequent performance on complex learning tasks supports the adaptive value of these structural changes. It is, however, important to realize that these two processes are not independent because the ability to extract reliable information from the environment relies on the presence of the appropriate perceptual capabilities. Thus, neurobiological development can be altered by factors that have an impact either by acting directly on the biochemistry of a particular pathway or by affecting the development of this pathway indirectly by influencing the nature of the interactions with the environment. From such a developmental systems perspective, specific behavioral attributes are the outcome of a complex system of interactions over the developmental period, in which causality arises from a cascade of effects that can be influenced in different ways at many levels (Gottlieb, 1998). Therefore, any factor, including nutrition, that affects the ability of the developing animal to engage appropriately with its environment during crucial developmental periods has the potential to affect brain development adversely (Pollitt, 2001).

11.2.3 METHODOLOGICAL IMPLICATIONS

This perspective on causality will necessarily influence the way in which questions related to brain and behavioral development are framed (Karmiloff-Smith, 1998). A common assumption in much developmental research has been that there will be a direct relationship between early and mature cognitive function. However, it is becoming evident that alterations in the functioning of specific neural systems may manifest themselves in different ways at different times during the developmental process. An example of this is a recent study that compares differences in cognitive profiles between individuals with Down syndrome and Williams syndrome, in which the characteristic patterns of differences between the adult populations are not consistent with those seen in infancy (Paterson et al., 1999). Thus, research designs that address developmental questions should include more extensive longitudinal measurements rather than relying on measurement at one particular age point. Furthermore, to understand whether accelerated (or retarded) development is beneficial (or problematic) it is important to study the entire developmental profile not only in terms of different timepoints but also in the context of related measures. For example, as has been shown in animals, manipulations that impair performance on complex learning tasks may actually improve performance on more simple tasks (Everitt and Robbins, 1997). It may also be that a transient effect on a function at one timepoint may have a longer-lasting effect on a different but related function measured at a later time. For example, delayed development of visual acuity during sensitive periods of development when visual input is important to the organization of the developing cortex could permanently alter other aspects of visual function at a later stage. This is similar to the situation of infantile cataracts, which, if not removed before 6 to 8 months of age, result in permanent deficits (Johnson and Mareschal, 2001). Another methodological implication of a systems perspective on development is a move away from global tests of function toward the use of measures more closely related to the operations of the neural substrates to which specific cognitive processes may be attributed. Tests such as the Bayley Scales of Infant Development, for example, were developed initially as screening tools in clinical situations (Bayley, 1969). Generally, on such scales, the infant's developmental status is represented by an aggregate score that represents the average of scores obtained across different developmental domains. Implicit in the use of this approach is the assumption that there is some general construct represented by this score such as a general intelligence quotient or IQ. However, it is possible for different infants to obtain the same score, but with very different developmental profiles. Moreover, although such scales have proved very useful in clinical situations, there is no strong correlation in a normal population between these scores and tests used later to assess intelligence in school-age children. Thus, recent developments in the field of infant cognition advocate alternatives to standardized scales that, rather than relying on tests of overall function, address questions directed at more specific aspects of cognitive development and for which there may be better predictive ability. These issues, discussed in depth in McCall and Mash (1997), Colombo (2001) and Rose et al. (2003), will be revisited in Section 11.4 in relation to the cognitive and behavioral effects of LCPUFAs. The present section concludes by addressing specific ways in which nutrition may be involved in these various interactive processes that constitute brain development.

11.2.4 NUTRITION AND BRAIN DEVELOPMENT

Without question, the biochemistry and function of both the mature and developing brain can be influenced by diet (Fernstrom, 2000). Much of the early research on the role of nutrition in brain and behavioral development addressed mainly the effects of undernutrition, particularly as this relates to protein and calories (Strupp and Levitsky, 1995). More recently, attention has focused on the role of specific nutrients, for example, the amino acids tyrosine and tryptophan that are the respective precursors of the neurotransmitters dopamine and serotonin or micronutrients such as vitamins and minerals that function as essential cofactors in specific biochemical pathways (Cooper et al., 1996). Generally, the concern has been with the effects of inadequate dietary supply. However, in this era of functional foods, the question often asked is whether, if deficiency is associated with a decline in function, would supplementation beyond what is considered adequate also have beneficial effects? In some cases, it may be possible to generalize from the deleterious effects of a nutritional deficiency to positive effects of supplementation, but this may not always be the case. There are many instances in the psychopharmacological literature in which, once an optimum response has been attained, more of a particular drug treatment is no longer advantageous and may sometimes even be disadvantageous in terms of the desired outcome. This is of particular concern when dealing with a process as intricate as the development of the nervous system. It is therefore important that recommendations for dietary supplementation be informed by dose–response studies in animals that not only include a wide range of doses but also incorporate other measures of physiological function that might be affected by the treatment together with cognitive outcomes. Furthermore, both the absolute and relative levels of dietary components may be important with respect to functional outcomes. For example, tyrosine and tryptophan fed as a high-protein meal compete for transport into the brain. However, a meal high in carbohydrate indirectly facilitates the access of tryptophan to the brain by raising insulin levels, and thus an effective way of raising brain tryptophan levels is to increase dietary carbohydrate (Fernstrom, 2000).

The condition known as phenylketonuria (PKU) provides an interesting example of the effects of protein imbalance on brain development in humans. This condition results from a recessive mutation in the gene coding for phenylalanine hydroxylase, which is the enzyme that converts dietary phenylalanine to tyrosine, the precursor of dopamine (Cooper et al., 1996). The resulting excess of phenylalanine impairs cognitive development in infants, which can be prevented to some extent by providing a diet low in phenylalanine started early in development. Notwithstanding such improvement, there remain subtle cognitive deficits that have been associated with dopaminergic function in the prefrontal cortex (Diamond et al., 1997). It is thought that, despite dietary intervention, there is a residual phenylalanine–tyrosine imbalance sufficient to limit access of tyrosine to the brain and thereby compromise the development of the dopaminergic pathways to the prefrontal cortex, which appear to be particularly sensitive to this type of dietary insult. The dopamine system is discussed in more detail in Section 11.4, which addresses the hypothesized relationship between brain n-3 FA composition and dopamine function.

PKU also illustrates the important contribution of nutrients to the trophic actions of neurotransmitters during development. In the mature nervous system, neurotransmitters are associated with well-defined pathways and mediate interneuronal communication by activating specific receptors and second messenger systems. In the developing system, however, they also function as trophic factors to regulate neurogenesis, neural migration, and synaptogenesis (Lauder, 1993; Levitt et al., 1997). Alterations in trophic factors that alter the nature of the interrelationships among developing neural systems can effect long-lasting change in developmental outcomes. It has been hypothesized, for example, that the relationship between maternal smoking behavior and sudden infant death syndrome may be mediated by nicotine-induced changes in the normal developmental interactions between the mechanisms underlying sympathoadrenal regulation (Slotkin, 1998). Similarly, deficiency, and also possibly supplementation, of essential dietary precursors during sensitive developmental periods can have far-reaching consequences. One example is the use of folic acid supplementation during pregnancy to prevent neural tube defects (Green, 2002). Another interesting story that is emerging from animal research is that of enhanced cognition in adult rats associated with choline supplementation during the prenatal period (Meck and Williams, 2003). Not only do the adult offspring show enhanced spatial maze learning ability, but there is also evidence that the dietary supplementation during the early developmental period increases their levels of nerve growth factor (NGF) in the cortex and hippocampus (Sandstrom et al., 2002), suggesting that it may offer some protection against pathology associated with aging. Interestingly, there is also a study showing that dietary n-3 FA deficiency decreases NGF in the hippocampus (Ikemoto et al., 2000). As regards other effects of dietary fats, an early study showed that deficiency of EFAs during prenatal development is associated with delayed maturation in brain ganglioside and glycoprotein content as well as subsequent learning impairment (Morgan et al., 1981). A more recent study conducted in our laboratory showed a reduction in dendritic branching of pyramidal neurons in the cortex of mice reared in an enriched environment and fed from conception on a saturated fat diet with marginal levels of n-6 FAs (Wainwright et al., 1998a). Other researchers have also demonstrated a relationship between dietary lipids and the FA content of neuronal growth cones in rats (Auestad and Innis, 2000; Innis and de la Presa Owens, 2001) as well as neuronal cell size in the hippocampus (Ahmad et al., 2002). Findings such as these support a potentially important relationship between the lipid content of diet and the developmental plasticity of the brain. Section 11.3 addresses the role of brain lipids in more detail, with an emphasis on LCPUFAs, particularly those of the n-3 family.

11.3 ROLE OF LONG-CHAIN POLYUNSATURATED FATTY ACIDS (LCPUFAs) IN THE BRAIN

11.3.1 BRAIN LIPIDS

The following brief overview of brain lipids is based on the more extensive treatment of this topic to be found in the chapters by Agranoff and Hajra (1994) and Clandinin and Jumpsen (1997) as well as the book by Jumpsen and Clandinin (1995). The reader interested in more detailed information should consult these sources.

The most important lipids in brain are the neutral lipids, phospholipids, and sphingolipids (glycolipids and sulfatides). All lipids contain FAs, typically with 12 to 26 carbon chains. These are esterified to a characteristic alcohol, for example, glycerol in phospholipids and sphingosine in sphingolipids. These FAs are saturated, monounsaturated, or polyunsaturated, and all double bonds are in the *cis* configuration. Any double bonds are usually nonconjugated and three carbons apart. In standard nomenclature of the FA, the first number refers to the length of the carbon chain; the second number, following the colon, refers to the number of double bonds; and the third number, after n-, represents the number of carbons from the methyl end of the molecule to the first double

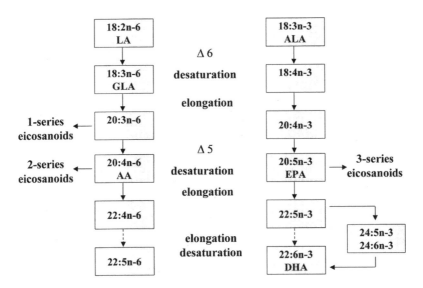

FIGURE 11.1 Shared biosynthetic pathway of n-6 and n-3 polyunsaturated fatty acids, whereby the essential fatty acid dietary precursors linoleic acid and α-linolenic acid are converted to their longer-chain metabolites through a series of desaturation and chain elongation reactions. Abbreviations: α-linolenic acid (ALA, 18:3 n-3); arachidonic acid (AA, 20:4 n-6); docosahexaenoic acid (DHA, 22:6 n-3); eicosapentaenoic acid (EPA, 20:5 n-3); γ-linolenic acid (GLA,18:3 n-6); linoleic acid (LA, 18:2 n-6).

bond. There are three series of unsaturated FAs: n-3, n-6, and n-9. Animals do not have the biosynthetic capacity to insert double bonds at the n-6 and n-3 positions and are therefore unable to synthesize LA (18:2 n-6) and ALA 1(8:3 n-3) *de novo*. Therefore these FAs must be provided by the diet, and consequently these 18-carbon FA are referred to as EFAs. As shown in Figure 11.1, they are then converted to their longer-chain compounds through a shared biosynthetic pathway involving desaturation and subsequent chain elongation.

In the generation of DHA, eicosapentaenoic acid (EPA, 20:5 n-3) is elongated to 22:5 n-3, followed by a second chain elongation and delta-6 desaturation that results in 24:6 n-3, which is converted subsequently to 22:6 n-3 through β-oxidation (Sprecher et al., 1995). The n-3 and n-6 FAs compete for the same desaturation enzymes as well as for the enzymes that esterify them into membrane phospholipids. The desaturation enzymes appear to have a higher affinity for n-3 FAs, but the relative activity of the enzymes involved in membrane remodeling are less well defined and are likely to vary among different tissues (Craig-Schmidt and Huang, 1998). A characteristic of neuronal phospholipids, compared to other tissues, is that the parent EFAs LA and ALA are found only in minor quantities. However, their long-chain derivatives constitute approximately one third of the dry weight of neuronal and retinal tissue, predominantly in the form of AA (20:4 n-6) and DHA (22:6 n-3), with smaller quantities of 20:3 n-6, 22:4 n-6, 22:5 n-6, and 20:5 n-3.

The lipid composition of the nervous system differs considerably between white matter and gray matter. The white matter is composed of myelin sheaths formed by the oligodendroglial cells. Although myelin composition varies with brain region, age, and species, a major lipid component is cholesterol, together with phospholipids and most of the sphingolipids. Very few cholesterol esters are found in adult brain. The predominant FAs in myelin are either very long chain saturated FAs or monounsaturated FAs. Cerebrosides, the simplest form of glycosphingolipids, are also highly prevalent in myelin. Gray matter, on the other hand, contains more gangliosides than the white matter does, and these play a crucial role in development, being involved in processes such as cell proliferation, recognition, migration, adhesion, and differentiation. Other major components of gray matter are phospholipids, which play an important role in cell signaling as well as in the formation and remodeling of synapses and dendrites. The role of neuronal membrane phospholipids is reviewed by Horrobin (1999) and is summarized briefly next.

Phospholipids consist of a three-carbon glycerol backbone with FAs at the *sn*-1 and *sn*-2 positions. The enzymes involved in phospholipid synthesis show a high degree of specificity with respect to the nature of the FAs attached at *sn*-1 and *sn*-2. Position *sn*-2 is often occupied by PUFAs with two or more double bonds, whereas saturated and monounsaturated FAs are more often located at *sn*-1. At the *sn*-3 carbon, there is a phosphorus atom to which a head group is attached; this head group usually consists of choline, ethanolamine, inositol, or serine. The quantitative distribution of FAs differs for each phospholipid class; for example, DHA constitutes a larger percentage of FAs in phosphatidylethanolamine (PE) than in phosphatidylcholine (PC). PE, including plasmalogens, is the most prevalent phospholipid in the adult human brain followed by PC and phosphatidylserine (PS). Although there are relatively much smaller quantities of phosphoinositides (PI), these play an important role in signal transduction. Plasmalogens are found in especially elevated concentrations in the brain, with levels increasing during myelination [reviewed in Farooqui et al. (2001)]. These are glycerophospholipids that have an α,β-unsaturated ether attached at *sn*-1 with either AA or DHA often located at *sn*-2. Interestingly, DHA induces their synthetic pathway and DHA, followed by AA, is also the most potent inhibitor of plamalogen-selective phospholipase A_2 (Farooqui et al., 2001).

Cell- and tissue-specific acyltransferase and phospholipase enzymes are involved in the synthesis and breakdown of phospholipids according to the needs of the membranes in specific sites within cells and tissues. Several classes of phospholipases, which can be activated by the interaction of agonists with receptors at the neural cell surface, mediate the breakdown of phospholipids. The FAs at *sn*-1 and *sn*-2 are removed by phospholipases A_1 and A_2, respectively, whereas phospholipase B is believed to have the ability to release the FAs at both *sn*-1 and *sn*-2. Phospholipase C cleaves the phosphorus head group from *sn*-3 to produce diacylglycerol (DAG), from which the FAs can be released by DAG lipases. Finally, phospholipase D splits the bond between the phosphorus group and the phospholipid head group to result in the formation of phosphatidic acid. The AA and DHA released can then be further metabolized to eicosanoids by cyclooxygenases or lipoxygenases. These are discussed further in the following section, which addresses the possible mechanisms by which these dietary-induced changes in membrane FA composition may affect brain function.

11.3.2 FUNCTIONAL EFFECTS OF LCPUFAs IN THE BRAIN

The types of head and acyl groups attached to the glycerol backbone of the phospholipids generate the physical and chemical environments within which receptors, channels, and other proteins function and which are influenced by the lipid environment. For instance, the presence of saturated FAs confers rigidity, whereas unsaturated FAs, due to their angled double bonds, give added flexibility. The elevated proportion of highly unsaturated FAs in brain phospholipids confers on the membrane the ability to quickly change shapes and allows membrane fusion. This fusing ability is important in creating and breaking synaptic connections and for dispersing neurotransmitters [reviewed by Horrobin (1999)]. FAs also have an impact on the function of membrane proteins. These effects may be mediated by a direct influence on protein conformation or by imposing limits on conformational changes through changes in membrane fluidity (Murphy, 1990). Unsaturated FAs can also influence the local order of the membrane because they are involved in determining the asymmetric distribution of sterols within the bilayer. Thus, by influencing membrane fluidity and order, FAs alter the actions of proteins that depend on mobility within the plane of the membrane, including receptor aggregation, interactions between receptors, regulatory proteins, and intracellular effectors, as well as receptor desensitization or downregulation [reviewed in Murphy (1990) and Salem et al. (2001)].

Within a phospholipid class, different pairs of FAs at *sn*-1 and *sn*-2 result in different molecular species, so that function may be related to not only the amount of FA present but also to its positional distribution. This is illustrated by the retinal lipids, in which DHA is particularly abundant in the rod outer segments (ROS), constituting up to 50% of all FAs in certain phospholipid classes [reviewed in Clandinin and Jumpsen (1997), Jeffrey et al. (2001), and Jumpsen and Clandinin (1995)]. Photoreceptor biosynthesis in the eye occurs at the same time as cellular differentiation and active synaptogenesis, and the outer segments of rod retinal membranes are thought to be extensions of nerve cell plasma membranes (Jumpsen and Clandinin, 1995). A characteristic of the phopholipids in these membranes is that PUFAs are located at both *sn*-1 and *sn*-2, and they are referred to as dipolyenes. The photopigment in the retina involved in the transduction of light to an electrical signal is rhodopsin. Dipolyenes increase the rate of rhodopsin activation (Jeffrey et al., 2001), and thus delays in activation because of a decreased availability of dipolyenes may result in a slower electroretinogram (ERG) response.

There is considerable evidence that rhodopsin function is influenced by alteration in the dietary n-6 to n-3 ratio [reviewed in Jumpsen and Clandinin (1995), Salem et al. (2001), and Jeffrey et al. (2001)]. Feeding diets with an imbalance of n-6 and n-3 FAs or with a deficiency of EFAs changes the FA profile of the retina and results in impaired visual cell renewal and abnormal visual function. For instance, in animal studies, declines of 50 to 80% in brain and retinal DHA have been associated with changes in neural function, including changes in ERG and visual acuity. The mechanisms involved in the changes in ERG and visual acuity that accompany a retinal deficiency of DHA have not been fully elucidated. It is known that despite decreased retinal DHA, there is no impact on rate of ROS disk synthesis, number of photoreceptors, ROS length, or width of the outer nuclear layer. Thus, the mechanism may involve altered rhodopsin activity and sensitivity due to low DHA in the membrane lipid environment. Alternatively, decreased DHA may have an impact on ion channel function. These and other potential mechanisms may explain the role of DHA and other FAs in vision [reviewed in Jeffrey et al. (2001)].

As described previously, changes in the FA composition of neuronal membranes can influence neurotransmission by altering the function of the receptors, enzymes, and ion channels embedded within or associated with the membrane phospholipids. In addition to their physical role in membranes, AA, and possibly DHA, may also act as second messengers themselves in cellular signal transduction pathways, therefore playing an important role in regulation of function. These signaling pathways are activated when neurotransmitters, hormones, or growth factors bind to specific membrane-bound receptors that activate G-proteins, resulting in the release of PLA_2, which cleaves AA and DHA from *sn*-2 of phospholipids. Phosphoinositides are particularly rich in AA, and PLC acts on *sn*-3 to catalyze the conversion of PI 4,5 bisphosphate to DAG and inositol triphosphate (IP_3). The generation of IP_3 causes the release of calcium ions from intracellular stores. Calcium and DAG activate protein kinase C (PKC). Through its phosphorylating activity, PKC regulates the activities of various neuronal proteins, including ion channels, and is also involved in regulating cell proliferation and differentiation (Kandel et al., 1995). Interesting in this regard is a recent report indicating that both DHA and EPA, but not AA, inhibit PKC *in vitro* (Seung Kim et al., 2001). AA is the n-6 FA that has been identified as necessary for growth, and this may be related to its signaling action in neuroendocrine cells, in which there is evidence for its involvement in the release of both growth hormone and prolactin (Roudbaraki et al., 1996). AA is also the precursor of anandamide, which has been identified as an endogenous ligand of the cannabinoid receptor; it has also been shown recently that dietary LCPUFAs can modulate levels of these compounds in the piglet brain (Berger et al., 2001).

LCPUFAs also play a very important functional role as precursors of eicosanoids. Eicosanoids are oxygenated 20-carbon compounds that include the prostaglandins (PGs), thromboxanes (TXs), leukotrienes (LTs), and some hydroxy and hydroperoxy FAs (Smith et al., 1991). Because they cannot be stored, eicosanoids are quickly synthesized following receptor stimulation and they play an important regulatory role by acting as modulators of cellular responses. AA is the main substrate for most eicosanoids, but they can also be produced from EPA, DHA, and DGLA. AA gives rise to series 1 eicosanoids, and DGLA and EPA give rise to series 2 and series 3 eicosanoids, respectively. There is competition among DHA, EPA, and AA for the production of their respective eicosanoids. Therefore, the quantitative and qualitative profile of eicosanoid production is a function of the balance of dietary n-3 and n-6 FAs (Broughton and Wade, 2002). In the presence of oxygen, AA is metabolized by the lipoxygenase, cyclooxygenase, or epoxygenase enzymatic pathways. LTs are synthesized by the lipoxygenase pathway and have been shown to constrict bronchial airway muscles, increase vascular permeability, and play a role in the interactions between endothelium and white blood cells (Clandinin and Jumpsen, 1997). They are also involved in immune activities mediated by cytokines. In addition, lipoxygenase products of AA and DHA have been identified in the pineal gland and their formation is affected by n-3 deficiency [reviewed in Salem et al. (2001)]. The cyclooxygenase pathway is associated with the synthesis of cell-specific prostanoids, including TXs and PGs such as PGD_2, PGE_2, PGF_2-α, PGI_2, and TXA_2. Whereas certain eicosanoids such as TXA_2 and PGI_2 are involved mostly in regulation of the circulatory system, the primary PGs, including PGD_2, PGE_2, and PGF_2-α, have direct neural activity. The formation of PGE_2 and PGF_2-α from AA results in altered release of neurotransmitters such as norepinephrine, serotonin, and vasoactive intestinal peptide. PGs have also been implicated in many other behavioral functions, including sedation, inhibition of locomotor and exploratory activities, and regulation of rapid eye movement (REM) sleep [reviewed in Wainwright (1997)]. Recently, some very interesting data have been published showing that in human infants higher maternal plasma DHA levels are associated with more mature neonatal sleep-state patterning (Cheruku et al., 2002). When such changes in basic regulatory function are considered in the overall context of the processes that contribute to brain development, it is conceivable that they may have important implications on ultimate functional capacity.

Particularly important is the role of LCPUFAs in excitotoxic or ischemic brain damage, which has been attributed to the uncontrolled accumulation of free PUFAs and related second messengers (Bazan et al., 1993). Some degree of protection is conferred by low concentrations of PUFAs, which has been associated with the opening of K^+ channels in the brain and subsequent inhibition of glutamatergic transmission (Lauritzen et al., 2000). Yavin et al. (2002) proposed that high levels of DHA accumulate during the period of rapid brain growth to function as a protective agent against oxidative stress. In support of this, they describe findings that associate elevated levels of DHA with a decrease in lipid peroxidation. This suggestion appears to be paradoxical at first, because high levels of PUFAs increase the potential for lipid peroxidation. However, they suggest that DHA may serve a dual role, being both prooxidant and antioxidant, the antioxidant effect being attributable to a reduction in free radicals mediated specifically by plasmalogen-associated DHA. Although apoptotic and antiapoptotic effects have been documented for both DHA and AA, Salem et al. (2001) also proposed that DHA may act as a trophic factor in the brain by supporting neuronal survival through an antiapoptotic effect. The apoptotic effects of DHA may be mediated by oxidative stress, whereas the antiapoptotic activity requires preincubation with DHA in the presence of antioxidants and is related to the extent of cellular accumulation of PS. Interestingly, cells supplemented with DHA showed decreased activity of caspase-3, an enzyme that mediates mammalian apoptosis. This was related in turn to modified gene expression, as seen in decreased levels of the mRNA coding for caspase-3 and an increase in the mRNA for Raf-1 (Kim et al., 2000).

The role of the dietary factors in regulating transcriptional activity is an area in which there is currently a great deal of interest. Dietary fat has been shown to have an important influence on metabolism, cell differentiation, and growth through its effects on gene expression in the liver [reviewed in Jump and Clarke (1999)]. These effects are seen in response to an excess or deficiency of fat or to an alteration in the FA profile of the diet. Certain nuclear receptors, such as the peroxisome proliferator-activated receptors (PPARs), are directly regulated by FAs, and a recent study describes increases in PPARβ transcripts in ocular tissue of rats consuming an n-3-deficient diet (Rojas et al., 2002). Another report identifies DHA as a ligand for the retinoid X receptor in the mouse brain (de Urquiza et al., 2000). A very interesting story in relation to the function of the brain will no doubt emerge as more studies address questions of this type.

11.3.3 Dietary Supply of LCPUFAs and Effects on Growth and Brain Fatty Acid Composition

The brain growth spurt includes maturation of the axons and dendritic trees and formation of the myelin sheath, and in humans occurs during the third trimester of pregnancy and the first 2 years postnatally (Dobbing and Sands, 1979). During this period, the greatest accretion of n-3 and n-6 FAs occurs in structural lipids, associated with increased cell size, cell type, and cell number. The n-3 FAs tend to accrue fastest prenatally, whereas the n-6 and n-9 FAs accumulate most rapidly during postnatal development [reviewed in Jumpsen and Clandinin (1995)]. Because of their low fat reserves, all neonates are at risk for FA deficiency. This risk is exaggerated in the premature or small-for-gestational-age infant, wherein infants born before week 32 have very low tissue reserves of DHA (Clandinin et al., 1981).

The fetus is largely dependent on the maternal supply of FAs, and therefore maternal FA status may be a significant player in determining the availability of these FAs to the developing fetus, particularly during early pregnancy. The human placenta allows free FAs to cross the placenta but not phosphoglycerides or triacylglycerols. The mechanisms that have been suggested to explain the increasing concentrations of AA and DHA in the fetus with gestational age include an increase in fetal or placental ability to form LCPUFAs from precursors or a preferential transfer of these FAs across the placenta from maternal to fetal circulation. Another potential source of LCPUFAs to the developing brain is that produced in the liver and transported in the blood as lipoprotein (Spector, 2001). Also, cerebromicrovascular endothelial cells and astrocytes have the potential to play a central role in the supply of LCPUFAs to the brain because they contain the necessary enzymes to desaturate and elongate LA and ALA (Moore, 2001). However, notwithstanding its capacity to desaturate and elongate the EFA precursors, it remains uncertain as to whether the human fetus and neonate are able to synthesize sufficient DHA for their needs from dietary precursors, particularly during the period of rapid brain growth when demand is high. This has led to the proposition that during these periods it is essential that preformed DHA be provided in the diet and hence be designated a conditional EFA (Cunnane, 2000).

There is evidence from many studies conducted in a variety of mammalian species that the developing brain responds to changes in dietary FA supply with characteristic changes in FA composition. For example, in both humans (e.g., Farquarson et al., 1992) and rodents (e.g., Wainwright et al., 1994) dietary n-3 deficiency decreases DHA in brain phospholipids, with a corresponding increase in 22:5 n-6. However, to produce a substantial reduction of DHA in the brain, particularly in rodents, it is necessary to raise animals for more than one generation on deficient diets in order to deplete maternal stores (Bourrè et al., 1989). Feeding EFA-deficient diets (i.e., deficient in both n-6 and n-3 FAs) results in growth retardation, accompanied by behavioral retardation and reproductive failure. An example of this is a recent study in our laboratory that showed neonatal mortality accompanied by growth and behavioral retardation in the surviving offspring of the second generation of mice maintained on a diet high in saturated fat and marginally deficient in EFAs (Fraser and Wainwright, 2001). Because the n-6 and n-3 families compete for the same biosynthetic pathways, both the absolute amount of each in the diet and their relative proportions are important (Craig-Schmidt and Huang, 1998). Generally n-3 deficiency does not affect growth, but increasing amounts of n-3 FAs in the diet have been associated with slightly decreased body weight (Heird, 1997). We addressed this question of dietary imbalance between the n-6 and n-3 FAs in a series of studies in which pregnant mice were fed the diets throughout gestation and lactation and the effects measured on the growth and brain and behavioral development of the pups (Wainwright et al., 1997, 1999a). A diet with a low n-6 to n-3 ratio (0.3) based on very high levels of n-3 FAs, provided as DHA (5% FAs) with marginal levels of n-6 FA as LA (1.6%), resulted in large increases in brain DHA. However, there was a corresponding decrease in AA, accompanied by growth and behavioral retardation in the pups. The growth retardation was offset partially if some of the LA was replaced by AA. This suggests that the high levels of DHA may have inhibited the activity of delta-6 desaturase in converting LA to γ-linolenic acid (GLA, 18:3 n-6), which is then converted to AA. These findings imply that diets supplemented with high levels of n-3 FAs should probably also provide additional n-6 FAs in the form of either GLA or AA. This is supported by a more recent study in which animals were fed a diet with a low n-6 to n-3 ratio (0.7). In this case, the n-3 FAs were provided as high levels of ALA (40% FAs), but with the n-6 FAs providing 1.6% GLA in addition to LA and here there were no effects on pup growth or behavioral development (Huang et al., 2002). This emphasizes the importance of considering not only the total n-6 FA content of the diet when supplementing with DHA but also the form in which this is provided, that is, as either LA, GLA, or AA.

These studies represent extreme dietary manipulations in terms of deficiency or excess. However, there are also questions associated with supplementation of DHA and AA in the context of a diet containing sufficient LA and ALA. Specifically, is there a limit to the incorporation of DHA into the developing brain and are these increased levels reflected in related functions? To address these issues, we used the artificial rearing model in rat pups, also known as "pups-in-cups," to systematically manipulate the DHA and AA content of diets containing adequate LA and ALA during the first two postnatal weeks. This is a period when the rat brain grows most rapidly and is approximately equivalent to the third trimester in humans (Dobbing and Sands, 1979). The advantage of this model is that it allows one to investigate directly the relationship between dietary supply and the pup brain, independent of factors related to maternal physiology. The pups were gastrostomized within the first few days of life and from postnatal day 5 to 18 were fed rat milk substitute containing oils providing 10% LA and 1% ALA and supplemented with AA and DHA from microbial cell oils. A 3×3 factorial design provided one of three levels of AA and DHA, 0.0, 0.4, and 2.4% of FAs (Ward et al., 1998). As measured on postnatal day 18, dietary supplementation resulted in a wide range of brain DHA, with a smaller range in AA. Again, these data show a strong reciprocal relationship between dietary levels of AA and DHA and those in the brain, with DHA in the diet raising DHA in the brain and decreasing AA, and vice versa for AA. Particularly noteworthy were the observations that the effects of supplementation of 0.4% DHA relative to the unsupplemented group were of approximately the same order of magnitude as were those of 2.4% relative to 0.4%. In addition, an exponential model provided the best fit of the relationship between DHA levels in the red blood cells with those in the brain. This suggests that there may be a threshold in terms of incorporation of DHA into brain phospholipids. It also implies that if the aim is to supplement diets with DHA to raise brain DHA, the optimal level of DHA is probably 0.5 to 1% of FAs, and that this should be accompanied by supplementation with either GLA or AA to maintain appropriate levels of AA. Our data suggest that <1% AA is probably sufficient at these levels of DHA; if GLA were used, the levels would need to be higher to accommodate the extra metabolic steps involved (Ward et al., 1999).

11.3.4 SENSITIVE PERIODS

An issue central to the discussion of the role of LCPUFAs in the development of the brain is the question of sensitive periods; that is, is there a specific period during which a deficiency or the imbalance of dietary FAs can result in potentially irreversible damages to structure and function? Studies in rats have shown that it is possible to rehabilitate DHA levels, albeit slowly, in the brain of animals deficient in n-3 through feeding diets enriched with fish oil (Youyou et al., 1986; Bourre et al., 1989). Nonetheless, although it may be possible to use dietary rehabilitation to restore normal membrane lipid composition, there may still be deficits in function. This is supported by some evidence from work in rhesus monkeys fed diets deficient in n-3 throughout development until the age of 22 months. Compared with controls, there was a large decrease in DHA in the retina and cortex whereas the proportion of 22:5 n-6 increased significantly. ERG recording at 21 months showed impaired recovery of the dark-adapted response to saturating flashes at short time intervals. From 4 to 12 weeks of age, the deficient monkeys also showed evidence of inferior acuity (Neuringer et al., 1986). A subsequent study examined the potential to rehabilitate function or behavior by using a fish oil diet rich in both DHA and EPA (20:5 n-3; Connor et al., 1990). Before repletion, biochemical n-3 deficiency, electroretinographic abnormalities, and visual acuity loss were again established in the juvenile monkeys. The repletion diet was administered starting between 10 and 24 months and feeding lasted between 43 and 129 weeks. Relatively rapid reversibility was observed in the cerebral cortex in this study, in that DHA increased significantly in all classes of phospholipids, whereas there were minor increases in EPA and reciprocal decreases in n-6 FAs. In showing that a severe FA deficiency of the brain could be reversed with an n-3-enriched diet, this study confirmed that phospholipids of the mammalian nervous system are not static entities. However, fish oil feeding did not alter the ERG, which remained abnormal even after biochemical repletion had been achieved (Connor and Neuringer, 1988). This suggests a permanent effect of n-3 deficiency on the function of the photoreceptor membrane and that for efficient visual transduction, high levels of retinal DHA are necessary. However, not only changes to the retina itself but also alterations to central visual pathways may be responsible for these visual impairments. It is also possible that repletion at an earlier stage might have offered a better chance of rehabilitating visual function.

Compared with studies in primates, studies in guinea pigs and rats are able to take advantage of the availability of large numbers of animals to implement cross-over designs in the investigation of sensitive periods in relation to dietary FAs and functional outcomes. Here animals in each dietary group are switched to the diet of the other group at a particular timepoint; a complete design also includes groups maintained on their respective diets. Such a design was used to address the issue of sensitive periods in the first generation of guinea pigs exposed to an n-3-deficient diet for 16 weeks from weaning, with rehabilitation beginning at 6 weeks for 5 or 10 weeks or at 11 weeks for 5 weeks (Weisinger et al., 1999). The results indicated that retinal function and brain DHA levels were similar to those of controls after 10 weeks of rehabilitation. Partial recovery was achieved after 5 weeks of rehabilitation, but, interestingly, despite similar retinal DHA levels, there were larger deficits in the ERG in the group in which rehabilitation began at 11 weeks compared with 6 weeks. This suggests that the age at which repletion is instituted may indeed be an important consideration with respect to ultimate functional outcome. Further support for the existence of a sensitive period for n-3 FA accretion in the retina and brain is provided by a preliminary report of a study conducted in rats by Armitage et al. (2000), in which the perinatal diet had the most significant impact on adult ERG function.

It is important to keep in mind that different regions of the brain mature at different rates (Rodier, 1980). It is therefore expected that the timing with respect to sensitive periods may differ with respect to different functions, depending on the neural systems involved. Thus, not all studies seeking evidence for such periods have focused on retinal changes. For example, as reported in Weisinger et al. (2001) and Armitage et al. (2003), ALA, when supplied during early development, can have an impact on blood pressure later in life in Sprague-Dawley rats, a strain that is not predisposed to hypertension. They used a double cross-over design that included groups maintained on their diets throughout the study. Both the rearing and post cross-over diets had significant effects on mean arterial pressure (MAP). The MAP of the def-def group was higher than MAPs for all other groups, followed by that of the def-con group, whereas the con-def and con-con groups were not significantly different. Thus, regardless of repletion attempts, early n-3 PUFA deficiency resulted in hypertension, whereas an adequate supply of n-3 PUFA during early development provided some degree of protection against increased blood pressure.

Only a few published studies appear to address this question in relation to behavior. One such study was conducted in our laboratory on behavioral development and visual learning. We used a cross-fostering design in pregnant and lactating mice to compare the effects of pre- and postnatal n-3 dietary deficiency on behavioral development and visual learning (Wainwright et al., 1991).

Animals were cross-fostered at birth to dams in either the opposite or the same dietary condition. At weaning, all pups were fed a control diet. Prenatal n-3 deficiency was associated with retardation of behavioral development. However, the adult animals did not differ in terms of learning a visual discrimination or visual acuity, suggesting that there were no long-term effects. Moriguchi and Salem (2003) examined the reversibility of the impairment seen in performance on a spatial learning task in the Morris water maze in the third generation of n-3-deficient Long-Evans rats. Deficient pups were switched to the n-3 adequate diet at birth, weaning, or at 7 weeks (adults); controls included both n-3-deficient and -adequate groups. Repletion from birth resulted in performance similar to that of the adequate group. Although the group rehabilitated at weaning was significantly improved compared to the n-3-deficient group, its performance was intermediate between this group and that of the adequate group, whereas the group that was rehabilitated as adults showed partial recovery of behavioral function. These findings suggest that with adequate supplementation, animals severely deprived during early development may regain brain DHA and improve some aspects of brain function.

11.4 RELATIONSHIP BETWEEN BRAIN DOCOSAHEXAENOIC ACID (DHA) AND COGNITIVE/BEHAVIORAL DEVELOPMENT

11.4.1 Studies in Infants

The high levels of DHA in the retina and brain have led much of the research on its functional effects to focus on outcomes associated with visual and cognitive development [reviewed in Gibson et al. (2001), Uauy et al. (2001) and Cockburn (2003)]. There are a number of studies in which human infants have been assigned randomly to be fed formulae supplemented with DHA only, or both DHA and AA, and then tested by standardized developmental scales. For example, recently there have been reports from two randomized clinical trials in which the formula of term infants was supplemented with DHA and AA during the first postnatal months and development measured on the Bayley scales at 18 months. In the large multicenter trial, there were no significant effects (Lucas et al., 1999). In the smaller study, however, the group of infants supplemented with AA and DHA showed a seven-point increase in the mental development index (MDI) relative to controls (Birch et al., 2000). They also showed enhanced visual maturation and a correlation between plasma and red blood cell DHA at 4 months and MDI scores at 18 months. Notwithstanding the limitations to the interpretation of such tests discussed previously, these findings provide a strong rationale for further research in relation to this question. Numerous studies on the effects of dietary supplementation with LCPUFAs have used the Fagan test of Infant Intelligence, which was developed to assess infant cognition by using paired comparisons of visual stimuli (Fagan, 1990). A consistent finding was that preterm infants fed DHA-supplemented diets showed shorter look durations on this test, but with no effects on visual recognition (Carlson and Werkman, 1996; Werkman and Carlson, 1996). Similar findings have been reported in rhesus monkeys (Reisbick et al., 1997), and it was suggested that the longer look durations in the n-3-deficient monkeys were due to their inability to disengage their attention from a visual stimulus (Reisbick and Neuringer, 1997). This dissociation in outcome between attention and recognition has led others to speculate that the effects of LCPUFAs may be selective with respect to one of the two separate neural substrates that have been related to different aspects of visual attention (Carlson, 2001). Specifically, the effect is not associated with the ventral stream, known as the "what" system, which projects to the inferior temporal lobe and mediates object recognition through the analysis of visual features. Rather, it is attributed to the dorsal stream, known as the "where" system, which projects to the posterior parietal lobe and mediates the direction of attention to the location of the object (Colombo, 2001). One other study in relation to dietary lipids that has used a well-controlled means–ends problem-solving task showed at 10 months that infants fed for 4 months with a formula containing AA and DHA were better than those fed a control formula that was low in n-3 FAs (Willatts et al., 1998). These findings encourage the use in future studies of tests that tap specific cognitive domains, with the choice of measure informed by a theoretical understanding of the neural substrates involved and the role of LCPUFAs in their development and function. Following we describe the studies in which we have used the rat as an animal model to address a possible relationship between n-3 FAs and performance on a dopamine-related behavioral task.

11.4.2 DOPAMINE HYPOTHESIS

One of the main neuromodulatory systems of the brain shown to be affected by dietary EFA deficiency is that involving the release of dopamine (Innis, 2000; Chalon et al., 2001). There are two major dopaminergic pathways in the brain, both originating in the midbrain (Smith and Kieval, 2000). The nigrostriatal pathway, which originates in the substantia nigra, projects to the basal ganglia and is involved in behavioral activation. The mesocorticolimbic system, which originates in the ventral tegmental area, projects to the frontal cortex as well as to limbic structures and the nucleus accumbens. The functions of dopamine in the prefrontal cortex have been identified as those associated with cognitive processes involving attention and working memory (Arnsten, 1997). The release of dopamine in the nucleus accumbens, on the other hand, has been associated with reinforcement. It is thought that here its role is to label environmental stimuli with appetitive value and to facilitate the initiation of the appropriate instrumental response (Kalivas and Nakamura, 1999). In rats, chronic dietary n-3 deficiency has been associated with neurochemical measures indicative of reduced frontal dopamine function, with a corresponding increase in limbic dopamine activity (Chalon et al., 2001). We therefore conducted a preliminary study in n-3-deficient rats to ascertain whether there were changes in behaviors that have been associated with the function of each of these systems. We first addressed effects on the pathway to the nucleus accumbens by testing the rats on conditioned place-preference in order to ascertain their response to the rewarding effects of amphetamine. This showed no consistent pattern of differences (Wainwright et al., 1998b). We then addressed the function of dopaminergic neurons projecting to the frontal cortex. It has been suggested that this system plays an important role in immediate short-term memory, sometimes called working memory (Arnsten, 1997). To measure short-term memory, we used a delayed-matching-to-place (DMP) version of the Morris water maze (described next). As predicted, the n-3-deficient rats showed impaired performance on this task (Wainwright et al., 1998b). We therefore used this task in further studies to investigate the functional effects of supplementation with DHA.

11.4.3 DELAYED-MATCHING-TO-PLACE AS A MEASURE OF SHORT-TERM MEMORY IN ANIMAL STUDIES BY USING THE MORRIS WATER MAZE

Short-term memory, also referred to as episode memory, is the type of memory necessary to solve a problem unique to a specific task. This is in contrast to reference memory, which involves learning generally the overall demands of the task. An analogy in humans would be a card game such as bridge, in which reference memory pertains to learning the rules of the game whereas working memory pertains to the capacity to hold in memory for the duration of one game the identity of cards that have already been played. This latter capacity is often assessed in animals by delayed-matching type tasks, in which the animal is required to hold information in memory for a specified period to make an appropriate behavioral choice (e.g., Murphy et al., 1996). In such tasks, it is also possible to further challenge mnemonic capabilities by increasing the length of the delay between the initial presentation information and the subsequent recall. In contrast with the place version, the DMP version of the Morris water maze is such a task. In the place version, rats learn to escape by learning the location of a fixed hidden platform relative to extramaze cues. In the DMP version they are presented with a series of problems, with the platform changing locations on each problem. Each problem comprises two trials, the first trial being a search trial that provides information regarding platform location and the second trial being the recall trial. Memory of the platform location from the first trial should be manifest in shorter distances swum on the second trial. In inferring cognitive effects in such studies, it is important to also take into account possible alterations in sensory, motor, or motivational attributes that may contribute to differences in performance.

We first used this task to show that, at a 60-sec delay, rats that had been reared and maintained on n-3-deficient diets showed a smaller difference between search and recall trials than controls did (Wainwright et al., 1998b). Then, having demonstrated the effects of deficiency, we used the same task to investigate effects of dietary supplementation with DHA and AA to diets already containing sufficient LA and ALA. In this study artificially reared rats were provided with either 0 or 2.5% AA or DHA, or both, and the dietary treatment was continued from weaning until behavioral testing at 6 weeks (Wainwright et al., 1999b). A fifth artificially reared group was fed a diet high in saturated fat (marginal n-6 and no n-3 FA), and a sixth group was suckled normally. Measures of brain FA composition were obtained at the conclusion of the behavioral testing, thereby allowing correlation of brain and behavioral measures. Although there were no differences among the supplemented groups on the DMP task, the saturated fat group, which had the lowest DHA levels, performed significantly more poorly than the suckled control group on this task. However, this difference cannot be attributed solely to differences in DHA status, as these two groups differed with respect to rearing condition, that is, artificially reared vs. dam reared, as well as diet. Furthermore, despite the considerable range in brain FA composition obtained in this study, subsequent analyses in the artificially reared animals did not support any correlations between forebrain values of either DHA or AA and performance on the DMP task. These findings suggest that once an optimum level of DHA is present in the diet, further supplementation does not improve short-term memory in rats as measured by the DMP task. There is, however, one caveat with respect to this interpretation: measures of total forebrain phospholipids may not be sufficiently sensitive to reflect changes in the individual brain region, that is, prefrontal cortex, that may be more salient with regard to this particular behavioral function.

11.4.4   THE SPONTANEOUSLY HYPERTENSIVE RAT AS AN ANIMAL MODEL OF ATTENTION DEFICIT HYPERACTIVITY DISORDER (ADHD)

In the supplementation study, we used the DMP task to assess whether DHA would improve short-term memory in a rat that has no known anomalies in terms of either behavior or FA metabolism. Other situations in which dietary supplementation may prove beneficial are pathological conditions, either hereditary or acquired, that lead to insufficient biosynthesis or utilization of LCPUFAs. Attention deficit hyperactivity disorder (ADHD) may be one such condition. ADHD is a behavioral disorder that affects 2 to 5% of grade-school children, with a higher prevalence in males (Paule et al., 2000). It is characterized by attention problems, impulsivity, and hyperactivity, and although learning deficits are not a diagnostic characteristic, many children with ADHD also experience learning difficulties. Neurobiological theories propose that ADHD is a neurodevelopmental disorder involving the brain dopamine systems. Specifically, it has been suggested that in this condition there is underactivity of frontal cortical dopamine systems, leading to cognitive deficits, with a corresponding overactivity in the subcortical regions, leading to hyperactivity (Swanson et al., 1998). Interestingly, these effects are similar to those described in relation to dietary FA deficiency. Although there has been some interest in diet, the preferred methods of treatment have been pharmacological interventions targeting dopamine systems, sometimes combined with behavioral strategies (Schachar et al., 1996). However, evidence of an inverse relationship between plasma concentrations of n-3 FAs and behavioral assessment scores in a subpopulation of children diagnosed with the disorder has led to the suggestion that there may be a therapeutic advantage to supplementation with LCPUFAs (Stevens et al., 1996). In the study described next, we addressed this in the spontaneously hypertensive rat (SHR), which is a widely accepted model of ADHD (Sagvolden, 2000), and again measured performance on the DMP task.

Comparisons of SHR with its progenitor strain, Wistar-Kyoto (WKY), indicate that it shares many of the behavioral characteristics of children with ADHD (Hendley, 2000). Neurochemical measures indicate that in addition it has decreased dopaminergic function in the prefrontal cortex (Russell et al., 1995). This is interesting because SHR has also been shown to have lower tissue levels of n-3 FAs than has WKY, possibly related to its propensity to oxidize a greater proportion of these compounds (Mills and Huang, 1993). This is a situation in which one might expect to see beneficial effects of dietary supplementation with DHA, in which provision of the preformed compound could to some extent offset the decrease due to increased oxidation and thereby improve related physiological and behavioral functions. In fact, dietary supplementation is an approach that has been used to control the increase in blood pressure that gives this strain its name, although the findings in this regard have been mixed (Mills and Huang, 1992; Frenoux et al., 2001).

We recently completed a study in which the objective was to compare the short-term memory capabilities of SHR with WKY rats, using the DMP version of the Morris water maze, and to ascertain whether these would be improved by dietary supplementation with AA and DHA (Clements et al., submitted). We also challenged the animals' mnemonic abilities by requiring them to find the hidden platform after both a short (60-sec) and long (60-min) delay between the search and recall trial. The performance of these animals in the water maze is independent of their status with respect to blood pressure (Gattu et al., 1997). Beginning at weaning, male SHR and WKY rats were fed either a control diet (AIN 93) in which the fat source was soybean oil or an LCPUFA diet in which soybean oil was supplemented with 0.5% AA and 1.0% DHA. The choice of levels of AA and DHA in this study was based on our previous work that used the artificial rearing model described previously, with the DHA increased to 1% to offset the expected increased oxidation. Behavioral testing began at 8 weeks. Interestingly, contrary to our predictions that SHR would be impaired, it was WKY that did not appear to use a short-term memory strategy to solve the task whereas SHR showed a dramatic improvement from the first to the second trial. Notwithstanding its short-term memory capabilities at a short delay, when challenged by the longer delay, SHR did show some memory impairment as indicated by longer path lengths on the recall trial after the 60-min compared with the 60-sec delay. However, contrary to expectation, there was no improvement related to dietary supplementation. This may be explained by the findings with respect to the data describing the FA composition of the brain phospholipids, which indicate that the effects of dietary supplementation also differed between the two strains. As expected, WKY supplementation increased DHA in the PE fraction, with no effect on AA. In contrast, the effects in SHR were extremely paradoxical, with supplemented groups either showing no significant change (PE) or decreases in both AA and DHA (PC and PS/PI). These findings emphasize the importance of taking the possibility of such gene-by-diet interactions into account when making recommendations based on studies of dietary supplementation in normal populations.

11.5 SUMMARY AND CONCLUSIONS

Nutritional factors play an integral role in the development of the brain. Because of the complex interactive processes involved, nutrition may influence function either directly by altering critical biochemical pathways or indirectly by altering patterns of input necessary to the formation of the integrated neural systems that mediate cognitive capabilities. The brain is high in lipids, of which a large proportion in the gray matter comprises LCPUFAs, particularly AA and DHA, which accrue rapidly during the brain growth spurt. There is growing evidence for an important role for n-3 FAs in brain function, including their involvement in mechanisms affecting neuronal survival and plasticity. There is some evidence in support of enhanced performance on standardized developmental tests in preterm and term infants fed DHA-supplemented formulae, but the effects of supplementation beyond what might be considered adequate dietary levels are less clear. Dose–response studies conducted in our laboratory suggest a threshold of incorporation of DHA into brain phospholipids in rats. Because of the reciprocity of effects of dietary AA and DHA on brain FA composition, our findings support inclusion of both in supplemented diets, with DHA being in the range of 0.5 to 1% of FAs and AA of 0.5%. These studies also show n-3 deficiency to be associated with a deficit in short-term memory, but with no indication that DHA supplementation improves performance beyond that of normal controls. Studies in the SHR rat, an animal model of ADHD, show this strain to be anomalous in terms of changes in brain phospholipids in response to dietary supplementation. An implication of this is that individual differences in metabolism may be an important factor influencing variability in response to dietary manipulations in neurodevelopmental disorders of this type.

ACKNOWLEDGMENTS

The preparation of this chapter was supported by a grant from the National Sciences and Engineering Research Council of Canada (NSERC) to P. Wainwight. Parts are based on material compiled by D. Martin during the tenure of an NSERC Undergraduate Fellowship in P. Wainwright's laboratory during the summer of 2002.

REFERENCES

Agranoff, B. W. and Hajra, A. K. (1994) Lipids. In *Basic Neurochemistry: Molecular, Cellular and Medical Aspects,* 5th ed. (B. W. Agranoff, R. W. Albers and P. B. Molinoff, Eds.), pp. 97–116. Raven Press, New York.

Ahmad, A., Murthy, M., Greiner, R. S., Moriguchi, T. and Salem, N., Jr. (2002) A decrease in cell size accompanies a loss of docosahexaenoate in the rat hippocampus. *Nutritional Neuroscience,* 5, 103–113.

Arnsten, A. F. T. (1997) Catecholamine regulation of the prefrontal cortex. *Journal of Psychopharmacology,* 11, 152–164.

Armitage, J. A., Pearce, A. D., Sinclair, A. J., Vingrys, A. J., Weisinger, R. S., and Weisinger, H. S. (2003) Increased blood pressure later in life may be associated with perinatal h-3 fatty acid deficiency. *Lipids,* 38, 459–464.

Armitage, J. A., Weisinger, J. S., Vingrys, A. J., Sinclair, A. J. and Weisinger, R. S. (2000) Perinatal omega-3 fatty acid deficiency alters ERG in adult rats irrespective of tissue fatty acid content. *Investigative Ophthalmology & Visual Science,* 41, S245 (abstract).

Auestad, N. and Innis, S. M. (2000) Dietary n-3 fatty acid restriction during gestation in rats: neuronal cell body and growth cone fatty acids. *American Journal of Clinical Nutrition,* 71, 312S–314S.

Bayer, S. (1989) Cellular aspects of brain development. *Neurotoxicology,* 10, 307–320.

Bayley, N. (1969) *Bayley Scales of Infant Development.* Psychological Corporation, New York.

Bazan, N. G., Allan, G. and Rodriguez de Turco, E. B. (1993) Role of phopholipase A_2 and membrane-derived lipid second messengers in membrane function and transcriptional activation of genes: implications in cerebral ischemia and neuronal excitability. In *Neurobiology of Ischemic Brain Damage* (K. Kogure, K.-A. Hossmann and B. K. Siesjo, Eds.), *Progress in Brain Research*, Vol. 96, pp. 247–257. Elsevier, Amsterdam.

Birch, E. E., Garfield, S., Hoffman, D. R., Uauy, R. and Birch, D. G. (2000) A randomized controlled trial of early dietary supply of long-chain polyunsaturated fatty acids and mental development in term infants. *Developmental Medicine and Child Neurology,* 42, 174–181.

Berger, A., Crozier, G. Bisogno, T., Cavaliere, P., Innis, S. and Di Marzo, V. (2001) Anandamide and diet: inclusion of dietary arachidonate and docosahexaenoate leads to increased brain levels of the corresponding *N*-acylethanolamines in piglets. *Proceedings of the New York Academy of Sciences,* 98, 6402–6406.

Black, J. E. (1998) How a child builds its brain: some lessons from animal studies of neural plasticity. *Preventive Medicine,* 27, 168–171.

Bourrè, J. M., Durand, G., Pascal, G. and Youyou, A. (1989) Brain cell and tissue recovery in rats made deficient in n-3 fatty acids by alteration of dietary fat. *Journal of Nutrition,* 119, 15–22.

Broughton, K. S. and Wade, J. W. (2002) Total fat and (n-3):(n-6) fat ratios influence eicosanoid production in mice. *Journal of Nutrition,* 132, 88–94.

Carlson, N. R. (2001) *The Physiology of Behavior,* 7th ed., pp 184–185. Allyn & Bacon, Boston.

Carlson, S. A. and Werkman, S. H. (1996) A randomized trial of visual attention of preterm infants fed docosahexaenoic acid until two months. *Lipids,* 31, 85–90.

Chalon, S., Vancassel, S., Zimmer, L., Guilloteau, D. and Durand, G. (2001) Polyunsaturated fatty acids and cerebral function: focus on monaminergic neurotransmission. *Lipids,* 36, 937–944.

Cheruku, S. R., Montgomery-Downs, H. E., Farkas, S. L., Thoman, E. B. and Lammi-Keefe, C. J. (2002) Higher maternal plasma docosahexaenoic acid during pregnancy is associated with more mature neonatal sleep-state patterning. *American Journal of Clinical Nutrition,* 76, 608–613.

Clandinin, M. T., Chappell, J. E., Heim, T., Swyer, P. R. and Chance, G. W. (1981) Fatty acid utilization in perinatal de novo synthesis of tissues. *Early Human Development,* 5, 355–366.

Clandinin, M. T., Chappell, J. E., Leong, S., Heim, T., Swyer, P. R. and Chance, G. W. (1980a) Intrauterine fatty acid accretion rates in human brain: implications for fatty acid requirements. *Early Human Development,* 4, 121–129.

Clandinin, M. T., Chappell, J. E., Leong, S., Heim, T., Swyer, P. R. and Chance, G. W. (1980b) Extrauterine fatty acid accretion in infant brain: implications for fatty acid requirements. *Early Human Development,* 4, 131–138.

Clandinin, M. T. and Jumpsen, J. (1997) Fatty acid metabolism in brain in relation to development, membrane structure, and signaling. In *Handbook of Essential Fatty Acid Biology* (S. Yehuda and D. I. Mostofsky, Eds.), pp. 15–65. Humana Press, Totowa, NJ.

Clements, K., Girard, T. A., Xing, H.-C. and Wainwright, P. E. (2003) Spontaneously hypertensive and Wistar Kyoto rats differ in terms of delayed-matching-to-place performance and their response to dietary long-chain polyunsaturated fatty acids. *Developmental Psychobiology,* 43, 57–69.

Cockburn, F. (2003) Role of infant dietary long chain polyunsaturated fatty acids, liposoluble vitamins, cholesterol and lecithin on psychomotor development. *Acta Paediatrica* Supplement 442, 19–33.

Colombo, J. (2001) Recent advances in infant cognition: implications for long-chain polyunsaturated fatty acid supplementation studies. *Lipids,* 36, 919–926.

Connor, W. E. and Neuringer, M. (1988) The effects of n-3 fatty acid deficiency and repletion upon the fatty acid composition and function of the brain and retina. In *Biological Membranes: Aberrations in Membrane Structure and Function* (M. L. Karnovsky, A. Leaf and L. C. Bolls, Eds.), pp. 275–294. A.R. Liss, New York.

Connor, W. E., Neuringer, M. and Lin, D. S. (1990) Dietary effects on brain fatty acid composition: the reversibility of n-3 fatty acid deficiency and turnover of docosahexaenoic acid in the brain, erythrocytes, and plasma of rhesus monkeys. *Journal of Lipid Research,* 31, 237–247.

Cooper, J. R., Bloom, F. E. and Roth, R. H. (1996) *The Biochemical Basis of Neuropharmacology,* 7th ed., pp. 126–193. Oxford University Press, New York.

Craig-Schmidt, M. C. and Huang, M.-C. (1998) Interaction of n-6 and n-3 fatty acids: implications for supplementation of infant formula with long-chain polyunsaturated fatty acids. In *Lipids in Infant Nutrition* (Y.-S. Huang and A. J. Sinclair, Eds.), pp. 63–84. AOCS Press, Champaign, IL.

Cunnane, S. C. (2000) The conditional nature of the dietary needs for polyunsaturates: a proposal to reclassify 'essential fatty acids' as 'conditionally-indispensable' or 'conditionally-dispensable' fatty acids. *British Journal of Nutrition,* 84, 803–812.

de Urquiza, A. M., Liu, S., Sjöberg, M., Zetterström, R. H., Griffiths, W., Sjövall, J. and Perlmann, T. (2000) Docosahexaenoic acid, a ligand for the reinoid-X receptor in mouse brain. *Science,* 290, 2140–2144.

Diamond, A., Prevor, M. B., Callender, G. and Druin, D. P. (1997) Prefrontal cortex cognitive deficits in children treated early and continuously for PKU. *Monographs of the Society for Research in Child Development,* 62, 1–208.

Dobbing, J. and Sands, J. (1979) Comparative aspects of the brain growth spurt. *Early Human Development,* 3, 79–83.

Everitt, B. J. and Robbins, T. W. (1997) Central cholinergic systems and cognition. *Annual Review of Psychology,* 48, 649–684.

Fagan, J. F. (1990) The paired-comparison paradigm and infant intelligence. *Annals of the New York Academy of Sciences,* 608, 337–364.

Farooqui, A. A. and Horrocks, L. A. (2001) Plasmalogens, phospholipase A_2 and docosahexaenoic acid turnover in brain tissue. *Journal of Molecular Neuroscience,* 16, 263–272.

Farquharson, J., Cockburn, F., Patrick, W. A., Jamieson, E. C. and Logan, R. W. (1992) Infant cerebral cortex fatty acid composition and diet. *Lancet,* 340, 810–813.

Fernstrom, J. (2000) Can nutrient supplements modify brain function? *American Journal of Clinical Nutrition,* 7, 1669S–1673S.

Fleming, A. H., O'Day, D. H. and Kraemer, G. W. (1999) Neurobiology of mother-infant interactions: experience and central nervous system plasticity across development and generations. *Neuroscience and Biobehavioral Reviews,* 23, 673–685.

Francis, D. D. and Meaney, M. J. (1999) Maternal care and the development of stress responses. *Current Opinion in Neurobiology,* 9, 128–134.

Fraser, M. and Wainwright, P. E. (2001) A study of the behavioural effects of prenatal ethanol exposure in mice fed a diet marginally deficient in essential fatty acids for two generations. *Nutritional Neuroscience,* 4, 445–459.

Frenoux, J.-M., Prost., E. D., Belleville, J. L. and Prost, J. L. (2001) A polyunsaturated fatty acid diet lowers blood pressure and improves antioxidant status in spontaneously hypertensive rats. *Journal of Nutrition,* 131, 39–45.

Gattu, M., Jr., Terry A., Pauly, J. and Buccafusco, J. (1997) Cognitive impairment in spontaneously hypertensive rats: role of central nicotinic receptors II. *Brain Research,* 771, 104–114.

Gibson, R. A., Chen, W. and Makrides, M. (2001) Randomized trials with polyunsaturated fatty acid interventions in preterm and term infants: functional and clinical outcomes. *Lipids,* 36, 873–883.

Gottlieb, G. (1998) Normally occurring environmental and behavioral influences on gene activity: from central dogma to probabilistic epigenesis. *Psychological Reviews,* 105, 792–802.

Green, N. S. (2002) Folic acid supplementation and prevention of birth defects. *Journal of Nutrition,* 132, 2356S–2360S.

Hall, W. H. and Oppenheim, R. W. (1987) Developmental psychobiology: prenatal, perinatal and early postnatal aspects of behavioural development. *Annual Review of Psychology,* 38, 91–128.

Heird, W. C. (1997) Statistically significant versus biologically significant effects of long-chain polyunsaturated fatty acids on growth. In *Developing Brain and Behaviour: The Role of Lipids in Infant Formula* (J. Dobbing, Ed.), pp. 169–198. Academic Press, London.

Hendley, E. (2000) WKHA rats with genetic hyperactivity and hyperreactivity to stress: a review. *Neuroscience and Biobehavioral Reviews,* 24, 41–44.

Horrobin, D. F. (1999) The phospholipid concept of psychiatric disorders and its relationship to the neurodevelopmental concept of schizophrenia. In *Phospholipid Spectrum Disorder in Psychiatry* (M. Peet, I. Glen and D. F. Horrrobin, Eds.), pp. 3–20. Marius Press, Carnforth, U.K.

Huang, E. and Reichardt, L. F. (2001) Neurotrophins: roles in neuronal development and function. *Annual Review of Neuroscience,* 24, 677–736.

Huang, Y.-S., Wainwright, P. E., DeMichele, S. J., Xing, H.-C., Biederman, J., Liu, J. W., Chuang, L.-T., Bobik, E., Jr. and Hastilow, C. (2002) Effects of feeding high levels of α-linolenic acid (ALA, 18:3n-3) and γ-linolenic acid (GAL, 8:3 n-6) to pregnant and lactating dams on tissue fatty acid distribution and behavioral development in pups. In *Fifth International Congress on Fatty Acids and Eicosanoids,* Taipei, Taiwan, August 2002.

Ikemoto, A., Nitta, A., Furukawa, S., Ohishi, M., Nakamura, A., Fujii, Y. and Okuyama, H. (2000) Dietary n-3 fatty acid deficiency decreases nerve growth factor content in rat hippocampus. *Neuroscience Letters,* 285, 99–102.

Innis, S. M. (2000) The role of dietary n-6 and n-3 fatty acids in the developing brain. *Developmental Neuroscience,* 22, 474–480.

Innis, S. M. and de la Presa Owens, S. (2001) Dietary fatty acid composition in pregnancy alters neurite membrane fatty acids and dopamine in newborn rat brain. *Journal of Nutrition,* 131, 118–122.

Jeffrey, B. G., Weisinger, H. S., Neuringer, M., and Mitchell, D. C. (2001) The role of docosahexaenoic acid in retinal function. *Lipids,* 36, 859–871.

Johnson, M. H. and Mareschal, D. (2001) Cognitive and perceptual development during infancy. *Current Opinion in Neurobiology,* 11, 213–218.

Jump, D. B. and Clarke, S. D. (1999) Regulation of gene expression by dietary fat. *Annual Review of Nutrition,* 19, 63–90.

Jumpsen, J. and Clandinin, M. T. (1995) *Brain Development: Relationship to Dietary Lipid and Lipid Metabolism.* AOCS Press, Champaign, IL.

Kalivas, P. W. and Nakamura, M. (1999) Neural systems for behavioral activation and reward. *Current Opinion in Neurobiology,* 9, 223–227.

Kandel, E. R., Schwartz, J. H. and Jessell, T. M (1995) *Essentials of Neural Science and Behavior,* pp. 243–267. Appleton-Lange, Norwalk, CT.

Karmiloff-Smith, A. (1998) Development itself is the key to understanding developmental disorders. *Trends in Cognitive Sciences,* 2, 389–398.

Keshavan, M. S. and Hogarty, G. E. (1999) Brain maturational processes and delayed onset in schizophrenia. *Development and Psychopathology,* 11, 525–543.

Kim, H. Y., Akbar, M., Lau, A. and Edsall, L. (2000) Inhibition of neuronal apoptosis by docosahexaenoic acid (DHA): role of phosphatidylserine in antiapoptotic effect. *Journal of Biological Chemistry,* 275, 35215–35223.

Klintsova, A. Y. and Greenough, W. T. (1999) Synaptic plasticity in cortical systems. *Current Opinions in Neurobiology,* 9, 203–208.

Kolb, B., Forgie, M., Gibb, R., Gorny, G. and Rowntree, S. (1998) Age, experience and the changing brain. *Neuroscience and Biobehavioral Reviews,* 22, 143–159.

Lauder, J. M. (1993) Neurotransmitters as growth regulatory signals role of receptors and second messengers. *Trends in Neurosciences,* 16, 233–240.

Lauritzen, I., Blondeau, N., Heurteaux, C., Widmann, C., Romey, G. and Lazdunski, M. (2000) Polyunsaturated fatty acids are potent neuroprotectors. *EMBO J.* 19, 1784–1793.

Lauritzen, L., Hansen, H. S., Jørgensen, M. H. and Michaelsen, K. F. (2001) The essentiality of long chain *n-3* fatty acids in relation to development and function of the brain and retina. *Progress in Lipid Research,* 40, 1–94.

Levitt, P., Harvey, J. A., Friedman, E., Simansky, K. and Murphy, E. H. (1997) New evidence for neurotransmitter influences on brain development. *Trends in Neurosciences,* 20, 269–274.

Lewis, D. A. and Levitt, P. (2002) Schizophrenia as a disorder of neurodevelopment. *Annual Review of Neuroscience,* 25, 409–432.

Lucas, A., Stafford, M., Morley, R., Abbott, R., Stephenson, T., MacFadyen, U., Elias-Jones, A. and Clements, H. (1999) Efficacy and safety of long-chain polyunsaturated fatty acid supplementation of infant-formula milk: a randomised trial. *Lancet,* 354, 1948–1954.

McCall, R. B. and Mash, C. W. (1997) Long-chain polyunsaturated fatty acids and the measurement and prediction of intelligence (IQ). In *Developing Brain and Behaviour: The Role of Lipids in Infant Formula* (J. Dobbing, Ed.), pp. 295–338. Academic Press, London.

Meck, W. H. and Williams, C. L. (2003) Metabolic imprinting of choline by its availability during gestation: implications for memory and attentional processing across the lifespan. *Neuroscience and Behavioral Reviews,* 27, 385–399.

Mills, D. E. and Huang, Y.-S. (1992) Metabolism of n-6 and n-3 polyunsaturated fatty acids in normotensive and hypertensive rats. In *Essential Fatty Acids and Eicosanoids* (A. Sinclair and R. Gibson, Eds.), pp. 345–348. AOCS Press, Champaign, IL.

Moore, S. A. (2001) Polyunsaturated fatty acid synthesis and release by brain-derived cells *in vitro. Journal of Molecular Neuroscience,* 16, 195–200.

Morgan, B. L. G., Oppenheimer, J. and Winick, M. (1981) Effects of essential fatty acid deficiency during late gestation on brain *N*-acetylneuramine acid content and behavior. *British Journal of Nutrition,* 46, 223–230.

Moriguchi, T. and Salem, N., Jr. (2003) Recovery of brain DHA leads to recovery of spatial task performance. *Journal of Neurochemistry,* 87, 297–309.

Murphy, M.G. (1990) Dietary fatty acids and membrane protein function. *Journal of Nutritional Biochemistry,* 1, 68–79.

Murphy, B., Arnsten, A., Goldman-Rakic, P. and Roth, R. (1996) Increased dopamine turnover in the prefrontal cortex impairs spatial working memory performance in rats and monkeys. *Proceedings of the National Academy of Sciences USA,* 93, 1325–1329.

Neuringer, M., Connor, W. E., Lin, D. S., Barstad, L. and Luck, S. (1986) Biochemical and functional effects of prenatal and postnatal 3 fatty acid deficiency on retina and brain in rhesus monkeys. *Proceedings of the National Academy of Sciences USA,* 83, 4021–4025.

Nowakowski, R. S. and Hayes, N. L. (1999) CNS development: an overview. *Development and Psychopathology,* 11, 395–417.

Pallas, S. L. (2001) Intrinsic and extrinsic factors that shape neocortical specification. *Trends in Neurosciences,* 24, 417–423.

Paterson, S. J., Brown, J. H., Gsödl, M. K., Johnson, M. H. and Karmiloff-Smith, A. (1999) Cognitive modularity and genetic disorders. *Science,* 286, 2355–2358.

Paule, M., Rowland, A., Ferguson, S., Chelonis, J., Tannock, R., Swanson, J. and Castellanos, F. (2000) Attention deficit/hyperactivity disorder: characteristics, interventions, and models. *Neurotoxicology and Teratology,* 22, 631–651.

Pollitt, E. (2001) The developmental and probabilistic nature of the functional consequencies of iron-deficiency anemia in children. *Journal of Nutrition,* 131, 669S–675S.

Quartz, S. R. and Sejnowski, T. J. (1997) The neural basis of cognitive development: a constructivist manifesto. *Behavioral and Brain Science,* 20, 537–596.

Reisbick, S. and Neuringer, M. (1997) Omega-3 fatty acid deficiency and behavior: a critical review and future directions for research. In *Essential Fatty Acid Biology: Biochemistry, Physiology and Behavioral Neurobiology* (S. Yehuda and D. I. Mostofsky, Eds.), pp. 397–426. Humana Press, Totowa, NJ.

Reisbick, S., Neuringer, M., Gohl, E., Wald, R. and Anderson, G. J. (1997) Visual attention in infant monkeys: effects of dietary fatty acids and age. *Developmental Psychology,* 33, 387–395.

Roudbaraki, M. M., Droulhault, R., Baqueart, T. and Vacher, P. (1996) Arachidonic acid and its lipoxygenase products on cytosolic calcium in GH3 cells. *American Journal of Physiology,* 63, 24–256.

Rodier, P. M. (1980) Chronology of neuron development: animal studies and their clinical implications. *Developmental Medicine and Child Neurology,* 22, 525–545.

Rojas, C. V., Greiner, R. S., Fuenzalida, L. C., Martinez, J. I., Salem, N., Jr. and Uauy, R. (2002) Long-term n-3 FA deficiency modifies peroxisome proliferator-activated receptor beta mRNA abundance in rat ocular tissues. *Lipids,* 37, 367–374.

Rose, S. A., Feldman, J. F. and Jankowski, J. J. (2003) The building blocks of cognition. *Journal of Pediatrics,* 143, 554–561.

Russell, V., Villiers, A. D., Sagvolden, T., Lamm, M. and Taljaard, J. (1995) Altered dopaminergic function in the prefrontal cortex, nucleus accumbens and caudate putamen of an animal model of attention deficit hyperactivity disorder — the spontaneously hypertensive rat. *Brain Research,* 676, 343–351.

Sagvolden, T. (2000) Behavioral validation of the spontaneously hypertensive rat (SHR) as an animal model of attention-deficit/hyperactivity disorder (AD/HD). *Neuroscience and Biobehavioral Reviews,* 24, 31–39.

Salem, N., Jr., Litman, B., Kim, H.-Y. and Gawrisch, K. (2001) Mechanisms of action of docosahexaenoic acid in the nervous system. *Lipids,* 36, 945–959.

Sandstrom, N. J., Loy, R. and Williams, C. L. (2002) Prenatal choline supplementation increases NGF levels in the hippocampus and frontal cortex of young and adult rats. *Brain Research,* 947, 9–16.

Sastry, P. S. (1985) Lipids of nervous tissue: composition and metabolism. *Progress in Lipid Research,* 24, 69–176.

Saugstad, L. F. (1998) Cerebral lateralisation and rate of maturation. *International Journal of Psychophysiology,* 28, 37–62.

Schachar, R., Tannock, R. and Cunningham, C. (1996) Treatment. In *Hyperactivity Disorders of Childhood* (S. Sandberg, Ed.), pp. 433–463. Cambridge University Press, Cambridge.

Seung Kim, H. F., Weeber, E. J., Sweatt, J. D., Stoll, A. L. and Marangell, L. B. (2001) Inhibitory effects of omega-3 fatty acids on protein kinase C activity *in vitro*. *Molecular Psychiatry,* 6, 246–248.

Shatz, C. (1992) The developing brain. *Scientific American,* 267, 61–67.

Slotkin, T. A. (1998) Fetal nicotine or cocaine exposure: which one is worse? *Journal of Pharmacology and Experimental Therapeutics,* 285, 931–945.

Smith, W. L., Borgeat, P. and Fitzpatrick, F. A. (1991) The eicosanoids: cyclooxygenase, lipoxygenase and epoxygenase pathways. In *Biochemistry of Lipids, Lipoproteins and Membranes* (D. E. Vance and J. Vance, Eds.), pp. 297–325. Elsevier, Amsterdam.

Smith, Y. and Kieval, J. Z. (2000) Anatomy of the dopamine system in the basal ganglia. *Trends in Neurosciences,* 23, S28–S33.

Spector, A. A. (2001) Plasma free fatty acid and lipoproteins as sources of polyunsaturated fatty acid for the brain. *Journal of Molecular Neuroscience,* 16, 159–165.

Sprecher, H., Luthria, D. L., Mohammed, B. S. and Baykousheva, S. P. (1995) Reevaluation of the pathways for the biosynthesis of polyunsaturated fatty acids. *Journal of Lipid Research,* 36, 2471–2477.

Strupp, B. J. and Levitsky, D. A. (1995) Enduring cognitive effects of early malnutrition: a theoretical appraisal. *Journal of Nutrition,* 125, 2221S–2232S.

Stevens, L. J., Zentall, S. S., Abate, M. L., Kuczek, T. and Burgess, J. R. (1996) Omega-3 fatty acids in boys with behaviour, learning and health problems. *Physiology and Behaviour,* 59, 915–920.

Swanson, J., Castellanos, F., Murias, M., LaHoste, G. and Kennedy, J. (1998) Cognitive neuroscience of attention deficit hyperactivity disorder and hyperkinetic disorder. *Current Opinion in Neurobiology,* 8, 263–271.

Tinoco, J. (1982) Dietary requirements and function of α-linolenic acid in animals. *Progress in Lipid Research,* 21, 1–45.

Uauy, R., Hoffman, D. R., Peirano, P., Birch, D. G. and Birch, E. E. (2001) Essential fatty acids in visual and brain development. *Lipids,* 36, 885–895.

Wainwright, P. E. (1997) Essential fatty acids and behavior: Is there a role for the eicosanoids? In *Essential Fatty Acid Biology: Biochemistry, Physiology and Behavioral Neurobiology* (S. Yehuda and D. I. Mostofsky, Eds.), pp. 299–341. Humana Press, Totowa, NJ.

Wainwright, P. E., Bulman-Fleming, M. B., Lévesque, S., Mutsaers, L. and McCutcheon, D. (1998a) Saturated-fat diet during development alters dendritic growth in mouse brain. *Nutritional Neuroscience,* 1, 49–58.

Wainwright, P. E., Huang, Y.-S., Bulman-Fleming, B., Mills, D. E., Redden, P. and McCutcheon, D. (1991) The role of n-3 essential fatty acids in brain and behavioral development: a cross-fostering study in the mouse. *Lipids,* 26, 37–45.

Wainwright, P. E., Huang, Y.-S, Bulman-Fleming, B., Lévesque, S. and McCutcheon, D. (1994) The effects of dietary fatty acid composition combined with environmental enrichment on brain and behavior in mice. *Behavioural Brain Research,* 60, 125–136.

Wainwright, P. E., Jalali, E., Mutsaers, L. M., Bell, R. and Cvitkovic, S. (1999a) An imbalance of dietary essential fatty acids retards behavioral development in mice. *Physiology and Behavior,* 66, 833–839.

Wainwright, P. E., Xing, H.-C., Girard, T., Parker, L. and Ward, G. R. (1998b) Effects of dietary n-3 deficiency on amphetamine-conditioned place preference and working memory in the Morris water-maze. *Nutritional Neuroscience,* 4, 281–293.

Wainwright, P. E., Xing, H.-C., Mutsaers, L., McCutcheon, D. and Kyle, D. (1997) Arachidonic acid offsets the effects on mouse brain and behavior of a diet with a low (n-6):(n-3) ratio and very high levels of docosahexaenoic acid. *Journal of Nutrition,* 127, 184–193.

Wainwright, P. E., Xing, H.-C., Ward, G. R., Huang, Y.-S., Bobik, E., Auestad, N. and Montalto, M. (1999b) Water maze performance is unaffected in artificially reared rats fed diets supplemented with arachidonic acid and docosahexaenoic acid. *Journal of Nutrition,* 129, 1079–1089.

Ward, G. R., Huang, Y.-S., Bobik, E., Xing, H.-C., Mutsaers, L., Auestad, N., Montalto, M. and Wainwright, P. E. (1998) Long-chain polyunsaturated fatty acid levels in formulae influence deposition of docosahexaenoic acid and arachidonic acid in brain and red blood cells of artificially reared neonatal rats. *Journal of Nutrition,* 128, 2473–2487.

Ward, G. R. Huang, Y.-S., Xing, H.-C., Bobik, E., Wauben, I., Auestad, N., Momtalto, M. and Wainwright, P. E. (1999) Effects of gamma-linolenic acid and docosahexaenoic acid in formulae on brain fatty acid composition in artificially reared rats. *Lipids,* 34, 1057–1063.

Weisinger, H. S., Vingrys, A. J., Bui, B. V. and Sinclair, A. J. (1999) Effects of dietary n-3 fatty acid deficiency and repletion in the guinea pig retina. *Investigative Opthalmology and Visual Science,* 40, 327–338.

Weisinger, H. S., Armitage, J. A., Sinclair, A. J., Vingrys, A. J., Burns, P. L. and Weisinger, R. S. (2001) Perinatal omega-3 fatty acid deficiency affects blood pressure later in life. *Nature Medicine,* 7, 258–259.

Werkman, S. W. and Carlson, S. E. (1996) A randomized trial of visual attention of preterm infants fed docosahexaenoic acid until nine months. *Lipids,* 31, 91–97.

Willatts, P., Forsyth, J. S., DiMondugno, M. K., Varma, S. and Colvin, M. (1998) Effect of long-chain polyunsaturated fatty acids in infant formula on problem solving at 10 months of age. *Lancet,* 352, 688–691.

Yavin, E., Brand, A. and Green, P. (2002) Docosahexaenoic acid abundance in the brain: a biodevice to combat oxidative stress. *Nutritional Neuroscience,* 5, 149–157.

Youyou, A. G., Durand, G., Pascal, M., Piccotti, M., Dumont, O. and Bourre, J. M. (1986) Recovery of altered fatty acid composition induced by a diet devoid of n-3 fatty acids in myelin, synaptosomes, mitochondria and microsomes of developing rat brain. *Journal of Neurochemistry,* 46, 224–227.

12 Dietary Modulation of the Behavioral Consequences of Psychoactive Drugs

Robin B. Kanarek, Kristen E. D'Anci, Wendy Foulds Mathes, Rinah Yamamoto, R. Todd Coy, and Monica Leibovici

CONTENTS

12.1 INTRODUCTION

Drug abuse represents one of the major health problems in the U.S. Exact information on the extent of the problem is difficult to obtain because of the illegal status of many drugs and the negative connotations associated with drug, alcohol, and tobacco use. However, data indicate that at any point in time 2–10% of adults in the U.S. are either abusing or addicted to illicit drugs, 5–10% regularly consume large amounts of alcohol, and 25–30% smoke tobacco (Doweiko, 2002).

It has been taken as a *sine qua non* that psychoactive drugs produce their effects by acting on receptors within the central nervous system (CNS). Although neuronal responses to a psychoactive drug are similar across members of a species, the behavioral consequences of these drugs differ widely among individuals. In our own species, some can drink alcohol, smoke cigarettes or marijuana, or even experiment with heroin and cocaine without developing problems of drug abuse or dependence, whereas others who initially engage in such behaviors go on to suffer the consequences of substance abuse (deWit, 1997). Individual differences in responses to psychoactive drugs are not unique to humans. Animals used as experimental models of drug abuse and to evaluate the clinical properties of psychoactive drugs also differ widely in their responses to these drugs (Koob and Le Moal, 2001).

Factors contributing to individual differences in responses to psychoactive drugs can be categorized into those intrinsic and extrinsic to the organism. Intrinsic factors are integral to the organism and include sex, age, and genetic background. With respect to the role of these factors in mediating the behavioral consequences of psychoactive drugs, recent work has shown that males and females display quantitative differences in their responses to the rewarding effects, abuse liability, expression of physical dependence, and relapse potential of a variety of psychoactive drugs (e.g., Cicero et al., 2000; Lynch and Carroll, 2000). Some of these differences may simply reflect that females typically weigh less than males. However, there is growing evidence that hormonal and other physiological factors also contribute to sex differences in responses to psychoactive drugs (e.g., Lynch and Carroll, 2000). Family, twin, and adoption studies indicate that although there are no genes uniquely associated with drug abuse, genetic factors play a crucial role in determining the consequences of drug use (Comings, 1997; Vanyukov et al., 2003). Similarly, research with animal models has demonstrated significant differences among rodent strains in their sensitivity to psychoactive drugs and that rats and mice can be selectively bred for high and low preferences for alcohol and other drugs of abuse (Comings, 1997; George et al., 1997; Mogil et al., 2000).

Extrinsic or environmental factors also are of critical importance when investigating individual differences in responses to psychoactive drugs. In animals, housing conditions, early social isolation, stress, contextual cues, availability of concurrent reinforcers, and prior drug experience can influence behavioral responses to psychoactive drugs. Similar variables modify the responses of our own species to drugs of abuse. For example, conditions that reinstate drug-seeking behavior in animals, such as stress or being returned to the environment in which drugs were experienced, also provoke drug craving and relapse in humans (DeVries and Shippenberg, 2002; Koob and LeMoal, 2001; O'Brian et al., 1992).

Nutrition is another important environmental variable to consider when determining the behavioral consequences of psychoactive drugs. In experimental animals, food deprivation increases drug-seeking behavior and can lead to the reinstatement of drug administration after a period of abstinence (Bell et al., 1996; Carroll, 1999; Carroll and Meisch, 1984; Shalev et al., 2003). In contrast, concurrent availability of palatable, sweet-tasting foods and fluids reduces drug self-administration (Carroll, 1999; Cosgrove and Carroll, 2003). Eating sweets may also moderate drug-taking behavior in humans. Recovering alcoholics and heroin addicts have been reported to consume large quantities of sweet-tasting foods and fluids, which may lessen the adverse consequences of drug withdrawal and thus help maintain drug abstinence (Morabia et al., 1989; Szpanowska-Wohn et al., 2000; Yung et al., 1983; Zador et al., 1996).

In addition to moderating drug intake, studies have shown that an organism's propensity to consume sweet-tasting substances can predict the quantities of alcohol, psychostimulants, and opiates an animal will use. For example, rats that typically eat large amounts of sucrose self-administer more alcohol, amphetamine, and morphine than those that, on average, take in lesser quantities of the sugar (DeSousa et al., 2000; Gosnell, 2000). In addition, rats selectively bred for having greater preferences for sweet tastes more rapidly attain stable rates of ethanol and cocaine self-administration than those bred for having lower preferences (Bell et al., 1994; Carroll et al., 2002; Gosnell, 2000). Sweet preferences may correlate with drug-taking behavior in humans as well as experimental animals. In a series of studies, Kampov-Polevoy et al. (2001, 2003) reported that the sweet preferences of alcoholics are greater than those of nonalcoholics. Similarly, recent work has shown that cocaine-abusing patients, like alcoholics, prefer higher concentrations of sweet-tasting fluids than those who refrain from drug-taking behavior (Janowsky et al., 2003). Additional support for a relation between sweet preferences and drug taking comes from the observation that individuals who consume large quantities of palatable high-carbohydrate and high-fat foods as a correlate of suffering from bulimia are more likely to have problems related to drug abuse than individuals without the disorder (Hardy and Waller, 1989).

This chapter reviews the role of nutrition in determining the behavioral and neurochemical consequences of psychoactive drugs. There is compelling evidence that subtle alterations in diet may significantly alter the rewarding, adverse, and therapeutic effects of drugs. As we will discuss, dietary variables affect many different classes of psychoactive substances including opiates, psychostimulants and hallucinogens. The ability of nutritional variables to influence such a broad spectrum of drugs may hold promise for a variety of therapeutic applications employing diet to decrease drug withdrawal symptoms, increase the success of drug-abuse treatment, and even facilitate pain management.

12.2 OPIATES

Opiates represent the Dr. Jekyll and Mr. Hyde of psychoactive agents. On their beneficent side, morphine and related opiate drugs are recognized as among the most effective drugs for treating pain. However, like the good doctor, these drugs have the potential to turn malicious. The ability of these drugs to stimulate reward pathways in the brain promotes drug abuse and its devastating consequences. Tolerance to the drugs' actions, physical dependence, and symptoms of drug withdrawal when drug taking ceases are not uncommon with prolonged use of opiates.

Opiate drugs act in the CNS on at least three different receptor subtypes, named μ, κ, and δ on the basis of the drugs with which they were originally found to bind. Each receptor type is characterized by a distinctive distribution within the brain and a specific pharmacological profile. Additional work conducted during the 1970s revealed that the body produces neuropeptides that bind selectively to these receptors and produce behavioral responses similar to those of morphine and other opiate agonists (DeVries and Shippenburg, 2002).

12.2.1 Nutrition and Opioid Effects on Feeding Behavior

Experiments conducted over the past 30 years have established that opiate drugs significantly alter both total food intake and selection of dietary components. In general, administration of opiate agonists such as morphine is followed by an increased food intake, whereas administration of opiate antagonists such as naloxone is associated with a reduction in feeding behavior. Opiates, however, do not uniformly affect intake of all types of foods. Rather, opiate agonists and antagonists alter intake of palatable foods to a greater degree than intake of less palatable items. For example, in experimental animals, morphine administration is associated with a greater enhancement and naloxone administration with a greater reduction in intake of palatable or preferred foods than of "ordinary" foods such as rodent chow (Bodnar, 2004; Kelley et al., 2002). Similarly, in humans, opiate antagonists reduce the rated pleasantness of foods and selectively decrease intake of the most preferred foods in a meal (Yeomans and Gray, 2002). On the basis of these findings, it is postulated that opiate agonists increase, whereas opiate antagonists decrease, the hedonic or rewarding properties of food.

Intake of a palatable substance can influence the effects of opiate drugs on food intake, even after the palatable item is no longer available. Rats that have previously consumed a palatable sweet-tasting fluid are more sensitive to the anorectic effects of the opiate antagonist naltrexone than rats that have no experience with the sweet-tasting solution (Kanarek et al., 1997; Yeomans and Gray, 2002). The results of these experiments have led to the hypothesis that the endogenous opioid system modulates the rewarding or hedonic aspects of feeding behavior.

12.2.2 PALATABLE FOODS ALTER THE PAIN-RELIEVING PROPERTIES OF OPIATE AGONISTS

Opiate drugs are an extremely important part of the physician's arsenal for treating chronic pain. However, these drugs are often administered in less than effective doses, as clinicians are concerned about the potential for the development of tolerance to the drugs' pain-relieving actions and addiction (Doweiko, 2002). Thus, it is important to find ways to increase the analgesic potency of opiate drugs that could simultaneously decrease their additive properties. Recent work suggests that intake of palatable foods might be one way to achieve this goal.

A growing body of research supports the conclusion that intake of palatable nutrients can moderate the pain-relieving properties of opiate agonists (D'Anci et al., 1997a, 1997b; Kanarek et al., 2001). When consumed for a short period of time (less than 24 h), rodents consuming nutritive sucrose solutions are less sensitive to the pain-relieving attributes of opiate agonists than animals not consuming the sugar. In contrast, chronic consumption (e.g., 3 weeks) of the same solutions is associated with an enhancement in the analgesic actions of opiate drugs (D'Anci et al., 1997a). Moreover, the development of tolerance to the pain-relieving properties of opiate agonists is delayed in rats chronically consuming a calorically dense, sweet-tasting solution relative to rats not given the palatable fluid (D'Anci, 1999).

The nutritional value of the palatable substances is important when examining the effects of these items on opiate-induced analgesia. Initially, novel nutritive and nonnutritive sweet-tasting substances similarly affect the pain-relieving qualities of opiate agonists. For example, short-term (less than 24-h) intake of either a sucrose or saccharin solution decreases the analgesic action of morphine. However, with chronic exposure, the effects of nutritive and nonnutritive sweeteners diverge, with intake of nutritive solutions generally enhancing opiate-induced analgesia and intake of nonnutritive solutions either decreasing or failing to have an effect on the pain-relieving actions of opiate drugs (D'Anci et al., 1997b). Recent work demonstrating that intraperitoneal injections of glucose, the final metabolic breakdown product of sucrose, can also potentiate morphine's pain-relieving properties suggests that the ability of sweet substances to alter the behavioral actions of opiate drugs is not solely a function of taste hedonics, but also depends, to some degree, on the physiological consequences of these substances (R. Yamamoto and R.B. Kanarek, unpublished results).

Intake of palatable foods and fluids can promote analgesia in humans and experimental animals. Human infants appear to be particularly sensitive to the pain-relieving properties of palatable nutrients. Numerous studies have demonstrated that newborns allowed to consume sweet-tasting wine or sucrose solutions display less indications of distress following painful procedures (e.g., blood drawing or circumcision) than infants not given these palatable fluids (Blass and Hoffmeyer, 1991). Although the analgesic actions of palatable foods and fluids have been reported to wane with age, there is evidence that consumption of subjectively palatable foods reduces pain sensitivity in adults as well as infants (Mercer and Holder, 1997).

12.2.3 NUTRITIONAL VARIABLES ALTER THE REWARDING ACTIONS OF OPIATE DRUGS

One method extensively used to investigate the rewarding effects of psychoactive drugs in experimental animals is drug self-administration. In this procedure, animals receive small quantities of a drug each time they perform an operant response. If the animals increase their rate of responding over time, it is assumed that the drug is serving as a positive reward.

Opiate drugs such as morphine and heroin are readily self-administered by both animals and humans (DeVries and Shippenberg, 2002). However, both food restriction and intake of palatable foods can moderate this behavior. Food restriction is often used in studies of drug abuse. Restricting animals' caloric intake to decrease body weight serves as a potent motivation for learning self-administration of drugs. However, the practice of food restriction can significantly alter drug self-administration. More specifically, with respect to opiates, it has been demonstrated that animals self-administered significantly more heroin and etonitazene when food deprived than when allowed free access to nutrients (Carroll, 1999; Carroll et al., 2001).

In contrast to food restriction, intake of palatable foods decreases self-administration of opiate drugs. When allowed to consume sugar, rats self-administered less morphine and etonitazene than when not given this sweet substance (Carroll, 1999; R.B. Kanarek, unpublished data). Eating sweets may also moderate the rewarding effects of opiates in our own species. Recovering heroin addicts who consume large quantities of sugar-containing foods and fluids are reported to remain drug abstinent for longer periods of time than those who consume smaller amounts of these palatable substances (Morabia et al., 1989).

Although self-administration studies are instructive for studying how a palatable diet can alter a drug's rewarding attributes, the effects of the drug on motor behavior can modify the rate of responding for the drug, independent of the drug's reinforcing effect. To decrease the potentially confounding effects of drug-induced alterations in motor behavior, conditioned place preference (CPP) procedures have been used to assess the rewarding effects of drugs of abuse (Bardo and Bevins, 2000). In a CPP procedure, animals receive injections of a rewarding drug in a compartment with distinct visual, tactile, and olfactory cues, and saline injections in another compartment with different cues. Evidence that a drug has rewarding consequences is provided by an increase in the amount of time spent in the drug-paired compartment after conditioning relative to that spent in this compartment before drug injections. In support of the idea that palatable foods alter the rewarding properties of opiate drugs, rats consuming a sucrose solution spent significantly more time on the side of the conditioning chamber paired with either morphine or the opiate agonist fentanyl than rats not given the sugar (Vitale et al., 2003). Evidence that the effect of diet is related to the rewarding effects of opiates and not from their effects on motor behavior comes from studies demonstrating that rats given highly palatable nutritive and nonnutritive fluids did not differ in overall locomotor activity following administration of morphine in comparison with chow-fed controls (K.E. D'Anci, unpublished results).

Preliminary data suggest that nutrient status may also play a role in determining the severity of opiate withdrawal. Morphine-addicted rats chronically consuming a sucrose solution displayed more severe symptoms of drug withdrawal when injected with the opiate antagonist naloxone than rats that had not received the sweet solution. In contrast, morphine-addicted rats given single intraperitoneal glucose injections showed less severe symptoms of drug withdrawal than animals not given the injections (Yamamoto and Kanarek, 2002).

12.2.4 EFFECTS OF PALATABLE FOODS AND FLUIDS ON THE ENDOGENOUS OPIOID SYSTEM

The results of the previous experiments have led to the hypothesis that intake of palatable substances modifies opiate-induced behaviors by acting directly on the endogenous opioid system. In support of this hypothesis, it has been shown that sucrose intake enhances the pain-relieving actions of centrally, as well as peripherally, administered opiate analgesics (Kanarek et al., 2001). On the more molecular side, short-term intake of palatable foods elevates plasma and brain levels of the endogenous opioid peptide β-endorphin, increases Fos-like immunoreactivity in a number of brain regions, and elevates brain levels of dynorphin peptide and proDYN mRNA (Dum et al., 1983; Park and Carr, 1998; Welch et al., 1996). When consumed for extended periods of time, sugar intake increases and saccharin intake decreases opiate receptor binding affinity (Colantuoni et al., 2001; Marks-Kaufman et al., 1989; Smith et al., 2002). Finally, chronic intake of a high-fat diet increases the density of hypothalamic μ-opioid receptors in rats (Barnes et al., 2003).

Taken together, the results of the preceding studies have led to the hypothesis that chronic intake of palatable nutritive substances potentiates the activity of the endogenous opioid system, which in turn leads to an enhancement in the behavioral actions of these drugs. As nonnutritive palatable substances are not as potent in producing these effects, it is assumed that both the taste and physiological consequences of palatable nutritive and nonnutritive solutions are important in determining their effects on neurochemistry and opiate-induced behaviors.

12.3 CENTRAL NERVOUS SYSTEM STIMULANTS: AMPHETAMINE AND COCAINE

Cocaine, amphetamine, and methamphetamine are the most frequently abused psychostimulants. Although there are differences in the behavioral consequences of these drugs, they all share the properties of leading to behavioral arousal, an increase in perceived energy and self-confidence, appetite suppression, and euphoria. Prolonged use of psychostimulants is associated with tolerance to the drugs' behavioral actions and physical as well as psychological dependence. In addition, psychostimulant abuse can precipitate mood swings, mental confusion, hallucinations, drug-induced psychoses, and damage to the cardiovascular system (Carroll, 2000; Doweiko, 2002).

Within the CNS, psychostimulants potentiate the activity of the monoamine neurotransmitters dopamine, norepinephrine, and serotonin. As with other drugs of abuse, the rewarding effects of psychostimulants result from the actions of these drugs on the release of dopamine from neurons in the ventral tegmental area, which project to a set of interconnected forebrain structures, including the nucleus accumbens, amygdala, and prefrontal cortex. Although all psychostimulants stimulate forebrain dopaminergic neurotransmission, the exact way in which this occurs differs among the drugs. Cocaine blocks the removal of dopamine from the synapse by binding to and competitively inhibiting the transmembrane protein that normally recycles the neurotransmitter from the synapse back into the presynaptic nerve terminal. As a result, dopamine remains for a longer period of time in the synapse, where it can stimulate postsynaptic receptors. In comparison, amphetamine is transported by the transmembrane protein into the presynaptic nerve terminal, where the drug then simulates the neuron to release dopamine (Nestler et al., 2001).

12.3.1 FOOD DEPRIVATION ALTERS THE BEHAVIORAL CONSEQUENCES OF PSYCHOSTIMULANTS

As for opiates, animals' responses to psychostimulants differ as a function of whether the animals have been deprived of food. For instance, many researchers have reported that both primates and rodents self-administered larger amounts of cocaine and amphetamine when food deprived to 80 to 90% of free-feeding body weight than when not deprived (de la Graza and Johanson, 1987; Macenski and Meisch, 1999). Short-term food restriction is also associated with increases in stimulant self-administration. Rats not allowed to eat for 24 h self-administered more of both cocaine and amphetamine than rats given *ad lib* access to food. However, food restriction increased cocaine self-administration nearly twice as much as it increased amphetamine self-administration (Glick et al., 1987; Shalev et al., 2003). These results demonstrate that both chronic and acute food deprivations alter psychostimulant self-administration.

Deprivation status can also alter drug-seeking behavior in rats. For example, deprived rats that receive their food after a self-administration session (acute deprivation) display increased responding during the drug withdrawal, or extinction, phase of the session compared to deprived rats that receive their food 30 min before the self-administration session (chronic deprivation) or nondeprived rats. In addition, acute food deprivation in chronically deprived rats leads to increases in the reinstatement of responding after extinction compared to chronically or nondeprived rats (Comer et al., 1995).

On the basis of the preceding data, it has been suggested that food restriction increases the reinforcement value of stimulant drugs (Macenski and Meisch, 1999). In support of this suggestion, it also has been noted that food deprivation increases cocaine-induced conditioned place preference in rats (Bell et al., 1996) and enhances the threshold-lowering effects of amphetamine on lateral hypothalamic self-stimulation (Cabeza de Vaca and Carr, 1998).

Food deprivation also increases the stimulatory effects of cocaine and amphetamine on loco-motor behavior. When amphetamine was injected into the lateral ventricles, food-deprived rats displayed greater increases in locomotor behavior than nondeprived animals (Cabeza de Vaca and Carr, 1998). Similarly, rats maintained at 80% of their free-feeding body weight through food restriction displayed greater cocaine-induced locomotor activity and a greater degree of behavioral sensitization to repeated cocaine injections than their nonrestricted counterparts (Bell et al., 1996).

12.3.2 SPECIFIC NUTRIENT DEFICITS CAN ALTER THE BEHAVIORAL ACTIONS OF PSYCHOSTIMULANT DRUGS

Recent work has demonstrated that the actions of psychostimulant drugs are modified not only by total food restriction but also by restriction of specific nutrients. In a series of studies, Galler and colleagues found that intake of low-protein diets enhances the behavioral consequences of cocaine in both young and adult rats (Galler and Tonkiss, 1998; Schultz et al., 1999; Shumsky et al., 1997). In young animals, prenatal exposure to cocaine was associated with greater developmental delays in rats whose dams had been maintained on a low-protein diet during pregnancy than in those whose dams received a high-protein diet (Galler and Tonkiss, 1998). Moreover, when tested later in life, male and female offspring of dams who had been maintained on a 6% protein diet during pregnancy displayed greater psychomotor sensitization following repeated cocaine injections when compared to offspring from nondeprived dams (Shultz et al., 1999). Interestingly, the male, but not female, offspring of protein-malnourished mothers showed increased stereotypy following a single cocaine injection compared to properly nourished controls. Provision of a low-protein diet also affects cocaine-mediated behavior in adult animals. Relative to rats given a high-protein diet or standard laboratory chow, female rats fed a 6% protein diet display increased levels of stereotyped behavior in response to repeated cocaine injections (Shumsky et al., 1997).

Dietary variables can also influence the behavioral outcomes of cocaine withdrawal. More specifically, rats consuming a high-protein or high-carbohydrate diet display less severe symptoms of withdrawal when cocaine administration ceases than rats fed a high-fat diet. The differences in withdrawal symptoms observed among animals fed diets varying in macronutrient content may reflect dietary-induced alterations in the serotonergic system (Loebens and Barros, 2003). In support of this possibility, intakes of specific amino acids such as tryptophan and tyrosine, the amino acid precursors for the neurotransmitters serotonin and dopamine, respectively, can also alter the effects of psychostimulant drugs. Animals maintained on a tryptophan-supplemented diet significantly reduce amphetamine and cocaine self-administration compared to rats not receiving tryptophan supplementation (Carroll et al., 1990; Smith et al., 1986). Furthermore, animals maintained on a tyrosine-supplemented diet are less sensitive to the locomotor stimulatory effects of amphetamine at low doses of the drug and more sensitive to the locomotor effects at high doses than animals consuming an unsupplemented diet (Thurmond et al., 1990).

Deficiencies in specific micronutrients can also alter the behavioral actions of psychostimulants. For example, magnesium deficiency attenuates whereas excess magnesium potentiates the effects of cocaine on aggressive behavior (Kantak, 1989). Changes in the behavioral consequences of psychostimulants have also been noted in iron-deficient animals. Male and female rats maintained on a low-iron diet for 4 weeks postweaning are significantly less active following cocaine injections compared to their nondeprived counterparts (Erikson et al., 2000). In addition, rats maintained on an iron-deficient diet are slower to acquire cocaine self-administration and administered less cocaine than rats maintained on an iron-rich diet (Jones et al., 2002). Further research has indicated that iron deficiency attenuates the effects of cocaine through decreased dopamine transporter function-ing, decreased dopamine receptor expression, and decreased dopamine reuptake (Erikson et al., 2000, 2001; Nelson et al., 1997).

12.3.3 INTAKE OF PALATABLE FOODS ALTERS THE BEHAVIORAL CONSEQUENCES OF PSYCHOSTIMULANTS

Consumption of palatable foods and fluid can both predict and moderate an animal's response to psychostimulant drug administration. For instance, rats that voluntarily consume large amounts of sucrose show greater locomotor activation and increased levels of drug-induced dopamine release following amphetamine administration than rats that consume lesser amounts of sucrose (Sills and Crawley, 1996; Sills et al., 1998). Individual preferences for sucrose also predict cocaine and amphetamine self-administration behavior in rats. Rats that consume greater amounts of sucrose acquire amphetamine self-administration faster and self-administer more of the drug than rats that less avidly eat the sugar (DeSousa et al., 2000). Similarly, rats that display a strong preference for sucrose acquire cocaine self-administration faster than rats that fail to show such preferences for the sugar. However, once self-administration behavior is established, cocaine intakes do not differ between high- and low-sucrose consumers (Gosnell, 2000).

Palatable foods and fluids not only predict an animal's propensity for drug abuse but also can act as alternative reinforcers, thus decreasing drug self-administration. Self-administration of psychostimulants is significantly decreased when palatable foods and fluids are presented concurrently. For example, animals self-administering cocaine will significantly decrease their cocaine consumption when they are given a glucose–saccharin cocktail compared to when they do not have access to the sweet fluid (Carroll et al., 1989). Similarly, rats orally consume about half as much amphetamine when given sucrose in addition to laboratory chow than when given only a chow diet. This effect is not specific to sweet solutions as intake of palatable high-fat diets also decreases oral amphetamine self-administration (Kanarek et al., 1996).

Evidence that intake of palatable substances can enhance the rewarding properties of psychostimulants comes from work demonstrating that rats chronically consuming a sucrose solution spent significantly more time on the side of a conditioning apparatus that had been paired with amphetamine injections than those not given the sugar. In addition, among rats given sucrose, those that consumed more than the median amount of the sugar had significantly greater conditioning scores than those that consumed less than the median amount (Vitale et al., 2002).

Even a brief exposure to sucrose may alter the behavioral effects of psychostimulant drugs. For example, amphetamine-sensitized rats that had drunk a sucrose solution for only a minute displayed increased motor activity compared to those that had drunk water (Avena and Hoebel, 2003a).

Results of the previous research indicate that intake of palatable foods can alter neuronal activity in the central dopamine system as well as the endogenous opioid system. More specifically, these results suggest that both intake of palatable foods and psychostimulant drugs reinforce behavior by stimulating the activity of dopamine in areas of the brain known to be involved in mediating reward, such as the nucleus accumbens.

The significance of this research to human problems of drug abuse remains to be elucidated. However, results of a study demonstrating that individuals who regularly use cocaine display greater preferences for sucrose than individuals who do not use the drug point to an interaction between intake of palatable substances and the behavioral consequences of psychostimulants in humans and experimental animals (Janowsky et al., 2003).

12.4 NICOTINE

Nicotine, the primary psychoactive ingredient in tobacco, is absorbed through the mucosal lining of the mouth and nose or by inhalation into the lungs. Once absorbed, nicotine, which is both water and lipid soluble, is rapidly distributed throughout the body, reaching the brain within 8 to 12 sec of inhalation. In the brain, nicotine initiates its actions by binding to nicotinic acetylcholine receptors. The primary consequence of this binding is to enhance the release of other neurotransmitters.

Nicotine predominantly affects the cholinergic system, but it is the drug's action on the dopaminergic system that most probably accounts for its addictive properties. Nicotine, like other drugs of abuse, elevates levels of the neurotransmitter dopamine in reward pathways in the brain. The increase in dopamine in areas such as the nucleus accumbens serves to reinforce drug-seeking behavior and ultimately leads to drug addiction (Dani and De Biasi, 2001).

12.4.1 NICOTINE, FOOD INTAKE, AND BODY WEIGHT

Individuals who smoke typically weigh less than those who do not smoke or those who have never smoked. Also, among the most commonly reported consequences of smoking cessation are the stimulation of appetite and subsequent weight gain. For many individuals, these are seen as adverse outcomes, which contribute to their return to tobacco use. Experimental studies support these anecdotal reports of increased appetite and weight gain following smoking cessation. In both rats and humans, nicotine administration decreases whereas nicotine cessation increases food consumption, particularly intake of palatable sweet-tasting items (Faraday et al., 2002; Guan et al., 2004).

12.4.2 EFFECTS OF PALATABLE FOOD INTAKE ON NICOTINE CRAVINGS

Tolerance to nicotine's behavioral effects is common. Furthermore, indications of physical withdrawal, including drowsiness, irritability, depressed mood, anxiety, difficulty concentrating, impairments in cognitive function, increased appetite, and craving for tobacco, begin within 24 h of the last cigarette. The addictive properties of nicotine cannot be underestimated. Only a small percentage of those who try to stop smoking succeed (Doweiko, 2002). Thus, it is important to find ways to help smokers successfully confront the negative consequences of nicotine withdrawal. Consumption of palatable food may be one way to reach this goal.

A number of studies have reported that food intake, especially that of palatable fare, activates neuronal reward circuits that are also mediated by various addictive drugs, including nicotine (Kelley et al., 2002). If this is the case, then intake of palatable foods by abstinent smokers could serve to attenuate some of the negative symptoms associated with drug withdrawal. In support of this possibility, Helmers and Young (1998) reported that sugar consumption reduced abstinence-induced anxiety and drowsiness. Moreover, West and colleagues (1999) observed that the desire to smoke after a day of refraining from smoking was significantly lower in individuals who had consumed glucose tablets than in those given a placebo. Also, individuals who consumed the glucose tablets were more likely to be abstinent at the end of a 4-week smoking cessation program than those given the placebo (West et al., 1999). These results support the idea that intake of palatable foods can lessen the adverse effects of nicotine withdrawal by stimulating dopaminergic neurons within brain areas involved with reward.

12.4.3 INTAKE OF PALATABLE FLUIDS INCREASES NICOTINE-INDUCED ANALGESIA

Nicotine, like opiates, is a potent analgesic (e.g., Mandillo and Kanarek, 2001; Pomerleau et al., 1984). Moreover, chronic intake of palatable nutritive fluids, but not nonnutritive fluids, significantly enhances the pain-relieving properties of nicotine as well as those of opiate drugs. More specifically, the analgesic potency of nicotine is significantly greater in rats that have consumed a sucrose solution, but not a saccharin solution, than in rats given only water to drink (Mandillo and Kanarek, 2001). Although sucrose increases nicotine-induced analgesia both in males and females, the effect of sugar intake is greater in females than in males.

Intake of a palatable substance can heighten the pain-relieving properties of nicotine in humans, as well as animals. In research using a cold-pressor test to assess pain sensitivity, college-aged smokers were able to keep their forearm in a cold-water bath ($2 \pm 0.2°C$) for longer periods of time before reporting that the experience was uncomfortable when allowed to smoke than when prohibited from doing so. Consumption of a sugar-containing beverage also increased the time until subjects found the cold-pressor test to be uncomfortable and also improved the pain-relieving actions of nicotine (Kanarek and Carrington, 2004).

12.4.4 NICOTINE, GLUCOSE, AND COGNITIVE BEHAVIOR

Administration of nicotine is associated with improvement in performance of cognitive tasks, particularly those with a memory component (Kumari et al., 2003; Newhouse et al., 2004). Studies, which demonstrate that administration of drugs blocking the nicotinic receptor reverse the memory-enhancing effects of nicotine and independently can lead to memory deficits, have provided evidence that central cholinergic systems are important for memory function (Levin and Rezvani, 2002).

As described in Chapter 5, glucose administration can also facilitate memory. It has been hypothesized that the memory-enhancing effects of glucose are a consequence of an increase in the synthesis of acetylcholine or a more general increase in energy-dependent cellular mechanisms within the brain, or both.

The effects of nicotine and glucose on memory clearly interact. Low doses of nicotine and glucose that fail to improve memory when given alone significantly improve performance on a memory task in mice when given together (Sansone et al., 2000). In addition, glucose administration can attenuate memory impairments resulting from the administration of nicotinic receptor antagonists and can enhance hippocampal release of glucose during memory testing in rats (Ragozzino and Gold, 1991; Ragozzino et al., 1996). Finally, in a recent study in which the effects of nicotine and sucrose on cognitive behavior in college-aged smokers were assessed, intake of the sugar enhanced the positive effects on nicotine on an attention task (Harte and Kanarek, 2004). Although it is believed that the effects of glucose and sucrose on the cholinergic system are important for the sugar's ability to enhance nicotine-induced improvements in memory, these sugars may affect other neurotransmitter systems as well.

12.5 MARIJUANA

Marijuana has been used for centuries for both its medicinal and rewarding consequences. The primary psychoactive ingredient of marijuana, Δ^9-tetrahydrocannabinol (THC), has recently been made available for pharmaceutical use in the U.K., and marijuana and THC are being considered as acceptable treatments for combating nausea and vomiting related to chemotherapy and as an appetite stimulant for individuals suffering from anorexia as a result of acquired immunodeficiency syndrome (AIDS)-related anorexia in the U.S. However, as a result of its ability to stimulate the reward system within the CNS, marijuana is more frequently used recreationally than medicinally. Although there is debate about the adverse effects of recreational use of the drug, it is clear that marijuana impairs motor coordination and visual perception, slows reflexes, causes modest deficits in short-term memory, and adversely affects the ability to accurately perceive the passage of time. Long-term marijuana use is associated with a number of detrimental side effects, such as damage to the lungs and reproductive system. Also, there is growing evidence that prolonged use of marijuana leads to tolerance to the drug effects and that cessation of use is associated with a mild withdrawal syndrome (Carroll, 2000; Haney et al., 1999).

In the early 1990s, two cannabinoid receptors, CB1 and CB2, were identified. CB1 receptors are located primarily on nerve endings in areas of the CNS involved with pain, memory, motor coordination, the hedonic aspects of eating, and reward, whereas CB2 receptors are found predominantly in the immune system. Following the discovery of the cannabinoid receptors, three endogenous cannabinoids (endocannabinoids), arachidonoyl ethanolamide (anandamide), 2-arachidonoylglycerol (2-AG), and 2-arachidonyl glyceryl ether, were indentified (Nestler et al., 2001).

One of the most commonly reported side effects of marijuana use is the stimulation of appetite (Berry and Mechoulam, 2002). Early research in human subjects confirmed these reports and further suggested that intake, particularly of sweet-tasting palatable foods, increased approximately 2 to 3 h after use (Abel, 1971). More recently, studies investigating dietary intake and nutritional status of adult marijuana users in the U.S. revealed that caloric intake of users was higher than that of nonusers and that intake increased directly as a function of drug use. When examined in more detailed, it was found that marijuana users consumed more soda, salty snacks, cheese, pork, and alcohol and less fresh fruits than non-drug users (Smit and Crespo, 2001). In contrast to the increase in food intake associated with marijuana use, results of recent studies indicated that in individuals who regularly use the drug, abstinence is associated with significant reductions in food intake (Haney et al., 1999).

Research in experimental animals has provided strong evidence for a role for endogenous cannabinoids in the regulation of appetite. Anandamide and the other endogenous cannabinoids, as well as THC, increase food intake by acting at CB1 receptor sites. In contrast, administration of the CB1 receptor antagonist SR 141716 is associated with reductions in food intake (Arnone et al., 1997; Kirkham et al., 2002; Koch, 2001b). Researchers have also reported a synergistic effect of opiate and cannabinoid antagonists on food intake, such that when administered together doses of SR141716 and the opiate antagonist naloxone, which when given alone have no effects, lead to significant reductions in food intake (Kirkham and Williams, 2001).

As noted previously, data from experiments examining nutritional intake in marijuana users suggest that cannabinoids do not affect intake of all foods in a similar manner. In experimental animals, cannabinoids stimulate intake of palatable, particularly sweet-tasting, foods to a greater degree than intake of less palatable fare (Koch, 2001a; Miller et al., in press). Moreover, administration of SR141716 at doses too low to modify consumption of a standard laboratory diet significantly reduces sucrose consumption (Arnone et al., 1997). These results have led to the suppositions that THC and other cannabinoids may promote feeding by enhancing the rewarding value of food and that the endogenous cannabinoid system may be involved in mediating the appetitive aspects of feeding behavior (Williams and Kirkham, 2002a, 2002b).

Recent work has suggested that the endocannabinoids may interact with leptin, the primary hormone through which the hypothalamus senses and modulates nutritional state and food intake. When leptin is administered to rats, food intake decreases and reduced levels of anandamide and 2-AG are observed. In addition, elevated endogenous cannabinoid levels have been noted in genetically obese mice and rats with deficient leptin signaling. These observations suggest that leptin may exert a negative control over endogenous cannabinoid levels and in this way moderate feeding behaviors (Di Marzo et al., 2001). Specifically, because of the relationship of the cannabinoid system with brain regions involved in reward and hedonics, feeding behaviors could be regulated by incentive salience. In other words, increases and decreases in cannabinoid levels by negative leptin control would change the state of desire for food or motivation to eat. More specifically, lower leptin levels would lead to increases in endogenous cannabinoid levels that interact with the brain's reward systems and brain areas involved in hedonics. This would be followed by an enhanced desire to consume food. Reciprocally, food consumption should be followed by increases in leptin levels that would decrease the levels of endocannabinoids and thereby reduce the desire for and incentive value of food.

The effects of dietary manipulations on the behavioral and neurochemical effects of THC have not been well studied. However, endogenous cannabinoid levels have been assessed relative to appetitive state. Kirkham et al. (2002) found that food restriction in rats significantly elevated anandamide and 2-AG levels in the limbic areas of the brain relative to control rats, satiated rats, and rats examined during feeding. More specifically, with respect to intake of palatable foods, intake of a sweet-tasting high-fat diet was associated with downregulation of CB1 receptor binding in extrahypothalamic brain regions closely related to the intake of palatable foods (Harrold et al., 2002).

12.6   3,4-METHYLENEDIOXYMETHAMPHETAMINE (MDMA)

3,4-Methylenedioxymethamphetamine (MDMA) is a catecholamine-like psychedelic drug. It is the active ingredient in the popular street drug commonly called "ecstasy," "x," or "Adam." The drug, first synthesized in 1914, began to be used in earnest in the late 1970s when case studies appeared suggesting that MDMA allowed patients undergoing psychotherapy to gain insights into their problems and enhanced communication between patients and therapists. These consequences of the drug coupled with reports that MDMA elevated mood, increased feelings of closeness, improved self-confidence, boosted physical energy, and altered sensory perception led to the drug becoming popular in social situations, particularly all-night dance parties, commonly called raves (Freese et al., 2002).

Although many individuals continue to take MDMA, in 1985, as a result of increasing reports of the adverse consequences of the drug, the Drug Enforcement Agency classified MDMA as a Schedule I substance and its use became illegal. Among the most frequently reported negative effects of acute drug use of MDMA are impaired decision making, depersonalization, anxiety, panic attacks, profuse sweating, blurred vision, headaches, jaw clenching, increased heart rate, high blood pressure, and nausea. With repetitive use, tolerance develops to the positive but not negative effects of MDMA. Also, there is evidence that prolonged use of the drug leads to permanent damage to the serotonergic neurotransmitter system (Doweiko, 2002; Freese et al., 2002).

The chemical structure of MDMA is reminiscent of both the monamine neurotransmitters and the psychostimulant amphetamine, which explains the drug's dimorphic actions as well as its appeal. People who have taken the drug report psychostimulant affects similar to those of amphetamine in conjunction with psychedelic affects, like taking lysergic acid diethylamide (LSD). These effects are consequences of MDMA's activation of the dopaminergic, serotinergic, and noradrenergic neurotransmitter pathways (Wilkerson and London, 1989). Although both amphetamine and MDMA stimulate the same neurotransmitter systems, on a quantitative basis MDMA leads to a greater release of serotonin and a lesser release of dopamine than amphetamine does. With respect to serotonin, MDMA causes the release, blocks the reuptake, and interferes with the synthesis of the neurotransmitter. Initially, MDMA acts as a serotonergic agonist, but with time the drug leads to a significant decrease in levels of the neurotransmitter and can result in the destruction of serotonin-containing neurons (Doweiko, 2002; Freese et al., 2002).

As MDMA is a relatively newly studied substance, there are few published works on the effects of either the drug on food intake or of nutrition on drug action. However, note that the drug was originally designed as an appetite suppressant and that decreased appetite is a commonly reported consequence of MDMA use (Freese et al., 2002). The effect of acute administration of MDMA on food intake is most probably the consequence of the drug's activation of serotonergic neurons in the brain, as numerous studies have shown that administration of serotonin agonists leads to significant reductions in feeding behavior, particularly intake of high-fat or high-carbohydrate foods (Curzon, 1990; Heisler et al., 1999).

In contrast to the short-term effects of MDMA on food intake, more chronic MDMA use has been associated with the onset of bulimic episodes (Schifano et al., 1998). For example, Schifano and colleagues (Schifano and Magni 1994; Schifano et al., 1998) found remarkably similar reports of carbohydrate and chocolate cravings amongst seven MDMA abusers seeking psychiatric treatment and 24% of individuals who regularly used MDMA developed bulimic behavior. As prolonged use of MDMA can lead to reductions in serotonergic activity in the brain, it is possible that the cravings serve as a way to replenish brain levels of serotonin. Intake of carbohydrates has been shown to increase brain serotonin levels and it has been suggested that individuals who crave carbohydrates do so because of low levels of the neurotransmitter.

In the long run, use of MDMA could decrease functioning of certain serotonin pathways because it is excitotoxic in serotonin axons. One consequence could affect behaviors relating to nutrition. However, Robinson et al. (1993) found that after repeated doses of MDMA, performance on certain memory tests was not impaired in rats, including tests that depended on food acquisition. In one such test in which rats pretreated with MDMA were tested on skilled reaching for food pellets, the MDMA rats performed even better at the farthest distance than control rats. In a second food-dependent task in which rats were tested on foraging for food, MDMA and control rats behaved similarly. Prior treatment with MDMA in rats did not seem to change their behaviors toward food in these experiments.

The acute behavioral effects of MDMA are similar to those of amphetamine with respect to nutrition. Psychostimulants activate the sympathetic nervous system, the "fight or flight," so appetite decreases and the feeling of energy increases. A one-time administration of various doses of MDMA in rats increased glucose metabolism in brain regions specific to serotonin pathways (Wilkerson and London, 1989). The glucose level before taking MDMA might affect how the serotonin system is activated. Perhaps an increase in carbohydrates high in tryptophan would intensify or prolong the enhancement of serotonin neurotransmission characteristic to MDMA use.

12.7 SUMMARY

The research described clearly demonstrates that dietary variables play a significant role in determining the behavioral consequences of a wide variety of psychoactive drugs, including opiates, psychostimulants, nicotine, MDMA, and THC. Taken together, current evidence argues for a powerful interaction at the neuronal level between drugs and diet. Although there are tantalizing hints of the underlying mechanism, much research remains to be done to determine the ultimate causes of this relationship. The interaction between nutrition and psychopharmacology is already being used in medical settings, in which infants are given sucrose-coated pacifiers to help alleviate discomfort from painful procedures. In addition, it has been reported that recovering addicts who increase intakes of sugary foods can help maintain drug abstinence. In the long term, nutritional factors may figure largely in the treatment of many common issues, such as postoperative pain, drug addition, and eating disorders.

REFERENCES

Abel, E.L. (1971) Effects of marijuana on the solutions of anagrams, memory and appetite. *Nature,* 231, 260–261.

Arnone, M., Maruani, J., Chaperon, F., Thiebot, M.-H., Poncelet, M., Sourbrie, G. and Le Fur, G. (1997) Selective inhibition of sucrose and ethanol intake by SR 141716, an antagonist of central cannabinoid (CB1) receptors. *Psychopharmacology,* 132, 104–106.

Avena, N.M. and Hoebel, B.G. (2003a) Amphetamine-sensitized rats show sugar-induced hyperactivity (cross-sensitization) and sugar hyperphagia. *Pharmacology Biochemistry and Behavior,* 74, 1635–1639.

Avena, N.M. and Hoebel, B.G. (2003b) A diet promoting sugar dependency causes behavioral cross-sensitization to a low dose of amphetamine. *Neuroscience,* 112, 17–20.

Bardo, M.T. and Bevins, R.A. (2000) Conditioned place-preference: what does it add to our preclinical understanding of drug reward? *Psychopharmacology,* 53, 31–43.

Barnes, M.J., Lapanowski, K., Conley, A., Rafols, J.A., Jen, K.L.C. and Dunbar, J.C. (2003) High fat feeding is associated with increased blood pressure, sympathetic nerve activity and hypothalamic mu opioid receptors. *Brain Research Bulletin,* 61, 511–519.

Bell, S.M., Gosnell, B.A., Krahn, D.D. and Meisch, R. (1994) Ethanol reinforcement and its relationship to saccharin preferences in Wistar rats. *Alcohol,* 11, 141–145.

Bell, S.M., Stewart, R.B., Thompson, S.C. and Meisch, R.A. (1996) Food-deprivation increases cocaine-induced conditioned place preference and locomotor activity in rats. *Psychopharmacology,* 131, 1–8.

Berry, E.M. and Mechoulam, R. (2002) Tetrahydrocannabinol and endocannabinoids in feeding and appetite. *Pharmacology and Therapeutics,* 95, 185–190.

Blass, E.M. and Hoffmeyer, L.B. (1991) Sucrose as an analgesic in newborn humans. *Pediatrics,* 87, 215–218.

Bodnar, R.J. (2004) Endogenous opioids and feeding behavior: a 30-year historical perspective. *Peptides*, 258, 697–725.

Cabeza de Vaca, S. and Carr, K.D. (1998) Food restriction enhances the central rewarding effects of abused drugs. *Journal of Neuroscience*, 18, 7502–7510.

Carroll, C.R. (2000) *Drugs in Modern Society*. McGraw Hill, New York.

Carroll, M.E. (1999) Interactions between food and addiction. In *Drugs of Abuse and Addiction: Neuorbehavioral Toxicology* (R.J.M. Niesink, R.M.A. Jaspers, L.M.W. Kornet and J.M. vanRee, Eds.), pp. 286–311. CRC Press, Boca Raton, FL.

Carroll, M.E., Campbell, U.C. and Heideman, P. (2001) Ketoconazole suppresses food restriction-induced increases in heroin self-administration in rats: sex differences. *Experimental and Clinical Psychopharmacology*, 9, 307–316.

Carroll, M.E., Lac, S.T. and Nygaard, S.L. (1989) A concurrently available non-drug reinforcer prevents the acquisition or decreases the maintenance of cocaine-reinforced behavior. *Psychopharmacology*, 97, 23–29.

Carroll, M.E., Lac, S.T., Asnecio, M. and Kragh, R. (1990) Intravenous cocaine self-administration in rats is reduced by dietary L-tryptophan. *Psychopharmacology*, 100, 293–300.

Carroll, M.E. and Meisch, R.A. (1984) Increased drug-reinforced behavior due to food deprivation. *Advances in Behavioral Pharmacology*, 4, 47–88.

Carroll, M.E., Morgan, A.D., Lynch, W.J., Campbell, U.C. and Dess, N.K. (2002) Intravenous cocaine and heroin self-administration in rats selectively bred for differential saccharin intake: phenotype and sex differences. *Psychopharmacology*, 161, 304–313.

Cicero, T.J., Ennis, T., Ogden, J. and Meyer, E.R. (2000) Gender differences in the reinforcing properties of morphine. *Pharmacology Biochemistry and Behavior*, 65, 91–96.

Colantuoni, C., Schwenker, J., McCarthy, J., Rada, P., Ladenheim, B., Cadet, J.-L., Schwartz, G. J., Moran, T.H. and Hoebel, B.G. (2001) Excessive sugar intake alters binding to dopamine and mu-opioid receptors in the brain. *Neuroreports*, 12, 3549–3552.

Comer, S.D., Lac, S.T., Wyvell, C.L., Curtis, L.K. and Carroll, M.E. (1995) Food deprivation affects extinction and reinstatement of responding in rats. *Psychopharmacology*, 121, 150–157.

Comings, D.E. (1997) Genetic factors in drug abuse and dependence. *NIDA Research Monograph*, 169, 16–38.

Cosgrove, K.P. and Carroll, M.E. (2003) Effects of a non-drug reinforcer, saccharin, on oral self-administration of phencyclidine in male and female rhesus monkeys. *Psychopharmacology*, 170, 9–16.

Curzon, G. (1990) Serotonin and appetite. *Annals of the New York Academy of Sciences*, 600, 521–531.

Dani, J.A. and De Biasi, M. (2001) Cellular mechanisms of nicotine addiction. *Pharmacology Biochemistry and Behavior*, 70, 439–446.

D'Anci, K.E. (1999) Tolerance to morphine-induced antinociception is decreased by chronic sucrose or polycose intake. *Pharmacology Biochemistry and Behavior*, 63, 1–11.

D'Anci, K.E., Kanarek, R.B. and Marks-Kaufman, R. (1997a) Duration of sucrose availability differentially alters morphine-induced analgesia in rats. *Pharmacology Biochemistry and Behavior*, 54, 693–697.

D'Anci, K.E., Kanarek, R.B. and Marks-Kaufman, R. (1997b) Beyond sweet taste: saccharin, sucrose, and Polycose differ in their effects upon morphine-induced analgesia. *Pharmacology Biochemistry and Behavior*, 56, 341–345.

de la Garza, R. and Johansion, C.E. (1987) The effects of food deprivation on the self-administration of psychoactive drugs. *Drug and Alcohol Dependence*, 19, 17–27.

DeSousa, N.J., Bush, D.E.A. and Vaccarino, F.J. (2000) Self-administration of intravenous amphetamine is predicted by individual differences in sucrose feeding in rats. *Psychopharmacology*, 148, 52–58.

DeVries, T.J. and Shippenberg, T.S. (2002) Neural systems underlying opiate addiction. *Journal of Neuroscience*, 22, 3321–3325.

de Wit, H. (1997) Individual differences in acute effects of drugs in humans: their relevance to risk for abuse. *NIDA Research Monograph*, 169, 176–187.

Di Marzo, V., Goparaju, S.K., Wang, L., Liu, J., Batkal, S., Jaral, Z., Fezza, F., Miura, G.L., Palmiter, R.D., Sugiura, T. and Kunos, G. (2001) Leptin-regulated endocannabinoids are involved in maintaining food intake. *Nature*, 410, 822–825.

Doweiko, H.E. (2002) *Concepts of Chemical Dependency*, 5th ed. Brooks Cole, Pacific Grove, CA.

Dum, J., Gramsch, C.H. and Herz, A. (1983) Activation of hypothalamic beta-endorphin pools by reward induced by highly palatable foods. *Pharmacology Biochemistry and Behavior*, 18, 443–448.

Erikson, K.M., Jones, B.J., Hess, E.J. and Beard, J.L. (2001) Iron deficiency decreases dopamine D1 and D2 receptors in rat brain. *Pharmacology Biochemistry and Behavior,* 69, 409–418.

Erikson, K.M., Jones, B.J. and Beard, J.L. (2000) Iron deficiency alters dopamine transporter function in rat striatum. *Journal of Nutrition,* 130, 2831–2837.

Faraday, M.M., Elliot, B.M. and Grunberg, N.E. (2002) Adult vs. adolescent rats differ in biobehavioral responses to chronic nicotine administration. *Pharmacology Biochemistry and Behavior,* 70, 475–489.

Freese, T.E., Miotto, K. and Reback, C. J. (2002) The effects and consequences of selected club drugs. *Journal of Substance Abuse Treatment,* 23, 151–156.

Galler, J.R. and Tonkiss, J. (1998) The effects of prenatal protein malnutrition and cocaine on the development of the rat. *Annals of the New York Academy of Sciences,* 846, 29–39.

George, F.R. (1997) Integrating genetic and behavioral models in the study of substance abuse mechanisms. *NIDA Research Monograph,* 169, 209–226.

Glick, S.D., Hinds, P.A. and Carlson, J.N. (1987) Food deprivation and stimulant self-administration in rats: differences between cocaine and D-amphetamine. *Psychopharmacology,* 91, 372–374.

Gosnell, B.A. (2000) Sucrose intake predicts rate of acquisition of cocaine self-administration. *Psychopharmacology,* 149, 286–292.

Guan, G., Kramer, S.F., Bellinger, L.L., Wellman, P.J. and Kramer, P.R. (2004) Intermittent nicotine administration modulates food intake in rats by acting on nicotine receptors localized to the brainstem. *Life Sciences,* 74, 2725–2737

Haney, M., Ward, A.S., Comer, S. D., Foltin, R.W., and Fischman, M.W. (1999) Abstinence symptoms following smoked marijuana in humans. *Psychopharmacology,* 141, 395–404.

Hardy, B.W. and Waller, D.A. (1989) Bulimia and opioids: eating disorder or substance abuse? In *Advances in Eating Disorders* (W. Johnson, Ed.), pp. 43–65. JAI Press, Greenwich, CT.

Harte, C.B. and Kanarek, R.B. (2004) The effects of nicotine and sucrose on spatial memory and attention. *Nutritional Neuroscience,* 7, 121–125.

Harrold, J.A., Elliott, J.C., King, P.J., Widdowson, P.S., and Williams, G. (2002) Down-regulation of cannabinoid-1 (CB-1) receptors in specific extrahypothalamic regions of rats with dietary obesity: a role for endogenous cannabinoids in driving appetite for palatable food? *Brain Research,* 952, 232–238.

Heisler, L.K., Kanarek, R.B. and Homoleski, B. (1999) Reduction of fat and protein intakes but not carbohydrate intake following acute and chronic fluoxetine in female rats. *Pharmacology Biochemistry and Behavior,* 63, 377–385.

Helmers, K.F. and Young, S.N. (1998) The effect of sucrose on acute tobacco withdrawal in women. *Psychopharmacology,* 139, 217–221.

Janowsky, D.S., Pucilowski, O. and Buyinza, M. (2003) Preference for higher sucrose concentrations in cocaine abusing-dependent patients. *Journal of Psychiatric Research,* 37, 35–41.

Jones, B.C., Wheeler, D.S., Beard, J.L. and Grigson, P.S. (2002) Iron deficiency in rats decreases acquisition of and suppresses responding for cocaine. *Pharmacology Biochemistry and Behavior,* 73, 813–819.

Kampov-Polevoy, A.B., Garbutt, J.C. and Khalitov, E. (2003) Family history of alcoholism and response to sweets. *Alcoholism-Clinical and Experimental Research,* 27, 1743–1749.

Kampov-Polevoy, A.B., Tsoi, M.V., Zvartau, E.E., Neznanov, N.G. and Khalitov, E. (2001) Sweet liking and family history of alcoholism in hospitalized alcoholic and non-alcoholic patients. *Alcohol and Alcoholism,* 36, 165–170.

Kanarek, R.B. and Carrington, C. (2004) Sucrose consumption enhances the analgesic effects of cigarette smoking in male and female smokers. *Psychopharmacology,* 173, 57–63.

Kanarek, R.B., Mandillo, S. and Wiatr, C. (2001) Chronic sucrose intake augments antinociception induced by injections of mu but not kappa opioid receptor agonists into the periaqueductal gray matter in male and female rats. *Brain Research,* 920, 97–105.

Kanarek, R.B., Mathes, W.F., Heisler, L.K., Lima, R.P. and Monfared, L.S. (1997) Prior exposure to palatable solutions enhances the effects of naltrexone on food intake in rats. *Pharmacology Biochemistry and Behavior,* 57, 377–381.

Kanarek, R.B., Mathes, W.F. and Przypek, J. (1996) Intake of dietary sucrose or fat reduces amphetamine drinking in rats. *Pharmacology Biochemistry and Behavior,* 54, 719–723.

Kantak, K.M. (1989) Magnesium alters the potency of cocaine and haloperidol on mouse aggression. *Psychopharmacology,* 99, 181–188.

Kelley, A.E., Bakshi, V.P., Haber, S.N., Steininger, T.L., Will, M.J. and Zhang, M. (2002) Opioid modulation of taste hedonics within the ventral striatum. *Physiology and Behavior,* 76, 365–377.

Kirkham, T.C. and Williams, C.M. (2001) Synergistic effects of opioid and cannabinoid antagonists on food intake. *Psychopharmacology,* 153, 267–270.

Kirkham, T.C., Williams, C.M., Fezza, F. and Di Marzo, V. (2002) Endocannabinoid levels in rat limbic forebrain and hypothalamus in relation to fasting, feeding and satiation: stimulation of eating by 2-arachidonoyl glycerol. *British Journal of Pharmacology,* 136, 550–557.

Koch, J.E. (2001a) Δ^9-THC stimulates food intake in Lewis rats: effects on chow, high-fat and sweet high-fat diets. *Pharmacology, Biochemistry and Behavior,* 68, 539–543.

Koch, J.E. (2001b) Δ^9-Tetrahydrocannabinol stimulates palatable food intake in Lewis rats: effects of peripheral and central administration. *Nutritional Neuroscience,* 4, 179–187.

Koob, G.F. and Le Moal, M. (2001) Drug addiction, dysregulation of reward, and allostasis. *Neuropsychopharmacology,* 24, 97–129.

Kumari, V., Gray, J.A., ffytche, D.H., Mitterschiffthaler, M.T., Das, M., Zachariah, E., Vythelingum, G.N., Williams, S.C.R., Simmons, A. and Sharma, T. (2003) Cognitive effects of nicotine in humans: an fMRI study. *Neuroimage,* 19, 1002–1013.

Levin, E.D. and Rezvani, A.H. (2002) Nicotinic involvement in cognitive function of rats. In *Nicotinic Receptors in the Nervous System* (E.D. Levin, Ed.), pp. 167–178. CRC Press, Boca Raton, FL.

Loebens, M. and Barros, H.M.T. (2003) Diet influences cocaine withdrawal behaviors in the forced swimming test. *Pharmacology Biochemistry and Behavior,* 74, 259–267.

Lynch, W.J. and Carroll, M.E. (2000) Reinstatement of cocaine self-administration in rats: sex differences. *Psychopharmacology,* 148, 196–200.

Macenski, M.J. and Meisch, R.A. (1999) Cocaine self-administration under conditions of restricted and unrestricted food access. *Experimental and Clinical Psychopharmacology,* 7, 324–337.

Mandillo, S. and Kanarek, R.B. (2001) Chronic sucrose intake enhances nicotine-induced antinociception in female but not male Long-Evans rats. *Pharmacology Biochemistry and Behavior,* 68, 211–219.

Marks-Kaufman, R., Hamm, M.W., and Barbato, G.F. (1989) The effects of dietary sucrose on opiate receptor binding in genetically obese (ob/ob) and lean mice. *Journal of the American College of Nutrition,* 8, 9–14.

Mercer, M.E. and Holder, M.D. (1997) Antinociceptive effects of palatable sweet ingesta on human responsivity to pressure pain. *Physiology and Behavior,* 61, 311–318.

Miller, C.C., Murray, T.F., Freeman, K.G. and Edwards, G.L. (2004) Cannabinoid agonist, CP 55,940, facilitates intake of palatable food when injected into the hindbrain. *Physiology and Behavior,* 80, 611–614.

Mogil, J.S., Chesler, E.J., Wilson, S.G., Juraska, J.M. and Sternberg, W.F. (2000) Sex differences in thermal nociception and morphine antinociception in rodents depend on genotype. *Neuroscience and Biobehavioral Review,* 24, 375–389.

Morabia, A., Fabre, J., Chee, E., Seger, S., Orsat, E. and Robert, A. (1989) Diet and opiate addiction: A quantitative assessment of the diet of non-institutionalized opiate addicts. *British Journal of Addiction,* 84, 173–180.

Nelson, C., Erikson, K., Pinero, D.J. and Beard, J.L. (1997) In vivo dopamine metabolism is altered in iron-deficient anemic rats. *Journal of Nutrition,* 127, 2282–2288.

Nestler, E.J., Hyman, S.E. and Malenka, R.C. (2001). *Molecular Neuropharmacology.* McGraw Hill, New York.

Newhouse, P.A., Potter, A. and Singh, A. (2004) Effects of nicotine stimulation on cognitive performance. *Current Opinion in Pharmacology,* 4, 36–46.

O'Brian, C.P., Childress, A.R., McLellan, A.T. and Ehrman, R. (1992) Classical conditioning in drug-dependent humans. *Annals of New York Academy of Sciences,* 654, 400–415.

Park, T.H. and Carr, K.D. (1998) Neuroanatomical patterns of Fos-like immunoreactivity induced by a palatable meal and meal-paired environments in saline- and naltrexone-treated rats. *Brain Research,* 805, 169–180.

Pomerleau, O.F., Turk, D.C. and Fertig, J.B. (1984) The effects of cigarette smoking on pain and anxiety. *Addictive Behaviors,* 9, 265–271.

Ragozzino, M.E. and Gold, P.E. (1991) Glucose effects on mecamylamine-induced memory deficits and decreases in locomotor behavior in mice. *Behavioral and Neural Biology,* 56, 271–282.

Ragozzino, M.E., Unick, K.E. and Gold, P.E. (1996) Hippocampal acetylcholine release during memory testing in rats: augmentation by glucose. *Proceedings of the National Academy of Sciences USA*, 93, 4693–4698.

Robinson, T.E., Castanedo, E. and Whishaw, I.Q. (1993). Effects of cortical serotonin depletion induced by 3,4 methylenedioxymethamphetamine (MDMA) on behavior, before and after additional cholinergic blockade. *Neuropsychopharmacology*, 8, 77–85.

Sansone, M., Battaglia, M. and Pavone, F. (2000) Shuttle-box avoidance learning in mice: improvement by glucose combined with stimulant drugs. *Neurobiology of Learning and Memory*, 73, 94–100.

Schifano, F. and Magni, G. (1994) MDMA ("ecstasy") abuse: psychopathological features and craving for chocolate — a case series. *Biological Psychiatry*, 36, 763–767.

Schifano, F., Di Furia, L., Forza, G., Minicuci, N. and Bricolo, R. (1998) MDMA ("ecstasy" consumption in the context of polydrug abuse: a report on 150 patients. *Drug and AlcoholDependence*, 52, 85–90.

Shalev, U., Marinelli, M., Baumann, M.H., Piazza, P.V. and Shaham, Y. (2003) The role of corticosterone in food deprivation-induced reinstatement of cocaine seeking in the rat. *Psychopharmacology*, 168, 170–176.

Shultz, P.L., Galler, J.R. and Tonkiss, J. (1999) Prenatal protein restriction increases sensitization to cocaine-induced stereotypy. *Behavioral Pharmacology*, 10, 379–387.

Shumsky, J.S., Shultz, P.L., Tonkiss, J. and Galler, J.R. (1997) Effects of diet on sensitization to cocaine-induced stereotypy in female rats. *Pharmacology Biochemistry and Behavior*, 58, 683–688.

Sills, T.L. and Crawley, J.N. (1996) Individual differences in sugar consumption predict amphetamine-induced dopamine overflow in nucleus accumbens. *European Journal of Pharmacology*, 303, 177–181.

Sills, T.L., Onalaja, A.O. and Crawley, J.N. (1998) Mesolimbic dopaminergic mechanisms underlying individual differences in sugar consumption and amphetamine hyperlocomotion in Wistar rats. *European Journal of Neuroscience*, 10, 1895–1902.

Smit, E. and Crespo, C.J. (2001) Dietary intake and nutritional status of US adult marijuana users: results from the Third National Health and Nutrition Examination Survey. *Public Health Nutrition*, 4, 781–786.

Smith, F.L., Yu, D.S.L., Smith, D.G., Lecesse, A.P. and Lyness, W.H. (1986) Dietary tryptophan supplements attenuate amphetamine self-administration in the rat. *Pharmacology Biochemistry and Behavior*, 25, 849–855.

Smith, S.L., Harrold, J.A. and Williams, G. (2002) Diet-induced obesity increases μ opioid receptor binding in specific regions of the rat brain. *Brain Research*, 953, 215–222.

Szpanowska-Wohn, A., Duzniewska, K., Groszek, B. and Lang-Mynarska, D. (2000) Nutrition disorders in persons qualified for the methadone treatment. Part II. Food choice and intake in diets of opiate addicts. *Przeglad Lekarski*, 57, 544–548.

Thurmond, J.B., Freeman, G.B., Soblosky, J.S., Ieni, J.R. and Brown, J.W. (1990) Effects of dietary tyrosine on L-dopa and amphetamine-induced changes in locomotor activity and neurochemistry in mice. *Pharmacology Biochemistry and Behavior*, 37, 259–266.

Vanyukov, M.M., Tarter, R.E., Kirisci, L., Kirillova, G.P., Maher, B.S. and Clark, D.B. (2003) Liability to substance use disorders. 1. Common mechanisms and manifestations. *Neuroscience and Biobehavioral Reviews*, 27, 507–515.

Vitale, M.A., Chen, D. and Kanarek, R. (2003) Chronic access to a sucrose solution enhances the development of conditioned place preferences for fentanyl and amphetamine in male Long-Evans rats. *Pharmacology Biochemistry and Behavior*, 74, 529–539.

Welch, C.C., Kim, E.M., Grace, M.K., Billington, C.J. and Levine, A.S. (1996) Palatability-induced hyperphagia increases hypothalamic dynorphin peptide and mRNA levels. *Brain Research*, 721, 126–131.

West, R., Hajek, P. and Burrows, S. (1990) Effect of glucose tablets on cravings for cigarettes. *Psychopharmacology*, 101, 555–559.

Wilkerson, G. and London, E.D. (1989) Effects of methylenedioxymethamphetamine on local cerebral glucose utilization in the rat. *Neuropharmacology*, 28, 1129–1138.

Williams, C.M. and Kirkham, T.C. (2002a) Observational analysis of feeding induced by Δ9-THC and anandamide. *Physiology and Behavior*, 76, 241–250.

Williams, C.M. and Kirkham, T.C. (2002b) Reversal of Δ^9-THC hyperphagia by SR141716 and naloxone but not dexfenfluramine. *Pharmacology Biochemistry and Behavior*, 71, 341–348.

Yamamoto, R.T. and Kanarek, R.B. (2002) Naloxone precipitated withdrawal from morphine is ameliorated by ip glucose administration. In *Abstracts of the Society for Neuroscience Annual Meeting,* Orlando, FL, Society for Neuroscience, Washington, D.C.

Yeomans, M.R. and Gray, R.W. (2002) Opioid peptides and the control of human ingestive behavior. *Neuroscience and Biobehavioral Reviews,* 26, 713–728.

Yung, L., Gordis, E. and Holt, J. (1983) Dietary choices and likelihood of abstinence among alcoholic patients in an outpatient clinic. *Drug and Alcohol Dependence,* 12, 355–361.

Zador, D., Wall, L. and Webster, I. (1996) High sugar intake in a group of women on methadone maintenance in south western Sydney, Australia. *Addiction,* 91, 1053–1061.

Part III

Micronutrients, Brain Function, and Behavior

13 Vitamins and Brain Function*

Jürg Haller

CONTENTS

13.1 INTRODUCTION

Vitamins are essential micronutrients involved in many aspects of brain function, from global processes such as energy metabolism of neurons and glia to specific functions such as coenzymes in the synthesis of neurotransmitters. The blood supply to the brain is indirectly influenced by the long-term vitamin status in so far as it influences pathological processes such as arteriosclerosis. One of the early findings in vitamin research in humans was that vitamin deficiency led to personality changes. This initiated a large body of work on the effects of vitamins on various aspects of mood and mental performance. Subsequent work showed that malnutrition could lead to cognitive impairment in children and adults that could be ameliorated by vitamin treatment. This chapter summarizes present knowledge on the direct actions of vitamins on neurophysiological function, their indirect actions on cerebral energy metabolism and blood supply, and the effects of vitamin deficiency and vitamin treatments on mental performance and mood in humans.

---

* Except antioxidants.

13.2 DIRECT ACTIONS OF VITAMINS

13.2.1 NEUROTRANSMITTER SYNTHESIS

Much of our knowledge of brain function derives from information on the mode of action of neurotransmitters and their neuroanatomical distribution. Hypotheses on the mechanisms involved in various functions such as attention, learning, memory, and psychopathological disorders such as anxiety, depression, and schizophrenia are based on the specific actions and interactions of the different excitatory and inhibitory neurotransmitter systems. Vitamins are essential cofactors in the synthesis of many of these neurotransmitters, and their concentrations can influence neurotransmitter levels (Table 13.1).

Pyridoxal 5'-phosphate (PLP) is required as a coenzyme by glutamic acid decarboxylase (GAD) for the conversion of glutamic acid to the inhibitory neurotransmitter γ-amino butyric acid (GABA). This coenzyme is also required by tyrosine decarboxylase for converting tyrosine to dopamine and noradrenaline, by tryptophan decarboxylase for converting tryptophan to 5-hydroxytryptamine, and by histidine decarboxylase for converting histidine to histamine. Nicotinamide in the form of nicotinamide adenine dinucleotide (NAD) is a cofactor for hepatic tryptophan pyrrolase, which is required for the synthesis of 5-hydroxytryptamine. Folic acid and vitamin B-12 are cofactors for catechol-O-methyl transferase, which breaks down noradrenaline and dopamine in the synaptic cleft. Ascorbic acid is a cofactor for dopamine β-hydroxylase, dihydropterin reductase, and dopamine-adenyl cyclase, which are involved in the synthesis of dopamine and noradrenaline (Ashley, 1986). Folate is involved in the synthesis of tetrahydrobiopterin, which is an essential cofactor for converting phenylalanine to tyrosine and for hydroxylating tyrosine and tryptophan to noradrenaline and dopamine (Anderson and Abou-Saleh, 1995). Glucose and thiamin are required for the synthesis of acetyl coenzyme A, which together with choline are required for the synthesis of acetylcholine in a reaction catalyzed by choline acetyltransferase (Heinrich et al., 1973).

13.2.2 NEURONAL RECEPTOR BINDING

Ionotrophic and metabotrophic neurotransmitters act by binding stereospecifically to pre- and postsynaptic receptors on neurons and initiate a cascade of reactions, which, respectively, lead to rapid short-term changes in membrane permeability or to long-term metabolic changes in the affected neuron. Vitamins can modulate these activities by enhancing or attenuating the receptor

TABLE 13.1
Vitamins and Synthesis of Neurotransmitters

Vitamin	Function	Site of Regulation	Neurotransmitters
Nicotinamide	Enzyme cofactor (NAD/NADH)	Tryptophan pyrrolase (in liver)	Serotonin (5-HT)
Pyridoxine	Enzyme cofactors (transamination, deamination, decarboxylation)	Tyrosine decarboxylase, tryptophan decarboxylase, histidine decarboxylase, glutamic acid decarboxylase	Noradrenaline, dopamine, serotonin (5-HT), histamine, GABA
Folic acid/B-12	Enzyme cofactor (transmethylation)	Catechol-O-methyl-transferase	Noradrenaline, dopamine
Ascorbic acid	Enzyme cofactor	Dopamine-β-hydroxylase, dihydropterin reductase, dopamine-adenyl cyclase	Noradrenaline, dopamine
Thiamin	Enzyme cofactor	Transketolase, pyruvate dehydrogenase, α-ketoglutarate dehydrogenase	Acetylcholine

binding of neurotransmitters. Vitamin B-6 modulates the binding of a number of neurotransmitters at postsynaptic receptors (Ebadi, 1981; Dakshinamurti, 1982; Dakshinamurti et al., 1990). Perinatal B-6 deficiency leads to long-lasting postnatal changes in dopaminergic function in rats, including changes in receptor number (Guilarte et al., 1987). Postsynaptic GABA binding is specifically inhibited by pyridoxal and PLP (Ebadi et al., 1980). Pyridoxine deficiency leads to increased GABAergic and benzodiazepine binding in deficient rat pups (Pilachowski Borek et al., 1990). Vitamin B-6 modulates the receptor binding of not only neurotransmitters but also of steroids to nuclear receptors (Allgood and Cidlowski, 1991).

13.2.3 NEURONAL MEMBRANE ION PUMP

The flow of information in the central nervous system is controlled not only by the activities of neurotransmitters but also by the rate at which action potentials can be generated and propagated along axons and dendrites. The integrity of the axons and their myelin sheaths is dependent on adequate levels of various vitamins such as cobalamin and folate. Thiamin is of functional importance for nerve conduction. It is present in the brain in different forms: ca. 4% as free thiamin; 9% as thiamin monophosphate; 84% as thiamin diphosphate, the active coenzyme form; and 3% as thiamin triphosphate. The last is found exclusively in the membrane fraction and appears to be involved in maintaining the transmembrane potential difference. Destruction of thiamin by UV light leads to loss of the membrane potential, and no action potentials can be generated in the isolated nerve preparation. These changes can be reversed by adding thiamin to the perfusate. These findings suggest that thiamin has a specific role in nerve conduction, independent of its metabolic role as a coenzyme in the form of the diphosphate. During thiamin deficiency in rats, the triphosphate is spared and changes little, whereas the free thiamin and the mono- and diphosphate decrease by ca. 20% within 4 weeks (Dreyfus, 1988; Bender, 1992).

13.3 INDIRECT ACTIONS OF VITAMINS

13.3.1 ENERGY METABOLISM OF THE BRAIN

The brain, although accounting for only 2 to 2.7% of the body weight, has a very high energy consumption, accounting for 20% of the body's oxygen supply, 25% of the glucose supply at rest, and 19% of the blood supply. Under normal physiological conditions, glucose is the primary source of energy for neurons in the adult brain. Astrocytes produce lactate anaerobically, which can be used by neighboring neurons. The metabolism of glucose in the brain requires thiamin for converting the pyruvate derived from glucose-6-phosphate to acetyl coenzyme A, which is converted to CO_2 and H_2O in reactions requiring the vitamins pantothenic acid, thiamin, riboflavin, and nicotinamide. Studies in humans have shown that during physiological neural activity evoked by visual or tactile stimulation, cerebral blood flow and glucose uptake increase by 50 and 51%, respectively, whereas oxygen consumption increases by only 5%, which would allow a maximum increase in aerobic and anaerobic energy in the form of ATP production of only 8% (Fox et al., 1988). This suggests that the increased neuronal uptake of glucose serves some function other than increased energy production, such as the synthesis of acetylcholine or other neurotransmitters. Alternatively, the disproportionate increase in glucose relative to oxygen uptake could reflect an increased anaerobic utilization either by neurons or glia cells. The high energy consumption of the brain and its dependence on glucose implies that the requirement for the vitamins involved in energy metabolism is also high.

13.3.2 CEREBRAL BLOOD SUPPLY

There is evidence to suggest that cerebral blood supply is protected by blood levels of folic acid, pyridoxine, and vitamin B-12 that are sufficient to reduce homocysteinemia (Selhub et al., 1995).

These three vitamins are involved in the metabolism of homocysteine to methionine and cystathionine, which when disturbed leads to elevated blood levels of homocysteine, a risk factor for vascular disease. Epidemiological studies have shown that there is an inverse relationship between the plasma levels of homocysteine and carotid artery stenosis and between the former and the levels of vitamins B-6, B-12, and folic acid (Selhub et al., 1995). Intervention studies have shown that these vitamins lower blood levels of homocysteine and methylmalonate and reduce the progression of atherosclerosis in the carotid arteries (Peterson and Spence, 1998). Epidemiological studies suggest that populations with higher plasma levels or higher dietary intake of folate and vitamin B-6 have a lower risk of carotid artery stenosis and coronary heart disease (Chasan-Taber et al., 1996; Morrison et al., 1996; Verhoef et al., 1996; Rimm et al., 1998). These vitamins of the B-complex therefore play an important role in maintaining an optimal blood supply to the brain throughout life.

13.4 BRAIN FUNCTION AND VITAMIN STATUS

13.4.1 DEPLETION STUDIES

Many studies have been carried out in a number of animal species and humans to examine the effects of vitamin deficiency on brain structure, function, and behavior. Deficiencies in vitamins B-1 and B-12, folic acid, and nicotinic acid lead to well-known pathological changes in humans, such as the Wernicke–Korsakoff syndrome, pellagra, subacute degeneration of the spinal cord, neuropathies, and neuropsychiatric symptoms.

Much work has been carried out to elucidate the effects of thiamin deficiency on brain function. Barclay et al. (1981) showed that in rats, thiamin deficiency induced by pyrithiamin, a centrally and peripherally active thiamin antagonist, led to a reduced incorporation of labeled $[^2H_4]$choline and labeled $[U^{-14}C]$glucose into acetylcholine after 12 days of treatment. Parallel with these changes, psychomotor performance decreased. These behavioral changes could be antagonized by physostigmine, an acetylcholinesterase inhibitor, or by arecoline, a direct muscarinic agonist. The effects of physostigmine could be blocked by the muscarinic blocker atropine, whereas nicotine or the nicotinic blocker mecamylamine had no effect. These data suggest that thiamin is directly involved in the synthesis of acetylcholine. Many other studies have confirmed the role of thiamin in the cholinergic system, and studies in humans have shown that it can antagonize the effects of scopolamine on memory and on P3 event-related potentials in healthy volunteers and patients (Meador et al., 1993; Easton and Bauer, 1997). However, thiamin deficiency not only leads to changes in acetylcholine levels but also affects the activity of various enzymes and of neurotransmitters such as dopamine and serotonin, the permeability of the blood-brain barrier, and can lead to neuronal apoptosis and necrosis. The lesions typically involve the periaqueductal grey, thalamus, mamillary bodies, vestibular nuclei, and colliculi. The effects of thiamin deficiency are intensified by a glucose load and attenuated by dietary restriction. An adequate dietary supply of thiamin is particularly critical for brain function because the active rate-limited process by which thiamin is transported across the blood-brain barrier has a maximum rate. This rate is approximately equal to the brain thiamin turnover and is therefore only just sufficient to meet the brain's requirements (Thomson, 2000). This, together with the low thiamin bioavailability of only ca. 5% and the small body stores, helps explain the high incidence of behavioral and cerebral deficits of the Wernicke–Korsakoff type in patients, whose thiamin supply has been interrupted for various reasons such as emesis gravidarum, glucose infusions without thiamin, gastric surgery, AIDS, anorexia nervosa, and alcoholism.

Pyridoxine deficiency can lead to developmental brain changes and abnormal behavior in neonatal rats and to decreases in serotonin levels and to decreased GABA receptor binding in adult rats (Guilarte, 1993). The decreases in serotonergic function lead to secondary decreases of thyroid function with decreased T_4 and T_3 but normal thyroid stimulating hormone (TSH) and thyroid releasing hormone (TRH) and decreased melatonin secretion from the pineal gland (Dakshinamurti et al., 1988). Abnormal electroencephalograms (EEGs) have been reported in healthy young women on a pyridoxine-deficient diet (Kretsch et al., 1991), and decreases in sensory nerve conduction velocity have been measured in pyridoxine-deficient rats (Claus et al., 1984).

Before most of these animal studies had been carried out, Brozek's group (Brozek et al., 1946; Guetzkow and Brozek, 1946) had examined the effects of restricted intakes of thiamin, riboflavin, and niacin on various psychological parameters in human volunteers. They found that personality changes were among the earliest symptoms of experimentally induced borderline and acute deficiencies. On the other hand, no deterioration of the intellectual or learning ability was found after partial restriction and only limited deterioration after acute restriction. The tests sensitive to the vitamin deficiency appeared to be those in which speed was the essential factor. Intellectual functions were among those that were most resistant to the vitamin deficiency. Similarly, in studies on acute thiamin restriction, the scores of intelligence tests were not adversely affected but there were large negative changes in the "psychoneurotic" scales of the Minnesota Multiphasic Personality Inventory (MMPI). Manual speed and complex reaction time were significantly affected by thiamin deficiency (Brozek and Guetzkow, 1957). In studies on vitamin C deprivation, the first symptoms that occurred were hypochondriasis, depression, and hysteria, as measured on the MMPI. Later, there were decrements in psychomotor performance associated with reduced arousal or motivation (Kinsman and Hood, 1971). A severe hyporiboflavinosis also led to significant changes in five personality subscales of the MMPI (hypochondriasis, depression, hysteria, psychopathic-deviate, and hypomania) as well as a long-lasting decrease in hand-grip strength, but intelligence as measured on various subtests of the Wechsler Adult Intelligence Scale (WAIS) was not affected (Sterner and Price, 1973). These studies on partial and acute vitamin deprivation suggest that the earliest impairments occur in measures of mood rather than of mental performance.

13.4.2 Epidemiological Studies

Many epidemiological studies have been carried out in different populations to examine the relationship between vitamin intake or vitamin blood (serum, plasma, or erythrocyte) levels and various aspects of mental function. In recent years, there has been an emphasis on the relationship between micronutrients with antioxidant properties and age-related cognitive changes and on interrelations between plasma concentrations of homocysteine and vitamins B-6, B-12, and folic acid.

Goodwin et al. (1983) tested the association between nutritional status and cognitive functioning in 260 free-living elderly volunteers (>60 years) who were healthy and not receiving any medication (Table 13.2). Their cognitive status was evaluated by the Halstead-Reitan Categories Test and by the Wechsler Memory Test. Subjects with low levels of vitamin C or B-12 scored significantly worse on both tests and those with low levels of riboflavin or folic acid scored significantly worse on the categories tests. These differences remained significant after controlling for age, gender, level of income, and amount of education.

Chomé et al. (1986) carried out a similar study in Germany in 60 subjects between 65 and 91 years. On the basis of their blood vitamin levels, they were allocated to either a deficient ($n = 34$) or a control ($n = 26$) group and their performance on various psychometric tests compared. On the Freiburg Personality Inventory the scores in the deficient group were significantly higher for emotional instability and depression. On the Scale of Attributes (Eigenschaftswörterliste), the deficient subjects had higher scores for excitability, fatigue, discouragement, and irritability. The scores for short-term memory measured with the Vienna test system were also significantly worse in the deficient group.

In the EURONUT SENECA study, the micronutrient status and its association with scores in the Minimental State Examination (MMSE) and the Geriatric Depression Scale (GDS) were examined in a randomized sample of 880 subjects of both sexes aged 75 to 80 years, stratified according to age and sex from 11 small towns in 9 European countries (Haller et al., 1996). There were highly significant but weak correlations between the total MMSE scores and the lycopene, β-carotene, β-cryptoxanthin, total carotene, α-carotene, α-tocopherol, cobalamin, and folate plasma

TABLE 13.2
Epidemiological Studies on Relationshps between Vitamin Status and Cognitive Function

Study	n	Subjects	Psychometric Parameters	Micronutrients and Other Blood Constituents	Outcome	Other
Goodwin et al., 1983	260	Free-living elderly (>60 years), who were healthy and not receiving any medication	Halstead–Reitan Categories Test, Wechsler Memory Test	Dietary intake of energy, protein, vitamins C, B-1, B-2, A, B-6, B-12, D, E, niacin, folic acid, Fe, Ca Zn, and P; plasma levels of vitamins C, B-12, folate, ETK, EGR, pyridoxal phosphate	Low levels of vitamin C or B-12 related to significantly worse scores on both tests and those with low levels of riboflavin or folic acid scored significantly worse on the categories tests	
Chomé et al., 1986	60	Elderly, 65–91 years	Freiburger Personality Inventory (FPI), Mood questionnaire, Vienna Test System	Plasma vitamins A, C, E, B-12, folate, carotenoids, RBC folate, α-ETK, α-EGR, α-EGOT	Deficient group had significantly worse scores on the FPI for emotional instability, depression, and on the mood scale for excitability, fatigue, discouragement, irritability, and short-term memory	
EURONUT Seneca Study: Haller et al., 1996	880	Free-living eldery, 75–80 years; 11 towns in 9 European countries; randomized sample of both sexes born from 1913 to 1918, stratified according to age and sex	MMSE, GDS	Plasma β-carotene, α-carotene, lycopene, lutein, zeaxanthin, β-cryptoxanthin, retinol, α-tocopherol, folate, cobalamin, PLP, cholesterol	Significant correlations between MMSE scores and plasma levels of carotenes, lutein, zeaxanthin, β-cryptoxanthin, α-tocopherol, folate and cobalamin; significant negative correlations between GDS scores and plasma levels of lycopene, lutein, zeaxanthin, β-cryptoxanthin and PLP	Education, ADL
DHSS Study Follow-up: Gale et al. 1996	921	Elderly, 65–85 years	Hodkinson Mental Test, mortality; stroke	Plasma leveles and daily intake of vitamin C	Plasma and intake of vitamin C correlated with cognitive function	Inverse relationship between vitamin C intake and plasma levels and mortality and stroke

Study	N	Population	Tests	Measures	Results	
Normative Aging Study: Riggs et al., 1996	70	Men, 54–81 years	Verbal fluency, Boston Naming test, vocabulary (WAIS-R), pattern comparison, Continuous Performance Test, Word List Memory Test, backward digit span, activity memory, pattern memory, Spatial Copying Test, Spatial Reasoning Test	Plasma B-6, B-12, folate, homocysteine	Lower levels of B-12 and folate related to poorer scores in copying skills; higher concentrations of B-6 asociated with better scores on two measures of memory	Raised homocysteine decreased scores
Hassing et al., 1999	71	Elderly, 90–101	Fuld Object Memory Evaluation, face and word recognition, word and object recall	Plasma vitamin B-12, folic acid	Low levels of folate related to poorer word and object recall and secondary memory; no relationship with B-12 levels	
New Mexico Elder Health Survey: Lindeman et al., 2000	783	Elderly, 60–85 years, non-Hispanic white, Hispanic	MMSE, Fuld WAIS-R Digits Forward, object memory, color trails, clock drawing, GDS 15 item, self-report of history of depression, list of current medication	Serum cobalamin, folate, vitamin C and dietary intake, use of multivitamins	Subjects with folate levels ≥ 11.1 nmol/l had significantly higher scores on the MMSE, Fuld object-memory, digits forward, color trails; history of depression reduced in those with B-12 > 221pmol/l and vitamin C > 57μmol/l	
Scottish Mental Surveys Follow-up: Duthie et al., 2002	334	Elderly born in 1921 or 1936, 64–78 years	MMSE, National Adult Reading Test (NART), Ravens Progressive Matrices, Auditory Verbal Learning Test (AVLT), Digit Symbol (DS), Block Design (BD)	Plasma vitamin B-12, folate and RBC folate, plasma homocysteine	Folate levels correlated with MMSE, NART, AVLT, DS, BD; B-12 correlated with MMSE	Negative correlation between plasma homocysteine and MMSE, RPM, DB, and DS

Note: α-ETK = activity coefficient erythrocyte transketolase; α-EGR = activity coefficient glutathione reductase; α-EGOT = activity coefficient erythrocyte glutamine oxaloacetic transaminase; CuZn-SOD = CuZn superoxide dismutase; DSST = Digit Symbol Substitution Test; GDS = Geriatric Depression Scale; MMSE = Minimental State Examination; PLP = pyridoxal 5'-phosphate; RBC = erythrocyte; TBARS = thiobarbituric acid reactive substances.

concentrations. There were weak negative but highly significant correlations between the GDS scores and the β-cryptoxanthin, cholesterol, PLP, and lycopene plasma levels. In a number of centers, there were many significant correlations between the MMSE and a number of carotenoids, such as β-cryptoxanthin, α-carotene, and lutein. In the case of the GDS, there were fewer significant correlations and they were with PLP and vitamin B-12 rather than with carotenoids.

In a 20-year follow-up study on 921 elderly subjects, the relationships between vitamin C plasma levels and from dietary intake and cognitive function, stroke, and all causes of mortality were examined (Gale et al., 1996). At the baseline, cognitive function determined by the Hodkinson Mental Test scores was poorest in those with the lowest dietary vitamin C intake or blood levels, independent of illness or social class. At the 20-year follow-up, those who had scored <7 on the Hodkinson Mental Test had a significantly higher risk of dying from ischemic disease and particularly of stroke, suggesting that ascorbic acid concentrations or intake were predictors of vascular pathology.

In the Boston Veterans Affairs Normative Aging Study, the relationship between plasma concentrations of homocysteine and vitamins B-6, B-12, and folate and a battery of psychological tests was examined in 70 men aged 54 to 81 years (Riggs et al., 1996). It was found that higher levels of homocysteine and lower concentrations of vitamin B-12 and folate were associated with poorer performance in the Spatial Copying Test. Also, higher plasma concentrations of PLP were associated with higher scores in two measures of memory (backward digit span test).

In another study carried out in a group of very elderly Swedish subjects aged 90 to 101 years, the relationship between levels of folate and vitamin B-12 on episodic memory was examined (Hassing et al., 1999). Lower folate levels were associated with decreased word and object recall and secondary memory, but there were no correlations with the cobalamin plasma levels (Table 13.2).

In the New Mexico Elder Health Survey, 783 elderly subjects aged 60 to 85 years took part in a study on the relationships between plasma levels of vitamins B-12, B-6, and C and various tests of cognitive function (Lindeman et al., 2000). It was found that subjects with folate levels >11.1 nmol/l had significantly higher scores on the MMSE, Fuld Object Memory, Digits Forward, and Color Trails tests. In those with vitamin B-12 serum levels ≥221 pmol/l or vitamin C serum levels ≥57 μmol/l, the incidence of depression was reduced.

In a Scottish study, the relationship between plasma folate, vitamin B-12, and homocysteine and cognitive performance measured by the MMSE, Ravens Progressive Matrices, National Adult Reading Test (NART), Auditory Verbal Learning Test (AVLT), Digit Symbol (DS), and Block Design (BD) was examined in two elderly cohorts born in 1921 and 1936, taking childhood IQ measured at the age of 11 years into account (Duthie et al., 2002). Folate levels correlated with MMSE, NART, AVLT, DS, and BD. Vitamin B-12 levels correlated with MMSE scores. Taking childhood IQ into account strengthened the partial correlations between plasma cobalamin, MMSE, folate, and NART scores, but reduced those between RBC folate and DS scores.

Statistically significant correlations between dietary intake and blood levels of vitamins and psychometric test scores have been found in many epidemiological studies in different types of populations, but the findings have not always been consistent. In most studies, many psychometric variables were measured and a small number were usually found to correlate significantly with either the blood levels or intake of certain micronutrients. Confounding can occur in studies of this type, as dietary variables frequently correlate with socioeconomic and health-related variables.

13.5 BRAIN FUNCTION AFTER VITAMIN TREATMENT

13.5.1 INTERVENTION STUDIES IN CHILDREN

The finding that malnutrition during pregnancy or during the neonatal period, or both, and early development can impair mental performance and reduce IQ scores in later life has led to many studies assessing the effects of multivitamin treatment on cognitive performance of children.

In one study, considerable gains in IQ were reported in anemic Indian children supplemented with iron and folate (Seshadri et al., 1982). Ninety-four 5- to 8-year-old children received either 20 mg iron and 100 mg folate daily or nothing for 60 days. The scores on the Wechsler Intelligence Scale for children of the experimental group increased by 7 to 13 points compared with those of the controls. This study illustrates that correcting vitamin and iron deficiency can lead to measurable changes in IQ scores and that iron and folate are important in mental function.

In another study (Benton and Roberts, 1988), 90 Welsh schoolchildren aged 12 to 13 years were treated for 8 months with a multivitamin and mineral preparation, a placebo, or no treatment under double-blind conditions and the nonverbal but not the verbal intelligence increased significantly in the verum group. The nutritional status of the children was estimated by a self-report dietary diary kept for 3 days. The 90 children were divided into three groups of 30 and matched on the basis of sex, school performance, and home background by a form master who knew them. The three groups were then randomly assigned to three treatment schedules: placebo tablets, multivitamin and mineral tablets (Table 13.3), or no treatment. The intelligence tests used were the verbal battery of the Cognitive Abilities Test (Thorndike) and the Calvert Nonverbal Test.

TABLE 13.3
Composition of the Dietary Supplement Used in Benton and Roberts (1988) and Crombie et al. (1990)

Vitamins		Minerals	
Bioflavanoids	50 mg	Calcium gluconate	100 mg
Biotin	100 µg	Chromium	0.2 mg
Choline bitartrate	70 mg	Magnesium	7.6 mg
Folic acid	100 µg	Manganese	1.5 mg
Inositol	30 mg	Molybdenum	0.1 mg
Niacin	50 mg	Iodine	50 µg
Pantothenic acid	50 mg	Iron	1.3 mg
Para-aminobenzoic acid	10 mg	Zinc	10 mg
Vitamin B-1	3.9 mg		
Vitamin B-2	5 mg		
Vitamin B-6	12 mg		
Vitamin A	375 µg		
Vitamin B-12	10 µg		
Vitamin C	500 mg		
Vitamin D	3 µg		
Vitamin E	70 IU		
Vitamin K	100 µg		

The dietary diary results showed that the energy intakes were low, with girls having a higher intake than boys. Some mean vitamin intakes were low, particularly vitamin D, which was 32 to 44% of the U.S. recommended daily allowances (RDAs), and folate, which was 60 to 62% of the U.S. RDA. Otherwise, only mean levels of vitamins B-6 and E were low, being, respectively, 85 to 91% and 79% of the U.S. RDA. Despite the seeming adequacy of the mean levels, many children had intakes that were below the RDAs. For example, 21% of the boys and 26% of the girls had less than 25% of the RDA for vitamin D and 27% of both sexes had less than 50% of the RDA for folate. Mineral intakes were generally low, ranging from 4% of the RDA for molybdenum to 59% for iodine. Two important minerals, iron and zinc, were low, with mean values of 23–51% and 41–46%, respectively, of the RDA.

In the verbal intelligence test, there was a nonsignificant increase of ca. four points in both the placebo and verum groups. In the test of nonverbal intelligence, there was a significant increase of nine points in the verum group and a nonsignificant increase of four points in the untreated group. There were four dropouts in the untreated group. Very little data were reported and no raw data of the IQ tests were provided for the placebo and verum group.

This study was repeated (Benton and Buts, 1990) in 167 thirteen-year-old school children in Belgium. Again, children of both sexes received a multivitamin and mineral (Table 13.4) supplement for five months and were tested at baseline and the end of treatment. The children completed dietary diaries for 15 days. The subjects were divided into those with a better or poorer diet on the basis of the dietary diaries. In the girls, there were no statistically significant differences in scores between the placebo and verum subgroups at the end of the treatment period regardless of their dietary status. However, in the boys, the subgroup receiving the active treatment improved their scores in both the better and diminished diet groups, only slightly in the former but very significantly in the latter.

In a third study, Benton and Cook (1991) examined the effects of a vitamin and mineral supplement (Table 13.5) on the mental performance of six-year-old children. This prospective randomized double-blind, placebo-controlled study was carried out in 47 schoolchildren with a mean age of 6.5 years who were attending two different schools. The treatment duration was 6 weeks in one school and 8 weeks in the other. The intelligence tests consisted of four subscales of the British Ability Scale: matrices, recall of digits, similarities, and naming vocabulary. Delayed reaction times were also measured. The children's reactions to frustration while playing a television video game, which was sufficiently difficult to ensure a high failure rate in all the children, were observed and scored. The dietary intake was assessed by a food frequency questionnaire, and

---

TABLE 13.4
Composition of the Dietary Supplement Used in Benton and Buts (1990)

Vitamins		Minerals	
Vitamin A	5000 IU	Calcium	1.6 mg
Vitamin B-1	1.5 mg	Copper	2 mg
Vitamin B-2	1.7 mg	Iron	18 mg
PP	20 mg	Magnesium	25 mg
Vitamin B-6	2 mg	Manganese	1 mg
Vitamin B-12	6 µg	Zinc	10 mg
Folic acid	400 µg		
Vitamin C	60 mg		
Vitamin E	15 IU		
Vitamin D-2	400 IU		

TABLE 13.5
Composition of the Dietary Supplement Used in Benton and Cook (1991)

Vitamins		Minerals	
Vitamin A	300 μg	Calcium	30 mg
Vitamin B-1	0.7 mg	Copper	0.03 mg
Vitamin B-2	0.9 mg	Chromium	0.1 mg
Nicotinamide	10 mg	Magnesium	8 mg
Vitamin B-6	1.3 mg	Manganese	1 mg
Vitamin B-12	2.5 μg	Molybdenum	0.1 mg
Vitamin C	70 mg	Iodine	50 μg
Vitamin D	10 μg	Iron	2.4 mg
Vitamin E	6 mg	Zinc	4 mg
Vitamin K	100 μg		
Biotin	10 μg		
Folic acid	200 μg		
Pantothenic acid	25 mg		
Choline bitartrate	35 mg		
Inositol	15 mg		
Para-aminobenzoic acid	5 mg		
Bioflavonoids	20 mg		

---

frequency scores were calculated for the food groups: dairy products, fruit and vegetables, grain products, and meat and fish. In addition, the sugar intake from confectionery, sweets, drinks, and jam was calculated. The heights and weights of the children were also measured.

At baseline, the intelligence scores of the placebo and active treatment groups did not differ significantly, but after treatment they increased significantly ($p < 0.001$) in the group receiving the vitamin/mineral supplement. The scores of the nonverbal measures of the British Ability Scale, recall of digits and matrices, improved significantly after the active treatment. The scores in the verbal measures of children receiving the active treatment, verbal similarities and naming vocabulary, improved in one school but not the other. In response to the frustrating television game in the children on active treatment, the scores for mental concentration were significantly better and those for fidgeting were significantly reduced. In the reaction time tests, vitamin/mineral supplementation treatment led to a significant reduction following a 3-sec and a 13-sec delay in the girls, whereas the placebo treatment did not produce any significant changes. In the boys, only the slower reaction times after placebo in the 3-sec-delay condition reached significance.

The relationships between dietary intake and changes in the intelligence scores after the vitamin/mineral treatment were also analyzed. The parents had been asked about their children's attitude to food, and 93% felt that their child always or frequently ate snacks or sweets in place of a large meal. There was a significant correlation between the amount of sugar in the diet and the change in intelligence scores and of the matrices and similarities scores after the active treatment. These results suggested that children with a high intake of refined carbohydrate had a low micronutrient intake and therefore responded positively to additional vitamins. The anthropometric data showed that these children were of average build and showed no obvious signs of malnutrition.

The many conflicting studies on the influence of nutrition on brain function prompted Eysenck (1991) and Schoenthaler et al. (1991) to test the hypothesis that specific micronutrient deficiencies in otherwise adequately nourished subjects may prevent optimal mental functioning. The hypothesis that they wanted to test was that a vitamin/mineral supplement could improve the intelligence of children, and, in addition, that there was an inverted U-shaped dose–response curve. A prospective

TABLE 13.6
Treatments Administered in the DRF Study (Schoenthaler et al.,1991)

Vitamin/Mineral	50% RDA	% RDA	100% RDA	% RDA	200% RDA	% RDA
Vitamin A	2500 IU	50	5000 IU	100	5000 IU	100
Vitamin B-1	0.75 mg	50	1.5 mg	100	3 mg	200
Vitamin B-2	0.85 mg	50	1.7 mg	100	3.4 mg	200
Vitamin B-6	1 mg	50	2 mg	100	4 mg	200
Vitamin B-12	3 μg	50	6 μg	100	12 μg	200
Niacin	10 mg	50	20 mg	100	40 mg	200
Folic acid	200 μg	50	400 μg	100	400 μg	100
Biotin	150 μg	50	300 μg	100	300 μg	100
Pantothenic acid	5 mg	50	10 mg	100	20 mg	200
Vitamin C	30 mg	50	60 mg	100	120 mg	200
Vitamin D	200 IU	50	400 IU	100	400 IU	100
Vitamin E	15 IU	50	30 IU	100	60 IU	200
Vitamin K	50 μg	*	50 μg	*	50 μg	*
Calcium	200 mg	20	200 mg	20	200 mg	20
Magnesium	80 mg	20	80 mg	20	80 mg	20
Iron	9 mg	50	18 mg	100	36 mg	200
Zinc	7.5 mg	50	15 mg	100	30 mg	200
Iodine	75 μg	50	150 μg	100	150 μg	100
Copper	1 mg	50	2 mg	100	3 mg	150
Manganese	1.25 mg	*	2.5 mg	*	5.0 mg	*
Chromium	0.05 mg	*	0.10 mg	*	0.20 mg	*
Selenium	0.05 mg	*	0.10 mg	*	0.20 mg	*
Molybdenum	0.12 mg	*	0.25 mg	*	0.50 mg	*

Note: * indicates that there is no U.S. RDA for these micronutrients.

parallel randomized double-blind study was therefore carried out in 610 U.S. schoolchildren aged 12 to 16 years divided into four treatment groups, which received either placebo or three different doses of multivitamin/mineral supplements (Schoenthaler et al., 1991; Schoenthaler, 1991). The entire study was completed by 411 children. The measures of intelligence and mental performance used at baseline and 10 to 13 weeks later were the Wechsler Intelligence Scale for Children-Revised (WISC-R), the Matrix Analogies Test (MAT), Reaction Time and Inspection Time (RT/IT), and the Comprehensive Test of Basic Skills (CTBS). The Raven Matrices (RM) was also given after 1 month of treatment but not at baseline. The treatments consisted of either placebo or three different vitamin/mineral supplements containing doses equivalent to one, two, or three times the U.S. RDA (Table 13.6).

The rationale for these three doses was that as it was unlikely that many children were eating less than 50% of the U.S. RDA, giving an additional 50% RDA should ensure that the subjects were receiving 100% RDA. To ensure that any vitamin deficiencies that might be present be completely eliminated and to allow comparison with other studies in which a 100% RDA dose was used, the second dose of 100% U.S. RDA was chosen. A third dose of 200% U.S. RDA was chosen because the RDAs had been attacked as being too low and the investigators wanted to test for any possible benefits on brain function at a higher dose.

The results for the WISC-R, MAT, and RM are given in Table 13.7. The increase of 3.6 in the WISC-R score for nonverbal intelligence in the group receiving the 100% RDA was significant ($p < 0.01$), as was the increase in the RM score ($p = 0.029$) after 4 weeks in the group receiving the 200% RDA treatment. The MAT results did not reach significance, because the MAT instructor's

TABLE 13.7
Results of the DRF Study (Schoenthaler, 1991)

Group	Number	Baseline Scores	Final Scores	Change	Difference from Placebo	p
Wechsler Intelligence Scale for Children — Revised (WISC-R)						
	410	104.8 (13.3)	115.4 (13.6)	10.6		
Placebo	100	104.7 (12.2)	113.7 (12.4)	9.0		
50% RDA	100	104.8 (14.0)	114.8 (14.0)	10.0	1.0	
100% RDA	105	104.5 (12.9)	117.1 (13.3)	12.6	3.6	0.01
200% RDA	105	105.4 (14.6)	115.8 (14.4)	10.4	1.4	
Matrix Analogies Test (MAT)						
	558	24.8 (5.2)	26.3 (6.4)	1.5		
Placebo	147	25.1 (4.4)	26.1 (6.7)	1.0		
50% RDA	138	24.6 (5.3)	25.9 (6.8)	1.3	0.3	
100% RDA	136	25.0 (5.2)	26.9 (5.7)	1.9	0.9	
200% RDA	137	24.6 (5.9)	26.2 (6.2)	1.6	0.3	
Raven Matrices (RM)						
	489					
Placebo	129		15.1 (5.8)			
50% RDA	119		15.5 (5.6)		0.4	
100% RDA	123		15.3 (5.7)		0.2	
200% RDA	118		16.6 (5.6)		1.5	0.029

manual suggested that 600 subjects would be required to obtain a significance when the standard deviation is 5.4, which lies within the range found at baseline (4.4–5.9) and is less than that at the final examination (5.7–6.7). The WISC-R and RM scores suggest that the hypothesis of an inverted U-shaped dose–response was correct.

Apart from these five studies showing positive effects in children, three studies found no effects of vitamins alone or in combination with minerals on intelligence scores. The study by Nelson et al. (1990) was carried out under double-blind conditions, and 194 children weighed and recorded their food intake for 7 days. The multivitamin preparation (Table 13.8) contained all the micronutrients that were present in the preparation used in the study of Benton and Roberts (1988), plus 2 mg copper and 150 mg selenium; however, the doses differed considerably for some vitamins such as retinol (1000 mg RE instead of 375 mg), nicotinic acid (24 mg instead of 50 mg), folic acid (450 mg instead of 100 mg), ascorbic acid (50 mg instead of 500 mg), vitamin E (10 mg instead of 70 mg), and iron (15 mg instead of 1.3 mg). The children were divided into two age groups, 7–10 years and 11–12 years, and given tests of verbal (only the older children) and nonverbal intelligence (AH1X, AH1Y, Heim AH4), digit-span, and coding tests from the WISC-R at the start of the study and after 28 days of treatment with either placebo or supplement.

In both the placebo and verum group, there were increases in the scores in all four tests, but there were no significant differences between the groups. The authors concluded, "results from our intervention study show clearly that no improvement in intelligence can be expected from the administration of vitamin-mineral supplements over 28 days to typical British schoolchildren."

This study differs from that by Benton in a number of important ways. First, the supplement was given for only 1 month instead of 8 months. Second, the children were younger, 7–12 years instead of 12–13. Also, the age range was much larger and the tests used were also different. The younger children did not carry out a test of verbal intelligence and did the AH1X test of nonverbal intelligence at the start of the study and the AH1Y, which is a different version of the AH1X test,

TABLE 13.8
Composition of the Dietary Supplement Used in Nelson et al. (1990)

Vitamins		Minerals	
Vitamin A	1000 RE	Calcium	100 mg
Vitamin B-1	2.2 mg	Iron	15 mg
Vitamin B-2	2.8 mg	Zinc	15 mg
Vitamin B-12	5 µg	Chromium	200 µg
Nicotinic acid equivalents	24 mg	Copper	2 mg
Vitamin B-6	2 mg	Iodine	100 µg
Folic acid	450 µg	Magnesium	25 mg
Biotin	100 vg	Manganese	1.5 mg
Pantothenic acid	50 mg	Selenium	150 µg
Vitamin C	50 mg		
Vitamin E	10 mg		
Vitamin D	15 µg		
Vitamin K	100 µg		

at the end, whereas the older children did the same AH4 test at both the start and end of the treatment. Therefore, the learning or carry-over effect would have been much greater in the older children and this is was confirmed by the scores: the differences in scores between the first and second test were, respectively, 7.5 and 7.6 in the supplement and placebo groups in the older children and 3.6 and 2.2 in the younger children. The treatment duration of 28 days was relatively short; it was 60 days to 1 year in studies in which a significant improvement in various intelligence tests was found. Most of the studies were longer than 6 months. Not only was the treatment period very short, but also the doses given were so low, in the order of one RDA, that they would hardly affect the blood levels of the water-soluble vitamins with short plasma elimination half-lives (e.g., B-1, B-2, B-6, folic acid, biotin; Table 3.15).

In common with most of the studies that have been published on the effects of vitamins on mental performance, no attempt was made to measure the vitamin status by determining the blood levels of the vitamins. Food intake data do not usually provide an accurate picture of the vitamin status, because of either measurement errors or inadequacies of the food tables (Southon et al., 1992).

Crombie et al. (1990) aimed to reproduce the original study by Benton and Roberts (1988) using the same treatment (Table 13.3) administered for 7 months under randomized double-blind conditions. They also used the same two tests, the Calvert Nonverbal Test and the verbal battery of the cognitive test. A 7-day weighed dietary survey was carried out. They found only a weak effect of supplementation on the scores for the Calvert Nonverbal Test. However, the age range of their subjects, 11–13 years, was higher than the 12–13 years of the population of the Benton and Roberts study.

Southon et al. (1994) carried out a study to further examine the relationship between children's diets, vitamin/mineral supplementation, and their IQ scores (Table 13.13). The study population consisted of schoolchildren aged 13 to 14 years. The diet was assessed by 7-day weighed dietary records over 7 weeks, one day per week. Duplicate portions of all food and drink were also collected on the days of dietary recording. The duplicate diets were analyzed for content of iron, zinc, copper, vitamins C, B-1, B-2, B-6, pteroylglutamates, and total energy. Fasting blood samples were collected to determine ferritin, zinc, copper, selenium, vitamins C, D, B-1, B-2, B-12, B-6, and plasma lipids. Erythrocyte transketolase (ETK) activity, the activity coefficient α-ETK, erythrocyte gutathione reductase (EGR) activity, and activity coefficient α-EGR were also measured. The children's IQ was measured by the WISC-R UK. The treatment consisted of vitamins and minerals (see Table 13.9)

TABLE 13.9
Composition of the Dietary Supplement Used in Southon et al. (1994)

Vitamins		Minerals	
Ascorbic acid	25 mg	Iron	12 mg
Thiamin	1.0 mg	Selenium	50 μg
Riboflavin	1.4 mg	Zinc	15 mg
Nicotinic acid equivalents	16 mg	Chromium	50 g
Pyridoxine	1.8 mg	Copper	50 μmg
Cyancobalamin	3 μg	Iodine	150 μg
Pteroylmonoglutamic acid	400 μg	Manganese	2.5 mg
Biotin	100 μg		
Pantothenic acid	4 mg		
Vitamin D	10 μg		

---

or placebo in the form of two capsules per day and was given for four months. The two treatment groups were matched for age, sex, height, weight, and overall IQ. It is not clear whether the treatment allocation was randomized; four girls in the micronutrient group developed nausea after taking the capsules and were transferred to the placebo group. The comparison of the calculated and analyzed food intake showed that the calculated energy intake for the boys was 4% less than the analyzed value, average calculated pteroylglutamate intake was 40% less than the analyzed value, and calculated thiamin intake was lower by 23 to 27%. Calculated copper, riboflavin, and pyridoxine intakes for girls were significantly higher than the analyzed values, but did not reach statistical significance. Calculated intakes for iron, zinc, and vitamin C were not significantly different from the values obtained by analysis. The WISC-R UK IQ verbal scores increased in the placebo group by 4.6 and 2.6 in the boys and girls, respectively, and the nonverbal scores increased by 5.6 and 5.7. In the treated groups, the verbal scores increased by 5.3 and 4.1 in the boys and girls, respectively, and the nonverbal scores by 5.9 and 6.0. There was a significant retest effect, but the differences between the treatments were not significant. The effects of the treatment on indices of the micronutrient status were inconsistent. There were no significant changes in the plasma levels of any of the vitamins in the treated group except cyanocobalamin and pteroylmono-glutamate, which increased significantly. The vitamin status of the children at baseline was good, with the ranges of vitamin C plasma concentrations more than 200 μmol/l and PLP more than 100 nmol/l, suggesting that these children were already taking supplements or eating food enriched with high levels of vitamins. As the multivitamin treatment only led to significant increases in the folate and vitamin B-12 plasma concentrations, which were already optimal at baseline, it is not surprising that no significant changes were found in the IQ scores after treatment. The treatment allocation was not randomized, as four girls who developed nausea were changed from the active to the placebo group at the start of the treatment phase. The reasons for the lack of an increase in plasma levels of the vitamins that were in the capsules are not clear. The doses were very low and the bioavailability of some vitamins such as thiamin is known to be low, but others such as pyridoxine are highly bioavailable (Table 13.15). Poor compliance or some inadequacy in the formulation of the capsules could account for the unchanged vitamin plasma levels, although the former is made less likely by the increases in pteroylmonoglutamate and cyanocobalamin. There was also a large unexplained increase in cyanocobalamin levels from 337 to 589 pmol/l in the boys in the placebo group. Overall, this study provided some very interesting data on the relationship between calculated dietary intake, vitamin content of food as determined by analysis, and corre-sponding plasma levels of vitamins, but for methodological reasons it did not provide any infor-mation on the effects of vitamin treatment on IQ in children. The study designs, study populations, and possible effects of growth rate of the children in these studies have been reviewed in detail by Haller (1995).

13.5.2 INTERVENTION STUDIES IN ADULTS

A number of studies have been carried out to examine the effects of vitamins on cognitive function and mood in healthy adults. One study that illustrated the efficacy of folate in restoring depressed IQ scores and mood was described by Botez et al. (1984). Forty-nine patients with folate deficiency presented with mild depression according to DSM-III criteria, and in addition some had mild neuropathies and mild neurological signs. Their CT scans showed that 68% had cerebral atrophy. After 7 to 11 months of 10 mg folate daily, their scores on the WAIS increased from 101.8 ± 2.20 to 111.2 ± 1.87 ($p < 0.001$), on the verbal scale from 103.0 ± 1.72 to 109.3 ± 1.21 ($p < 0.001$), and on the performance scale from 98.3 ± 2.21 to 106.5 ± 2.01 ($p < 0.001$). In this study, there was again a larger increase in the nonverbal than in the verbal intelligence scores. Polyneuropathies, fatigue, depression, and distractibility improved under folate treatment.

A very interesting study on the psychological changes occurring in subjects with low vitamin blood levels and the effects of supplementation with a low-dose multivitamin supplement was carried out by Heseker et al. (1990). Initially, 1228 subjects aged 17 to 29 years were screened for low vitamin blood levels (vitamins B-1, B-2, B-6, C) and 197 selected. This group, together with a random group of 884, were entered into the study and subjected to a psychometric test battery at baseline and blood collected 1 week later. They were then given placebo or a multivitamin treatment (Table 13.10) under randomized double-blind conditions for 8 weeks, after which they were retested and blood was again drawn. The psychometric test battery consisted of the Freiburger Personality Inventory (FPI), which is the Eysenck Personality Inventory normalized for the South German population; the Eigenschaftswörterliste (EWL) (adjective checklist), which is similar to the Profile of Mood States (POMS); and the Vienna Test System for measuring reaction time, attention, and concentration. A comparison of the baseline scores between subjects with vitamin levels below the fifth percentile and those with vitamin levels above the median for that particular vitamin and above the twenty-fifth percentile for the other vitamins was made. There were significant differences in only a few psychometric scales. As it was found that the vitamin status at baseline and at week 8 in the placebo group varied considerably in individual subjects, it was decided to compare those whose blood levels of a particular vitamin were above the median at both baseline and week 8 and whose other vitamin levels were at least adequate with those whose blood levels of that particular vitamin were below a defined level at both times. The results of this

TABLE 13.10
Composition of the Vitamin Supplement Used in Heseker et al. (1990)

Vitamin B-1	3.0 mg
Vitamin B-2	3.5 mg
Vitamin B-6	4.5 mg
Niacin	20.0 mg
Folic acid	800 μg
Vitamin B-12	6.0 μg
Vitamin C	150 mg
Vitamin A	1.5 mg (RE)
Vitamin E	20.0 mg

comparison, despite the small numbers ($n = 10$), were many highly significant differences on most of the scales of the EWL and FPI for vitamins B-1, B-2, B-6, B-12, C, and E and folate. These results are very interesting as they seem to show that differences in psychological parameters become measurable when a chronically borderline group is compared with an adequately supplied group.

In the analysis of the treatment effects, the placebo and active subgroups, which had low vitamin levels at baseline, were compared at baseline and after 8 weeks. Significant differences were found for a number of scales of the EWL and FPI for vitamins B-1 and C and folate. The effects of supplementation were more limited, presumably because the verum subgroup contained subjects with both chronic and acute low blood levels, as this could not be determined from their blood levels at the end of the treatment period as in the case of the placebo group. This study was particularly good because it attempted to address the problem of low vitamin stores and the consequences of chronic vitamin deficiency on mental performance. The overall placebo verum comparison would presumably have led to more significant results if the doses used had been higher and if the treatment duration had been longer than 8 weeks.

A prospective parallel randomized placebo-controlled study on the effects of long-term high-dose vitamin supplementation used a study design intended to deal with some of the problems specific to this type of trial (Benton et al., 1995a, 1995b). The study population consisted of 207 young adults of both sexes aged 17 to 27 years, and anthropometric, smoking, drinking, and food intake data were collected and vitamin blood levels measured at baseline and 3, 6, 9, and 12 months when the psychometric test batteries were applied. The composition of the supplement given is listed in Table 13.11. The test battery consisted of the POMS, the General Health Questionnaire (GHQ), simple (SR) and complex reaction time (CRT), Digit Symbol Substitution (DSST), and the Continuous Attention Task (CAT). In the active treatment group, the blood levels of all the vitamins included in the supplement, except vitamin A, increased significantly. The scores of the GHQ and POMS improved significantly only after 12 months of treatment and only in the females. The improvements in mood were correlated with increases in the blood levels of vitamins B-2 and B-6 and in the GHQ scores with vitamins B-1, B-2, B-6, biotin, and vitamins C and E. Those having the lowest thiamin levels, with an erythrocyte transketolase activation coefficient (ETKAC) more than 1.20, had significantly worse POMS scores for composure, confidence, and elation (Benton et al., 1995a). In the results of the test of cognitive function, the only significant findings were in the women after 12 months of treatment. There were significant improvements in the CRT

TABLE 13.11
Composition of the Vitamin Supplement Used in Benton et al. (1995a)

Vitamin	Daily Dose
Vitamin A	3334 IU (1000 RE)
Vitamin B-1	14 mg
Vitamin B-2	16 mg
Vitamin B-6	22 mg
Vitamin B-12	0.03 mg
Vitamin C	600 mg
DL-α-Tocopherol acetate	100 mg
Folic acid	4 mg
D-Biotin	2 mg
Nicotinamide	180 mg

reaction and decision times and the CAT associated with increases in the blood levels of vitamins B-1, B-6, biotin, and vitamin B-12 (Benton et al., 1995b). The vitamin status of this study population was poor: 36% of the women and 28% of the men had a high risk of a biochemical vitamin B-6 deficiency, 29% of women and 20% of men for vitamin B-2, 1.8% of women and 4.6% of men for vitamin B-1, less than 1% of both men and women for vitamin C, 1.8% of women and 1.5% of men for vitamin B-12, and 4.4% of women and 4.6% of men for biotin (Benton et al., 1995b).

This study showed that there was a strong correlation between baseline thiamin status and improvement in thiamin status and mood scores in women. Therefore, it was decided to carry out a second study to examine the effects of high-dose thiamin treatment in women (Benton et al., 1997). Female university students (120) with a mean age of 20.3 years were treated with 50 mg thiamin or placebo for 2 months with random allocation to treatment under double-blind conditions. Fasted blood samples were drawn before and at the end of the treatment period to determine thiamin, riboflavin, and pyridoxine status. Mood was assessed using the bipolar POMS, psychopathology was assessed by the 30-item GHQ, reaction time with the CRT, and memory with the Familiar Faces test and a word recall list. The study was completed by 117 subjects, and in the verum group the treatment led to a significant increase in ETK and decrease in α-ETK, whereas there were no significant changes in the placebo group. None of the subjects had a thiamin deficiency at baseline, riboflavin status remained unchanged in both groups, and pyridoxine status, which was moderately deficient (α-EAST = 1.71), worsened significantly during the study to 1.78. The POMS scores improved significantly in the thiamin group, but the GHQ scores did not change. The memory scores did not change under treatment, but the decision times in the CRT improved significantly in the thiamin group. Changes in thiamin status correlated significantly with change in mood for the factors Agreeable, Clearheaded, Composed, Elated, Energetic, and Total Mood.

In a study examining the effects of B-vitamins, calcium, magnesium, and zinc (Table 13.12), 80 healthy men, aged 18 to 42 years, were treated for 28 days and the following psychometric instruments were used to assess the effects of treatment: GHQ-28; Hospital Anxiety and Depression Scale (HADS); Perceived Stress Scale (PSS); self-rating scales for anxiety, depression, tension, tiredness, and concentration; a physical symptoms checklist; and a treatment evaluation (Carroll et al., 2000). Plasma zinc was measured at baseline and at the end of the study. The composition of the supplement is listed in Table 13.12. After 28 days of treatment, the total GHQ-28 scored

TABLE 13.12
Composition of the Vitamin Supplement Used in Carroll et al. (2000)

Vitamin	Daily Dose
Vitamin B-1	15 mg
Vitamin B-2	15 mg
Vitamin B-12	10 µg
Vitamin C	500 mg
Folic acid	400 µg
Biotin	150 µg
Niacin	50 mg
Pantothenic acid	23 mg
Calcium	100 mg
Magnesium	100 mg
Zinc	10 mg

decreased significantly in the vitamin-treated group and somatic symptoms increased significantly in the placebo group. In the HADS subscale for anxiety, there was a significant reduction in the verum group. In the PSS, the verum group had significantly lower levels of perceived stress than the placebo group. In the self-rating scales, the subjects of the vitamin group rated themselves as being significantly less anxious, tired, and unable to concentrate, but there were no significant treatment effects. Physical symptoms increased significantly in the placebo group, but remained unchanged in the vitamin group.

13.5.3 INTERVENTION STUDIES IN THE ELDERLY

A number of studies have been carried out to examine the effects of vitamins on the mental performance and mood of normal elderly, elderly subjects with age-related impairments, and patients suffering from various forms of dementia. (Table 13.14) The rationales and study designs of these studies have been critically reviewed in detail in Haller (1996).

The hypothesis that high doses of thiamin can increase acetylcholine levels and cholinergic function in the brains of Alzheimer patients was tested in two small studies (Blass et al., 1988; Nolan et al., 1991). In both studies, outpatients with a diagnosis of probable senile dementia of the Alzheimer type (SDAT) according to the National Institute of Neurological and Communicative Disorders and Stroke — Alzheimer's Disease and Related Disease Association (NINCDS-ADRDA) criteria were treated orally with 1 g thiamin/day for either 3 or 12 months. The vitamin status was not measured. Both studies included the MMSE as a measure of efficacy as well as the Haycox Behavioral Rating Scale, the Blessed Dementia Scale, and the Consortium to Establish a Registry for Alzheimer's disease (CERAD) neuropsychological battery. In the shorter study the MMSE total score improved significantly but not the Haycox or Blessed Dementia scales, whereas in the 12-month study both the MMSE and CERAD scores decreased in all groups. Burns et al. (1989) treated 15 inpatients with multiinfarct dementia or SDAT with 100 mg thiamin, 10 mg riboflavin, 10 mg pyridoxine, and 400 mg nicotinamide for 6 weeks. The status of vitamins B-1 and B-6 was measured and found to be marginal or deficient. The main outcome measure, the MMSE, remained unchanged whereas the behavioral disturbance score decreased.

Smidt et al. (1991) examined the relationship between thiamin status and effects of thiamin treatment (10 mg/day p.o.) for 6 weeks in 80 healthy, free-living Irish women aged 65 to 92 years. Of the 80 subjects, 48% percent had a thiamin deficiency. After treatment with thiamin, there were significant improvements in subjective ratings of appetite, fatigue, and general well-being.

The effects of a multivitamin also containing magnesium, L-tryptophan, trace elements, and intrinsic factor on the MMSE scores were examined in geriatric inpatients (Abalan et al., 1992). After 105 days of treatment the MMSE improved significantly. The vitamin status was not determined.

Deijen et al. (1992) examined the effects on mood and mental performance of treating 74 elderly men with 20 mg pyridoxine or placebo daily for 3 months. On the basis of the baseline test session scores, the subjects were matched on age, vitamin B-6 status, and IQ scores and then randomly allocated to a 12-week treatment with either vitamin B-6 or placebo and then tested again. Mood was assessed with the adjective checklist of Janke and Debus. Different aspects of memory were assessed by a battery of tests: short-term verbal memory by the associate learning task, the sensory register by the Sperling whole report task, long-term verbal memory by the associate recognition test, long-term memory storage by a Forget score derived from trial 3 of the associate learning task and from the associate recognition test, a long-term visual and visual memory task, attention and performance by the Cognitrone Form Perception Test, and perceptual motor skill by the Vienna Determination Unit. The pyridoxine treatment led to better verbal long-term memory. The relationship between the memory score differences and plasma PLP levels had an inverted U-shaped curve, showing that there was an optimal vitamin B-6 level and that both high and low levels had detrimental effects.

TABLE 13.13
Results of Studies on the Effects of Vitamins on Intelligence or Other Aspects of Mental Performance

Study	n	Treatment	Dose (U.S. RDA)	Duration	Outcome[a]	Subjects
Benton and Roberts, 1988	90	Multivitamins/minerals	0.25–8.3	8 months	Positive	Children, 12–13 years
Benton and Buts, 1990	167	Multivitamins/minerals	0.5–1.0	5 months	Positive	Children, 13 years
Nelson et al., 1990	194	Multivitamins/minerals	0.3–1.0	1 month	NS	Children, 7–12 years
Crombie et al., 1990	86	Multivitamins/minerals	0.25–8.3	7 months	NS	Children, 11–13 years
Benton and Cook, 1991	47	Multivitamins/minerals	0.2–1.0	1.5–2 months	Positive	Children, 6 years
Schoenthaler et al., 1991	411	Multivitamins/minerals	1.0	3.25 months	Positive	Children, 12–16 years
Southon et al., 1994	54	Multivitamins/minerals	1.0	4 months	NS	Children, 13–14 years
Heseker et al., 1990	197	Multivitamins	0.67–2.5	2 months	Positive	Adults, male, 17–29 years
Benton et al., 1995a	127	Multivitamins	10	12 months	Positive	Adults, 17–27 years
Benton et al., 1996	117	Thiamin	45	2 months	Positive	Women, 20.3 years
Botez et al., 1984	49	Folate	25	7–11 months	Positive	Adults
Carroll et al., 2000	80	B-complex, C, Zn, Ca, Mg	10	28 days	Negative	Adults, male, 18–42 years

[a] Statistically significant improvement in some measure of cognitive function or mood; NS = not significant.

TABLE 13.14
Results of Studies on the Effects of Vitamins on Mental Performance in the Elderly

Study	n	Treatment	Median Dose[a]	Duration	Outcome[b]	Subjects/Patients
Blass et al., 1988	11	B-1	2490	3 months	Positive	SDAT patients, 59–83 years
Burns et al., 1989	15	B-1, B-2, B-6, C, nicotinamide	3.3–83	6 weeks	Negative	SDAT, MID patients, 81 ± 5.3 years
Nolan et al., 1991	15	B-1	2490	12 months	Negative	SDAT patients, 59–88 years
Smidt et al., 1991	80	Thiamin	6.7	6 weeks	Positive	Women, 65–92 years
Abalan et al., 1992	29	Multivitamins, etc.	≅1	105 days	Positive	Geriatric inpatients, ≅85 years
Deijen et al., 1992	82	Pyridoxine	10	3 months	Positive	Adults, male, 70–79 years
Cockle et al., 2000	139	Multivitamins	10	24 weeks	Negative	Elderly, 60–83 years
Cockle et al., 1998	239	Multivitamins	10	12 months	Negative	Elderly, 60–82 years

[a] Expressed in U.S. RDAs.
[b] Positive statistically significant improvement in some measure of cognitive function or mood.

The effects of high doses of multivitamins on a battery of psychometric tests were examined in 139 healthy elderly volunteers aged 60 to 83 years in a randomized placebo-controlled parallel group study (Cockle et al., 2000). The vitamin supplement had the same composition as that used by Benton et al. (1995a) (see Table 13.11). The vitamin status assays showed that at baseline 22% were at high risk of a biochemical riboflavin deficiency, 63% a pyridoxine deficiency, 4.3% a thiamin deficiency, 6.8% a vitamin B-12 deficiency, and 4.8% a vitamin C deficiency. The baseline vitamin status correlated positively with many measures of mental health. Higher scores in the MMSE correlated with thiamin status overall, higher levels of vitamin B-12 in men, and β-cryptoxanthin in women. The status of vitamin B-1 was correlated with a number of factors in the POMS questionnaire, such as composure, energy, clearheadedness, and positive mood in women and negatively with elation in men. Riboflavin status was associated with confidence, energy, clearheadedness, and overall positive mood in men, with few correlations with mood in women. Feelings of composure, agreeability, elation, and overall positive mood were related to higher biotin levels in men. In both sexes, pyridoxine status was associated with energy and clearheadedness. Vitamin A plasma levels were associated with greater energy in women and vitamin E with less energy and less confidence in both genders. There were also a number of significant associations between plasma levels of β-cryptoxanthin in females and folic acid in males for total and motor reaction times from the choice reaction time task. Higher vitamin C and B-2 stati were correlated with faster reaction times in women. However, a higher pyridoxine status was associated with slower performance on the Sternberg Memory Scanning Task in males. Lower critical flicker fusion (CFF) was correlated with higher retinol in females and vitamin B-12 in males. The multivitamin treatment significantly raised plasma levels or decreased activity coefficients for most of the vitamins except retinol in the verum group, and the vitamin C and folic acid levels increased considerably in the placebo group. The 24-week multivitamin treatment led to significant improvements in CFF in males; total reaction time improved in the CRT in both sexes, but had no significant effect on mood.

This study was repeated in a larger population of 239 elderly subjects, using the same multivitamin or placebo treatment for 12 months (Cockle et al., 1998; Cockle et al, in press; Table 13.11). The subjects were tested on a battery of psychometric tests and their vitamin status measured at baseline and after 4, 8, and 12 months of treatment. The vitamin status of this population of elderly was also not optimal, with 30% at high risk of a vitamin B-6 deficiency, 10% of a riboflavin deficiency, and 2% of a cobalamin deficiency. At baseline, there were a small number of significant correlations between the vitamin status and the Hospital Anxiety Scale (HAS) for pyridoxine in females and riboflavin overall. In males, biotin correlated with higher scores on the HAS. Higher vitamin E levels in men and lower vitamin C levels in women, correlated with lower CFF. Poorer performance in the Sternberg Memory Scanning Task was associated with poorer thiamin status in women. In contrast, higher levels of vitamin C were associated with longer reaction times. In the CRT, higher pyridoxine status was correlated with longer total reaction and recognition times and higher levels of retinol with longer motor reaction time in females. In males, higher vitamin E levels were associated with shorter motor reaction times.

In the verum group, the levels of vitamins A, B-12, C, E, biotin, and folic acid increased significantly and the activity coefficients for thiamin, riboflavin, and pyridoxine decreased significantly during the treatment period, indicating improvements in status, but one subject was still at high risk of a cobalamin deficiency. In the placebo group, levels remained constant or decreased, and after 48 weeks 19% had a high risk of a biochemical riboflavin deficiency, 43% a pyridoxine deficiency, 4% a cobalamin deficiency, and 4% a vitamin E deficiency.

In the vitamin group, CFF improved relative to baseline but not in comparison to the placebo group. There were, however, no significant differences between the verum and placebo group for the CRT, the Sternberg Memory Scanning Task, the Cognitive Failures Questionnaire, the Syndrom Kurtz Test, the HADS, or the Ravens Progressive Matrices.

13.6 DISCUSSION

Vitamins are essential for the energy metabolism of neurons and glia cells as well as for the synthesis of many neurotransmitters. Vitamins also modulate the receptor binding of several neurotransmitters, influence the structural integrity of neurons, and affect conduction velocity. Dose–response effects of vitamins have been studied on various aspects of brain function in animals and humans with and without vitamin deficiencies. The first studies addressed the question of minimal vitamin requirements and types of symptoms and adverse effects that occurred during various degrees of vitamin deficiencies. The symptoms of vitamin deficiency are complex and depend on the degree of deficiency and its duration. The signs and symptoms of acute deficiency can differ considerably from those due to marginal long-term deficiency. Deficiency in a single vitamin is rare and is also difficult to attain under experimental conditions. More recent work has examined the prophylactic effects of vitamins and the potential benefits of vitamin doses above those required to prevent deficiency. Studies in countries where malnutrition is endemic have shown that giving children additional vitamins and minerals can improve their school and mental performance. Similar studies have also been carried out in prosperous Western countries, because it was felt that many school-children were consuming food having a high energy but low micronutrient density. The baseline micronutrient intake of the children has been determined in a number of studies before the start of the treatment, but the blood levels have rarely been measured, and thereby very little is known about the true vitamin status of these subjects. The randomized double-blind controlled studies on healthy children and adults are listed in Table 13.13 together with the factors that might be expected to influence their outcome, namely, the type of study population, its size, its age, the dose, expressed as U.S. RDAs to simplify comparison, and the treatment duration. In two studies in children, the study population, doses, treatment duration, and composition of the supplement were almost identical, yet the outcomes were different (Benton and Roberts, 1988; Crombie et al., 1990). The only major difference between these studies was in the age range of the children, 12 to 13 years in the Benton and Roberts (1988) study and 11 to 13 years in the Crombie et al. (1990) study. This might affect the outcome because the growth rate of girls reaches a maximum at 12 years and then declines. Therefore, in the Crombie et al. (1990) study the girls would still have been growing very rapidly and perhaps have had a very high vitamin demand, which was only just met by the food intake and supplement. The other studies are difficult to compare because of the many differences in study parameters, including the doses of the individual vitamins in the supplements, which ranged from 25 to 830% of the U.S. RDA in some studies, and the different outcome measures, which included IQ, memory, reaction times, and mood. The vitamin stores of the subjects before treatment are not known, although the vitamin blood concentrations or coenzyme activation coefficients were measured in three studies and the dietary intake in six studies, providing some information about the vitamin status. The dosage regimens have rarely taken into account the biokinetic properties of the vitamins used in the treatments (Table 13.15). Also, the baseline vitamin status of the subjects has not been sufficiently characterized. Cut-offs for defining acute vitamin deficiency and for defining optimal levels having a protective action against degenerative processes have been proposed, but little is known about their true validity (Table 13.16). In some studies in which the vitamin status indicated that the subjects had some vitamin deficiencies at baseline, the treatments led to significant improvements, as might be expected. However, in the study with high-dose thiamin, the participants did not have a thiamin deficiency, yet there were clear-cut improvements in mood and reaction times. In the few studies in which they were measured, baseline vitamin levels or activity coefficients correlated significantly with various measures of mental performance, but raising the blood levels did not necessarily improve test scores.

In the studies on elderly people, most participants were impaired by dementia or other age-related factors, except for three studies involving healthy volunteers. As in the studies on adults, neither the baseline vitamin status nor the dosage, duration, or composition of the supplement are able to predict the outcome of the study. It is surprising that in the studies in which vitamin

TABLE 13.15
Biokinetic Parameters of Vitamins A, B-1, B-2, B-6, E, and K and of Carotene in Humans[a]

Vitamin	T_{max}	Bioavailability (%)	$T_{1/2}$	Depletion Time
Carotene	24–48 h	40–60	5–6 days	
Vitamin A	3–4 h	80–90	30 h	1–2 years
α-Tocopherol	12–13 h		73–80 h	6–12 months
Pyridoxine		75–100	2.5 h	2–6 weeks
Riboflavin		60–90		2–6 weeks
Ascorbic acid		80–90 (dose dependent)	3.4 h	2–6 weeks
Thiamin	1 h	5	20 min/1 h/2 days	1–2 weeks
Vitamin K		50	1.5–3 h	

Note: T_{max} = time to reach maximum plasma concentration; $T_{1/2\beta}$ = plasma elimination half-life; depletion time = duration of vitamin deficiency required before body stores are completely depleted.

[a] Biokinetic data from studies with micronutrients administered in the form of oral dosage (tablets or capsules).

TABLE 13.16
Cut-Offs for Defining Risk of Acute Deficiency and Optimal Levels to Prevent Chronic Diseases (Haller, 1999)

Vitamin	Acute Levels			Chronic Optimal Level
	High Risk	Moderate Risk	Low Risk	
α-ETK	>1.25	1.15–1.25	<1.15	
α-EGR	>1.30	1.2–1.3	<1.2	
α-EAST	>1.8	1.7–1.8	<1.7	
Plasma biotin	<122 ng/l	122–244	>244 ng/l	
	<0.5 nM/l	0.5–1.0	>1.0 nM/l	
Plasma folic acid	<3.0 µg/l	3.0–5.9	>5.9 µg/l	6.6 µg/l
	<6.8 nM/l	6.8–13.4	>13.4 nM/l	15 nM/l
Plasma vitamin B-12	<150 ng/l	150–200	>200 ng/l	349 ng/l
	<111 pM/l	111–147	>147 pM/l	258 pM/l
Plasma ascorbic acid	<2 mg/l	2–3 mg/l	>3 mg/l	>8.8–10.6 mg/l
	<11 µM/l	11–16.5 µM/l	>16.5 µM/l	>50–60 µM/l
Plasma α-tocopherol	<5 mg/l	5–7	>7 mg/l	>12.9 mg/l
	<11.6 µM/l	11.6–16.3	>16.3 µM/l	>30 µM/l
Vitamin E:cholesterol ratio	<2.5 mg/g		>2.5 mg/g	
	<2.2 µmol/mmol		>2.2 µmol/mmol	>5.2 µmol/mmol
Plasma retinol	<100 µg/l	100–300	>300 µg/l	>716 µg/l
	<0.35 µM/l	0.35–1.05	>1.05 µM/l	>2.5 µM/l
Plasma total carotene	<200 µg/l	200–400 µg/l	>400 µg/l	>215 µg/l
	<0.37 µM/l	0.37–0.75 µM/l	>0.75 µM/l	>0.40 µM/l

deficiencies were present at baseline, successfully correcting it still did not lead to improvements in cognitive performance or mood. One explanation could be that these healthy elderly volunteers were still able to perform at such a high level that there was a ceiling effect and no further improvement in scores was possible. It is also possible that the geriatric inpatients in some other studies were irreversibly mentally impaired due to lesions of various etiologies, such as β-amyloid toxicity or ischemia.

The outcomes measures used in these different types of studies were also very heterogeneous. Some were designed to measure pathology in patients and others to measure psychological scores for assessing normal children or adults. In some studies, only one sex responded to treatment. The inconsistent results suggest that the treatment effects are minor compared with the variance of the outcome measures.

13.7 CONCLUSIONS

Vitamins are one of many factors that can affect mental performance. Manipulating their intake has been shown to influence mental performance and mood in some populations, particularly children and young adults. They are essential to maintain normal brain function and deficiencies lead to measurable deficits, particularly in children. Some aspects of age- or pathology-related cognitive impairments are due to vitamin deficiencies. It remains to be seen from the results of ongoing research whether the severity or progress of Alzheimer's disease or vascular dementias can be attenuated by long-term treatment with certain vitamins or combinations of vitamins. More needs to be known about the pathological mechanisms in these diseases before rational treatment strategies that use vitamins can be developed. Future studies should examine the effects of individual vitamins in well-defined homogenous study populations and should use well-validated and appropriate outcome measures. Study design must take into account the rationale for the choice of dose and the time course of the expected changes in mental performance. A consistent finding of studies on the vitamin status of various populations is that a certain proportion have a high or moderate risk of a biochemical vitamin deficiency, which raises the question of whether their mental performance is suboptimal. Case histories in which neurological or psychiatric deficits are finally found to have a nutritional etiology demonstrate that vitamin deficiencies due to idiosyncratic diets or pathology affecting absorption or utilization lead to various forms of impairment.

REFERENCES

Abalan, F., Mancier, G., Dartigues, J.-F., Décamps, A., Zapata, E., Saumally, B., Galley, P. (1992) Nutrition and SDAT. *Biol. Psychiatry*, 31, 99–105.

Allgood, V.E., Cidlowski, J.A. (1991) Novel role for vitamin B6 in steroid hormone action: a link between nutrition and the endocrine system. *J. Nutr. Biochem*, 2(10), 523–534.

Anderson, D.N., Abou-Saleh, M.T. (1995) Tetrahydrobiopterin, folate and depression: a modified amine hypothesis. *Human Psychopharmacol.*, 10, 1–5.

Ashley, V.M. (1986) Nutritional control of brain neurotransmitter synthesis and its implications. *Bibl. Nutr. Dieta.*, 38, 39–53.

Barclay, L.L., Gibson, G.E., Blass, J.P. (1981) Impairment of behavior and acetylcholine metabolism in thiamin deficiency. *Pharmacol. Exp. Therapeut.*, 217(3), 537–543.

Bender, D.A. (1992). *Nutritional Biochemistry of the Vitamins*. Cambridge University Press, Cambridge, MA.

Benton, D, Buts, J.P. (1990) Vitamin/mineral supplementation and intelligence. *Lancet*, 335, 1158–1160.

Benton, D., Cook, R. (1991) Vitamin and mineral supplements improve the intelligence scores and concentration of six-year-old children. *Person. Individ. Diff.*, 12(11), 1151–1158.

Benton, D., Fordy, J, Haller, J. (1995a) The impact of long-term vitamin supplementation on cognitive functioning. *Psychopharmacology*, 117(3), 298–305.

Benton, D., Griffiths, R., Haller, J. (1997a) Thiamin supplementation, mood and cognitive functioning. *Psychopharmacology*, 129, 66–71.

Benton, D., Haller, J., Fordy, J. (1997b) The vitamin status of young British adults. *Int. J. Vitam. Nutr. Res.*, 67, 34–40.

Benton, D., Haller, J., Fordy, J. (1995b) Vitamin supplementation for 1 year improves mood. *Neuropsychobiology*, 3, 98–105.

Benton, D., Roberts, G. (1988) Effect of vitamin and mineral supplementation on intelligence of a sample of schoolchildren. *Lancet*, 1(8578), 140–143.

Blass, J.P., Gleason, P., Brush, D., DiPonte, P., Thaler, H. (1988) Thiamine and Alzheimer's disease. *Arch. Neurol.*, 45, 833–835.

Botez, M.I., Botez, T., Maag, U. (1984) The Wechsler subtests in mild organic brain damage associated with folate deficiency. *Psychol. Med.*, 14, 431–437.

Brozek, J., Guetzkow, H. (1957) Psychological effects of thiamin restriction and deprivation in normal young men. *Am. J. Clin. Nutr.*, 5(2), 109–120.

Brozek, J., Guetzkow, H., Keys, A., Cattell, R.B., Harrower, M.R., Hathaway, S.R. (1946) A study of personality of normal young men maintained on restricted intakes of vitamins of the B complex. *Psychosomat. Med.*, 8, 98–109.

Burns, A., Marsh, A., Bender, D.A. (1989) A trial of vitamin supplementation in senile dementia. *Int. J. Geriatr. Psychiatry*, 4, 333–338.

Carroll, D., Ring, C., Suter, M., Willemsen, G. (2000) The effects of an oral multivitamin combination with calcium, magnesium, and zinc on psychological well-being in healthy young male volunteers: a double-blind placebo-controlled study. *Psychopharmacology*, 150, 220–225.

Chasan-Taber, L., Selhub, J., Rosenber, I.H., Malinow, R., Terry, P., Tishler, P.V., Willett, W., Hennekens, C.H., Stampfer, M.J. (1996) A prospective study of folate and vitamin B6 and risk of myocardial infarction in US physicians. *J. Am. Coll. Nutr.*, 15, 136–143.

Chomé, J., Paul, T., Pudel, V., Bleyl, H., Heseker, H., Hüppe, R., Kübler, W. (1986) Effects of suboptimal vitamin status on behavior. *Bibl. Nutr. Dieta.*, 38, 94–103.

Claus, D., Neundorfer, B., Warecka, K. (1984) The influence of vitamin B6 deficiency on somatosensory stimulus conduction in the rat. *Eur. Arch. Psychiatr. Neurol. Sci.*, 234, 102–105.

Cockle, S.M., Dawe, R., Robinson, L., Haller, J., Hindmarch, I. (1998) Repeated multivitamin supplementation and cognitive function in healthy elderly volunteers. In *Eight Congress of the CENP*, Vienna, 13–17 September 1998.

Cockle, S.M., Haller, J., Kimber, S., Dawe, R.A., Hindmarch, I. (2000) The influence of multivitamins on cognitive function and mood in the elderly. *Aging Mental Health*, 4(4), 339–353.

Cockle, S.M., Dawe, R., Robinson, L., Haller, J., Hindmarch, I. (1999) The effects of repeated doses of vitamins on cognitive function in healthy elderly volunteers. *Int. J. Neuropsychopharm.*, 2(1), 245.

Crombie, J.K., Todman, J., McNeill, G., Florey, C., Du, V., Menzies, J., Kennedy, R.A. (1990) Effect of vitamin and mineral supplementation on verbal and non-verbal reasoning of schoolchildren. *Lancet*, 335, 744–747.

Dakshinamurti, K. (1982) Neurobiology of pyridoxine. In *Advances in Nutritional Research*, Vol 4 (H.H. Draper, Ed.), pp. 143–179. Plenum Press, New York.

Dakshinamurti, K., Paulose, C.S., Viswanathan, M., Siow, Y.L. (1988) Neuroendocrinology of pyridoxine deficiency. *Neurosci. Biobehav. Rev.*, 12, 189–193.

Dakshinamurti, K., Sharma, S.K., Bonke, D. (1990) Influence of B vitamins on binding properties of serotonin receptors in the CNS of rats. *Klin. Wochenschr.*, 68, 142–145.

Deijen, J.B., van der Beek, E.J., Orlebeke, J.F., van den Berg, H. (1992) Vitamin B-6 supplementation in elderly men: effects on mood, memory, performance and mental effort. *Psychopharmacology*, 109, 489–496.

Dreyfus, P.M. (1988) Vitamins and neurological dysfunction. In *Nutritional Modulation of Neural Function* (J.E. Morley, M.B. Sterman, J.H. Walsh, Eds.), pp. 155–164. Academic Press, London.

Duthie, S.J., Whalley, L.J., Collins, A.R., Leaper, S., Berger, K., Deary, I.J. (2002) Homocysteine, B vitamin status, and cognitive function in the elderly. *Am. J. Clin. Nutr.*, 75, 908–913.

Easton, C.J., Bauer, L.O. (1997) Beneficial effects of thiamin on recognition memory and P300 in abstinent cocaine-dependent patients. *Psychiatry Res.*, 70, 165–174.

Ebadi, M. (1981) Regulation and function of pyridoxal phosphate in CNS. *Neurochem. Int.*, 3(3/4), 181–206.

Ebadi, M., Klangkalya, B., Deupree, J.D. (1980) Inhibition of GABA binding by pyridoxal and pyridoxal phosphate. *Int. J. Biochem.*, 11, 313–317.

Eysenck, H.J. (1991) Raising I.Q. through vitamin and mineral supplementation: an introduction. *Person. Individ. Diff.*, 12(4), 329–333.

Fox, P.T., Raichle, M.E., Mintun, M.A., Dence, C. (1988) Nonoxidative glucose consumption during focal physiological neural activity. *Science,* 241, 462–464.

Gale, C.R., Martyn, C.N., Cooper, C. (1996) Cognitive impairment and mortality in a cohort of elderly people. *Br. Med. J.,* 312, 608–611.

Goodwin, J.S., Goodwin, J.M., Garry, P.J. (1983) Association between nutritional status and cognitive functioning in a healthy elderly population. *JAMA,* 249(21), 2917–2921.

Guetzkow, H., Brozek, J. (1946) Intellectual functions with restricted intakes of B-complex vitamins. *Am. J. Psychol.,* 59, 358–381.

Guilarte, T.R. (1993) Vitamin B6 and cognitive development: recent research findings from human and animal studies. *Nutr. Rev.,* 51(7), 193–198.

Guilarte, T.R., Wagner, H.N., Frost, J.J. (1987) Effects of perinatal vitamin B6 deficiency on dopaminergic neurochemistry. *J. Neurochem.,* 48, 432–439.

Hassing, L., Wahlin, A., Winblad, B., Bäckman, L. (1999) Further evidence on the effects of vitamin B12 and folate levels on episodic memory functioning: a population-based study of healthy very old adults. *Biol. Psychiatry,* 45, 1472–1480.

Haller, J. (1995) The actions of vitamins and other nutrients on psychological parameters. In *Human Psychopharmacology,* Vol. 5 (Hindmarch, I. and Stonier, P.D., Eds.), pp. 229–261. John Wiley & Sons, Chichester, U.K.

Haller, J. (1996) The actions of vitamins on mental function in the elderly. *Age Nutr.,* 7(1), 16–25.

Haller, J. (1999) The vitamin status and its adequacy in the elderly: an international overview. *Int. J. Vitam. Nutr. Res.,* 69(3), 160–168.

Haller, J., Weggemans, R.M., Ferry, M., Guigoz, Y. (1996) Mental health: minimental state examination and geriatric depression score of elderly Europeans in the SENECA study of 1993. *Eur. J. Clin. Nutr.,* 50(Suppl. 2), 112–116.

Heinrich, C.P., Stadler, H., Weiser, H. (1973) The effect of thiamin deficiency on the acetylcoenzyme A and acetylcholine levels in the rat brain. *J. Neurochem.,* 21, 1273–1281.

Heseker, H., Kübler, W., Westenhöfer, J, Pudel, V. (1990) Psychische vränderungen als frühzeichen einre suboptimalen vitaminversorgung. *Ernährungsumschau,* 37(3), 87–94.

Kinsman, R.A., Hood, J. (1971) Some behavioral effects of ascorbic deficiency. *Am. J. Clin. Nutr.,* 24, 455–464.

Kretsch, M.J., Sauberlich, H., Newbrun, E. (1991) Electroencephalographic changes and periodontal status during short-term vitamin B-6 depletion in young, nonpregnant women. *Am. J. Clin. Nutr.,* 53, 1266–1274.

Lindeman, R.D., Romero, L.J., Koehler, K.M., Lian, H.C., LaRue, A., Baumgartner, R.N., Garry, P.J. (2000) Serum vitamin B12, C and folate concentrations in the New Mexico Elder Health Survey: correlations with cognitive and affective functions. *J. Am. Coll. Nutr.,* 19(1), 68–76.

Meador, K.J., Nickols, M.E., Franke, P., Durkin, M.W., Oberzan, R.L., Moore, E.E., Loring, D.W. (1993) Evidence for a central cholinergic effect of high-dose thiamin. *Ann. Neurol.,* 34, 724–726.

Morrison, H.I., Schaubel, D., Desmeules, M., Wigle, D.T. (1996) Serum folate and risk of fatal coronary heart disease. *JAMA,* 275, 1893–1896.

Nelson, M., Naismith, D.J., Burely, V., Gatenby, S., Geddes, N. (1990) Nutrient intakes, vitamin/mineral supplementation, and intelligence in British schoolchildren. *Br. J. Nutr.,* 64, 13–22.

Nolan, K.A., Black, R.S., Sheu, K.F.R., Langberg, J., Blass, J.P. (1991) A trial of thiamin in Alzheimer's disease. *Arch. Neurol.,* 48, 81–83.

Peterson, J.C., Spence, J.D. (1998) Vitamins and progression of atherosclerosis in hyper-homocyst(e)inaemia. *Lancet,* 351, 263.

Pilachowski Borek, J., Guilarte, T.R. (1990) Effects of vitamin B-6 nutrition on benzodiazepine receptor binding in the developing rat brain. *FASEB J.,* 4(3), A674.

Riggs, K.M., Spiro, A., Tucker, K., Rush, D. (1996) Relations of vitamin B-12, vitamin B-6, folate, and homocysteine to cognitive performance in the Normative Aging Study. *Am. J. Clin. Nutr.,* 63, 306–314.

Rimm, E.B., Willett, W.C., Hu, F.B., Sampson, L., Colditz, G.A., Manson, J.E., Hennekens, C., Stampfer, M.J. (1998) Folate and vitamin B6 from diet and supplements in relation to risk of coronary heart disease among women. *JAMA,* 279, 358–364.

Schoenthaler, S. (1991) Brains and vitamins. *Nature,* 337, 728–729.

Schoenthaler, S.J., Amos, S.P., Eysenck, H.J., Peritz, E., Yudkin, J. (1991) Controlled trial of vitamin-mineral supplementation: effects on intelligence and performance. *Person. Individ. Diff.,* 12(4), 351–362.

Selhub, J., Jacques, P.F., Bostom, A.G., D'Agostino, R.B., Wilson, P.W., Belanger, A.J., O'Leary, D.H., Wolf, P.A., Schaefer, E.J., Rosenberg, I.H. (1995) Association between plasma homocysteine concentrations and extracranial carotid-artery stenosis. *N. Engl. J. Med.,* 332(5), 286–291.

Seshadri, S., Hirode, K., Naik, P., Malhotra, S. (1982) Behavioural responses of young Indian children to iron-folic acid supplements. *Br. J. Nutr.,* 48, 233–240.

Smidt, L.J., Cremin, F.M., Grivetti, L.E., Clifford, A.J. (1991) Influence of thiamin supplementation on the health and general well-being of an elderly Irish population with marginal thiamin deficiency. *J. Gerontol.,* 46(1), M16–M22.

Southon, S., Wright, A.J.A., Finglas, P.M., Bailey, A.L., Belsten, J.L. (1992) Micronutrient intake and psychological performance of schoolchildren: consideration of the value of calculated nutrient intakes for the assessment of micronutrient status in children. *Proc. Nutr. Soc.,* 51, 315–324.

Southon, S., Wright, A.J.A., Finglas, P.M., Bailey, A., Loughridge, J.M., Walker, A.D. (1994) Dietary intake and micronutrient status of adolescents: effect of vitamin and trace element supplementation on indices of status and performance in tests of verbal and non-verbal intelligence. *Br. J. Nutr.,* 71, 897–918.

Sterner, R.T., Price, W.R. (1973) Restricted riboflavin: within-subject behavioral effects in humans. *Am. J. Clin. Nutr.,* 26, 150–160.

Thomson, A.D. (2000) Mechanisms of vitamin deficiency in chronic alcohol misusers and the development of the Wernicke-Korsakoff syndrome. *Alcohol Alcoholism,* 35(Suppl. 1), 2–7.

Verhoef, P., Stampfer, M.J., Buring, J.E., Gaziano, M., Allen, R.H., Stabler, S.P., Reynolds, R.D., Kok, F.J., Hennekens, C.H., Willet, W.C. (1996) Homocysteine metabolism and risk of myocardial infarction: relation with vitamins B6, B12, and folate. *Am. J. Epidemiol.,* 143, 845–859.

14 Iron and Brain Function

Domingo J. Piñero and James R. Connor

CONTENTS

14.1 INTRODUCTION

Iron is essential for normal neurological function because of its role in oxidative metabolism and because it is a cofactor in the synthesis of neurotransmitters and myelin. In the past several years, increased attention has been given to the importance of oxidative stress in the central nervous system (CNS). Because iron is the most potent inducer of reactive oxygen species, the relation of iron to neurodegenerative processes is more appreciated at present than it was a few years ago. Nevertheless, in spite of this increased attention and awareness, our knowledge of iron metabolism in the brain at the cellular and molecular levels is still limited.

Iron is distributed in a heterogeneous fashion among different regions and cells of the brain. This regional and cellular heterogeneity is preserved across many species. Brain iron concentrations are not static: they increase with age and in many diseases and decrease when iron is deficient in the diet. In infants and children, dietary iron deficiency is associated with decreased brain iron and with changes in behavior and cognitive functioning. Abnormal iron accumulation in diseased brain areas and alterations in iron-related proteins have been reported in many neurodegenerative diseases, including neurodegeneration with brain iron accumulation type 1 (NBIA-1), Alzheimer's disease (AD), Parkinson's disease (PD), and Friedreich's ataxia (FRDA). There is strong evidence for iron-mediated oxidative damage as a primary contributor to cell death in these disorders. Demyelinating diseases such as multiple sclerosis (MS) especially warrant study in relation to iron availability. Myelin synthesis and maintenance have a high iron requirement; thus, oligodendrocytes must have a relatively high and constant supply of iron. However, high oxygen utilization, high density of lipids, and high iron content of white matter combine to increase the risk of oxidative damage. We review here the current knowledge of the normal metabolism of iron in the brain, the alterations in brain iron status as a result of dietary manipulation, and the suspected contribution of iron in neuropathology.

14.2   IRON METABOLISM

Iron is essential for all living organisms, with the exception of certain members of the bacterial genera *Lactobacillus* and *Bacillus*. Iron is involved in a series of very important biochemical functions, including oxygen transport, electron transport, glucose metabolism, synthesis of neu-rotransmitters and myelin, and DNA replication in all organisms. The compounds that carry out all these reactions can be can be grouped as iron-containing proteins, heme proteins, iron-sulfur enzymes, and iron-containing enzymes. The deficiency of iron causes alterations in these different proteins and their functions, but iron excess is harmful because of its involvement in the formation of highly reactive hydroxyl radicals.

To prevent either excess or insufficiency in iron status, the homeostasis of iron is regulated at the levels of absorption, storage, and release in relation to its availability and need. Iron is present in the diet as heme iron and nonheme iron (ionic iron). Heme is absorbed intact by the enterocyte by a mechanism still unclear and its iron released by heme oxygenase to become part of the intracellular pool of iron. The first step in the uptake of nonheme iron by the enterocyte involves ferrireductase activity on the luminal side, seemingly by a membrane-bound protein named Dcytb, capable of reducing iron from the ferric to the ferrous state (McKie et al., 2001). The divalent metal transporter 1 (DMT1, previously known as Nramp2) on the apical membrane of the enterocytes transports the ferrous iron into the cell (Figure 14.1; Fleming et al., 1998; Gunshin et al., 1997). The iron transported by DMT1 and the iron released from heme can be stored in ferritin or transported across the cell by some unidentified iron chaperon proteins. Another protein, the metal transporter protein 1 (MTP1, also known as IREG1 and ferroportin 1), found on the basolateral membrane of the duodenal epithelial cell appears to be involved in the export of iron from the cell (Abboud and Haile, 2000). The multicopper oxidase protein hephaestin expressed in the basolateral

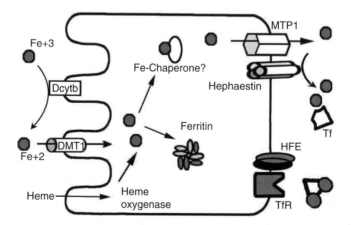

FIGURE 14.1 Intestinal absorption of nonheme iron. The proposed mechanisms involve the presence of ferrireductase activity by Dcytb acting in conjunction with the divalent metal transporter 1 (DMT1) for iron uptake on the apical membrane, an elusive intracellular iron chaperone for the intracellular transport of iron, and the metal transporter protein 1 (MTP1) and hephaestin on the basolateral membrane for the export of iron into the circulation. Transferrin receptor (TfR1) and the HFE protein, and probably some other factors, interact on the basolateral membrane to regulate the absorption of nonheme iron by mechanisms still unclear.

membrane is required for export of iron from the enterocyte, where it becomes bound to circulating transferrin (Tf, Vulpe et al., 1999).

The absorption of iron at the intestinal level increases in iron deficiency and decreases when iron stores augment. Regulation of the amount and rate of iron absorption by the enterocytes is thought to involve the interaction of HFE (the hemochromatosis protein) with β_2-microglubulin and transferrin receptor (TfR) on the basolateral membrane of the enterocyte (Feder et al., 1998). The current thinking is that the TfR serves as a sensor of the whole-body iron status, regulating the uptake of iron from circulating Tf and, as a result, the iron status of the enterocyte. A genetic mutation in the gene for HFE is associated with hereditary hemochromatosis (HH), a disease characterized by systemic iron overload (Feder et al., 1996). The alteration in the protein prevents the binding of HFE to β_2-microglubulin, altering HFE protein trafficking and cell surface expression (Feder et al., 1997). Disruption of the TfR–HFE–β_2-microglubulin interaction results in a state of low iron in the enterocyte and iron absorption is not downregulated by the iron stores of the individual despite the high saturation of circulating Tf.

Although iron can be transported in plasma bound to different compounds (heme-hemopexin, ferritin, lactoferrin, etc.), the most significant iron transport molecule is Tf. Tf is responsible for the delivery of iron from the liver to peripheral tissues and the protection of iron from glomerular filtration. Tf is a single-chain 80-kDa protein that contains two iron-binding motifs. Tf is on average 25 to 50% saturated with iron (Bothwell et al., 1979), so that under physiological circumstances there is an excess of iron-binding capacity in relation to the levels of iron in plasma. Tf transports not only the iron absorbed from the intestine (9–35 mmol/day) but also the iron recycled from the degradation of hemoglobin (350–450 mmol/day).

For most cells in the body, the uptake of iron involves the interaction between Tf and the transferrin receptor 1 (TfR1), a 180-kDa transmembrane glycoprotein composed of two identical subunits linked by two disulfide bridges (Jing and Trowbridge, 1987). Each TfR1 subunit binds one Tf with high affinity. The Tf–TfR1 complex is taken up by cells through endocytosis by a clathrin-mediated process. On acidification of the endosome, iron is released from Tf and exported into the cytoplasm by DMT1. Transferrin receptor and apotransferrin are recycled back to the cellular surface, where apo-Tf dissociates from the receptor (Dautry-Varsat et al., 1983). In the cytoplasm, iron can be incorporated into the storage pool (ferritin), the functional pool (component of enzymes), the regulatory pool [iron regulatory proteins (IRPs)], and the labile iron pool (LIP, defined as the metabolically and catalytically reactive iron of cells).

Another Tf receptor, transferrin receptor 2 (TfR2), has been recently identified and cloned (Kawabata et al., 1999). TfR2 is expressed at high levels in the liver and in epithelial cells of the duodenum (Deaglio et al., 2002) and its expression was shown to be independent of iron status in a murine model (Fleming et al., 2000). Patients with mutations in the TfR2 gene develop a condition of iron overload called hemochromatosis type 3 (HFE3; Roetto et al., 2002), and it has been proposed that TfR2 functions as an iron sensor and that its regulatory function is linked to TfR1.

Ferritin, the main iron storage protein, is composed of 24 polypeptide subunits. These subunits exist in two isoforms, H and L, which allows considerable heterogeneity in the structure of the full protein. The H-chain possesses ferroxidase activity and is capable of self-loading iron, and the L-chain lacks the ferroxidase site. The H-chain releases iron more rapidly and predominates in the heart and brain, where iron turnover is higher, and the L-form is more abundant in the liver, an organ typically associated with the storage of iron (Levi et al., 1988, 1989, 1992). Ferritin can store up to 4500 Fe^{3+} ions (Fishbach and Andreregg, 1965) in the form of ferrihydrite, and iron can be rapidly released from the iron core by reduction. In addition to its intracellular role, ferritin is also secreted from cells and may be an iron delivery protein. Receptors for ferritin have been identified in a number of systemic cells and cell lines (Fargion et al., 1988, 1991; Moss et al. 1992a, 1992b) and more recently in both the mouse and human brain (Hulet et al., 1999a, 1999b). Ferritin has also been observed in cell nuclei, prompting yet another new line of investigation on the possibility of ferritin as a DNA protectant (Thompson et al., 2002; Cai and Linsenmayer, 2001).

Iron regulates its own intracellular concentration by mechanisms that involve cytoplasmic IRPs that interact with specific mRNA stem-loop structures called iron responsive elements (IREs) in response to cellular iron status (Leibold and Munro, 1988; Müllner et al., 1989). mRNAs for ferritin, TfR, erythroid-D-aminolevulinate synthase, mitochondrial aconitase, ferroportin 1, and an isoform of DMT1 contain IREs (Aziz and Munro, 1987; Müllner and Kühn, 1988; Zheng et al., 1992; Melefors et al., 1993). Ferritin mRNA contains an IRE in the 5′ untranslated region, and the TfR mRNA has five IREs in the 3′ untranslated region. Two distinct IRPs, IRP1 and IRP2, have been identified. In situations of normal iron supply, IRP1 contains an iron-sulfur cluster (4Fe-4S) and has cytosolic aconitase activity, whereas in iron-depleted conditions the apo form of IRP1 has high affinity for the IRE (Haile et al., 1992). In contrast, IRP2 does not have aconitase activity, does not contain an iron-sulfur cluster, and it is rapidly degraded when iron supply is adequate (Kühn, 1998). In iron-depleted situations, IRPs bind with high affinity to the 5′-UTR of ferritin mRNA, preventing the initiation of translation, and to the 3′-UTR of the TfR mRNA, stabilizing the mRNA for translation of the receptor. In the presence of iron, IRP1 is inactivated and released from the ferritin mRNA, allowing the attachment of ribosomes and the translation of ferritin. On the release of IRP from the TfR1 mRNA, rapid degradation of the mRNA occurs (Rouault and Klausner, 1997). Although other factors such as H_2O_2 and nitric oxide are able to activate IRP1 (Hentze and Kühn, 1996), the coordinated control of iron uptake and storage by the IRE–IRP interaction is a very elegant system to keep intracellular levels of iron at a stable level and simultaneously prevent iron excess and deprivation in a process regulated by iron itself. For a more extensive appraisal of the intracellular iron regulatory system, readers are referred to some reviews (Beard et al., 1996; Kühn, 1998; Wood and Han, 1998; Wessling-Resnick, 2000; Eisenstein, 2000).

14.3 IRON IN NEUROBIOLOGY

Our understanding of the role of iron in normal brain functioning has greatly improved during the past decade, and the publication of reviews and numerous papers focusing on nutritional status of iron and brain iron, iron status and cognition, iron status and neurotransmitters metabolism, and brain iron and neuropathologies is a clear sign of the interest this topic attracts. Even so, many basic questions remain unanswered regarding the normal metabolism of iron and the consequences of alterations in the homeostasis of iron in the CNS, as well as the relationship between dietary iron and brain iron.

The brain requirement for iron is relatively high, consistent with its high-energy needs. There is a high demand for ATP to maintain membrane ionic gradients, synaptic transmission, and axonal transport, all highly active processes in the brain. Iron is an essential component of cytochromes a, b, and c, cytochrome oxidase, and the iron-sulfur complexes of the oxidative chain and hence has an important role in ATP production. Iron is also a cofactor of the enzymes tyrosine hydroxylase (Tyr to L-DOPA) and tryptophan hydroxylase (Trp to 5-hydroxytryptophan), which are involved in the synthesis of neurotransmitters (Glinka et al., 1997; Waldemeier et al., 1993). Ribonucleoside reductase, the rate-limiting enzyme of the first metabolic reaction committed to DNA synthesis, as well as succinate dehydrogenase and aconitase of the TCA cycle, are also Fe-dependent enzymes (Crowe and Morgan, 1992). Finally, iron is essential for the biosynthesis of lipids and cholesterol, which are important substrates in the synthesis and metabolism of myelin (Larkin and Rao, 1990).

The mechanisms by which iron enters the brain are not completely understood. Important features of the acquisition of iron by the brain are the regulatory control of both the blood-brain barrier (BBB) and the blood-cerebrospinal fluid barrier (Figure 14.2). The BBB prevents the brain from having direct access to iron in the plasma pool. Iron is transferred to the brain by plasma Tf through an interaction between circulatory Tf and TfR1 in the brain microvasculature (Morris et al., 1992b). Capillary endothelial cells have a high density of TfR1, 6 to 10 times higher than the amount on the brain parenchyma (Kalaria et al., 1992; Jefferies et al., 1984; Connor, 1994), and there is an increase in the number of TfR1s in response to iron depletion both on the cell surface

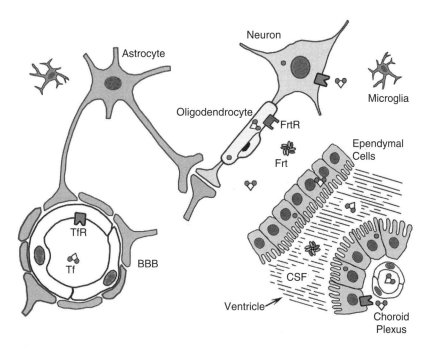

FIGURE 14.2 Brain iron uptake and mobilization. The main areas of iron entry to the brain are the blood-brain barrier (BBB) and the blood-cerebrospinal fluid barrier (choroid plexuses). At the BBB level, iron is transferred to the brain by plasma Tf through an interaction between circulatory Tf and transferrin receptors (TfR1) in the brain microvasculature. The release of iron from the endothelial cell is presumed to occur by a process similar to that in the basolateral side of the enterocyte, but direct evidence is lacking. Some studies have shown transcytosis of transferrin, suggesting that Tf may transport iron through the BBB. Once across the BBB, iron is transported by Tf (synthesized by the choroid plexus or by oligodendrocytes) and taken up by different cell types. The choroid plexus, which secretes the cerebrospinal fluid (CSF), forms another barrier to circulating plasma iron and the brain. Endothelial cells of the choroid plexus express TfR1. Secretion of Tf by the choroid plexus into the ventricles seems to be an important mechanism for delivery of iron to the rest of the brain, but iron may also be transported from the CSF in a non-transferrin-bound state. The CSF also contains ferritin. Neurons express transferrin receptors, whereas mature oligodendrocytes have ferritin receptors for the uptake of iron.

and participating in the recycling of Tf (van Gelder et al., 1998). Once Tf is endocytosed into the endothelial cell, little is known about how iron is transferred to the brain. The presence of DMT1 in the endothelial cells suggests that iron is released there, and supporting this is the report that Belgrade rats, which have a defect in DMT1, have a decrease in brain iron uptake (Burdo et al., 2001). The release of iron from the endothelial cell does not appear to occur by a process similar to that in the basolateral side of the enterocyte because MTP1 has not been detected in brain endothelial cells. Some studies have shown transcytosis of Tf across the BBB (Descamps et al., 1996; Fishman et al., 1987), suggesting Tf may transport iron trough the BBB; however, a mechanism for release of Tf from the Tf receptor at the abluminal membrane has not been identified. The current understanding is that once across the BBB, iron is transported by Tf made in the brain and taken up by different cell types, but this basic concept of iron entry into the brain is likely to be flawed. For example, hypotransferrinemic mice, which do not make Tf, accumulate iron in the brain and in the appropriate cell type when the animal is given systemic injections of Tf (Dickinson and Connor, 1998). Also, recent reports indicate that the Tf made in the brain by oligodendrocytes is not secreted (de Arriba Zerpa et al., 2000).

Transport of iron on Tf-like molecules may also take place at the BBB. Receptor-mediated transcytosis of lactoferrin, an iron-binding glycoprotein normally associated with milk, has been demonstrated in primary cultures of bovine vascular endothelial cells (Fillebeen et al., 1999). Although there are contradictory reports about its presence in the normal human brain, lactoferrin has been detected in neurodegenerative diseases such as PD (Faucheux et al., 1995; Leveugle et al., 1994, 1996). Many of these diseases, as discussed later, are associated with an alteration in iron and iron management proteins.

Iron transport to the brain also occurs through the ventricular system. The choroid plexus, which secretes the cerebrospinal fluid (CSF), forms another barrier between circulating plasma iron and the brain. Endothelial cells of the choroid plexus express TfRs, although at a lower density than the capillary endothelia (Moos, 1996), and when rats received peripheral injections of ^{59}Fe, the majority of radioactivity after 1 h was detected in the choroid plexus (Moos and Morgan, 1998). Secretion of Tf by the choroid plexus into the ventricles may be an important mechanism for delivery of iron to the rest of the brain, but a direct contribution of this source of iron to brain iron requirements has not been demonstrated. Other compounds in the CSF, for instance, citrate and ascorbate, could be responsible for iron transport if Tf is fully saturated in the CSF as some studies suggest (Bradbury, 1997).

The levels of CSF ferritin are roughly 10% of those found in serum (Campell et al., 1986), ca. 3 ng/ml in normal CSF, with elevated ferritin concentration occurring during infectious meningoencephalitis, CNS vascular diseases, dementia without vascular pathology, and chronic progressive active MS patients (Sindic et al., 1981; LeVine et al., 1999). In contrast to serum ferritin, which is highly glycosylated (>70%), ferritin within the CSF is largely nonglycosylated (<20%). Nonglycosylated ferritin is usually considered as tissue ferritin, leading some to believe that CSF ferritin is derived from cell death and subsequent ferritin release (Zappone et al., 1986). However, the concentration of ferritin within the CSF of normal patients is significantly higher than what would be expected to occur by passive diffusion across the blood-CSF barrier, suggesting local synthesis and release of ferritin by cells of the choroid plexus (Keir et al., 1993).

Some brain regions, such as the posterior pituitary and median eminence, are not protected by a blood barrier. These are known as the circumventricular organs and generally contain fenestrated capillaries lacking TfR1, presenting the possibility of paracellular Tf transport or non-Tf-bound iron transport across these fenestrated vessels. However, free diffusion of iron into the CSF from the circumventricular organs is prevented by the tight junctions that tanycytes, modified glial cells that line the ventricles, form between the circumventricular organs and the ventricles. Tanycytes, which stain robustly for iron and ferritin (Benkovic and Connor, 1993; Burdo et al., 1999) but do not contain TfR1 (Dickinson and Connor, 1998; Moos, 1996), project to discrete nuclei in the thalamus and could be a potential mechanism for the transport of iron into the brain.

The mechanisms of iron efflux from the brain are poorly understood, but recent studies have shown that on par with the calculated iron influx there is also high iron efflux (Moos and Morgan, 1998). Although the natural tendency is for the brain to accumulate iron with age, the importance of these efflux mechanisms deserves further investigation.

14.3.1 BRAIN IRON AND IRON PROTEINS DURING DEVELOPMENT

In humans, brain levels of nonheme iron are low at birth, increase rapidly for the next two decades, and then increase at a slower rate for the rest of the life (Hallgren and Sourander, 1958; Sourkes, 1982; Bartzokis et al., 1997). In terms of macrodivision of the brain, white matter has the higher concentration of iron (Connor et al., 1992a; Larkin and Rao, 1990). Regionally, iron concentration is highest in the globus pallidus, red nucleus, substantia nigra (pars reticulata), caudate putamen, and dentate nucleus (Aoki et al., 1989; Connor et al., 1995a; Roskams and Connor, 1994). The concentration of iron in basal ganglia is similar to that of liver tissue (Hallgren and Sourander, 1958). The amount and the distribution of iron in the brain change with age. Using rat as a model, it has been shown that the total amount is lower at birth and increases with age and that some regions that are rich in iron in adult rat brains are not rich in iron for the first 60 days of life

(Benkovic and Connor, 1993). This observation is also true for humans (Aoki et al., 1989). However, when expressed per gram of protein, iron levels are higher at birth, decrease during the first 20 to 30 days of life, and then increase with maturation at around 60 days of life in the rat and 12 to 15 years in humans (Connor et al., 1995a). The uptake of iron into the rat brain increases rapidly after birth, peaks at postnatal day 15, and decreases thereafter (Erikson et al., 1997). These changes are associated with rapid brain growth up to postnatal day 15 and a decrease in the concentration of nonheme iron. The high concentration and requirement for iron during brain development is consistent with the devastating and long-term neurological and cognitive deficits associated with iron deficiency during development.

The concentration of Tf is also higher at birth than any other time, falls until weaning, and then increases to a plateau that appears to be different for each region (Roskams and Connor, 1994; Erikson et al., 1997; Chen et al., 1995). Brain Tf is assumed to be of systemic origin before the closure of the BBB, but in the rat brain there is an increase in Tf synthesis from the choroid plexus and oligodendrocytes after postnatal day 5, coinciding with the closure of the BBB (Levin et al., 1984; Barlett et al., 1991). Tf appears first in the white matter, associated with oligodendrocytes, and in the corpus callosum and the optic nerve; it then appears in the gray matter (striatum and also spinal cord; Connor, 1994). The distributions of Tf and Fe are closely correlated in the gray matter, whereas in the white matter Tf-positive cells are evenly distributed, whereas Fe-positive cells are clustered and possibly associated with blood vessels (Connor, 1994).

In rats, brain ferritin levels are high at birth, decrease with age (until day 17), and then increase again (Erikson et al., 1997). During development, iron and ferritin are initially found in microglia, and then as myelination is initiated, ferritin, as well as iron and Tf mRNA, are found within oligodendrocytes (Connor et al., 1995a; Connor and Menzies, 1996; Cheepsunthorn et al., 1998). The expressions of ferritin and Tf mRNAs are early events in differentiating oligodendrocytes (Barlett et al., 1991; Bloch et al., 1985; Sanyal et al., 1996). The brain is richer in the H- than the L-chain of ferritin, and the isoform distribution is cell specific: H-ferritin is predominantly found in neurons, whereas L-ferritin is found in oligodendrocytes. Alterations in the ratio of H:L subunits in the brain could imply different functional roles of the ferritin subunits in the brain cells. Not much is known of the H:L ratio alterations in the brain associated with changes in iron supply, but it has been reported that oligodendrocytes in 1-month-old-piglets express H-chain ferritin but not L-chain ferritin (Blissman et al., 1996) whereas there is more L-ferritin in the adult pig brain (Erikson et al., 1998). This difference may imply a developmental stage in which the growing brain tissue expresses ferritin consisting primarily of the H-chain ferritin (Blissman et al., 1996). Sanyal et al. (1996) found upregulation of the expression of the H-chain ferritin gene associated with the differentiation of oligodendrocytes, showing that H-chain ferritin transcription is modulated by factors that control cell growth and differentiation. A recently developed genetically altered mouse model in which the levels of H-ferritin have been reduced by 80% while maintaining normal brain iron status could prove invaluable in the study of the contribution of iron imbalance to neurode-generative diseases (Thompson et al., 2003).

The brains of humans and monkeys appear more conditioned for iron storage than do rodent brains. In rodents, the ratio of Tf to ferritin is 4:1, indicating a high level of mobility relative to storage of iron. However, in humans and monkeys, there is 10 times more ferritin than Tf, indicating that considerably more iron can be stored in these brains. Nonetheless, the cellular and regional distribution of ferritin and iron among the various regions of the brain is very similar between the two species.

The TfR is a key player in the acquisition of iron by the brain. The concentration of TfR1 is highest in the cortex, amygdala, hippocampus, and brainstem in the rat brain (Hill et al., 1985). Immunohistochemical studies on the adult brain show that neurons are the predominant cells that stain for TfR1 (Moos, 1996), although immunopositive astrocytes and oligodendrocytes are also found (Giometto et al., 1990; Connor and Menzies, 1995). It is interesting to note that the nucleus accumbens and the caudate putamen (low Fe areas) have high TfR1 densities. These regions connect with the globus pallidus and the substantia nigra (high Fe areas); thus, there is the possibility of an uptake of iron in one area and delivery to storage or function in another through axonal transport (Mash et al., 1990). A recent report that ferritin receptors are predominantly expressed in white matter (Hulet et al., 1999a, 1999b) indicates two nonoverlapping receptor systems for iron in the brain.

The proteins involved in iron transport in the gut can be expected to play a role in brain iron transport. For example, DMT1 has been reported in the brain tissue, and animals with a defect in this endosomal iron transport protein have decreased concentrations of iron in their brain (Burdo et al., 1999). The iron transporter MTP1, responsible for iron efflux from cells, is also present in the brain (Abboud and Haile, 2000; Burdo et al., 2001), and its expression in different brain regions is influenced by age (Jiang et al., 2002). Clearly, the characterization of iron transport proteins in the brain is critical to understand the mechanisms by which iron accumulates in specific brain regions in neurodegenerative diseases. Table 14.1 lists the location of iron and some proteins involved in its metabolism by brain region and cell type.

14.3.2 DISRUPTION OF THE NORMAL PATTERN OF IRON ACQUISITION AND EXPRESSION OF IRON MANAGEMENT PROTEINS IN THE BRAIN

The establishment of a normal brain iron profile may be affected in a number of ways. A genetic defect in oligodendrocyte development, exposure to alcohol in the uterus (Connor and Menzies, 1996), perinatal hypoxic or ischemic insult (Cheepsunthorn et al., 2002), and dietary iron manipulation can have long-term effects on brain iron status. For this discussion, we focus on the influence of dietary manipulation.

Brain iron is decreased by dietary iron deficiency in the rat. Although both iron uptake and Tf uptake are reportedly increased in iron deficient (ID) rats compared with normal control animals, the total amount of iron entering the brain may not be increased because of the diminished pool of systemic iron (Crowe and Morgan, 1992; Morris et al., 1992). ID in rat models, both pre- and postweaning, results in increased Tf levels in the rat brain (Crowe and Morgan, 1992; Chen et al., 1995, Levin et al.., 1984; Piñero et al., 2000b). It has been proposed that ID may alter the BBB (Ben-Shachar et al., 1988), but there is no clear evidence for such an alteration, and it is reasonable to assume that the upregulation of TfR1 is responsible for this increase in iron uptake in ID.

TABLE 14.1
Distribution of Iron and Some Proteins Involved in Iron Metabolism by Brain Region and Cell Type

	Region	Cell Type
Iron	Ubiquitous, high in motor areas	Oligodendrocytes, tanycytes, ependymal cells, neurons
Transferrin	Ubiquitous	Oligodendrocytes, vasculature, choroid plexus, ependymal cells
Transferrin receptor 1	Cortex, hippocampus, striatum	Neurons, vasculature, astrocytes, oligodendrocytes
H-ferritin	White matter, cortex	Oligodendrocytes, neurons
L-ferritin	White matter	Oligodendrocytes, microglia
Ferritin receptor	White matter	Oligodendrocytes
Divalent metal transporter 1	Striatum, thalamus, choroid plexus, cerebellum	Astrocytes, neurons, ependymal cells, vasculature, Purkinje cells
Metal transporter protein 1	Cortex, cerebellum	Neurons, oligodendrocytes

Early studies of the effects of iron deficiency and repletion on the brain of young rats led to the belief that if iron deficiency happened during the critical period of brain growth spurt, it was impossible to normalize brain iron afterward (Dallman et al., 1975; Dallman and Spirito, 1977; Weinberg et al., 1980, 1981). Recent studies suggest that this notion should be reconsidered. Piñero et al. (2000) showed that late lactational iron deficiency or iron supplementation in the rat, between postnatal days 10 and 21, produced alterations in iron as well as Tf, TfR, and ferritin in different brain regions. Furthermore, those changes were different from the ones found in models of postweaning iron deficiency, implying a role for the level of neuromaturation on the effects of iron deficiency. In their study, the overall brain iron concentration was corrected by 2 weeks of iron repletion, although the larger amounts of iron provided by an iron-supplemented diet were necessary to fully normalize iron and iron-related proteins in some brain regions. Using a rat model of postweaning iron deficiency, Erikson et al. (1997) also found heterogeneous changes in the concentrations of iron and iron regulatory proteins in the brain as a consequence of iron deficiency, and brain iron was normalized on repletion with a diet adequate in iron content.

Both human studies and animal models have been used to study the effects of iron deficiency on behavior. Studies in adults and children have demonstrated that iron deficiency is associated with apathy, irritability, lethargy, lack of concentration, hypoactivity, and decreased cognitive and attentional processes. Similar results have been found from early studies of Oski and Honig (1978) and Oski et al. (1983) to subsequent ones by different groups [reviewed in Granthan-McGregor and Ani (2001) and Yager and Hartfield (2002)], despite some major differences in their design. Although the biological bases of these cognitive and behavioral alterations are unclear, an elegant conceptual model of mechanisms for poorer development in iron-deficient anemic infants put forward by Lozoff (1998) includes environmental as well as biological factors, mechanisms that are not mutually exclusive. The first group of variables includes disadvantaged environment, poor feeding practices, and limited support for child development. The biological variables are all related to the decrease in brain iron associated with iron deficiency, and include hypomyelination, impaired dopaminergic function, and delayed neuromaturation. Although direct studies measuring myelination changes in iron-deficient humans are lacking, studies in animals have shown a negative impact of iron deficiency on myelination (Larkin and Rao, 1990; Yu et al., 1986). Studies in humans have used proxy indicators of myelination to evaluate the effects of iron deficiency. Researchers found on measuring auditory brainstem-evoked responses in 6-month-old iron-deficient infants during spontaneous naps that absolute and interpeak latency values were longer among iron-deficient infants at several stimulus conditions, which is correlated with delayed neuromaturation (Rocangliolo et al., 1998). Central conduction times were even higher at follow-up 12 and 18 months after iron therapy, indicating a probable irreparable damage to the normal neuromaturation process. Another study by the same group reports longer evoked responses in 4-year-old children treated for iron-deficiency anemia when compared with children who were nonanemic as infants (Algarin et al., 2003). These results support and strengthen the hypothesis that myelination is altered by iron-deficiency anemia in infancy and provide evidence that the effects can be long lasting and a possible core mechanism for all the other developmental delays associated with iron-deficiency anemia.

Higher urinary levels of catecholamines have been reported in iron-deficient subjects, and these levels normalize with the recovery from iron deficiency (Beard, 1987; Oski et al., 1983). In the rat, normal levels of catecholamines and neurotransmitters but decreases in the dopamine (DA) receptors D1 and D2 as well as the dopamine transporter have been reported in a postweaning model of iron deficiency (Erikson et al., 2000, 2001). Using microdialysis *in vivo*, an increase in extracellular DA and a blunted DA reuptake mechanism were found in ID animals when compared with controls (Beard et al., 1994; Nelson et al., 1997). This alteration disappeared when the animals were iron repleted, evidence that the process is reversible in postweaning rats. The functional importance of these differences has been more difficult to prove. Even though the neural mechanisms governing exploratory behavior in animals are very complex, it is accepted that dopaminergic neurons mediate many forms of motivated behavior and motor function (Robbins and Everitt, 1982; Izquierdo, 1989). It is then to be expected that these mentioned ID-related changes in DA metabolism would result in some type of alteration in exploratory behavior in animals; indeed, experiments have shown that ID rats have altered motor activity when compared with control animals. When confronted with a novel open-field environment, ID rats were less explorative in a hole board apparatus (Weinberg et al., 1980), and after 6 weeks of iron repletion there was still difference in their open-field exploratory behavior compared with control rats. Stereotypy, an excellent measure of dopamine-related behavior, was reduced in ID rats treated with dopamine agonists (Youdim et al., 1980). A reversal in the physical activity pattern of ID rats, normalized within 2 weeks with iron repletion, has also been reported (Youdim et al., 1981). Hunt et al. (1994) reported decreased motor activity and exploratory behavior in ID rats, but could not replicate the described phenomenon of reversed motor activity. Felt and Lozoff (1996) also found altered behavior in rats 3 months after the iron deficiency had been reversed. In the study by Piñero et al. (2000b), the rats were also tested for changes in physical activity. There was a decrease in physical activity for both the ID and, surprisingly, iron-supplemented rats, and these behavioral alterations persisted even though iron repletion normalized the levels of brain iron (Piñero et al., 2001). Beard et al. (2002) showed that iron concentration in dopamine-containing brain regions and densities of dopamine receptors and the transporter are significant predictors of measures of activity and reactivity in ID rats, and are in favor of the iron–dopamine link to explain the neurobehavioral abnormalities found in ID children. These studies taken together strongly support the idea that the behavioral alterations found in iron deficiency are irreversible when the deficiency occurs during critical periods of neural development. Although there may be problems when using the rat as a model for ID-related behavioral alterations, the use of a developmentally sensitive model that takes into account species differences would take care of most of those problems. An important contribution from these studies on the relationship between iron and dopamine has been the implication for related research; for example, the findings of reduced CSF ferritin and elevated CSF Tf and the known changes in dopamine metabolism in patients with restless legs syndrome led to the hypothesis that an alteration in brain iron could be involved in the etiology of the disease (Earley et al., 2000a).

The issue of the reversibility of cognitive and behavioral deficits in ID children is still a troublesome one. Although the now classic study by Idjradinata and Pollitt (1993) showed reversibility of the delay in cognitive development in Indonesian infants after iron repletion, a follow-up study of Costa Rican children who were treated for iron deficiency as infants by Lozoff et al. (2000) showed that more than 10 years after treatment these children are still at behavioral and developmental risk. The latter study makes a strong argument for the irreversibility of the deleterious effects of iron deficiency on the developing CNS.

14.4 IRON AND NEUROPATHOLOGY

Many neuropathologies are associated with increased iron deposition and iron-related oxidative damage in the brain. We present a synopsis of the relationship of iron to some of these neuropathologies (Table 14.2).

14.4.1 NEURODEGENERATION WITH BRAIN IRON ACCUMULATION TYPE 1 (NBIA-1)

NBIA-1, previously known as the Hallervorden–Spatz syndrome and pantothenate kinase-associated neurodegeneration (PKAN), is a rare autosomal recessive neurodegenerative disease characterized by extrapyramidal dysfunction confirmed by rigidity, dystonia, and choreathetosis. NBIA-1 is most commonly of childhood onset, with progressive signs and symptoms. The association between excessive accumulation of brain iron and NBIA-1 was made with the discovery and description of the disease more than 70 years ago. Massive iron accumulation is found in the globus

TABLE 14.2
Alterations in Iron and Iron Metabolism Proteins in Some Neurological Diseases

Neurological Disease	Alteration
Alzheimer's disease	Increase in iron in amygdala, hippocampus, inferior parietal lobe, frontal lobe, globus pallidus, motor cortex, nucleus basalis, and amygdala; decrease in iron in occipital cortex and substantia nigra; increase in iron in amyloid plaques; decrease in ferritin in areas that accumulate iron; decrease in transferrin in the hippocampus; altered distribution and activity of iron regulatory protein (IRP); increase in p97 in plaques and plasma
Parkinson's disease	Increase in iron in striatum, substantia nigra, pars compacta, caudate, and putamen; decrease in ferritin in substantia nigra, caudate, and putamen; altered transferrin, transferrin receptor 1, and lactoferrin receptor
Neurodegeneration with brain iron accumulation type 1 (NBIA-1)	Increase in iron in globus pallidus and substantia nigra
Friedreich's ataxia	Increase in iron in striatum and cerebellum
Restless legs syndrome	Increase in iron in striatum, substantia nigra, and putamen; decrease in ferritin in plasma and cerebrospinal fluid; increase in transferrin in cerebrospinal fluid
Huntington's disease	Decrease in iron in striatum (presymptomatic)
Multiple sclerosis	Loss of ferritin receptor in periplaque white matter

pallidus and the pars reticulata of the substantia nigra in patients with NBIA-1 (Swaiman, 1991). The genetic defect for NBIA-1 has been mapped to the gene encoding pantothenate kinase 2 (PANK2) in chromosome 20 (Taylor et al., 1996; Zhou et al., 2001; Hayflick et al., 2003). It has been proposed that a defect in PANK2, essential for the synthesis of coenzyme A from pantothenate, produces alterations in energy metabolism, fatty acid synthesis, and defective membranes (Zhou et al., 2001). Phosphopantothenate, the product of PANK2, condensates with cysteine, and a decrease in phosphopantothenate could also produce the reported high cysteine concentration in the basal ganglia and, as a result, iron accumulation and oxidative stress in these areas (Perry et al., 1985).

There is accumulation of synucleins (α-, β-, and γ-) in Lewy bodies in NBIA-1 patients (Arawaka et al., 1998; Galvin et al., 2000), which suggests a common pathological mechanism with AD and PD. This common mechanism could also involve iron, because iron has been shown to promote synuclein aggregation (Golts et al., 2002).

14.4.2 ALZHEIMER'S DISEASE (AD)

Iron homeostasis is altered in the AD brain. Increased iron content has been reported in the amygdala, hippocampus, frontal lobe, inferior parietal lobe, and temporal cortex (Cornett et al., 1998; Bartzokis et al., 2000) as well as in the nucleus basalis, globus pallidus, and motor cortex (Thompson et al., 1998). In regions such as the occipital cortex and substantia nigra, iron levels are reportedly decreased in comparison with age-matched controls (Connor et al., 1992b). At the microscopic level, increased iron has also been noted in the neurofibrillary tangle containing neurons (Good et al., 1992b) and in senile plaques and their surrounding cells (Connor et al., 1992a).

In AD, there are alterations in the levels and distribution of iron-related proteins, which could affect the storage as well as the mobilization and delivery of iron in the brain. In regions with a relatively modest age-related accumulation of iron (e.g., occipital cortex), the levels of ferritin and iron are proportionally adequate. However, in regions with a higher age-related accumulation of iron (e.g., the superior temporal gyrus and frontal cortex) the levels of ferritin are inadequate for the amount of iron (Connor et al., 1992b). This is a powerfully suggestive finding, given that the histopathology of AD is more severe in these latter regions whereas regions such as the occipital cortex remain relatively unaffected. The increase in iron without a concomitant increase in ferritin could put the cells at an increased risk for oxidative stress, accepted to be a major contributor to neurodegeneration in AD. Additional evidence for abnormal iron accumulation in AD is the report of decreased number of TfRs in AD hippocampus (Kalaria et al., 1992). This reported decrease in Tf receptors is more profound when considering the observation that cells surrounding neuritic plaques stain robustly for TfR (Connor, 1994; Connor and Menzies, 1995). This altered distribution of TfR could result in the targeting of iron away from neurons, which would further limit their metabolic abilities.

The synthesis of ferritin and TfR is inversely regulated by the IRP/IRE system. A more stable IRP/IRE complex, possibly resulting from increased RNase activity, has been found in AD brains (Piñero et al., 2000a). Such a complex could increase the stability of the TfR mRNA and inhibit ferritin synthesis, which would explain the increase in iron accumulation without an increase in ferritin found in AD brains. The distribution of IRP is also reportedly altered in AD brains (Smith et al., 1998).

An iron imbalance may also affect amyloid in AD. Iron reportedly promotes the deposition of the β-amyloid peptide (Mantyh et al., 1993). Iron may influence the production of pathogenic amyloid from the amyloid precursor protein (APP) by affecting α-secretase activities (Bodovitz et al., 1995). Iron may also regulate the processing of APP mRNA through the presence of an iron-responsive element on the 5'-UTR of the APP transcript (Rogers et al., 2002). Thus, increased intracellular iron levels or a dysfunctional IRP could potentially affect APP translation.

Another iron-related protein, melanotransferrin, also known as p97, a homologue of Tf that is found as a secreted form and a membrane-bound form, is expressed in capillary endothelium and in microglia associated with amyloid plaques in AD (Jefferies et al., 1996). Although the function and mechanisms of action of p97 are not totally clear, it appears that the soluble form of p97 donates iron to the cell through nonspecific internalization of the protein (Food et al., 2002). p97 is elevated in the serum of AD patients and has been proposed as an biomarker to diagnose the disease (Jefferies et al., 2001; Feldman et al., 2001).

Excess accumulation of iron in AD brains suggests that diseases such as hemochromatosis could be a comorbidity factor in AD. Individual who are heterozygotes for the HFE mutation have been shown to have either increased risk for AD (Moalem et al., 2000) or earlier onset of AD (Sampietro et al., 2001). Although the brain has traditionally been considered protected from iron overload in hemochromatosis, this view has been challenged by the presence of the HFE protein in the brain microvasculature (Connor et al., 2001). Individuals who are homozygous may not be at increased risk for AD because they will either die before they get old enough for AD or if they are treated for hemochromatosis, the repeated phlebotomies would likely have the effect of minimizing brain iron accumulation as it does with all the other organs. It is important to identify the possible connection between carrying the HFE mutation and AD, because intervention, including limiting dietary intake of iron, could prove effective.

In summary, the extent of iron dysregulation in the AD brain increases the potential for oxidative stress. Data such as those showing that desferrioxamine (an iron chelator) and vitamin E are effective in decreasing the progression of the disease support the theory of an iron-induced oxidative stress component of AD (Subramaniam et al., 1998). However, before embarking on an aggressive iron chelation therapy in AD or any disease involving excess iron, one must also consider the effects of limiting iron availability on cellular metabolism and function, including loss of energy, decreased neurotransmitter synthesis, and loss of synapses.

14.4.3 PARKINSON'S DISEASE (PD)

PD is a neurodegenerative disease characterized by rigidity, tremor, bradykinesia, and the death of dopaminergic neurons. Although the precise mechanisms that result in nigral neuronal death in PD are still unknown, oxidative stress and iron mismanagement appear to be important factors. Abnormal levels of iron have been reported in the substantia nigra and striatum (Dexter et al., 1989, 1991; Reiderer et al., 1989) as well as in association with Lewy bodies and neuromelanin-containing cells. Most studies have found increased iron in the substantia nigra in individuals with severe PD, more specifically in the substantia nigra pars compacta (Sofic et al., 1991; Bartzokis et al., 1999a), the primary site of neurodegeneration.

Iron accumulation occurs primarily in the neuromelanin granules of pigmented substantia nigra pars compacta neurons (Good et al., 1992a). Data on iron in other regions involved in PD, such as the caudate nucleus, putamen, and globus pallidus, are inconsistent. Several studies suggest the absence of iron accumulation in the PD caudate and putamen relative to age-matched controls (Dexter et al., 1989; 1991; Reiderer et al., 1989). In the globus pallidus, both decreased (Dexter et al., 1989, 1991; Reiderer et al., 1989) and increased (Loeffler et al., 1995) iron levels have been reported in PD. The reasons for the discrepancy are unclear and likely reflect problems associated with the study of tissue collected at end-stage disease as well as medical treatment received before death. A recently developed genetically altered mouse model, in which the levels of H-ferritin have been reduced by 80% while maintaining normal brain iron status, could prove invaluable in the study of the contribution of iron imbalance to neurodegenerative diseases (Thompson et al., 2003). Alterations in the distribution and levels of iron-associated proteins, such as ferritin, Tf, TfR1, and lactoferrin receptor, have also been reported in the PD brain (Faucheux et al., 1995; Kalaria et al., 1992). As in AD, the normal compensatory increases in ferritin expression to iron accumulation appear to be altered in PD. Ferritin expression is decreased in comparison to elderly controls on a per weight basis in the substantia nigra of the PD brain (Dexter et al., 1990). Specifically, H- and L-ferritin levels are reportedly 75% and 37% of the normal, respectively (Connor et al., 1995b). The normal age-related increase in ferritin was not found in the caudate and putamen, suggesting a dysregulation in the normal age-related response to iron accumulation. Thus, there is a potential for oxidative stress secondary to inadequate iron storage that may cause or contribute to the pathogenesis of PD. The evidence on the use of antioxidants in the treatment of PD is inconclusive; and the use of iron chelators to treat PD must consider the dependency of dopamine synthesis on iron (Waldemeier et al., 1993).

14.4.4 DEMYELINATING DISORDERS

It is now accepted that iron acquisition by oligodendrocytes is essential for myelination. Iron concentrations in the brain white matter consistently exceed those of the corresponding gray matter (Connor, 1994), suggesting a role for iron in myelin maintenance. In support of this concept, studies on rats have demonstrated that iron deficiency is connected with hypomyelination (Larkin and Rao, 1990) through a decrease in myelin-associated lipids (Oloyede et al., 1992). As mentioned before, oligodendrocytes are the principal cells in the brain that stain following iron histochemistry (Benkovic and Connor, 1993; Dwork et al., 1988; Morris et al., 1992). Iron-positive oligodendrocytes are present at birth and appear first in white matter tracts in the brain; in the developing white matter, these cells are clustered in areas reported to be myelinogenic foci (Connor et al., 1995a) and peak iron uptake in the brain coincides with the onset of myelination (Crowe and Morgan, 1992). Iron modulates the differentiation of precursor cells into oligodendrocytes during development (Morath and Mayer-Proschel, 2001), and it has been shown that holotransferrin but not apotransferrin prevents dedifferentiation of Swchann cells *in vitro* (Salis et al., 2002). If oligodendrocytes are not functional (myelin-deficient rats), brain iron uptake continues (Gocht et al., 1993) but accumulates in astrocytes and microglia (Connor and Menzies, 1990). If oligodendrocytes reach mature stages and produce myelin, even if the myelin is altered (shiverer mice) iron accumulates in oligodendrocytes (Connor et al., 1993; Levine, 1991). Taken together, these studies provide clear evidence that brain iron uptake, iron accumulation in oligodendrocytes, and myelination are coincident temporally and spatially.

There is evidence that the expression of Tf and ferritin in oligodendrocytes, as well as the cellular and regional distribution of iron, is altered in MS (Craelius et al., 1982) and other demyelinating disorders such as Pelizaeus–Merzbacher disease (Koeppen et al., 1988), central pontine myelinolysis (Gocht and Lohler, 1990), and progressive rubella panencephalitis (Valk and van der Knaap, 1991). In animal models of demyelinating disorders (EAE), iron chelation therapy decreases both the severity of the histopathology and the clinical symptoms (Bowern et al., 1984). Hence, knowledge of how iron is managed and the consequence of loss of iron management may reveal potential targets for intervention in demyelinating disorders. Increased lipid peroxidation has been found in the MS brain (Hunter et al., 1985) as has increased production of superoxide radicals and hydrogen peroxide (Fisher et al., 1988) by macrophages from MS patients, suggesting that oxidative stress and demyelination are causally related. Oligodendrocytes, because of their high iron content, may be especially susceptible to this type of oxidative stress secondary to macrophage-mediated inflammation. The presence of substances such as nitric oxide (NO), which accompany inflammatory responses, might compromise the iron-sequestering capabilities of ferritin. NO has been shown to remove iron from ferritin (Reif and Simmons, 1990) and kill oligodendrocytes (Mitrovic et al., 1995).

What is the role of iron in myelin production? Iron is required for the biosynthesis of lipids and cholesterol, which are abundant and key components of myelin (Larkin and Rao, 1990). The demyelination associated with tellurium toxicity is thought to result from blockage of an iron-requiring step in cholesterol biosynthesis (Wagner-Recio et al., 1991). Oligodendrocytes reportedly have a twofold higher metabolic activity rate than other cell types in the brain (Hamberger, 1961). Also, specific iron-requiring enzymes involved in maintaining a high rate of metabolic activity, such as glucose-6-phosphate dehydrogenase, dioxygenase, succinic dehydrogenase, NADH dehydrogenase, and the cytochrome oxidase system, are elevated in oligodendrocytes relative to other cells in the brain (Cammer, 1984). Thus, iron involvement in myelination is both direct, through synthetic pathways, and indirect, through its essential role in oxidative metabolism.

Because of the high iron requirements in oligodendrocytes, these cells appear to have developed a specific mechanism for iron uptake. Traditionally, Tf has been considered the protein to deliver iron to cells by binding to the TfR, and TfRs are found predominantly in the gray matter in the brain. Recently, ferritin receptors have been reported to be expressed in white matter tracts in rodent and human brains (Hulet et al., 1999a, 1999b), and the appearance of ferritin binding in the postnatal brain follows the onset and progression of myelination (Hulet et al., 2002). In addition, ferritin receptors are found specifically on oligodendrocytes in mixed glial cultures and addition of ferritin to the medium increases the intracellular iron content of these cells (Hulet et al., 2000), leading to the hypothesis that high iron requirements of oligodendrocytes are met by delivery through ferritin. In MS, ferritin receptors are lost in the periplaque region in the white matter, whereas Tf binds to immature oligodendrocytes found in the same region (Hulet et al., 1999b), suggesting that ferritin receptors could be an autoimmune target in MS.

14.4.5 FRIEDREICH'S ATAXIA (FRDA)

FRDA is an autosomal recessive disorder characterized by progressive ataxia, motor weakness, proprioceptive loss, dysarthria, and cardiac dysfunction. It has an early age of onset, and most individuals become symptomatic between the first and second decade of life (Harding, 1993). FRDA is primarily a disease of the spinal cord and peripheral nerves. There is neuronal loss in the dorsal root ganglia, dorsal nuclei of Clarke, and deep cerebellar nuclei (dentate nuclei). There is also degeneration and sclerosis of the posterior columns, spinocerebellar tracts, and corticospinal tracts, with mild degenerative changes in the pontine and medullary nuclei as well as in the Purkinje cells of the cerebellum (Harding, 1993). Peripherally, there is selective loss of large myelinated neurons.

The gene responsible for FRDA has been identified on chromosome 9q13 (Campuzano et al., 1996). FRDA is due to an intronic expansion of a guanine-adenine-adenine (GAA) trinucleotide repeat in the gene encoding the mitochondrial protein frataxin (Priller et al., 1997; Koutnikova et al., 1997). This leads to reduced synthesis of frataxin and results in overload of mitochondrial iron and oxidative stress. Losses in mitochondrial DNA and deficiencies in Fe-S-dependent enzyme activities as well as a decrease in cytochrome complexes I, II, and III and cytosolic aconitase in the heart have also been reported (Babcock et al., 1997; Wilson and Roof, 1997; Rotig et al., 1997). It has been suggested that these decreases in enzyme activities as well as the losses in mitochondrial DNA are the result of oxidative stress secondary to mitochondrial iron overload (Koutnikova et al., 1997; Wilson and Roof, 1997). These data suggest that frataxin may function as a protector against iron-induced oxidative stress, or alternatively, that frataxin may function in mitochondrial iron transport (Babcock et al., 1997).

Finally, it is particularly noteworthy that the organs of greatest pathology in FRDA (the heart and brain) share in common that their cells are predominantly in the postmitotic state as well as the predominant expression of H-ferritin. Although the significance of this connection is uncertain, it is possible that postmitotic cells, which are limited to the expression H-ferritin as opposed to L-ferritin, have a decreased capacity for long-term iron storage. As such, neurons and cardiac myocytes represent vulnerable targets for oxidative stress, given their dependence on a continuous demand for iron uptake coupled with a relatively high level of mitochondrial energy production.

14.4.6 HUNTINGTON'S DISEASE (HD)

Huntington's disease (HD) is an autosomal dominant neurodegenerative disorder caused by a (CAG)n repeat expansion in the HD gene, which results in the expression of a polyglutamine chain in the N-terminus of the protein huntingtin (The Huntington's Disease Collaborative Research Group, 1993). The cleavage and subsequent generation of a toxic N-terminus fragment may be crucial steps in the pathogenesis of the disease (Leavitt et al., 1999). HD is characterized by degeneration in the striatum and motor, psychiatric, and cognitive disturbances. It has been hypothesized to be a disease of mitochondrial metabolism (Beal, 1998; Schapira, 1999). Significantly increased striatal iron levels have been found in HD patients (Dexter et al., 1991; Bartzokis et al., 1999b) and the involvement of free radical generation has been suggested to contribute to neurodegeneration (Schapira, 1999). In cell culture models the, overexpression of huntingtin altered iron metabolism (Hilditch-Maguire, 2000), but it is not known whether this effect is specific to iron management proteins or to changes in the cell's protein general trafficking.

14.4.7 RESTLESS LEGS SYNDROME (RLS)

Restless legs syndrome (RLS) is a disorder of sensation, characterized by an uncontrollable need or urge to move the legs (referred to as akathisia), with a clear circadian pattern in which the symptoms dominate during nighttime (Walters, 1995). RLS is highly responsive to treatment with dopaminergic agents, suggesting a relative or absolute decrease in dopaminergic activity. Iron deficiency has been considered a significant contributing cause of RLS (Norlander, 1953; Ekbom, 1960), and decreased plasma ferritin correlates with increased severity of RLS symptoms (O'Keeffe et al., 1994; Sun et al., 1998). Oral iron therapy improved the symptoms of some RLS patients (Ekbom, 1960; O'Keeffe et al., 1994), and high doses of intravenous iron therapy produced a complete relief of symptoms in almost all patients treated, even though the majority had normal iron status before treatment (Norlander, 1953). The CSF of RLS patients is 65% lower in ferritin and threefold higher in Tf compared with control individuals (Earley et al., 2000b), which may reflect brain iron deficiency, a notion supported by the lower iron concentrations found in the substantia nigra and putamen of RLS patients by magnetic resonance imaging (MRI) analysis (Allen et al., 2001). The idea that the substantia nigra is iron deficient in RLS has been recently supported by an autopsy analysis. The autopsy analysis did not identify any histopathological changes in the RLS brains, but did observe a lack of Tf receptor expression on neuromelanin cells despite the rest of the iron management proteins being consistent with iron deficiency. Decreased binding potential for the dopamine receptor and transporter have also been reported in basal ganglia (Staedt et al., 1995; Turjanski et al., 1999), similar to the alterations shown in ID animals. Thus, although only speculative at present, we have proposed that iron acquisition by neuromelanin cells is part of the underlying pathogenesis of RLS.

14.4.8 ACERULOPLASMINEMIA

Aceruloplasminemia is an autosomal recessive disorder of iron metabolism caused by mutations in the ceruloplasmin gene (Harris et al., 1995). It is characterized by diabetes, retinal and basal ganglia degeneration, and neurological symptoms. The brain is a main site of iron deposition (Yoshida et al., 1995), and CSF from aceruloplasminemic patients showed a threefold increase in iron as well as increased lipid peroxidation products and superoxide dismutase activity (Miyajima et al., 1998). Expression of a surface-anchored form of ceruloplasmin by astrocytes is assumed critical for iron metabolism and neuronal survival in the basal ganglia and retina (Gitlin, 1998; Patel and David, 1997). A clear indicator of the involvement of iron in the neurological problems associated with aceruloplasminemia is the report that treatment with the iron chelator desferriox-amine decreased brain iron stores and prevented progression of the neurological symptoms and reduced plasma lipid peroxidation in a patient with aceruloplasminemia (Miyajima et al., 1997).

14.5 CONCLUSION

Both iron deficiency and iron excess have deleterious consequences for the nervous system. The cognitive and developmental problems associated with iron deficiency require further investigation with better-controlled studies. There is a heterogeneous distribution and accumulation of iron in the human brain, but the reason for these differences between areas of the normal brain is still unknown. There appear to exist predetermined set points for iron accumulation for the different brain regions; thus, in normal brains substantia nigra reaches a plateau for iron content with age. The mechanisms for these set points and the implications for the abnormal iron accumulation in the diseases discussed need to be studied. The propensity for the striatum to accumulate excessive iron in many neuropathological states, as well as accumulation in any other brain area, presumably involves a loss of concordant regulation between uptake and storage and perhaps transport within the brain. This could as well be associated with a change in the set points.

Iron could also accumulate in the brain because of an alteration in iron transport across the brain barriers. Although poorly investigated to date, iron efflux is suspected to be an important component in brain iron management. There is the possibility of RLS being an example of brain iron deficiency by means of an increased iron efflux from the brain (brain-CSF barrier?). Conversely, it is easy to understand that a decrease in iron efflux would lead to regional brain iron accumulation. Increased iron influx caused by alterations in the BBB attributable to conditions such as inflammation, as in MS, seems to be one means for iron accumulation in the brain. Cytokines secreted during inflammation alter normal IRP functioning, which could also lead to iron accumulation. A dysfunctional IRP–IRE interaction is also a suspected cause for elevated iron without concomitant increase in ferritin found in PD and AD.

The field of iron research is very active and new components have been added in the recent past to the already complex metabolism of iron. Several of these components, such as DMT1, IRP1, and IRP2, are being actively investigated in relation to iron mismanagement in the brain. Some of the newest ones, the Tfr2, the novel mitochondrial ferritin (Drysdale et al., 2002), and hepcidin, a circulating antimicrobial peptide recently proposed as a key component in the interaction between body stores and dietary iron absorption (Pigeon et al., 2001; Nicolas et al., 2002), require special attention in relation neurodegenerative processes associated with iron deposition.

Independent of the mechanisms to explain the accumulation of iron in the brain, the overwhelming evidence is that iron induces cell and tissue damage through the generation of reactive oxygen species and oxidative stress. Iron mismanagement, both regionally and at the cellular level, is a component of many neurodegenerative diseases. It is clear that the assessment of brain iron status by techniques such as MRI and CSF analysis should be included in discussions of diagnosis and treatment of neurological disorders. The development of iron chelators and iron chelation therapy should include neurological disorders as target pathologies for their use.

REFERENCES

Abboud, S. and Haile, D.J. (2000) A novel mammalian iron-regulated protein involved in intracellular iron metabolism. *J Biol Chem* 275(26):19906–19912.

Algarin, C., Peirano, P., Garrido, M., Pizarro, F. and Lozoff, B. (2003) Iron deficiency anemia in infancy: long-lasting effects on auditory and visual system functioning. *Pediatr Res* 53(2):217–223.

Allen, R.P., Barker, P.B., Wehr, F., Song, H.K. and Earley, C.J. (2001) MRI measurement of brain iron in patients with restless legs syndrome. *Neurology* 56(2):263–265.

Aoki, S., Okada, Y., Nishimura, K., Barkovich, A.J., Kjos, B.O., Brasch, R.C. and Norman, D. (1989) Normal deposition of brain iron in childhood and adolescence: MRI imaging at 1.5T. *Radiology* 172:381–385.

Arawaka, S., Saito, Y., Murayama, S. and Mori, H. (1998) Lewy body in neurodegeneration with brain iron accumulation type 1 is immunoreactive for alpha-synuclein. *Neurology* 51(3):887–889.

Aziz, N. and Munro, H.N. (1987) Iron regulates ferritin mRNA translation through a segment of its 5' untranslated region. *Proc Natl Acad Sci USA* 84:8478–8482.

Babcock, M., de Silva, D., Oaks, R., Davis-Kaplan, S., Jiralerspong, S., Montermini, L., Pandolfo, M. and Kaplan, J. (1991) Regulation of mitochondrial iron accumulation by Yfh1p, a putative homolog of frataxin. *Science* 276(5319):1709–1712.

Bartlett, W.P., Li, X.-S. and Condor, J.R. (1991) Expression of transferrin mRNA in the CNS of normal and jimpy mice. *J Neurochem* 57(1):318–322.

Bartzokis, G., Beckson, M., Hance, D.B., Marx, P., Foster, J.A. and Marder, S.R. (1997) MRI evaluation of age-related increase of brain iron in young adult and older normal males. *Magn Reson Imag* 15(1):29–35.

Bartzokis, G., Cummings, J.L., Markham, C.H., Marmarelis, P.Z., Treciokas, L.J., Tishler, T.A., Marder, S.R. and Mintz, J. (1999a) MRI evaluation of brain iron in earlier- and later-onset Parkinson's disease and normal subjects. *Magn Reson Imag* 17(2):213–222.

Bartzokis, G., Cummings, J., Perlman, S., Hance, D.B. and Mintz, J. (1999b) Increased basal ganglia iron levels in Huntington disease. *Arch Neurol* 56(5):569–574.

Bartzokis, G., Sultzer, D., Cummings, J., Holt, L.E., Hance, D.B., Henderson, V.W. and Mintz, J. (2000) In vivo evaluation of brain iron in Alzheimer disease using magnetic resonance imaging. *Arch Gen Psychiatry* 57(1):47–53.

Beal, M.F. (1998) Mitochondrial dysfunction in neurodegenerative diseases. *Biochim Biophys Acta* 1366(1/2): 211–223.

Beard, J.L. (1987) Feed efficiency and norepinephrine turnover in iron deficiency. *Proc Soc Exp Biol Med* 184:337–344.

Beard, J.L., Chen, Q., Connor, J. and Jones, B.C. (1994) Altered monoamine metabolism in caudate-putamen of iron-deficient rats. *Pharmacol Biochem Behav* 48(3):621–624.

Beard, J.L., Dawson, H. and Piñero, D.J. (1996) Iron metabolism: a comprehensive review. *Nutr Rev* 54:295–317.

Beard, J.L., Erikson, K.M. and Jones, B.C. (2002) Neurobehavioral analysis of developmental iron deficiency in rats. *Behav Brain Res* 21:134(1/2):517–524

Benkovic, S.A. and Connor, J.R. (1993) Ferritin, transferrin, and iron in selected regions of the adult and aged rat brain. *J Comp Neurol* 338:97–113.

Ben-Shachar, D., Yehuda, S., Finberg, J.P.M., Spanier, I. and Youdim, M.B.H. (1988) Selective alteration in blood brain barrier and insulin transport in iron-deficient rats. *J Neurochem* 50:1434–1437.

Blissman, G., Menzies, S., Beard, J., Palmer, C. and Connor, J.R. (1996) The expression of ferritin subunits and iron in oligodendrocytes in neonatal porcine brains. *Dev Neurosci* 18:274–281.

Bloch, B., Popovici, T., Levin, M.J., Tuil, D. and Kahn, A. (1985) Transferrin gene expression visualized in oligodendrocytes of the rat brain by using in situ hybridization and immunohistochemistry. *Proc Natl Acad Sci USA* 82:6706–6710.

Bodovitz, S., Falduto, M.T., Frail, D.E. and Klein, W.L. (1995) Iron levels modulate α-secretase cleavage of amyloid precursor protein. *J Neurochem* 64:307–315.

Bothwell, T.H., Charlton, R.W., Cook, J.D. and Finch, C.A. (1979) *Iron Metabolism in Man.* Blackwell Scientific Publications, Oxford.

Bowern, N., Ramshaw, I.A., Clark, I.A. and Doherty, P.C. (1984) Inhibition of autoimmune neuropathological process by treatment with an iron-chelating agent. *J Exp Med* 160:1532–1543.

Bradbury, M.W. (1997) Transport of iron in the blood-brain-cerebrospinal fluid system. *J Neurochem* 69(2):443–454.

Burdo, J.R., Martin, J., Menzies, S.L., Dolan, K.G., Romano, M.A., Fletcher, R.J., Garrick, M.D., Garrick, L.M. and Conner, J.R. (1999) Cellular distribution of iron in the brain of the Belgrade rat. *Neuroscience* 93:1189–1196.

Burdo, J.R., Menzies, S.L., Simpson, I.A., Garrick, L.M., Garrick, M.D., Dolan, K.G., Haile, D.J., Beard, J.L. and Conner, J.R. (2001) Distribution of divalent metal transporter 1 and metal transport protein 1 in the normal and Belgrade rat. *J Neurosci Res* 66(6):1198–1207.

Cai, C.X. and Linsenmayer, T.F. (2001) Nuclear translocation of ferritin in corneal epithelial cells. *J Cell Sci* 114(Pt. 12):2327–2334.

Cammer, W. (1984) Oligodendrocyte associated enzymes. In *Oligodendroglia* (Norton, W.T. Ed.), pp. 199–232. Plenum Press, New York.

Campbell, D.R., Skikne, B.S. and Cook, J.D. (1986) Cerebrospinal fluid ferritin levels in screening for meningism. *Arch Neurol* 43(12): 1257–1260.

Campuzano, V., Montermini, L., Moltò, M.D., Pianese, L., Cossée, M., Cavalcanti, F., Monros, E., Rodius, F., Duclos, F., Monticelli, A., Zara, F., Cañizares, J., Koutnikova, H., Bidichandani, S.I., Gellera, C., Brice, A., Trouillas, P., DeMichele, G., Filla, A., DeFrutos, R., Palau, F., Patel, P.I., DiDonato, S., Mandel, J.-L., Cocozza, S., Koenig, M. and Pandolfo, M. (1996) Friedreich's ataxia: autosomal recessive disease caused by an intronic GAA triplet repeat expansion. *Science* 271(5254):1423–1427.

Cheepsunthorn, P., Palmer, C. and Connor, J.R. (1998) Cellular distribution of ferritin subunits in postnatal rat brain. *J Comp Neurol* 400:73–86.

Cheepsunthorn, P., Palmer, C., Menzies, S., Roberts, R.L. and Connor, J.R. (2001) Hypoxic/ischemic insult alters ferritin expression and myelination in neonatal rat brains. *J Comp Neurol* 19:431(4):382–496.

Chen, Q., Connor, J.R. and Beard, J.L. (1995) Brain iron, transferrin and ferritin concentrations are altered in developing iron-deficient rats. *J Nutr* 125(6):1529–1535.

Connor, J.R. (1994) Iron regulation in the brain at the cell and molecular level. *Adv Exp Med Biol* 356:229–238.

Connor, J.R. and Menzies, S.L. (1990) Altered cellular distribution of iron in the CNS of myelin deficient rats. *Neuroscience* 34:265–271.

Connor, J.R. and Menzies, S.L. (1995) Cellular management of iron in the brain. *J Neurol Sci* 134(Suppl):33–44.

Connor, J.R. and Menzies, S.L. (1996) Relationship of iron to oligodendrocytes and myelination. *Glia* 17:83–93.

Connor, J.R., Menzies, S.L., St. Martin, S.M. and Mufson, E.J. (1992a) A histochemical study of iron, transferrin, and ferritin in Alzheimer's diseased brains. *J Neurosci Res* 31:75–83.

Connor, J.R., Milward, E.A., Moalem, S., Sampietro, M., Boyer, P., Percy, M.E., Vergani, C., Scott, R.J. and Chorney, M. (2001) Is hemochromatosis a risk factor for Alzheimer's disease? *J Alzheimers Dis* 3(5):471–477.

Connor, J.R., Pavlick, G., Karli, D., Menzies, S.L. and Palmer, C. (1995a) A histochemical study of iron-positive cells in the developing rat brain. *J Comp Neurol* 355:111–123.

Connor, J.R., Roskams, A.J.I., Menzies, S.L. and Williams, M.E. (1993) Transferrin in the central nervous system of the shiverer mouse myelin mutant. *J Neurosci Res* 36:501–507.

Connor, J.R., Snyder, B.S., Arosio, P., Loeffler, D.A. and LeWitt, P. (1995b) A quantitative analysis of isoferritins in select regions of aged, parkinsonian, and Alzheimer's diseased brains. *J Neurochem* 65:717–724.

Connor, J.R., Snyder, B.S., Beard, J.L., Fine, R.E. and Mufson, E.J. (1992b) Regional distribution of iron and iron-regulatory proteins in the brain in aging and Alzheimer's disease. *J Neurosci Res* 31:327–335.

Cornett, C.R., Markesbery, W.R. and Ehmann, W.D. (1998) Imbalances of trace elements related to oxidative damage in Alzheimer's disease brain. *Neurotoxicology* 19(3):339-346.

Craelius, W., Migdal, M.W., Luessenhop, C.P., Sugar, A. and Mihalakis, I. (1982) Iron deposits surrounding multiple sclerosis plaques. *Arch Pathol Lab Med* 106:397–399.

Crowe, A. and Morgan, E.H. (1992) Iron and transferrin uptake by brain and cerebrospinal fluid in the rat. *Brain Res* 592:8–16.

Dallman, P.R., Siimes, M.A. and Manies, E.C. (1975) Brain iron: persistent deficiency following short term iron deprivation the young rat. *Br J Haematol* 31:209–215.

Dallman, P.R. and Spirito, R.A. (1977) Brain iron in the rat: extremely slow turnover in normal rats may explain the long lasting effects of early iron deficiency. *J Nutr* 107:1075–1081.

Dautry-Varsat, A., Ciechanover, A. and Lodish, H.F. (1983) pH and the recycling of transferrin during receptor-mediated endocytosis. *Proc Natl Acad Sci USA* 80(8):2258–2262.

de Arriba-Zerpa, G.A., Saleh, M.C., Fernandez, P.M., Guillou, F., Espinosa de los Monteros, A., de Vellis, J., Zakin, M.M. and Baron, B. (2000) Alternative splicing prevents transferrin secretion during differentiation of a human oligodendrocyte cell line. *J Neurosci Res* 61(4):388–395.

Deaglio, S., Capobianco, A., Cali, A., Bellora, F., Alberti, F., Righi, L., Sapino, A., Camaschella, C. and Malavasi, F. (2002) Structural, functional, and tissue distribution analysis of human transferrin receptor-2 by murine monoclonal antibodies and a polyclonal antiserum. *Blood* 100(10):3782–3789.

Descamps, L., Dehouck, M.P., Torpier, G. and Cecchelli, R. (1996) Receptor-mediated transcytosis of transferrin through blood-brain barrier endothelial cells. *Am J Physiol* 270(4, Pt. 2):H1149–H1158.

Dexter, D.T., Carayon, A., Javoy-Agid, F., Agid, Y., Wells, F.R., Daniel, S.E., Lees, A.J., Jenner, P. and Marsden, C.D. (1991) Alterations in the levels of iron, ferritin and other trace metals in Parkinson's disease and other neurodegenerative diseases affecting the basal ganglia. *Brain* 114:1953–1975.

Dexter, D.T., Carayon, A., Vidailhet, M., Ruberg, M., Agid, F., Agid, Y., Lees, A.J., Wells, F.R., Jenner, P. and Marsden, C.D. (1990) Decreased ferritin levels in brain in Parkinson's disease. *J Neurochem* 55(1):16–20.

Dexter, D.T., Wells, F.R., Lees, A.J., Agid, F., Agid, Y., Jenner, P. and Marsden, C.D. (1989) Increased nigral iron content and alterations in other metal ions occurring in brain in Parkinson's disease. *J Neurochem* 52:1830–1836.

Dickinson, T.K. and Connor, J.R. (1998) Immunohistochemical analysis of transferrin receptor: regional and cellular distribution in the hypotransferrinemic (hpx) mouse brain. *Brain Res* 801:171–181.

Drysdale, J., Arosio, P., Invernizzi, R., Cazzola, M., Volz, A., Corsi, B., Biasiotto, G. and Levi, S. (2002) Mitochondrial ferritin: a new player in iron metabolism. *Blood Cells Mol Dis* 29(3):376–383.

Dwork, A.J., Schon, E.A. and Herbert, J. (1988) Nonidentical distribution of transferrin and iron in rat brain. *Neuroscience* 27:333–335.

Earley, C.J., Allen, R.P., Beard, J.L. and Connor, J.R. (2000a) Insight into the pathophysiology of restless legs syndrome. *J Neurosci Res* 62(5):623–628.

Earley, C.J., Connor, J.R., Beard, J.L., Malecki, E.A., Epstein, D.K. and Allen, R.P. (2000b) Abnormalities in CSF concentrations of ferritin and transferrin in restless leg syndrome. *Neurology* 54(8):1698–1700.

Eisenstein, R.S. (2000) Iron regulatory proteins and the molecular control of mammalian iron metabolism. *Annu Rev Nutr* 20:627–662.

Ekbom, K.A. (1960) Restless legs syndrome. *Neurology* 10:868–873.

Erikson, K.M., Beard, J.L. and Connor, J.R. (1998) Distribution of brain iron, ferritin, and transferrin in the 28-day-old piglet. *Nutr Biochem* 9:276–284.

Erikson, K.M., Jones, B.C. and Beard, J.L. (2000) Iron deficiency alters dopamine transporter functioning in rat striatum. *J Nutr* 130(11):2831–2837.

Erikson, K.M., Jones, B.C., Hess, E.J., Zhang, Q. and Beard, J.L. (2001) Iron deficiency decreases dopamine D1 and D2 receptors in rat brain. *Pharmacol Biochem Behav* 69(3/4):409–418.

Erikson, K.M., Piñero, D.J., Connor, J.R. and Beard, J.L. (1997) Regional brain, iron, ferritin, and transferrin concentrations during iron deficiency and iron repletion in developing rats. *J Nutr* 127:2030–2038.

Fargion, S., Arosio, P., Fracanzani, A.L., Cislaghi, V., Levi, S., Cozzi, A., Piperno, A. and Fiorelli, G. (1988) Characteristics and expression of binding sites specific for ferritin H-chain on human cell lines. *Blood* 71:753–757.

Fargion, S., Fracanzani, A.L., Cislaghi, V., Levi, S., Cappellini, M.D. and Fiorelli, G. (1991) Characteristics of the membrane receptor for human H-ferritin. *Curr Stud Hematol Blood Transfus* 58:164–170.

Faucheux, B.A., Nillesse, N., Damier, P., Spik, G., Mouatt-Prigent, A., Pierce, A., Leveugle, B., Kubis, N., Hauw, J.J., Agid, Y. and Hirsch, E.C. (1995) Expression of lactoferrin receptors is increased in the mesencephalon of patients with Parkinson's Disease. *Proc Natl Acad Sci USA* 92:9603–9607.

Feder, J.N., Gnirke, A., Thomas, W., Tsuchihashi, Z., Ruddy, D.A., Basava, A., Dormishian, F., Domingo, Jr., R., Ellis, M.C., Fullan, A., Hinton, L.M., Jones, N.L., Kimmel, B.E., Kronmal, G.S., Lauer, P., Lee, V.K., Loeb, D.B., Mapa, F.A., McClelland, E., Meyer, N.C., Mintier, G.A., Moeller, N., Moore, T., Morikang, E., Prass, C.E., Quintana, L., Starnes, S.M., Schatzman, R.C., Brunke, K.J., Drayna, D.T., Risch, N.J., Bacon, B.R. and Wolff, R.K. (1996) A novel MHC class I-like gene is mutated in patients with hereditary haemochromatosis. *Nat Genet* 13(4):399–408.

Feder, J.N., Penny, D.M., Irrinki, A., Lee, V.K., Lebron, J.A., Watson, N., Tsuchihashi, Z., Sigal, E., Bjorkman, P.J. and Schatzman, R.C. (1998) The hemochromatosis gene product complexes with the transferrin receptor and lowers its affinity for ligand binding. *Proc Natl Acad Sci USA* 95(4):1472–1477.

Feder, J.N., Tsuchihashi, Z., Irrinki, A., Lee, V.K., Mapa, F.A., Morikang, E., Prass, C.E., Starnes, S.M., Wolff, R.K., Pakkila, S., Sly, W.S. and Schatzman, R.C. (1997) The hemochromatosis founder mutation in HLA-H disrupts beta2-microglobulin interaction and cell surface expression. *J Biol Chem* 272(22):14025–14028.

Feldman, H., Gabathuler, R., Kennard, M., Nurminen, J., Levy, D., Foti, S., Foti, D., Beattie, B.L. and Jeffries, W.A. (2001) Serum p97 levels as an aid to identifying Alzheimer's disease. *J Alzheimer's Dis* 3(5):507–516.

Felt, B.T. and Lozoff, B. (1996) Brain iron and behavior of rats are not normalized by treatment of iron deficiency anemia during early development. *J Nutr* 126(3):693–701.

Fillebeen, C., Descamps, L., Dehouck, M.P., Fenart, L., Benaissa, M., Spik, G., Cecchelli, R. and Pierce, A. (1999) Receptor-mediated transcytosis of lactoferrin through the blood-brain barrier. *J Biol Chem* 274(11):7011–7017.

Fishbach, F.A. and Andreregg, J.W. (1965) An x-ray scattering study of ferritin and apoferritin. *J Mol Biol* 14:458–473.

Fisher, M., Levine, P.H., Weiner, B.H., Vaudreuil, C.H., Natale, A., Johnson, M.H. and Hoogasian, J.J. (1988) Monocyte and polymorphonuclear leukocyte toxic oxygen metabolite production in multiple sclerosis. *Inflammation* 12(2):123–131.

Fishman, J.B., Rubin, J.B., Handrahan, J.V., Connor, J.R. and Fine, R.E. (1987) Receptor-mediated transcytosis of transferrin across the blood-brain barrier. *J Neurosci Res* 18(2):299–304.

Fleming, M.D., Romano, M.A., Su, M.A., Garrick, L.M., Garrick, M.D. and Andrews, N.C. (1998) Nramp2 is mutated in the anemic Belgrade (b) rat: evidence of a role for Nramp2 in endosomal iron transport. *Proc Natl Acad Sci USA* 95:1148–1153.

Fleming, R.E., Migas, M.C., Holden, C.C., Waheed, A., Britton, R.S., Tomatsu, S., Bacon, B.R. and Sly, W.S. (2000) Transferrin receptor 2: continued expression in mouse liver in the face of iron overload and in hereditary hemochromatosis. *Proc Natl Acad Sci USA* 97(5):2214–2219.

Food, M.R., Sekyere, E.O. and Richardson, D.R. (2002) The soluble form of the membrane-bound transferrin homologue, melanotransferrin, inefficiently donates iron to cells via nonspecific internalization and degradation of the protein. *Eur J Biochem* 269(18):4435–4445.

Galvin, J.E., Giasson, B., Hurtig, H.I., Lee, V.M. and Trojanowski, J.Q. (2000) Neurodegeneration with brain iron accumulation, type 1 is characterized by alpha-, beta-, and gamma-synuclein neuropathology. *Am J Pathol* 157(2):361–368.

Giometto, B., Bozza, F., Argentiero, V., Gallo, P., Pagni, S., Piccinno, M.G. and Tavolato, B. (1990) Transferrin receptor in rat central nervous system. An immunohistochemical study. *J Neurol Sci* 98:81-90.

Gitlin, J.D. (1998) Aceruloplasminemia. *Pediatr Res* 44(3):271–276.

Glinka, Y., Gassen, M. and Youdim, M.B.H. (1997) Iron and neurotransmitter function in the brain. In *Metals and Oxidative Damage in Neurological Disorders* (Connor, J.R., Eds), pp. 1–22. Plenum Press, New York.

Gocht, A., Keith, A.B., Candy, J.M. and Morris, C.M. (1993) Iron uptake in the brain of the myelin deficient rat. *Neurosci Lett* 154:187–190.

Gocht, A. and Lohler, J. (1990) Changes in glial cell markers in recent and old demyelinated lesions in central pontine myelinolysis. *Acta Neuropathol (Berl)* 80(1):46–58.

Golts, N., Snyder, H., Frasier, M., Theisler, C., Choi, P. and Wolozin, B. (2002) Magnesium inhibits spontaneous and iron-induced aggregation of alpha-synuclein. *J Biol Chem* 277(18):16116–16123.

Good, P.F., Olanow, C.W. and Perl, D.P. (1992a) Neuromelanin-containing neurons of the substantia nigra accumulate iron and aluminum in Parkinson's disease: a LAMMA study. *Brain Res* 593(2):343–346.

Good, P.F., Perl, D.P., Bierer, L.M. and Schmeidler, J. (1992b) Selective accumulation of aluminum and iron in the neurofibrillary tangles of Alzheimer's disease: a laser microprobe (LAMMA) study. *Ann Neurol* 31:286–292.

Grantham-McGregor, S. and Ani, C. (2001) A review of studies on the effect of iron deficiency on cognitive development in children. *J Nutr* 131(25-2):649S–666S.

Gunshin, H., Mackenzie, B., Berger, U.V., Gunshin, Y., Romero, M.F., Boron, W.F., Nussberger, S., Gollan, J.L. and Hediger, M.A. (1997) Cloning and characterization of a mammalian proton-coupled metal-ion transporter. *Nature* 388(6641):482–428.

Haile, D.J., Rouault, T.A., Hartford, J.B., Kennedy, M.C., Blondin, G.A., Beinert, H. and Klausner, R.D. (1992) Cellular regulation of the iron-responsive element binding protein: disassembly of the cubane iron-sulfur cluster results in high affinity RNA binding. *Proc Natl Acad Sci USA* 89:11735–11739.

Hallgren, B. and Sourander, P. (1958) The effect of age on the non-haemin iron in the human brain. *J Neurochem* 3:41–45.

Hamberger, A. (1961) Oxidation of tricarboxylic acid cycle intermediates by nerve cell bodies and glial cells. *J Neurochem* 8:31–35.

Harding, A.E. (1993) Clinical features and classification of inherited ataxias. *Adv Neurol* 61:1–14.

Harris, Z.L., Takahashi, Y., Miyajima, H., Serizawa, M., MacGillivray, R.T. and Gitlin, J.D. (1995) Aceruloplasminemia: molecular characterization of this disorder of iron metabolism. *Proc Natl Acad Sci USA* 92(7):2539–2543.

Hayflick, S.J., Westaway, S.K., Levinson, B., Zhou, B., Johnson, M.A., Ching, K.H. and Gitschier, J. (2003) Genetic, clinical, and radiographic delineation of Hallervorden-Spatz syndrome. *N Engl J Med* 348(1):33–40.

Hentze, M.W. and Kühn, L. (1996) Molecular control of vertebrate iron metabolism: mRNA-based regulatory circuits operated by iron, nitric oxide, and oxidative stress. *Proc Natl Acad Sci USA* 93:8175–8182.

Hilditch-Maguire, P., Trettel, F., Passani, L.A., Auerbach, A., Persichetti, F. and MacDonald, M.E. (2000) Huntingtin: an iron-regulated protein essential for normal nuclear and perinuclear organelles. *Hum Mol Genet* 9(19):2789–2797.

Hill, J.M., Ruff, M.R., Weber, R.J. and Pert, C.B. (1985) Transferrin receptors in rat brain: neuropeptide-like pattern and relationship to iron distribution. *Proc Natl Acad Sci USA* 82:4453–4557.

Hulet, S.W., Hess, E.J., Debinski, W., Arosio, P., Bruce, K., Powers, S. and Conner, J.R. (1999a) Characterization and distribution of ferritin binding sites in the adult mouse brain. *J Neurochem* 72:868–874.

Hulet, S.W., Heyliger, S.O., Powers, S. and Connor, J.R. (2000) Oligodendrocyte progenitor cells internalize ferritin via clathrin-dependent receptor-mediated endocytosis. *J Neuroscience Res* 61:52–60.

Hulet, S.W., Menzies, S. and Connor, J.R. (2002) Ferritin binding in the developing mouse brain follows a pattern similar to myelination and is unaffected by the jimpy mutation. *Dev Neurosci* 24(2/3):208–213.

Hulet, S.W., Powers, S. and Connor, J.R. (1999b) Distribution of transferrin and ferritin binding in normal and multiple sclerotic human brains. *J Neurol Sci* 165:48–55.

Hunt, J.R., Zito, C.A., Erjavec, J. and Johnson, L. (1994) Severe or marginal iron deficiency affects spontaneous physical activity in rats. *Am J Clin Nutr* 59:413–418.

Hunter, M.I., Nlemadim, B.C. and Davidson, D.L. (1985) Lipid peroxidation products and antioxidant proteins in plasma and cerebrospinal fluid from multiple sclerosis patients. *Neurochem Res* 10(12):1645–1652.

Idjradinata, P. and Pollitt, E. (1993) Reversal of developmental delays in iron-deficient anaemic infants treated with iron. *Lancet* 341:1–4.

Izquierdo, I. (1989) Different forms of post-training memory processing. *Behav Neur Biol* 51:171–202.

Jefferies, W.A., Brandon, M.R., Hunt, S.V., Williams, A.F., Gatter, K.C. and Mason, D.Y. (1984) Transferrin receptor on endothelium of brain capillaries. *Nature* 312(5990):162–163.

Jefferies, W.A., Dickstein, D.L. and Ujiie, M. (2001) Assessing p97 as an Alzheimer's disease serum biomarker. *J Alzheimer's Dis* 3(3):339–344.

Jefferies, W.A., Food, M.R., Gabathuler, R., Rothenberger, S., Yamada, T., Yasuhara, O. and McGeer, P.L. (1996) Reactive microglia specifically associated with amyloid plaques in Alzheimer's disease brain tissue express melanotransferrin. *Brain Res* 712(1):122–126.

Jiang, D.H., Ke, Y., Cheng, Y.Z., Ho, K.P. and Qian, Z.M. (2002) Distribution of ferroportin1 protein in different regions of developing rat brain. *Dev Neurosci* 24(2/3):94–98.

Jing, S. and Trowbridge, I.S. (1987) Identification of the intermolecular disulfide bonds of the human transferrin receptor and its lipid attachment site. *EMBO J* 6:327–331.

Kalaria, R.N., Sromek, S.M., Grahovac, I. and Harik, S.I. (1992) Transferrin receptors of rat and human brain and cerebral microvessels and their status in Alzheimer's disease. *Brain Res* 585:87–93.

Kawabata, H., Yang, R., Hirama, T., Vuong, P.T., Kawano, S., Gombart, A.F. and Koeffler, H.P. (1999) Molecular cloning of transferrin receptor 2. A new member of the transferrin receptor-like family. *J Biol Chem* 274(30):20826–20832.

Keir, G., Tasdemir, N. and Thompson, E.J. (1993) Cerebrospinal fluid ferritin in brain necrosis: evidence for local synthesis. *Clin Chim Acta* 216(1/2):153–166.

Koeppen, A.H., Barron, K.D., Csiza, C.K. and Greenfield, E.A. (1988) Comparative immunocytochemistry of Pelizaeus-Merzbacher disease, the jimpy mouse, and the myelin-deficient rat. *J Neurol Sci* 84(2/3):315–327.

Koutnikova, H., Campuzano, V., Foury, F., Dolle, P., Cazzalini, O. and Koenig, M. (1997) Studies of human, mouse and yeast homologues indicate a mitochondrial function for frataxin. *Nat Genet* 16(4):345–351.

Kühn, L.C. (1998) Iron and gene expression: molecular mechanisms regulating cellular iron homeostasis. *Nutr Rev* 56:S11–S19.

Larkin, E.C. and Rao, G.A. (1990) Importance of fetal and neonatal iron: adequacy for normal development of central nervous system. In *Brain, Behaviour and Iron in the Infant Diet* (Dobbing J., Ed.), pp. 43–63. Springer-Verlag, London.

Leavitt, B.R., Wellington, C.L. and Hayden, M.R. (1999) Recent insights into the molecular pathogenesis of Huntington disease. *Semin Neurol* 19(4):385–395.

Leibold, E.A. and Munro, H.N. (1988) Cytoplasmic protein binds *in vitro* to a highly conserved sequence in the 5′ untranslated region of ferritin heavy- and light-subunit mRNAs. *Science* 241:1207–1210.

Leveugle, B., Faucheux, B.A., Bouras, C., Nillesse, N., Spik, G., Hirsch, E.C., Agid, Y. and Hof, P.R. (1996) Cellular distribution of the iron-binding protein lactotransferrin in the mesencephalon of Parkinson's disease cases. *Acta Neuropathol (Berl)* 91(6):566–572.

Leveugle, B., Spik, G., Perl, D.P., Bouras, C., Fillit, H.M. and Hof, P.R. (1994) The iron-binding protein lactotransferrin is present in pathologic lesions in a variety of neurodegenerative disorders: a comparative immunohistochemical analysis. *Brain Res* 650(1):20–31.

Levi, S., Franceschinelli, F., Cozzi, A., Doerner, M.H. and Arosio, P. (1989) Expression and structure and functional properties of human ferritin L-chain from *Escherichia coli*. *Biochemistry* 28:5179–5185.

Levi, S., Luzzago, A., Cesareni, G., Lozzi, A., Franceschinelli, F., Albertini, A. and Arosio, P. (1988) Mechanism of ferritin iron uptake: activity of the H-chain and deletion mapping of the ferro-oxidase site. *J Biol Chem* 263:18086–18092.

Levi, S., Yewdall, S.J., Harrison, P.M., Santambrogio, P., Cozzi, A., Robida, E., Albertini, A. and Arosio, P. (1992) Evidence that H- and L-chains have cooperative roles in the iron-uptake mechanism of human ferritin. *Biochem J* 288:591–596.

Levin, M.J., Tuil, D., Uzan, G., Dreyfus, J.C. and Kahn, A. (1984) Expression of the transferrin gene during development of non-hepatic tissues: high levels of transferrin mRNA in fetal muscle and adult brain. *Biochem Biophys Res Commn* 122:212–217.

LeVine, S.M., Lynch, S.G., Ou, C.N., Wulser, M.J., Tam, E. and Boo, N. (1999) Ferritin, transferrin and iron concentrations in the cerebrospinal fluid of multiple sclerosis patients. *Brain Res* 821(2):511–515.

Levine, S.M. (1991) Oligodendrocytes and myelin sheaths in normal, quaking and shiverer brains are enriched in iron. *J Neurosci Res* 29:413–419.

Loeffler, D.A, Connor, J.R., Juneau, P.L., Snyder, B.S., Kanaley, L., DeMaggio, A.J., Nguyen, H., Brickman, C.M. and LeWitt, P.A. (1995) Transferrin and iron in normal, Alzheimer's disease, and Parkinson's disease brain regions. *J Neurochem* 65:710–716.

Lozoff, B. (1998) Exploratory mechanisms for poorer development in iron-deficient anemic infants. In *Nutrition, Health, and Child Development. Research Advances and Policy Recommendations*, Scientific Publication No. 556, pp.162–178. Pan American Health Organization, Tropical Metabolism Research Unit of the University of the West Indies, and the World Bank, Washington, D.C.

Lozoff, B., Jimenez, E., Hagen, J., Mollen, E. and Wolf, A.W. (2000) Poorer behavioral and developmental outcome more than 10 years after treatment for iron deficiency in infancy. *Pediatrics* 105(4):e51.

Mantyh, P.W., Ghilardi, J.R., Rogers, S., DeMaster, E., Allen, C.J., Stimson, E.R. and Maggio, J.E. (1993) Aluminum, iron and zinc ions promote aggregation of physiological concentrations of beta-amyloid peptide. *J Neurochem* 61:1171–1174.

Mash, D.C., Pablo, J., Flynn, D.D., Efange, S.N.B.M. and Weiner, W.J. (1990) Characterization and distribution of transferrin receptors in the rat brain. *J Neurochem* 55:1972–1979.

McKie, A.T., Barrow, D., Latunde-Dada, G.O., Rolfs, A., Sager, G., Mudaly, E., Mudaly, M., Richardson, C., Barlow, D., Bomford, A., Peter, T.J., Raja, K.B., Shirali, S., Hediger, M.A., Farzaneh, F. and Simpson, R.J. (2001) An iron-regulated ferric reductase associated with the absorption of dietary iron. *Science* 291(5509): 1755–1759.

Melefors, Ö., Goossen, B., Johansson, H.E., Stripecke, R., Gray, N.K. and Hentze, M.W. (1993) Translational control of 5-aminolevulinate synthase mRNA by iron-responsive elements in erythroid cells. *J Biol Chem* 268:5974–5978.

Mitrovic, B., Ignarro, L.J., Vinters, H.V., Akers, M.A., Schmid, I., Uittenbogaart, C. and Merrill, J.E. (1995) Nitric oxide induces necrotic but not apoptotic cell death in oligodendrocytes. *Neuroscience* 65(2):531–539.

Miyajima, H., Fujimoto, M., Kohno, S., Kaneko, E. and Gitlin, J.D. (1998) CSF abnormalities in patients with aceruloplasminemia. *Neurology* 51(4):1188–1190.

Miyajima, H., Takahashi, Y., Kamata, T., Shimizu, H., Sakai, N. and Gitlin, J.D. (1997) Use of desferrioxamine in the treatment of aceruloplasminemia. *Ann Neurol* 41(3):404–407.

Moalem, S., Percy, M.E., Andrews, D.F., Kruck, T.P., Wong, S., Dalton, A.J., Mehta, P., Fedor, B. and Warren, A.C. (2000) Are hereditary hemochromatosis mutations involved in Alzheimer disease? *Am J Med Genet* 93(1):58–66.

Moos, T. (1996) Immunohistochemical localization of intraneuronal transferrin receptor immunoreactivity in the adult mouse central nervous system. *J Comp Neurol* 375:675–692.

Moos, T. and Morgan, E.H. (1998) Kinetics and distribution of [59Fe-125I]transferrin injected into the ventricular system of the rat. *Brain Res* 790(1/2):115–128.

Morath, D.J. and Mayer-Proschel, M. (2001) Iron modulates the differentiation of a distinct population of glial precursor cells into oligodendrocytes. *Dev Biol* 237(1):232–243.

Morris, C.M., Candy, J.M., Oakley, A.E., Bloxham, C.A. and Edwardson, J.A. (1992a) Histochemical distribution of non-haem iron in the human brain. *Acta Anatomica* 144:235–257.

Morris, C.M., Keith, A.B., Edwardson, J.A. and Pullen, R.G.L. (1992b) Uptake and distribution of iron and transferrin in the adult rat brain. *J Neurochem* 59:300–306.

Moss, D., Fargion, S., Fracanzani, A.L., Levi, S., Cappellini, M.D., Arosio, P., Powell, L.W. and Halliday, J.W. (1992a) Functional roles of the ferritin receptors of human liver, hepatoma, lymphoid and erythroid cells. *J Inorg Biochem* 47(3/4):219–227.

Moss, D., Powell, L.W., Arosio, P. and Halliday, J.W. (1992b) Characterization of the ferritin receptors of human T lymphoid (MOLT-4) cells. *J Lab Clin Med* 119(3):273–279.

Müllner, E.W. and Kühn, L.C. (1988) A stem-loop in the 3′ untranslated region mediates iron-dependent regulation of transferrin receptor mRNA stability in the cytoplasm. *Cell* 53:815–825.

Müllner, E.W., Neupert, B. and Kühn, L.C. (1989) A specific mRNA-binding factor regulates the iron-dependent stability of cytoplasmic transferrin receptor mRNA. *Cell* 58:373–382.

Nelson, C., Erikson, K.M., Pinero, D.J. and Beard, J.L. (1997) In vivo dopamine is altered in iron-deficient anemic rats. *J Nutr* 127:2282–2288.

Nicolas, G., Viatte, L., Bennoun, M., Beaumont, C., Kahn, A., Vaulont, S., Hepcidin, A. (2002) New iron regulatory peptide. *Blood Cells Mol Dis* 29(3):327–335.

Norlander, N.B. (1953) Therapy in restless legs syndrome. *Acta Med Scand* 145:453–457.

O'Keeffe, S.T., Gavin, K. and Lavan, J.N. (1994) Iron status and restless legs syndrome in the elderly. *Age Ageing* 23:200–203.

Oloyede, O.B., Folayan, A.T. and Odutauga, A.A. (1992) Effects of low-iron status and deficiency of essential fatty acids on some biochemical constituents of rat brain. *Biochem Int* 27:913–922.

Oski, F.A. and Honig, A.S. (1978) The effects of therapy on the developmental scores of iron-deficient infants. *J Pediatr* 92:21–25.

Oski, F.A., Honig, A.S., Helu, B. and Howanitz, P. (1983) Effects of iron therapy on behavior performance in nonanemic, iron-deficient infants. *Pediatrics* 71:877–880.

Patel, B.N. and David, S. (1997) A novel glycosylphosphatidylinositol-anchored form of ceruloplasmin is expressed by mammalian astrocytes. *J Biol Chem* 272(32):20185–20190.

Perry, T.L., Norman, M.G., Yong, V.W., Whiting, S., Crichton, J.U., Hansen, S. and Kish, S.J. (1985) Haller-vorden-Spatz disease: cysteine accumulation and cysteine dioxygenase deficiency in the globus pallidus. *Ann Neurol* 18(4):482–489.

Pigeon, C., Ilyin, G., Courselaud, B., Leroyer, P., Turlin, B., Brissot, P. and Loreal, O. (2001) A new mouse liver-specific gene, encoding a protein homologous to human antimicrobial peptide hepcidin, is overexpressed during iron overload. *J Biol Chem* 276(11):7811–7819.

Piñero, D.J., Hu, J. and Connor, J.R. (2000a) Alterations in the interaction between iron regulatory proteins and their iron responsive element in normal and Alzheimer's diseased brains. *Cell Mol Biol (Noisy-Le-Grand)* 46(4):761–776.

Piñero, D.J., Jones, B. and Beard, J. (2001) Variations in dietary iron alter behavior in developing rats. *J Nutr* 131(2):311–318.

Piñero, D.J., Li, N.Q., Connor, J.R. and Beard, J.L. (2000b) Variations in dietary iron alter brain iron metabolism in developing rats. *J Nutr* 130(2):254–263.

Priller, J., Scherzer, C.R., Faber, P.W., MacDonald, M.E. and Young, A.B. (1997) Frataxin gene of Freidreich's ataxia is targeted to mitochondria. *Ann Neurol* 42:265–269.

Reiderer, P., Sofic, E., Rausch, W.-D, Schmidt, B., Reynolds, G.P. and Jellinger, K. (1989) Transition metals, ferritin, glutathione, and ascorbic acid in parkinsonian brains. *J Neurochem* 52:515–520.

Reif, D.W. and Simmons, R.D. (1990) Nitric oxide mediates iron release from ferritin. *Arch Biochem Biophys* 283:537–541.

Robbins, T.W. and Everitt, B.J. (1982) Functional studies of the central catecholamines. *Int Rev Neurobiol* 23:303–365.

Roetto, A., Daraio, F., Alberti, F., Porporato, P., Cali, A., De Gobbi, M. and Camaschella, C. (2002) Hemo-chromatosis due to mutations in transferrin receptor 2. *Blood Cells Mol Dis* 29(3):465–470.

Rogers, J.T., Randall, J.D., Cahill, C.M., Eder, P.S., Huang, X., Gunshin, H., Leiter, L., McPhee, J., Sarang, S.S., Utsuki, T., Greig, N.H., Lahiri, D.K., Tanzi, R.E., Bush, A.I., Giordano, T. and Gullans, S.R. (2002) An iron-responsive element type II in the 5'-untranslated region of the Alzheimer's amyloid precursor protein transcript. *J Biol Chem* 277(47):45518–45528.

Roncagliolo, M., Garrido, M., Walter, T., Peirano, P. and Lozoff, B. (1998) Evidence of altered central nervous system development in infants with iron deficiency anemia at 6 mo: delayed maturation of auditory brainstem responses. *Am J Clin Nutr* 68(3):683–690.

Roskams, A.J. and Connor, J.R. (1994) Iron, transferrin, and ferritin in the rat brain during development and aging. *J Neurochem* 63(2):709–716.

Rotig, A., de Lonlay, P., Chretien, D., Foury, F., Koenig, M., Sidi, D., Munnich, A. and Rustin, P. (1997) Aconitase and mitochondrial iron-sulfur protein deficiency in Friedreich ataxia. *Nat Genet* 17(2):215–217.

Rouault, T. and Klausner, R. (1997) Regulation of iron metabolism in eukaryotes. *Curr Top Cell Reg* 35:1–19.

Salis, C., Goedelmann, C.J., Pasquini, J.M., Soto, E.F. and Setton-Avruj, C.P. (2002) HoloTransferrin but not ApoTransferrin prevents Schwann cell de-differentiation in culture. *Dev Neurosci* 24(2/3):214–221.

Sampietro, M., Caputo, L., Casatta, A., Meregalli, M., Pellagatti, A., Tagliabue, J., Annoni, G. and Vergani, C. (2001) The hemochromatosis gene affects the age of onset of sporadic Alzheimer's disease. *Neurobiol Aging* 22(4):563–568.

Sanyal, B., Polak, P.E. and Szuchet, S. (1996) Differential expression of the heavy chain ferritin gene in non-adhered and adhered oligodendrocytes. *J Neurosci Res* 46:187–197.

Schapira, A.H. (1999) Mitochondrial involvement in Parkinson's disease, Huntington's disease, hereditary spastic paraplegia and Friedreich's ataxia. *Biochim Biophys Acta* 1410(2):159–170.

Sindic, C.J.M., Collet-Cassart, D., Cambiaso, C.L., Masson, P.L. and Laterre, E.C. (1981) The clinical relevance of ferritin concentration in the cerebrospinal fluid. *J Neurol Neurosurg Psychiatr* 44:329–333.

Smith, M.A., Wehr, K., Harris, P.L.R., Siedlak, S.L., Connor, J.R. and Perry, G. (1998) Abnormal localization of iron regulatory protein in Alzheimer's disease. *Brain Res* 788(1/2):232–236.

Sofic, E., Paulis, W., Jellinger, K., Riederer, P. and Youdim, M.B. (1991) Selective increase of iron in substantia nigra pars compacta of Parkinsonian brains. *J Neurochem* 56(3):978–982.

Sourkes, T.L. (1982) Transition elements and the nervous system. In *Iron Deficiency: Brain Biochemistry and Behavior* (Pollitt, E. and Leibel, L.R., Eds.), pp. 1–30. Raven Press, New York.

Staedt, J., Stoppe, G., Kogler, A., Riemann, H., Hajak, G., Munz, D.L., Emrich, D. and Rustin, P. (1995) Nocturnal myoclonus syndrome (periodic movements in sleep) related to central dopamine D2-receptor alteration. *Eur Arch Psychiatry Clin Neurosci* 245(1):8–10.

Subramaniam, R., Koppal, T., Green, M., Yatin, S., Jordan, B., Drake, J. and Butterfield, D.A. (1998) The free radical antioxidant vitamin E protects cortical synaptosomal membranes from amyloid beta-peptide(25-35) toxicity but not from hydroxynonenal toxicity: relevance to the free radical hypothesis of Alzheimer's disease. *Neurochem Res* 23(11):1403–1410.

Sun, E.R., Chen, C.A., Ho, G., Earley, C.J. and Allen, R.P. (1998) Iron and the restless legs syndrome. *Sleep* 21:371–377.

Swaiman, K.F. (1991) Hallervorden-Spatz syndrome and brain iron metabolism. *Arch Neurol* 48:1285–1293.

Taylor, T.D., Litt, M., Kramer, P., Pandolfo, M., Angelini, L. Nardocci, N., Davis, S., Pineda, M., Hattori, H., Flett, P.J., Cilio, M.R., Bertini, E. and Hayflick, S.J. (1996) Homozygosity mapping of Hallervorden-Spatz syndrome to chromosome 20p12.3-p13. *Nat Genet* 14(4):479–481.

The Huntington's Disease Collaborative Research Group. (1993) A novel gene containing a trinucleotide repeat that is expanded and unstable on Huntington's disease chromosomes. *Cell* 72(6): 971–983.

Thompson, C.M., Markesbery, W.R., Ehmann, W.D., Mao, Y.-X and Vance, D.E. (1998) Regional brain trace-element studies in Alzheimer's disease. *Neurotoxicology* 9:1–9.

Thompson, K., Menzies, S., Muckenthaler, M., Torti, F.M., Wood, T., Torti, S.V., Hentze, M.W., Beard, J. and Conner, J.R. (2003) Mouse brains deficient in H-ferritin have normal iron concentration but a protein profile of iron deficiency and increased evidence of oxidative stress. *J Neurosci Res* 71(1):46–63.

Thompson, K.J., Fried, M.G., Ye, Z., Boyer, P. and Connor, J.R. (2002) Regulation, mechanisms and proposed function of ferritin translocation to cell nuclei. *J Cell Sci* 115(Pt. 10):2165–2177.

Turjanski, N., Lees, A.J. and Brooks, D.J. (1999) Striatal dopaminergic function in restless legs syndrome: 18F-dopa and 11C-raclopride PET studies. *Neurology* 52(5):932–937.

Valk, J. and van der Knaap, M.S. (1991) White matter disorders. *Curr Opin Neurol Neurosurg* 4(6):843–851.

van Gelder, W., Huijskes-Heins, M.I., Cleton-Soeteman, M.I., van Dijk, J.P. and van Eijk, H.G. (1998) Iron uptake in blood-brain barrier endothelial cells cultured in iron-depleted and iron-enriched media. *J Neurochem* 71(3):1134–1140.

Vulpe, C.D., Kuo, Y.M., Murphy, T.L., Cowley, L., Askwith, C. Libina, N., Gitschier, J. and Anderson, G.J. (1999) Hephaestin, a ceruloplasmin homologue implicated in intestinal iron transport, is defective in the sla mouse. *Nat Genet* 21(2):195–199.

Wagner-Recio, M., Toews, A.D. and Morell, P. (1991) Tellurium blocks cholesterol synthesis by inhibiting squalene metabolism: preferential vulnerability to this metabolic block leads to peripheral nervous system demyelination. *J Neurochem* 57(6):1891–1901.

Waldemeier, P.C., Buchle, A.M. and Steulet, A.F. (1993) Inhibition of catechol-*O*-methyltransferase (COMT) as well as tyrosine and tryptophan hydroxylase by the orally active iron chelator, 1,2-dimethyl-3-hydroxypyridin-4-one (L1, CP20), in rat brain *in vivo*. *Biochem Pharmacol* 45:2417–2424.

Walters, A.S. (1995) Toward a better definition of the restless legs syndrome. The International Restless Legs Syndrome Study Group. *Mov Disord* 10(5):634–642.

Weinberg, J., Bert, L.P., Levine, S. and Dallman, P.R. (1981) Long term effects of early iron deficiency on consummatory behavior in the rat. *Pharmacol Biochem Behav* 14:493–502.

Weinberg, J., Dallman, P.R. and Levine, S. (1980) Iron deficiency during development in the rat: behavioral and physiological consequences. *Pharmacol Biochem Behav* 12:493–502.

Wessling-Resnick, M. (1999) Biochemistry of iron uptake. *Crit Rev Biochem Mol Biol* 34(2):285–314.

Wessling-Resnick, M. (2000) Iron transport. *Annu Rev Nutr* 20:129–151.

Wilson, R.B. and Roof, D.M. (1997) Respiratory deficiency due to loss of mitochondrial DNA in yeast lacking the frataxin homologue. *Nat Genet* 16(4):352–357.

Wood, R.J. and Han, O. (1998) Recently identified molecular aspects of intestinal iron absorption. *J Nutr* 128(11):1841–1844.

Yager, J.Y. and Hartfield, D.S. (2002) Neurologic manifestations of iron deficiency in childhood. *Pediatr Neurol* 27(2):85–92.

Yoshida, K., Furihata, K., Takeda, S., Nakamura, A., Yamamoto, K., Morita, H., Hiyamuta, S., Ikeda, S., Shimizu, N. and Yanagisawa, N. (1995) A mutation in the ceruloplasmin gene is associated with systemic hemosiderosis in humans. *Nat Genet* 9(3):267–272.

Youdim, M.B.H., Green, A.R., Bloomfield, M.R., Mitchell, B.D., Heal, D.J. and Grahame-Smith, D.G. (1980) The effects of iron deficiency on brain biogenic monoamine biochemistry and function in rats. *Neuropharmacology* 19:259–267.

Youdim, M.B.H., Yehuda, S. and Ben-Uriah, Y. (1981) Iron deficiency-induced circadian rhythm reversal of dopaminergic-mediated behaviours and thermoregulation in rats. *Eur J Pharmacol* 74:295–301.

Yu, G.S., Steinkirchner, T.M., Rao, G.A. and Larkin, E.C. (1986) Effect of prenatal iron deficiency on myelination in rat pups. *Am J Pathol* 125:620–624.

Zappone, E., Bellotti, V., Cazzola, M., Ceroni, M., Meloni, F., Pedrazzoli, P. and Perfetti, V. (1986) Cerebrospinal fluid ferritin in human disease. *Haematologica* 71(2):103–107.

Zheng, L., Kennedy, M.C., Blondin, G.A., Beinert, H. and Zalkin, H. (1992) Binding of cytosolic aconitase to the iron responsive element of porcine mitochondrial aconitase mRNA. *Arch Biochem Biophys* 299:356–360.

Zhou, B., Westaway, S.K., Levinson, B., Johnson, M.A., Gitschier, J. and Hayflick, S.J. (2001) A novel pantothenate kinase gene (PANK2) is defective in Hallervorden-Spatz syndrome. *Nat Genet* 28(4):345–349.

15 Iodine and Brain Function

John H. Lazarus

CONTENTS

15.1 INTRODUCTION

The element iodine, a member of the seventh column of the periodic group of elements, was discovered in 1811 by Courtois. It was not recognized to be relevant to thyroid physiology until 1895, when its presence in the thyroid gland was first recognized. Marine in 1915 found that thyroid tissue from dogs who were fed iodine contained a large amount of the element, especially if the dog was goitrous. During the 20th century, following the introduction of radioisotopes and techniques of chemical analysis, the physiological role of iodine was extensively investigated. This chapter reviews the thyroidal and extrathyroidal physiology of iodine before discussing its critical role in nutrition and particularly in regard to the maturation of the developing brain.

15.2 NATURE AND GEOLOGY OF IODINE

15.2.1 NATURE

Iodine is an essential element for normal growth and development in animals and humans. The healthy human contains 15–20 mg of iodine, of which 70–80% is in the thyroid gland. The normal daily requirement for dietary intake is 100–150 µg of iodine, but this requirement is increased in pregnancy to 200 µg/day. Following dietary ingestion, iodine is absorbed mainly in the jejunum and circulates in the plasma as inorganic iodine. The thyroid gland may be regarded as a factory utilizing iodine in synthesizing thyroid hormones (www.Thyroidmanager.org; Figure 15.1). Iodide is actively concentrated by the thyroid 20 to 40 times compared to the plasma concentration. The mechanism of the concentrating process (sometimes known as the iodide trap) is through an iodide symporter situated on the basolateral membrane of the follicular cell. The symporter gene was cloned in 1996, but more recently other iodide transporters have been described in the apical follicular membrane, which transport the anion into the follicular lumen, thus making it available for incorporation into tetraiodothyronine, that is, thyroxine (T4) (Spitzweg and Morris, 2002). This process occurs on thyroglobulin, a 660-kDa protein situated in the thyroid follicular lumen, whose structure may be adversely affected by alterations in iodine status. Once synthesized, T4 can enter the follicular cell, thereby reaching the peripheral circulation when required. T4 is essentially a prohormone that is peripherally converted to triiodothyronine (T3) by deiodinase enzymes. This deiodination is noted in many tissues, such as heart, kidney, liver and, importantly, the brain (Leonard and Koehrle, 2000). As discussed later, T4 is critically important for fetal central nervous system development and maturation and there is strong evidence for placental transfer of maternal T4 into the fetus.

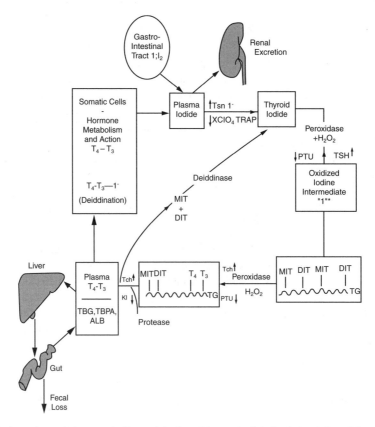

FIGURE 15.1 Overview of the metabolism of iodine. Plasma iodide is derived from absorption from the gastrointestinal tract and concentrated by the thyroid gland (thyroid iodide). It is then incorporated into thyroid hormones, which circulate in the plasma. Note the role of deiodinase in maintaining thyroid iodide. Plasma thyroid hormones are metabolized by the liver and excreted in the feces whereas iodide is excreted in the urine. (From Rousset, BA, Dunn, JT In *Thyroid Hormone Synthesis and Secretion*. Thyroid Disease Manager. With permission.)

15.2.2 GEOLOGY

Most iodine is found in the oceans. It is believed that large amounts of the element have been leached from the surface soil by glaciation, snow, and rain resulting in carriage to the sea by rivers, floods, and wind. Older soils and those at high altitude are more likely to be iodine deplete. Iodine occurs in the soil and sea as iodide. Iodide ions are oxidized to the volatile elemental iodine, resulting in evaporation from the sea to the air. However, airborne iodine is subsequently returned to the soil, although in lesser quantities, thereby leading to iodine deficiency in the soil. This traditional view of the iodine cycle has been challenged by recent geochemical studies, which broadly indicate the presence of complex molecules in seawater that bind iodine. There are also inconsistencies in iodine concentrations between iodine-deficient populations and the soil (sampled by soil bores) on which they live.

The 1993 World Health Organization (WHO) report on iodine deficiency disorders (IDDs) noted that "soil and inland bodies of water may become deficient of iodine due to the leaching effects of glaciation, snow, high rainfall and floods" (World Health Organization, Micronutrient Deficiency Information System, 1993). The accuracy of some of these assertions has been questioned by environmental scientists who have indicated that environmental levels have not been measured (Maberly et al., 1981; Fuge and Johnson, 1986). For example, IDDs have few or no borders in common with glaciation, as determined from a comprehensive global survey (Kelly and Snedden, 1958).

The concentration and organification of iodine by algae occur in localized hot spots around the world's oceans (Moore and Tokarczyk, 1992). These compounds (CH_3I, C_2H_5I, CH_2ICl, CH_2IBr, CH_2I_2, C_3H_7I, and CH_3CHICH_3) are released into the atmosphere, possibly by volatilization but more likely under biological control. There is seasonal variation of atmospheric iodine, in association with the growth of marine algae. Iodine has been found in the atmosphere wherever it has been sought. This indicates that, although concentration decreases above the mean boundary layer (the line of separation above the surface-influenced atmosphere), iodine is transported easily around the globe. Once in the atmosphere, organic iodine in the gaseous phase interacts with ozone and other compounds in a complex of reactions. Iodine is more reactive than chlorine in ozone degradation. These reactions determine the transit time in the atmosphere of iodine, because different compounds have different weights and, accordingly, different depositional velocities due to gravity. Furthermore, it is likely that wet deposition (in rain and snow) and not dry deposition by gravity controls the amount of iodine delivered to the surface of the earth. There is a direct relationship between rainfall and depositional amounts of iodine (Truesdale and Jones, 1996). First rain contains more iodine than later rain in the same shower, indicating some form of cleansing from the atmosphere of iodine in gas and dust.

Atmospheric input of iodine to the soil is more important than any input resulting from the degradation of the underlying rock, although it is not clear how the presumed initial high soil iodine came about or why it has changed. Iodine distribution in soils is governed both by the supply of iodine and the ability of the soil to retain it (Fuge and Johnson, 1986). Leaching of soil iodine is also unlikely as it is remarkably resistant to removal by hot or cold water. The relationship between soil and plant concentrations of iodine is not a straightforward one (Whitehead, 1979). Plants reject the large iodine ion by methylation and release into the local air, where it is moved and redeposited. Fuge (1996) suggests that this is the main mechanism for transporting iodine long distances inland despite evidence of iodine even in polar air. It is probable that if environmental iodine is related to the incidence of IDDs, then there would be a relationship between the distribution of IDDs and the atmospheric deposition and distribution across the earth's surface of iodine.

15.3 IODINE DEFICIENCY

15.3.1 THYROIDAL ADAPTATION TO IODINE DEFICIENCY

The thyroidal response to iodine deficiency involves adjustment of all the physiological processes of thyroid hormone production to maximize iodine use (Ingbar, 1985). The thyroid enlarges in response to iodine deficiency and large goiters may occur. The iodide-concentrating mechanism is stimulated, resulting in an increased thyroidal iodine uptake. Although thyroid stimulating hormone (TSH) is primarily responsible for thyroidal iodine uptake, its plasma concentration is often not elevated, suggesting a degree of thyroid autoregulation in this situation. There is a significantly increased thyroidal production of T3 as opposed to T4, as the former requires less iodine and is the biologically active circulating thyroid hormone in terms of action in peripheral tissues. In addition, very little iodine is stored in thyroglobulin and there is an increased iodine turnover.

15.3.2 IODINE AND THE DEVELOPING NERVOUS SYSTEM

During the early 20th century, studies of endemic cretinism had suggested that the fetal developing thyroid was dependent on factors that may impair the maternal thyroid reserve (Hetzel, 1994). Subsequently, a cause–effect relationship was shown between maternal iodine deficiency and the birth of neurological cretins. Furthermore, there was evidence that the degree of maternal hypothyroxinemia correlated with the CNS damage of the progeny. These data were difficult to explain at the time because it was not thought that transplacental thyroid hormone transport took place. The view was that the placenta was impermeable to iodothyronines and that small amounts possibly transferred would be of no physiological importance. It has now been shown convincingly that this does occur not only before the fetal thyroid starts to synthesize thyroid hormones (i.e., up to 12 weeks gestation) but right through pregnancy (Vulsma et al., 1989).

15.3.2.1 Animal Data Relating Thyroid Hormone to Brain Development

Table 15.1 lists the availability of thyroid hormone to the fetal brain. Although fetal thyroid function does not commence until the equivalent of 12 weeks gestation in humans, the presence of functional fetal nuclear receptors for T3 are noted in early pregnancy, indicating that triiodothyronine exerts an action at this time (De Nayer and Dozin, 1989). Maternal thyroid hormone is necessary before the onset of fetal thyroid function, as shown by interference of cortical cell migration and the cortical expression of several genes in the fetuses of mothers rendered hypothyroid by goitrogens.

TABLE 15.1
Physiology of Thyroid Hormone Availability to Fetal Brain

Before Onset of Fetal Thyroid Function:
- T4 and T3 are present in embryonic and fetal fluids and tissues.
- T4 and T3 are of maternal origin.
- Nuclear receptors are present and occupied by T3.
- D2 and D3 are expressed in the brain.

Between Onset of Fetal Thyroid Function and Birth:
- Maternal transfer continues.
- Brain T3 is dependent on T4 and D2 and 3 not on systemic T3.
- Normal maternal T4 protects fetal brain from T3 deficiency.
- Normal T3 in low T4 mother does not prevent cerebral T3 deficiency.

After fetal thyroid function is present, maternal thyroxine still contributes to the thyroid hormone available to the fetal tissues at term. Maternal T4 is sufficient to prevent fetal cerebral T3 deficiency until birth in a hypothyroid fetus. The T3 at this stage is locally produced in cerebral structures by deiodination of T4 by type 2 iodothyronine deiodinase — hence the requirement for T4. Both type 2 (D2) and type 3 (D3) deiodinases are critical in producing and modulating the supply of T4 to the fetus as well as producing locally derived T3. The type 3 enzyme is mainly placentally located and inactivates T4 and T3, thereby regulating the influx of T4 to the fetus.

Corroborative data have been obtained by extensive studies of iodine deficiency in sheep (Hetzel and Mano, 1994). In these animals, induced iodine deficiency results in reduced brain weight, reduction in brain DNA, and retarded myelination. In keeping with earlier observations, morphological changes in the cerebellum were noted accompanied by delayed maturation (Hetzel et al., 1989).

15.3.2.2 Human Data Relating Thyroid Hormone to Brain Development

The situation in humans is similar to that observed in animals, although placentation is different in some anatomical respects. Nuclear receptors have been demonstrated in brains of 10-week-old fetuses; low amounts of T4 have been found in coelomic fluid, and this provides enough for transport into the fetal nervous system. Table 15.1 gives an overview of thyroid hormone supply to the fetal brain.

15.3.3 EPIDEMIOLOGY OF IODINE DEFICIENCY

The epidemiological demonstration of the association of cretinism with goiter has been known for ca. 150 years and goiter was also found in ca. 30% of cases. In a seminal study in New Guinea (Pharoah and Connolly, 1994), it was discovered that cretinism could be prevented by iodine administration but only if given (as iodized oil in this case) before the onset of pregnancy. The reduction in cretinism in Europe in the early 20th century was due to the increase in iodine intake from a variety of foodstuffs as well as specific supplementation programs.

Endemic goiter and cretinism have occurred widely in the world (Kelly and Snedden, 1958). Table 15.2 lists some of the most notable areas. These areas have been intensively studied and treatment regimes with iodine supplementation have been carried out by several different methods.

An example of the complexity of the iodine deficiency situation is that which has pertained in Europe during the 20th century and even into the 21st century. In fact, iodine deficiency was under control in only five countries (Austria, Switzerland, Finland, Norway, and Sweden). This was due to the introduction of iodized salt in varying concentrations and also to milk consumption. The situation in other European countries is variable, being characterized by discrete areas of significant iodine deficiency in some countries (e.g., Spain and Italy) and other areas being iodine sufficient. In some countries, mild iodine deficiency persists and this is evidenced by studies of thyroid physiology in pregnancy (e.g., Belgium).

TABLE 15.2
Examples of Iodine-Deficient Countries

Continent	Countries/Region
Latin America	Peru, Bolivia, Chile
Africa	Zaire, Mali, Algeria, Nigeria
Asia	India, Pakistan, Southeast Asia
Europe	Switzerland, Spain, Poland

TABLE 15.3
Recommended Daily Iodine Intake

Group	Intake (μg)
Infants (first 12 months)	50
Children (1–6 years)	90
Schoolchildren (7–12 years)	120
Adults (more than 12 years)	150
Pregnant and lactating women	200

The WHO, United Nations Children's Fund, and the International Council of the Control of Iodine Deficiency Disorders (ICCIDD) have published recommended iodine levels in salt and guidelines for monitoring their adequacy and effectiveness. Table 15.3 gives the recommended daily iodine intakes for different groups.

The generally accepted method of supplying iodine to large populations is by consumption of appropriately iodized salt (Delange and Burgi, 1989). Iodine lost from salt is 20% from production site to household and another 20% during cooking before consumption. Assuming an average daily salt intake of 10 g, iodine concentration in salt at the point of production should be 20–40 mg of iodine (i.e., 34–66 mg potassium iodate) per kilogram of salt.

15.3.4 Assessment of Iodine Deficiency

Ideally, an accurate measure of iodine status should be obtained by estimating iodine in all foodstuffs ingested during a given time period. This has proved impractical for population studies, as has estimation by dietary questionnaire, although some useful data have been generated by the latter painstaking method.

As it is very difficult or impossible to obtain data on iodine intake in most populations, iodine nutrition is classically evaluated by indirect indices. The urinary excretion of iodine can be readily measured on a random sample of urine obtained at any time of the day. It would be more accurate to determine the iodine concentration in a 24-h urine collection, but this is usually impractical, especially in developing countries. Casual urinary iodine estimations should be determined in at least 50 to 100 persons to allow for statistical variations due to dilution. There has been much debate as to whether the concentration of urinary iodine should be expressed as an iodine:creatinine ratio or just as micrograms of iodine per unit volume. The excretion of creatinine is usually very constant, although in areas of low protein nutrition it may vary considerably within a population and be significantly lower than that observed in a population with normal protein calorie nutrition. As endemic goiter generally occurs in developing countries and mainly in areas with a low socioeconomic level, a relative protein deficiency might be present without clinical signs of severe protein calorie malnutrition. If this is the case, then iodine:creatinine ratios may not be representative of the true iodine excretion (Knudsen et al., 2000).

Advances in ultrasound technology have enabled thyroid volumetric ultrasound to be used as a reliable and valid index of endemic goiter and inferentially of iodine deficiency. The relevance of thyroid volume measurements by thyroid ultrasound for assessing iodine deficiency has been established by comparing thyroid volumes in schoolchildren from iodine-deficient and iodine-sufficient areas (Vitti et al., 1994). Thyroid volume determination by ultrasound in children provides more reliable quantitative and reproducible data than palpation, particularly in mild goiter endemia (Vitti and Rago, 2002). Thus, thyroid volume measurement is a marker for iodine deficiency and may be used to assess the effect of iodination programs.

Other thyroid parameters may also reflect iodine status in an indirect way. Thyroidal radioiodine uptake is significantly increased in iodine deficiency. The practicalities of this procedure now make

TABLE 15.4
Health Indicators of Iodine Status

Variable	Normal	Mild IDDs	Moderate IDDs	Severe IDDs
% Goiter in SAC	<5	5–19.9	20–29.9	>30
% Thy vol in SAC >97th centile	<5	5–19.9	20–29.9	>30
Median urinary I in SAC and adults (µg/l)	100–200	50–99	20–49	<20
% Neonatal TSH >5 mU/l	<3	3–19.9	20–39.9	>40

Note: SAC, school-age children; IDD, iodine deficiency disorders.

Source: Adapted from WHO-UNICEF-ICCIDD (1994). Indicators for assessing iodine deficiency disorders and their control through salt iodization. WHO/NUT/94.6. Geneva: World Health Organization. With permission.

it less suitable for field studies, and results may be difficult to compare with others due to variations in methodology. Serum thyroglobulin concentration has been shown to be elevated in endemic goiter due to thyroidal stimulation consequent to iodine deficiency. In practice, thyroglobulin assays are expensive and subject to error if thyroglobulin antibodies are present. More recently, it has been shown that prevalence of iodine deficiency may be estimated from the incidence of congenital hypothyroidism as determined by newborn screening of TSH and T4. In addition, the finding that a definite shift of TSH and T4 levels in cord blood toward high and low values, respectively, indicates that newborns are particularly sensitive to iodine deficiency. As this group is a target population in iodine-deficient areas, this method of screening is the most sensitive index of iodine deficiency. Unfortunately, use of this strategy had not been as widespread as expected.

The severity of iodine status in a population can be estimated by reference to Table 15.4, which documents three different stages of iodine deficiency.

15.3.5 Impact of Iodine Deficiency on Development

15.3.5.1 Animal Studies

The studies on animals complement data obtained in humans. In particular, animal observations allow studies to be made on regional areas of the brain as well as the whole organ under controlled experimental conditions. Clearly, these kinds of data cannot be sought in humans. Three animal models have been extensively studied in relation to iodine deficiency: the sheep, the marmoset, and the rat.

Iodine deficiency was known to be a problem with sheep grazing in many parts of Australia. Hetzel and others (Hetzel and Mano, 1994) took advantage of this and conducted many experiments to examine in detail the effects of iodine deficiency on lamb development in general and the brain in particular. The brains from fetuses removed at various stages of pregnancy showed a lowered brain weight associated with fewer brain cells. Further observations showed a denser brain than normal because of failure of arborization of axons and dendrites in the cerebellum and also in the cerebral hemisphere (Hetzel et al., 1989). In addition, there was evidence of retarded myelination in the cerebral hemispheres and the brain stem. Experiments involving iodine-deficient ewes and surgical thyroidectomy at varying stages of gestation suggested that normal brain development required the availability of both maternal and fetal thyroid hormones, a most important physiological finding not previously recognized. Similar data relating to brain weight were obtained in fetuses of iodine-deficient marmosets and histological changes were very evident in the cerebellum.

The biochemical mechanisms of thyroid control of brain development are the subject of continuing investigations, but advances in the understanding of thyroid hormone action at the cellular and subcellular level have helped provide an overall view (Bernal, 2002). Thyroid hormone receptors are encoded by two genes, TRα and TRβ, located in different chromosomes (17 and 3, respectively). In the rat fetus, T3 receptor in the brain can be detected at low concentrations several days before the onset of thyroid gland function. The presence of receptors in the neural tissue at these early stages of development suggests that the fetal brain is a target for thyroid hormone even before the onset of fetal thyroid gland function.

Thyroid hormones reach the brain mainly from the blood through the capillaries, although a significant quantity is also transported through the choroid plexus. The regional cerebral distribution of thyroid hormones is affected by the deiodinase enzymes, and the type 3 enzyme in particular is rate limiting in the placenta.

Thyroid hormone regulates the expression of many genes in the brain, as evidenced by studies on hypothyroid neonatal rats. These genes may be primary targets of thyroid hormone or contain thyroid hormone response elements. Most genes are thyroid dependent only during a critical period in development, although a few genes are thyroid dependent in adult life. During brain maturation, thyroid hormone influences the expression of most myelin genes that have been studied, including those encoding myelin basic protein and other myelin-associated proteins. Thyroid hormone influences the expression of nuclear-encoded and mitochondrial-encoded mitochondrial RNAs in the brain. The hormone also controls the expression of nerve growth receptors, which ensures cooperation between neurotrophic factors in the growth and maintenance of different cell types in different brain regions (e.g., cholinergic neurons in the forebrain and of factors controlling Purkinje cell differentiation). Thyroid hormone controls the expression and cell distribution of cytoskeletal components such as specific tubulin isotypes and microtubule-associated proteins. Other developmental genes whose expressions are known to be controlled by thyroid hormone include transcription factors, splicing regulators, and proteins involving intracellular signaling and adhesion molecules. Detailed attention has been paid to thyroid control of cerebellar genes in view of the clear histological changes observed in that region in thyroid deficiency. There are many such genes, the details of which are not given here. It is therefore apparent that thyroid hormone is a critical factor in brain development, but must act at specific times during gestation in precise concentrations that have a regional variation.

Extensive studies of the effect of thyroid hormone on the developing brain have been conducted by Spanish investigators, using a rat model (Morreale de Escobar et al., 1994). Their conclusions are a valuable contribution to the problem and are reproduced in relation to rats and humans. Research has confirmed that maternal thyroid hormone reaches the early developing fetus before the onset of fetal thyroid function. Transfer of maternal thyroid hormone to the fetus continues after fetal thyroid function commences and constitutes a significant fraction of the hormonal requirements of the fetus throughout pregnancy. Maternal thyroid failure during early pregnancy may result in irreversible effects on fetal development, but the fetus is able, through its own thyroid hormone production, to compensate in later gestation for the maternal lack of hormone. The developing brain is highly dependent on T4, because it is dependent on production of T3 by intracellular deiodination. In iodine deficiency during pregnancy, maternal T4 is low and the fetus is unable to compensate for the hormone deficiency and is severely deficient in brain triiodothyronine. The human phenotypic examples of iodine deficiency are characterized at one extreme by the neurologically severely damaged infants of mothers who have had low levels of T4. Cretins with lesser or no obvious neurological damage are associated with only modest reduction in maternal T4. The group from Madrid has synthesized the animal data pictorially (Figure 15.2) to explain the temporal importance of maternally derived T4 to the fetus in the context of iodine deficiency, the different types of cretinism, and the brain structures affected.

15.3.5.2 Human Studies

Although iodine deficiency had long been suspected as the cause of cretinism in areas of endemic goiter, it was the demonstration of the prevention of cretinism in a double-blind controlled trial

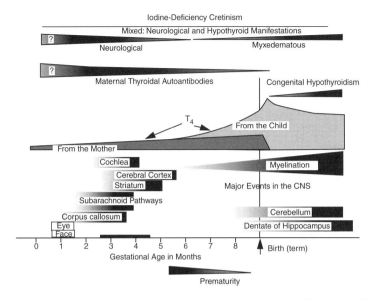

FIGURE 15.2 Thyroxine delivery to the developing fetus and neonate. The middle graph shows the approximate proportion of fetal T4 derived from the mother and from fetal thyroid synthesis. The top lines indicate the role of hormone delivery in relation to different types of cretinism. The brain areas affected are shown at the bottom at their developmental stages. A possible role for thyroid antibodies in neurodevelopment is also shown. (From Morreale de Escobar G et al. (2000) *J. Clin. Endocrinol. Metab.,* 85, 3975–3987. With permission.)

with iodized oil in Papua New Guinea (Pharoah and Connolly, 1987) that established the causal role of iodine deficiency in cretinism by an effect on the developing fetus. In this situation, cretinism could not be prevented unless the iodized oil was given before pregnancy.

Cretinism is associated with endemic goiter and iodine deficiency. Clinically, cretins are mentally deficient (IQ ca. 40) and may have a predominant neurological syndrome, comprising hearing and speech defects, and disorders of stance and gait, or they may have predominant hypothyroidism with impaired growth or a mixture of the two (Delange, 1986). The clinical picture of cretinism has been appreciated for hundreds of years, but the realization that minor degrees of iodine deficiency could be associated with a spectrum of disorders has only been documented during the past 20 to 30 years. Hetzel (1983) suggested the term *iodine deficiency disorders* to replace the term *goiter*. The clinical manifestations of IDDs range from mild mental impairment all the way to myxedematous cretinism and neurological cretinism. The full spectrum of IDDs is shown in Table 15.5.

Two clinical syndromes of cretinism have been observed in many parts of the world. The neurological type is characterized by mental retardation, deaf mutism, cerebral diplegia, no physical signs of hypothyroidism, and with no clinical response to thyroid hormone therapy. In contrast, hypothyroid cretinism shows features of severe growth retardation (growth is normal in neurological cretinism), coarse dry skin, but no deaf mutism or cerebral diplegia. There may be a clinical response to thyroid hormone therapy (Boyages, 2002).

Results of epidemiological field studies have shown that apart from iodine deficiency, other factors may impact on the pathogenesis of endemic cretinism. In parts of Africa (e.g., Zaire) the consumption of cassava (manioc) has been shown to result in an increase in serum thiocyanate concentration, which itself is goitrogenic and enhances the effect of iodine deficiency (Bourdoux et al., 1978). The element selenium is a vital component of the thyroid deiodinase enzyme, and selenium deficiency plays an important role in some areas of iodine deficiency.

TABLE 15.5
Spectrum of Iodine Deficiency Disorders

Fetus	Abortions
	Stillbirths
	Congenital anomalies
	Increased perinatal mortality
	Increased infant mortality
	Cretinism, neurological, myxedematous
Neonate	Goiter
	Hypothyroidism
Child/Adolescent/Adult	Goiter
	Hypothyroidism
	Impaired mental function
	Iodine induced hyperthyroidism
	(following I administration)

Knowledge of the development of the fetal brain makes it possible to correlate the developmental defects in neurological cretinism with specific brain structures. The major defects in neurological cretinism relate to intellectual deficiency, deafness, and motor rigidity, which correlate to the cerebral neocortex, the cochlea, and the basal ganglia, respectively, being involved (Morreale de Escobar et al., 2000). Experiments using knockout mice lacking specific T3 receptors have shown cochlear defects, suggesting that thyroid hormone is essential for the developing auditory apparatus *in utero*.

Although it is now accepted that iodine deficiency, acting alone or with other nutritional factors, may impair brain development, which results in a wide range of clinical effects, it should not be forgotten that this is due to lack of adequate thyroid hormone concentrations in the developing fetal brain. During the past 20 to 30 years, it has become apparent that fetal T4 deficiency may occur in the presence of iodine sufficiency. Thus, maternal thyroid dysfunction (hypothyroidism or subclinical hypothyroidism) during pregnancy results in neurointellectual impairment of the child. Two studies (Man et al., 1991; Pop et al., 1999) have shown that low thyroid hormone concentrations in early gestation can be associated with significant decrements of IQ of the children when tested at 7 years and 10 months, respectively. Pop et al. (1995) have also shown a significant decrement in IQ in children aged 5 years whose mothers were known to have circulating anti-TPO antibodies at 32 weeks gestation and were biochemically euthyroid. Moreover, as shown by Haddow et al. (1999), the 7-year-old children of women known to be hypothyroid during gestation showed impaired psychological development compared to children of the same age from carefully matched control mothers whose thyroid function was known to be normal during pregnancy. In this report, it is interesting that a subgroup of children whose mothers had been receiving thyroxine for hypothyroidism during pregnancy (albeit at inadequate doses as evidenced by high TSH during pregnancy) also showed some impairment of psychological performance, although not as much as the other children. The neurodevelopmental impairment is similar to that seen in iodine-deficient areas and implies that iodine status should be normalized in regions of deficiency. However, much of the U.S. and parts of Europe are not iodine deficient, which raises the question of routine screening of thyroid function during early pregnancy or even at preconception. It will now be realized that similar mental impairment characteristics can be seen in iodine-deficient areas as well as iodine-sufficient ones, thus providing further evidence for the role of thyroid hormone in fetal neurodevelopment.

15.4   IODINE EXCESS

Chronic excessive iodine intake may result from ingestion of large amounts of seaweed (seen in Japan), kelp, marine fish, and iodine-containing supplements. In some areas, water may contain large amounts of iodine. An important cause of excess iodine is the consumption of salt that is adulterated with abnormally large quantities of the element. In the clinical setting, several drugs such as amiodarone, expectorant preparations, and radiodinated contrast agents have led to acute iodine excess.

The thyroidal response to iodine excess includes the development of goiter, hypothyroidism, or hyperthyroidism with or without goiter (Roti and Braverman, 1996; Dremier et al., 1996). The particular response depends on the preexisting iodine status as well as thyroid status before iodine exposure. One of the earliest and best documented outbreaks of iodine-induced hyperthyroidism was seen in Tasmania as a result of iodination of bread and the use of iodophors by the milk industry. More recently, the iodine excess problem was highlighted by an outbreak of iodide-induced hyperthyroidism in Zimbabwe (Todd et al., 1995). Here, the iodine intake had increased 100% since 1991 and by 1994 a sudden increase in cases of thyrotoxicosis was noted such that the incidence increased threefold from 1991 to 1995. A multicenter study conducted in seven African countries found that acute iodine overload was due to insufficient monitoring in the salt industry (WHO, 1997). As some deaths occurred in Zimbabwe, where treatment facilities were deficient, the experience highlighted the importance of monitoring salt iodization processes and supplementation programs.

During pregnancy, the daily iodine requirement increases to 200 µg/day. If excess iodine occurs, the fetal thyroid may enlarge and hypothyroidism may develop in the mother. Hypothyroid infants have been observed as a result of excess topical application of iodine in pregnancy with a presumed risk to neonatal brain function.

From the foregoing discussion, it is apparent that IDDs, including subclinical hypothyroidism, goiter, and brain damage, are prevented only when the iodine supply to pregnant and lactating women is ca. 200 µg/day (Zoeller et al., 2002). However, WHO, UNICEF, and the ICCIDD recognize that when populations that have previously been chronically iodine deficient increase their iodine intake to normal or above normal, there is a risk that a small number of individuals may suffer from iodine-induced hyperthyroidism. This applies particularly to people with long-standing nodular and potentially autonomous goiters, usually more than 40 years old. As indicated, this problem must be taken very seriously by those responsible for micronutrient fortification strategies.

15.5   IODINE AND PUBLIC HEALTH

During 1980 to 1990, it was realized that the effects of iodine deficiency were apparent at the individual, social, and national levels (Dunn, 1994). The term *iodine deficiency disorders* became used to indicate the various effects of iodine deficiency on growth and development. It was also apparent that this condition was serious enough to threaten the social and economic development of many developing countries as well as still being observed to a lesser degree in areas such as Europe. To help address this problem, ICCIDD was formed in 1986. This nongovernmental organization set out to collaborate with WHO and UNICEF to develop national IDD control programs in countries with significant IDD problems. A major function of ICCIDD was and still is to communicate the IDD message and awareness to national governments, decision makers, international agencies, as well as all relevant health professionals and planners. ICCIDD has operated by organizing meetings and contributing to seminars and workshops in all continents. A quarterly newsletter disseminates information about progress in improving iodine status in particular countries and reviews relevant abstracts from the world scientific literature. More recently, a Web site (www.iccidd.org) has been set up that gives details about ICCIDD and current developments in the control of iodine deficiency. Membership of ICCIDD is all inclusive and not just confined to medical personnel. In addition to endocrinologists and public health workers, members include salt producers, management specialists, communicators, laboratory analysts, and researchers. Universal salt iodization, according to which all salt used in agriculture, food processing, catering, and households should be iodized, is the agreed strategy for achieving iodine sufficiency. During the past 20 years or so, great progress has been made toward eliminating worldwide IDDs (Hetzel 2002). However, IDDs are still a major public health problem worldwide and WHO estimates that 740 million people are at present affected by goiter. Nevertheless, the mobilization of WHO, UNICEF, and ICCIDD has resulted in remarkable progress in IDD control, especially in Africa and in Southeast Asia, where the endemia is the most severe. It is estimated that 68% of the populations of affected countries currently have access to iodized salt compared to less than 10% a decade ago (Delange et al., 2001). Unfortunately, of the 130 affected countries, ca. 30 have no iodization program. In addition, salt quality control and monitoring of population iodine status are still weak in many countries, thus exposing the population to an excessive iodine intake and subsequently to the risk of iodine-induced hyperthyroidism. Furthermore, IDDs are reemerging in some countries, especially in Eastern Europe. To reach the goal of IDD elimination, it is important to insist on the sustainability of salt iodization programs, which implies an increased commitment of both health authorities and representatives of the salt industry. It is critical that managers and sponsors of programs of iodized salt appreciate the continuing need for greatly improved monitoring and quality control. A recent evaluation of iodine nutrition in ca. 35,000 schoolchildren has shown that many previously iodine-deficient parts of the world now have median urinary iodine concentrations well above 300 µg/l, which is excessive and carries the risk of adverse health consequences (Delange et al., 2002). The elimination of iodine deficiency is within reach, but major additional efforts are required to cover the whole population at risk and to ensure quality control and sustainability. The progress made to reduce IDDs worldwide is a great achievement, a public health success unprecedented in the field of noncommunicable diseases (Benoist and Delange, 2002).

REFERENCES

Benoist BD, Delange F (2002) Iodine deficiency: current situation and future prospects. *Sante,* 12, 9–17.
Bernal J (2002) Action of thyroid hormone in brain. *J. Endocrinol. Invest.,* 25, 268–288.
Bourdoux P, Delange F, Gerard M, Mafuta M, Hanson A, Ermans AH (1978) Evidence that cassava ingestion increases thiocyanate formation: a possible etiologic factor in endemic goiter. *J. Clin. Endocrinol. Metab.,* 46, 613–621.
Boyages S (2002) Iodine deficiency disorders. In *Oxford Textbook of Endocrinology and Diabetes* (JAH Wass, SM Shalet, Eds.), pp. 372–380. Oxford University Press, Oxford.
Delange FM (1986) Anomalies in physical and intellectual development associated with severe endemic goiter. In *Towards the Eradication of Endemic Goiter, Cretinism and Iodine Deficiency* (JT Dunn, EA Pretell, CH Daza, FE Viteri, Eds.), pp. 49-65. Pan American Health Organization, Washington, D.C.
Delange F, Burgi H (1989) Iodine deficiency disorders in Europe. *Bull. World Health Org.,* 67, 317–325.
Delange F, de Benoist B, Burgi H (2002) Determining median circulary iodine concentrations that indicates adequate iodine intake at population level. *Bull. World Health Org.,* 80, 663–666.
Delange F, de Benoist B, Pretell E, Dunn JT (2001) Iodine deficiency in the world: where do we stand at the turn of the century? *Thyroid,* 11, 437–447.
De Nayer P, Dozin B (1989) Thyroid hormone receptors in the developing brain. In *Iodine and the Brain* (GR Delong, J Robbins, PD Condliffe, Eds.), pp. 51–58. Plenum Press, New York.
Dremier S, Coppee F, Delange F, Vassart G, Dumont JE, Van Sande J (1996) Thyroid autonomy: mechanism and clinical effects. *J. Clin. Endocrinol. Metab.,* 81, 4187–4193.
Dunn JT (1994) Societal implications of iodine deficiency and the value of its prevention. In *The Damaged Brain of Iodine Deficiency* (JB Stanbury, Ed.), pp. 309–314. Cognizant Communication Corporation, New York.

Fuge R (1996) Geochemistry of iodine in relation to iodine deficiency diseases. In *Environmental Geochemistry and Health* (JD Appleton, R Fuge, GJH McCall, Eds.), Sp. Pub. 113, pp. 201–211. The Geological Society, London.

Fuge R, Johnson CC (1986) The geochemistry of iodine — a review. *Environ. Geochem. Health,* 8, 31–54.

Haddow JE, Palomaki GE, Allan WC, Williams JR, Knight GJ, Gagnon MA, O'Heir CE, Mitchell ML, Hermos RJ, Waisbren SE, Faix JD, Mein RZ (1999) Maternal thyroid deficiency during pregnancy and subsequent neuropsychological development of the child. *N. Engl. J. Med.,* 341, 549–555.

Hetzel BS (1983) Iodine deficiency disorders (IDD) and their eradication. *Lancet,* 2, 1121.

Hetzel BS (1989) The evidence from Papua New Guinea. In *The Story of Iodine Deficiency* (BS Hetzel, Ed.), pp. 52–69. Oxford University Press, New York.

Hetzel BS (1994) Historical development of the concepts of the brain–thyroid relationships. In *The Damaged Brain of Iodine Deficiency* (JB Stanbury, Ed.), pp. 1–7. Cognizant Communication Corporation, New York.

Hetzel BS (2002) Eliminating iodine deficiency disorders: the role of the International Council in the global partnership. *Bull. World Health Org.,* 80, 410–412.

Hetzel BS, Mano ML (1994) Hormone nurturing of the developing brain: sheep and marmoset models. In *The Damaged Brain of Iodine Deficiency* (JB Stanbury, Ed.), pp. 123–130. Cognizant Communication Corporation, New York.

Hetzel, BS, Mano ML, Chevadev J (1989) The fetus and iodine deficiency: marmoset and sheep models of iodine deficiency. In *Iodine and the Brain* (GR Delong, J Robbins, PG Condliffe, Eds.), pp.177–186. Plenum Press, New York.

Ingbar SH (1985) Autoregulation of the thyroid: the effects of thyroid iodine enrichment and depletion. In *Thyroid Disorders Associated with Iodine Deficiency and Excess* (R Hall, J Kobberling, Eds.), Vol. 22, pp. 153–161. Raven Press, New York.

Kelly FC, Snedden WW (1958) Prevalence and geographical distribution of endemic goitre. *Bull. World Health Org.,* 18, 5–173.

Knudsen N, Christiansen E, Brandt-Christensen M, Nygaard B, Perrild H (2000) Age and sex-adjusted iodine/creatinine ratio — a new standard in epidemiological surveys? Evaluation of three different estimates of iodine excretion based on casual urine samples and comparison to 24h values. *Eur. J. Clin. Nutr.,* 54, 361–363.

Leonard JL, Koehrle J (2000) Intracellular pathways of iodothyronine. In *The Thyroid* (LE Braverman, RD Utiger, Eds.), 8th ed., pp. 136–173. Lippincott, Philadelphia, PA.

Maberly GF, Eastman CF, Corcoran JM (1981) Effect of iodination of a village water supply on goitre size and thyroid function. *Lancet,* 2, 1270–1272.

Man EB, Brown JF, Serunian SA (1991) Maternal hypothyroxinemia: psychoneurological deficits of progeny. *Ann. Clin. Lab. Sci.,* 21, 227–239.

Moore RM, Tokarczyk R (1992) Chloro-iodomethane in North Atlantic waters: a potentially significant source of atmospheric iodine. *Geophys. Res. Lett.,* 19, 1779–1782.

Morreale de Escobar G, Obregon MJ, Calvo R, Escobar del Rey F (1994) Hormone nurturing of the developing brain: the rat model. In *The Damaged Brain of Iodine Deficiency* (JB Stanbury, Ed.), pp.103–122. Cognizant Communication Corporation, New York.

Morreale de Escobar G, Obregon MJ, Escobar del Rey F (2000) Is neuropsychological development related to maternal hypothyroidism or to maternal hypothyroxinemia? *J. Clin. Endocrinol. Metab.,* 85, 3975–3987.

Pharoah POD, Connolly KC (1987) A controlled trial of iodinated oil for the prevention of endemic cretinism: a long term follow-up. *Int. J. Epidem.,* 16, 68–73.

Pharoah POD, Connolly KC (1994) Iodine deficiency in Papua New Guinea. In *The Damaged Brain of Iodine Deficiency* (JB Stanbury, Ed.), pp. 299–305. Cognizant Communication Corporation, New York.

Pop VJ, deVries E, van Baar AL, Waelkens JJ, de Rooy HA, Horsten M, Donkers MM, Kimpro IH, van Son MM, Vader HL (1995) Maternal thyroid peroxidase antibodies during pregnancy: a marker of impaired child development? *J. Clin. Endocrinol. Metab.,* 80, 3561–3566.

Pop V, Kuipens JL, van Baar AL, Verkerk G, van Son MM, de Vijlder JJ, Vulsma T, Wiersinga WM, Drexhage HA, Vader HL (1999) Low maternal free thyroxine concentrations during early pregnancy are associated with impaired psychomotor development in infancy. *Clin. Endocrinol.,* 50, 147–148.

Roti E, Braverman LE (1996) Iodine excess and thyroid function. In *The Thyroid and Iodine* (J Nauman, D Glinoer, LE Braverman, U Hostalek, Eds.), pp. 7–17. F.K. Schattauer Verlagsgesellschaft mbH, Germany.

Rousset BA, Dunn JT (2004) Thyroid hormone synthesis and secretion. In *Thyroid Disease Manager* (LJ DeGroot, G Henneman, Eds.), chapter 2, www.thyroidmanager.org accessed October 1, 2004.

Spitzweg C, Morris JC (2002) The sodium iodide symporter: its pathophysiological and therapeutic implications. *Clin. Endocrinol. (Oxf.),* 57, 559–574.

Todd CH, Allain T, Gomo ZAR, Hasler JA, Ndiweni, M, Oken E (1995) Increase in thyrotoxicosis associated with iodine supplements in Zimbabwe. *Lancet,* 346, 1563–1564.

Truesdale VW, Jones SD (1996) The variation of iodate and total iodine in some UK rainwaters during 1980-1981. *J. Hydrol.,* 179, 67–86.

Vitti P, Rago T (2002) Thyroid volume in children: role of iodine intake. *J. Endocrinol. Invest.,* 25, 666–667.

Vitti P, Martino E, Aghini-Lombardi F (1994) Thyroid volume measurement by ultrasound in children as a tool for the assessment of mild iodine deficiency. *J. Clin. Endocrinol. Metab.,* 79, 600–603.

Vulsma T, Gons MH, De Vijlder JJM (1989) Maternal-fetal transfer of thyroxine in congenital hypothyroidism due to a total organification defect or thyroid agenesis. *N. Engl. J. Med.,* 321, 13–16.

Whitehead DC (1979) Iodine in the UK environment with particular reference to agriculture. *J. Appl. Ecol.,* 16, 269–279.

WHO-UNICEF-ICCIDD (2004) Indicators for assessing iodine deficiency disorders and their control through salt iodization. WHO/NUT/94.6 Geneva: World Health Organization.

World Health Organization, Micronutrient Deficiency Information System (1993) Global prevalence of iodine deficiency disorders, MDIS Working Paper 1. World Health Organization/United Nations Children's Fund [UNICEF]/International Consultative Council for Iodine Deficiency Disorders, Geneva.

Zoeller TR, Dowling AL, Herzig CT, Iannacone EA, Gauger KJ, Bansal R (2002) Thyroid hormone, brain development, and the environment. *Environ. Health Perspect.,* 110, 355–361.

16 Dietary Zinc in Brain Development, Behavior, and Neuropathology

Tammy M. Bray and Mark A. Levy

CONTENTS

16.1 INTRODUCTION

Zinc is best characterized as an essential trace element that displays relatively benign toxicity properties, but exhibits a broad spectrum of deficiency syndromes that range from mild to severe and that may even be lethal if not corrected (Prasad, 1988). In humans, zinc was first demonstrated to be essential in the 1960s, when delayed sexual development and arrested growth were reversed with zinc supplementation (Prasad, 1963). Since then, a myriad of zinc deficiency syndromes, including immunological and neurological abnormalities, have been described in infants, adolescents, and the elderly (Vallee and Falchuk, 1993). Indeed, the field of zinc research, and our understanding of the physiological importance of zinc in human health and disease in particular, has increased dramatically as a result of this landmark study.

Zinc is ubiquitous in cells across the plant and animal kingdoms. It is required for processes of gene transcription, formation of zinc finger proteins that act as DNA-binding transcription factors, and activity of more than 300 different enzymes involved in every aspect of cellular metabolism (Falchuk, 1998). Zinc exerts its biological activity through its association with proteins, in which it may serve as a catalytic, structural, or regulatory element (Cousins, 1996). The catalytic function of zinc can be found in all six classes of enzymes (oxidoreductases, ligases, lyases, isomerases, transferases, and hydrolases; Vallee and Galdes, 1984). Removal of zinc from these enzymes leaves a catalytically inactive apoenzyme, although they retain their tertiary structure (Vallee and Galdes,

1984). Structurally, zinc forms a tetrahedral complex with cysteine and histidine residues in, for example, zinc finger proteins (Klug and Schwabe, 1995). Zinc has also been proposed to play a structural role in maintaining the integrity and hence functionality of biomembranes (Bettger and O'Dell, 1981). The regulatory role of zinc is best characterized by its influence on metallothionein (MT) gene expression, in which zinc is required for metal transcription factor (MTF) binding to the metal response element (MRE) in the promoter region of the MT gene (Cousins, 1994). Yet, despite this research, the physiological consequences of zinc deficiency are incompletely understood and have yet to be attributed to specific biochemical functions. This is particularly true in regard to our understanding of the roles of zinc in neurobiology. Although zinc has been implicated in numerous neurological disorders, for example, Alzheimer's, Lou Gehrig's disease, as well as behavioral abnormalities, at present most evidence of a relationship between zinc and neuropathology is circumstantial. Thus, there is currently considerable debate on whether the role of zinc in specific neuropathologies is primary and therefore causal in nature or secondary and hence a consequence of these disease processes.

16.2 ZINC HOMEOSTASIS

Zinc-related diseases have been attributed primarily to a loss of homeostasis, either locally or systemically, in which deficient or excess levels of the mineral alter biochemical or physiological processes. Indeed, it was generally assumed that the relatively benign properties of zinc stem from efficient processes that tightly regulate absorption and excretion and hence homeostasis of this metal. However, there is increasing evidence that many factors, including dietary, hereditary, and environmental factors, may adversely affect zinc homeostasis and have profound effects on human health and behavior. Thus, a critical understanding of the physiology of zinc homeostasis is essential to preventing and treating zinc-related pathologies. The brain is unique among all organs in that it is encased within the brain-barrier system, a protective barrier that regulates the uptake and export of nutrients and other essential compounds. It therefore serves as a regulatory interface critical for maintaining homeostasis within the brain. Thus, zinc homeostasis within the brain is dependent not only on zinc absorption and excretion in the gastrointestinal (GI) tract but also on regulatory mechanisms that control zinc transport through the brain-barrier system.

16.2.1 WHOLE-BODY ZINC HOMEOSTASIS

Homeostatic control of whole-body zinc metabolism is regulated primarily through a balance between dietary zinc absorption through enterocytes in the small intestine and endogenous excretion into the GI tract. In animal models and in humans, dietary zinc is absorbed and transported through intestinal epithelial cells into the blood, where most becomes bound and transported by albumin and α_2-macroglobulin. However, many factors influence zinc absorption, including competition with other metal species and the presence of dietary zinc-binding ligands that may facilitate or inhibit zinc absorption (Sandstrom, 1989). The fraction of zinc absorption increases as the dietary zinc content declines (Taylor et al., 1991; Hoadley et al., 1988).

   At the molecular level, significant progress has been made in our understanding of the physiological pathways that govern zinc absorption and transport. During periods of high zinc intake both paracellular and transcellular diffusion pathways may predominate, whereas during periods of low zinc intake carrier-mediated processes account for a major proportion of zinc absorption. Several mechanisms have been proposed to facilitate zinc uptake in enterocytes, including endocytotic vesicle formation, transport through membrane-spanning zinc channels, or via cotransport as a peptide–zinc complex by a peptide carrier system (Cousins, 1996). A family of transport proteins, the ZRT/IRT-related proteins (ZIP), has been identified in prokaryotes and eukaryotes that function in zinc transport into cells (Gaither and Eide, 2001; Guerinot, 2000). Most notable in this family is the gene encoding the putative human ZIP-4 (hZIP-4), one of twelve members thus far identified in the human genome (Gaither and Eide, 2001). In studies of transgenic mice, this gene is expressed and localizes to the apical membrane of enterocytes. Moreover, several different mutations in this gene have been identified in human patients with acrodermatitis enteropathica, a disease characterized by defective zinc absorption (Wang et al., 2002). Taken together, these results provide convincing evidence that the hZIP-4 transporter functions in intestinal absorption of zinc.

After absorption, zinc becomes bound within the enterocyte to several proteins, including MT and cysteine-rich intestinal peptide (CRIP; Richards and Cousins, 1977; Hempe and Cousins, 1991). MT is regarded as an intracellular buffering agent that modulates the transfer and homeostasis of zinc and other metal ions. Its expression increases with increasing dietary zinc intake (Cousins and Lee-Ambrose, 1992), and increased MT expression decreases zinc absorption (Hoadley et al., 1988). In this way, MT functions to regulate intracellular levels of free zinc through intracellular binding (Davis and Cousins, 2000). CRIP, on the other hand, does not exhibit regulatory properties, but rather may function in cytokine production, cellular differentiation, or host defense (Levenson et al., 1993; Lanningham-Foster et al., 2002).

Just as zinc uptake is not fully characterized, our understanding of zinc export from enterocytes is still in its infancy. However, enterocytes typically contain significantly lower concentrations of zinc than serum, suggesting that zinc efflux is an energy-dependent process. Several candidate genes have been recognized that may function in this multistep process. The most prominent of these is zinc transporter-1 (ZnT-1). Transfection studies and *in vivo* work have demonstrated that ZnT-1 functions in cellular zinc export and that it localizes to the basolateral membrane of enterocytes (Palmiter and Findley, 1995; McMahon and Cousins, 1998). Another zinc transport protein, ZnT-2, is also expressed in the small intestine. However, this protein is thought to be located on the surface of intracellular vesicles, where it may function in vesicular zinc storage (Harris, 2002). ZnT-1 and ZnT-2 are just two of a six-member family of zinc transporters (ZnT-1 to ZnT-6). Although research thus far indicates that they function to regulate cellular zinc status either through zinc transport out of the cell or through intracellular trafficking to various intracellular compartments, much more research is required to define the specific characteristics of each, such as factors that regulate expression, tissue specificity, and subcellular location (Guerinot, 2000; Huang et al., 2002; Kambe et al., 2002).

Following transport across the basolateral membrane of the enterocyte, studies in human and animal models have demonstrated that zinc is transported in the plasma to peripheral tissues and organs bound to either albumin (~70%) or α_2-macroglobulin (~30%), although neither is essential for zinc transport into the brain (Takeda et al., 1997). The remainder, typically 1 to 2% of plasma zinc, is bound to low-molecular weight ligands such as the amino acids histidine and cysteine. Free zinc is thought to be almost negligible, with concentrations as low as 10^{-9} to 10^{-10} M (Magneson et al., 1987).

16.2.2 BRAIN ZINC HOMEOSTASIS

As our understanding of dietary zinc uptake and transport continues to improve, so does our understanding of the various aspects of zinc uptake, transport, and metabolism in the brain. Research to date demonstrates that there are two routes of zinc entry into the central nervous system (CNS), the blood-brain barrier and the brain-cerebrospinal fluid (CSF) barrier, that collectively constitute the brain-barrier system. Indeed, the brain-barrier system is critical to brain zinc homeostasis, regulating brain zinc turnover in a manner considerably slower than in peripheral tissues (Kasarskis, 1984). Brain zinc levels are usually not affected by dietary zinc, although changes have been observed (Noseworthy and Bray, 2000; O'Dell et al., 1989; Wallwork et al., 1983; Takeda et al., 2001). It has been proposed that the brain is not responsive to dietary zinc deprivation (Golub et al., 1995). However, neuromotor and cognitive dysfunction associated with zinc deficiency seems to argue against this view (Sandstead et al., 2000).

Although the brain contains significant quantities of zinc, very little is known about the molecular mechanisms that underlie zinc transport and homeostatic mechanisms. As with intestinal zinc absorption, carrier-mediated processes have been proposed to function in the brain. For example, brain uptake of zinc is enhanced by L-histidine provided either through infusion or diet (Keller et al., 2000; Aiken et al., 1992). Zinc and histidine may cotransport across the brain-barrier system through the peptide/histidine transporter PHT1 (Yamashita et al., 1997) or histidine may act as a shuttle and present zinc to divalent metal transporter 1 (DMT1), a transport protein expressed in brain capillary endothelial cells throughout the brain-barrier system (Gunshin et al., 1997).

In addition, brain zinc homeostasis is almost certainly dependent on three families of zinc-binding proteins: the ZIP proteins, the ZnT proteins, and the MTs (MT 1–4). The ZIP family of proteins functions in zinc uptake into cells, transporting zinc from the outside to the inside of the cell. ZIP transporters have also been shown to mobilize compartmental zinc from within organelles or vesicles, releasing it into the cytoplasm (Gaither and Eide, 2001). However, our understanding of ZIP transporters is still in the early stages. Twelve members have thus far been identified in humans, but functional data are available for only two. Human ZIP-1 (hZIP-1) is proposed to be the major zinc transporter, as it is expressed in 24 tissues, including the brain. In contrast, hZIP-2 has thus far been identified in prostate and uterus tissues only, thus suggesting that it may have a very specialized tissue-specific role (Gaither and Eide, 2001).

Three members of the ZnT family have thus far been identified in the brain. ZnT-1, the first of these to be cloned, has been proposed to transport zinc out of cells and thus protect against potentially toxic build-up of intracellular zinc (Palmiter and Findley, 1995). Increases in ZnT-1 expression have recently been documented to parallel increases in the level of chelatable zinc in neurons of the postnatal mouse brain, suggesting that it may function to protect cells against the potentially toxic effects of zinc (Nitzan et al., 2002). ZnT-3 and ZnT-4 have also been identified in mammalian brain tissue. Specifically, ZnT-3 transports zinc into synaptic vesicles and is therefore considered to play an integral role in glutamatergic neurotransmission (Cole et al., 1999). The function of ZnT-4 in the brain remains undefined. However, mutation of this gene is responsible for zinc deficiency in the lethal milk mouse syndrome, a finding consistent with its proposed role as a zinc efflux protein (Huang and Gitschier, 1997).

The MTs are a family of low-molecular-weight cysteine-rich proteins. Two isoforms, MT-1 and MT-2, are expressed in virtually all tissues, but a third isoform, MT-3, is expressed primarily in the CNS. Functionally, MT-1 and MT-2 exhibit a broad spectrum of roles, but their high affinity for zinc indicates an integral role for the homeostasis and metabolism of this metal. Indeed, similar to their role in intestinal epithelia, MT-1 and MT-2 are potent zinc acceptors and are thus proposed to serve as storage reservoirs that limit intracellular free zinc levels. On the other hand, although MT-3 shares zinc binding-properties similar to those of MT-1 and MT-2, it exhibits distinctly different biological properties. Whereas MT-1 and MT-2 expression is reduced during zinc deficiency, MT-3 expression is unchanged and may in fact compete for zinc binding and thus exacerbate the zinc deficiency. In fact, MT-3 was originally named growth inhibitory factor (GIF), because it was found to reduce survival of neuronal cells in culture (Palmiter, 1995). Histochemical studies have demonstrated abundant MT-3 levels in the hippocampus, prompting speculation that it may play an important regulatory role in the functioning of zinc-enriched neurons (Hidalgo et al., 2001). Thus, although MTs are generally acknowledged to figure prominently in zinc metabolism and homeostasis in the brain, the exact nature of their roles remains to be defined.

16.3 ZINC IN THE BRAIN

The human brain contains a significant quantity of zinc. In fact, it is second only to iron in terms of total concentration and is three times more abundant than copper (Prasad, 1988). During postnatal growth, there is a gradual rise in brain zinc content, increasing from approximately 5 µg/g wet tissue in the first year of life (Volkl et al., 1974), to a relatively constant 10 µg/g in the adult

(Markesbery et al., 1984). However, there is a dramatic shift in the relative distribution of zinc throughout the brain during aging. In neonatal rats the concentration of zinc is highest in the cerebellum, whereas in the adult it is higher in the cerebral cortex and the hippocampus (Sawashita et al., 1997). These changes in distribution and concentration may reflect the particular functional role of zinc in each brain region. During the early postnatal stage, cerebellar growth and development is rapid, reflecting its central role in the rapid acquisition of motor skills, balance, and posture during the early postnatal period (Willis, 1998). Rapid cerebellar growth may also reflect more recent findings of the importance of the cerebellum in directing cerebral cortex growth and development (Molinari et al., 2002). Hence, elevated zinc levels in the cerebellum in the early postnatal stage may be indicative of its central role in DNA and RNA synthesis as well as cell growth and division. In contrast, the more gradual increase in zinc content in the cerebral cortex and the hippocampus may be more indicative of the role of zinc in neurotransmission, memory, and learning, processes associated with the development and maturation of cognitive circuits in these brain regions (Zola-Morgan et al., 1986; Valente et al., 2002; Adolphs et al., 1994).

Within the brain, 80 to 90% of all zinc is protein bound, located within metalloproteins, where it functions in metabolic reactions, or within biomembranes, where it provides structural support. The remaining zinc is localized within synaptic vesicles of nerve terminals and is referred to as vesicular zinc. It is histochemically reactive and therefore considered ionic zinc (Frederickson and Danscher, 1990). Although its precise role has yet to be defined, vesicular zinc appears to be released into the synapse during neural activity of glutamatergic neurons, where it may serve to modulate neurotransmitter receptors (Assaf and Chung, 1984; Howell et al., 1984). In addition to its functions in metalloproteins and neural activity, zinc may serve a more global but equally vital role as an antioxidant in the brain. Although it has only recently been proposed that zinc functions as an antioxidant (Bray and Bettger, 1990), there is unequivocal documentation that this is indeed a critical function of this mineral (Powell, 2000). Zinc is a component of the free radical defense system, and zinc deficiency induces oxidative damage both *in vitro* and *in vivo* (Xu et al., 1994; Oteiza et al., 1995; Bagchi et al., 1998).

Zinc is essential for brain development. Zinc deficiency can lead to a variety of malformations during embryogenesis (Hurley and Swenerton, 1966; Swenerton et al., 1969), can lead to behavioral deficits in both animals (Halas et al., 1986) and humans (Sandstead et al., 1978), and has been proposed to be a risk factor for age-associated dementias and neuropathologies (Burnet, 1981; Tully et al., 1995). Thus, zinc is not only critical for structural aspects of brain development, but may also play a significant role in behavioral and cognitive aspects of brain function. In general terms, there are three different stages to be considered when evaluating the function of zinc in brain pathology: the role of zinc in (1) CNS development in embryonic and early postnatal stages, (2) the function of the developed brain in the young, and (3) neurodegenerative disorders of the old. The mechanisms by which zinc deficiency affects the CNS in these three stages resulting in clinical pathology are probably distinctively different.

16.3.1 ZINC AND STRUCTURAL BRAIN DEVELOPMENT IN THE EMBRYONIC AND EARLY POSTNATAL STAGE

The importance of zinc in embryonic CNS development was first established by nutritional studies demonstrating that dietary zinc deficiency in dams throughout pregnancy resulted in brain malformations such as spina bifida and exencephaly in rat pups (Hurley and Swenerton, 1966). Similarly, zinc deficiency during suckling resulted in decreased brain size as well as reduced brain DNA, RNA, and protein concentration (Sandstead et al., 1975). These adverse neurological effects during the early brain development stage appear to be permanent, showing little or no improvement from zinc repletion therapy (Hurley and Swenerton, 1966). In humans, a causal relationship between maternal zinc deficiency and abnormal fetal brain development has not been established, in part because zinc status is difficult to ascertain, particularly during pregnancy (Shah and Sachdev, 2001).

However, congenital CNS malformations have been observed in infants of mothers who exhibit abnormalities of zinc metabolism (Bergmann et al., 1980), and a series of studies conducted in Turkey demonstrated that indices of zinc status were much lower in mothers of anencephalic children than in control subjects (Cavdar et al., 1983). In addition, a review of pregnancy outcomes in women with acrodermatitus enteropathica suggested that human fetal development is also prone to the teratogenic effects of zinc deficiency (Hambidge et al., 1975). Early postnatal zinc deficiency and brain development in humans has not been well documented. However, one case study reported that supplementation of zinc sulfate improved indices of cortical atrophy in infants that developed acrodermatitis enteropathica during breastfeeding (Ohlsson, 1981).

The mechanism by which zinc deficiency causes abnormal brain formation and development has been attributed to impaired cell division during embryonic development (Dvergsten et al., 1983). Because zinc plays a critical role in replication, transcription, and translation, compromised zinc status may lead to abnormal cell cycle regulation and organ development.

16.3.2 Zinc and Cognitive Development in the Young

By age two, the human brain reaches 75% of its adult size and no new neurons are created after birth (Frederickson, 1989). Thus, the involvement of zinc in the developed brain is proposed to be mostly functional. Various biochemical indices have been used to detect and demonstrate that zinc deficiency can impair brain function. For example, it is known that zinc metalloenzymes are essential to the normal biology of the developed brain. Many investigators have measured the activity of various zinc metalloenzymes in zinc-deficient animals, expecting that the results would provide some information on the mechanisms behind impaired neurological function. For instance, glutamate dehydrogenase, dopamine-β-hydroxylase, phenyethanolamine-N-methyl transferase, thymidine kinase, and alkaline phosphatase have shown reduced activity in brains of rats exposed to prolonged zinc deficiency (Burnet, 1981). However, direct causal consequences of the reduced activity of these enzymes in the developed brain have not been clearly identified. It is not likely that the observed changes in zinc metalloenzyme activity can mediate the rapid effects that zinc can have on brain function.

Evidence of behavioral abnormalities in both animals and humans suggests that zinc deficiency is most likely involved in abnormal neurochemical processes. Zinc has been shown to inhibit N-methyl D-aspartate (NMDA) and γ-amino butyric acid (GABA) receptors and to regulate opiodergic, glycinergic, and sigma receptors. It has also been suggested that zinc may function to stabilize secretory vesicles (Burnet, 1981). Colocalization of zinc with glutamate in presynaptic terminals may indicate that zinc regulates neurotransmission by modulating postsynaptic gluatmate receptors. However, there is not yet any direct evidence to demonstrate the effect of dietary zinc deficiency on the neurochemical processes.

In humans, there is evidence that zinc deficiency may interfere with cognitive performance. Supplementation of zinc and a micronutrient mixture improved tests of memory, reasoning, and psychomotor functioning in comparison with subjects given the micronutrient mixture only (Sandstead et al., 1998). Zinc deficiency may also increase lethargy. Spontaneous activity levels were significantly increased in 6- to 17-month-old Indian infants supplemented with 10 mg/day of zinc and a multivitamin mixture for 6 to 7 months compared with infants given a multivitamin mixture only (Sazawal et al., 1996). Similarly, in a trial conducted in Guatemala, 10 mg zinc/day for 7 months significantly elevated activity in 6- to 9-month-old infants compared with infants given a placebo (Bentley et al., 1997).

Premature infants are a unique population susceptible to neurological disorders related to cognitive and motor function, including cerebral palsy (Graziani et al., 1992; Volpe, 1990). They are at risk of developing zinc deficiency, because two thirds of total body zinc is acquired during the last trimester (Widdowson, 1974). Moreover, they are often exposed to external factors such as hyperoxia exposure and drug therapy, which increases oxidative stress. Recently, we have examined the effects of zinc deficiency in an animal model of premature infancy and demonstrated that zinc deficiency increased glutathione oxidation in the brain tissue (Noseworthy and Bray, 1998). Moreover, exposure of animals to hyperoxia increased the extent of glutathione oxidation. Similarly, magnetic resonance imaging (MRI) analysis revealed that zinc deficiency induced a loss in blood-brain barrier integrity and that this was exacerbated following hyperoxia exposure (Noseworthy and Bray, 2000). These results suggested that the neurological deficits of premature infants may in part be attributable to zinc deficiency and the resulting decrease in antioxidant protection.

16.3.3 ZINC IN NEUROLOGICAL DEGENERATIVE DISORDERS

The role of dietary zinc in age-related degenerative disorders of the brain is an area of research that has grown enormously in the past two decades. Recently, a correlation between low zinc levels and some clinical neurological disorders such as Parkinson's disease and Alzheimer's disease has been established (Cuajungco and Lees, 1997). Aging is associated with increased oxidative stress and diminished antioxidant defenses that could account for the delayed onset and progressive nature of age-related diseases, particularly neurodegenerative diseases (Coyle and Puttfarcken, 1993). Moreover, zinc has been proposed to function as an antioxidant, and there is evidence that marginal and clinically manifested zinc deficiency may be common in the elderly population (Bray and Bettger, 1990; Bales et al., 1994). As a result, the possible role of zinc in neurodegenerative diseases has garnered considerable attention, particularly with regard to its role as an antioxidant.

The proposed function of zinc as an antioxidant, in addition to its roles as a catalytic, structural and regulatory element, has arisen primarily from a body of work that has demonstrated antioxidant functions of zinc *in vitro* (Gibbs et al., 1985; Washabaugh and Collins, 1986). Zinc functions as an antioxidant through two primary mechanisms. First, it protects sulfhydryl groups from oxidation either through binding directly to the sulfhydryl group and thereby reducing its reactivity, by binding near the sulfhydryl group and through steric hindrance reducing its reactivity, or by binding to sulfhydryl-containing proteins and inducing conformational changes that reduce sulfhydryl group reactivity. Second, zinc functions as an antioxidant by effectively competing with redox active metals (e.g., iron and copper) for binding sites within macromolecules, thereby preventing their production of $OH^\bullet$ and $O_2^{\bullet-}$ (Bray and Bettger, 1990). Indeed, sulfhydryl group oxidation and redox metal-catalyzed tissue damage are two hallmark features of neurodegeneration (Aksenov and Markesbery, 2001; Sayre et al., 1999).

16.3.3.1 Alzheimer's Disease

Alzhemier's disease is the most common form of senile dementia. It is characterized by progressive neuronal cell death in various regions of the brain, particularly the temporal lobe and the hippocampus (Price et al., 1982). Senile plaques, composed of insoluble β-amyloid protein, and neurofibrillary tangles containing the microtubule-associated protein tau, are the anatomically distinguishing features of Alzheimer's disease. Although the causes of Alzheimer's have yet to be identified, oxidative stress has been implicated, either through diminished antioxidant defenses or increased reactive oxygen species production (Olanow, 1993; Coyle and Puttfarcken, 1993). Moreover, the role of trace metals, particularly zinc, is being intensively investigated (Bush, 2000).

A link between zinc and Alzheimer's initially emerged through a series of papers demonstrating that β-amyloid, soluble in the absence of metal binding, forms an insoluble protease-resistant aggregate in the presence of zinc (Bush et al., 1992, 1994; Seubert et al., 1992). Not surprisingly, zinc and other metals were subsequently proposed to be risk factors for Alzheimer's (Frederickson and Danscher, 1990; Connor et al., 1992). Indeed, evidence suggests that zinc, copper, and iron may contribute to the aggregation of β-amyloid and the accumulation of plaques. First, postmortem studies reveal that the concentration of zinc and copper is elevated in Alzheimer's disease brain tissue, particularly in regions of dense plaque formation (Deibel et al., 1996). Second, both zinc and copper have been shown to bind to and render β-amyloid insoluble in *in vitro* aqueous environments (Miura et al., 2000; Bush et al., 1994). Third, metal chelation reverses zinc- and copper-induced β-amyloid precipitation (Atwood et al., 1998), releases soluble β-amyloid from Alzheimer's disease brain homogenates (Cherny et al., 1999), and inhibits plaque formation in a mouse model of Alzheimer's disease (Cherny et al., 2001). However, the most compelling evidence of the potential involvement of zinc in Alzheimer's comes from a murine model in which mice that overexpress the human amyloid precursor protein (hAPP$^+$) were mated with zinc transporter-3 (ZnT-3) knockout mice (ZnT-3$^{-/-}$). Three genotypes thus resulted in the offspring: hAPP$^+$:ZnT-3$^{+/+}$ (ZnT-3 wild-type), hAPP$^+$:ZnT-3$^{+/-}$ (ZnT-3 hemizygous), and hAPP$^+$:ZnT-3$^{-/-}$ (ZnT-3 knockout). Hence, in the last genotype, zinc transport into synaptic vesicles and release into the synapse was completely abolished. Compared with hAPP$^+$:ZnT-3$^{+/+}$, hAPP$^+$:ZnT-3$^{+/-}$ mice had significantly lower insoluble β-amyloid plaque levels in brain tissue, with still larger reductions observed in the hAPP$^+$:ZnT-3$^{-/-}$, thus demonstrating a prominent role of zinc, and specifically synaptic zinc, in β-amyloid aggregation and plaque formation (Lee et al., 2002).

Nevertheless, there is equally compelling evidence that zinc may play a protective role in β-amyloid plaque formation and development of Alzheimer's disease. Neuronal cell death in Alzheimer's has been proposed to be mediated, in part, through β-amyloid-induced lipid peroxidation, 4-hydroxynonenal production, and H_2O_2 generation (Hensley et al., 1994). However, zinc, which is redox inert, inhibits β-amyloid-mediated H_2O_2 production and neurotoxicity *in vitro* and may account for an inverse correlation between β-amyloid levels and DNA oxidation in brain tissue from Azheimer's patients (Cuajungco et al., 2000). More recently, zinc has been proposed to protect against β-amyloid toxicity through its effects on the structural characteristics of β-amyloid aggregation. By analyzing secondary structure, zinc was revealed to promote nonfibrillar or granular β-amyloid aggregation, a conformation that exhibits substantially reduced cytotoxicty compared to fibrillar β-amyloid (Yoshiike et al., 2001). Hence, although zinc is well documented to reside in β-amyloid plaques and is thus proposed to be a contributing factor in β-amyloid neurotoxicity (Cherny et al., 2001), there is growing evidence that zinc may in fact be neuroprotective in Alzheimer's disease and thus occupy a much more intricate role in the pathogenesis of this disease than initially suspected.

16.3.3.2 Amyotrophic Lateral Sclerosis

Amyotrophic lateral sclerosis (ALS) is a neurodegenerative disease that results in the progressive loss of motor neurons in the motor cortex, brain stem, and spinal column (Brown, 1995). ALS progresses rapidly, leading to paralysis and death within 2 to 3 years of diagnosis. It occurs in both a sporadic (SALS) and familial (FALS) form, the latter accounting for ~10% of all cases. Within the FALS subset, ca. 20% of the cases have been associated with more than 90 different missense mutations of the antioxidant enzyme copper-zinc superoxide dismutase (CuZnSOD). Hence, mutations of CuZnSOD account for 2 to 3% of individuals with ALS. Nevertheless, it thus far remains the only proven cause of the disease (Beckman et al., 2001). Curiously, mutated CuZnSOD retains full activity *in vitro*, and knockout of the endogenous CuZnSOD gene in animal models does not lead to motoneuron disease. In addition, motoneuron disease develops in transgenic mice that express mutated CuZnSOD, regardless of the presence or absence of endogenous CuZnSOD (Gurney et al., 1994). Thus, the toxicity of mutant CuZnSOD in ALS appears to be independent of its antioxidant role and is proposed to arise from a toxic gain of function of this enzyme (Brown, 1996).

The question therefore emerges: what common gain of function could be shared by the many different CuZnSOD mutations that result in ALS? Excitotoxicity, mitochondrial defects, and toxic protein aggregates have been implicated in the etiology of ALS (Julien, 2001). However, evidence that mutant CuZnSOD exhibits decreased thermal stability (Rodriguez et al., 2002) as well as increased susceptibility to proteolysis and accelerated turnover *in vivo* (Ratovitski et al., 1999) has provided some interesting clues. Indeed, the decreased stability of mutant CuZnSOD has been attributed to a decreased affinity of this enzyme for zinc (Deng et al., 1993). Moreover, loss of zinc from this enzyme (zinc-deficient SOD) alters its redox properties, effectively switching its function from a superoxide dismutase to a superoxide synthase (Estevez et al., 1999). Thus, in the presence of nitric oxide, superoxide generation by zinc-deficient SOD may function to produce peroxynitrite, a strong oxidant that induces apoptosis in motor neurons (Estevez and Jordan, 2002). However, generation of peroxinitrite by zinc-deficient SOD may not fully account for ALS disease progression, as various neuroprotective mechanisms may inhibit apoptotic neuronal cell death *in vivo*. Therefore, a model has been proposed in which neurofilaments, abundant structural proteins within motoneurons, may play a key role in establishing a viscious cycle of disease progression (Crow et al., 1997). In this model, neurofilaments are proposed to outcompete zinc-deficient SOD for binding of zinc *in vivo* and thus serve as a zinc sink, enhancing formation of both zinc-deficient SOD and peroxynitrite. Moreover, the neurofilament subunit L is particularly susceptible to peroxynitrite-mediated tyrosine nitration. Nitration of this subunit may then disrupt neurofilament synthesis and lead to its erroneous aggregation (Beckman et al., 2001). As neurofilament aggregates accumulate they simultaneously bind zinc and increase zinc-deficient SOD levels, thereby facilitating further peroxinitrite formation until enough is generated to activate apoptosis. Thus, in this model, the decreased zinc-binding affinity of mutant CuZnSOD is proposed as the central mechanism to explain how so many different mutations could manifest as a singular disease entity. Moreover, it predicts that altered zinc metabolism may be a significant feature not only of FALS but also of SALS (Beckman et al., 2001).

16.4 CONCLUSION

Zinc is ubiquitous across the plant and animal kingdoms and is essential for an impressive array of biochemical functions involved in every aspect of cellular metabolism. In general terms, zinc functions as a constituent of metalloproteins, where it serves a structural, catalytic, regulatory, or antioxidant role. Indeed, incorporation of zinc into these metalloproteins is critical for their proper function.

There has been intense speculation regarding the role of zinc in the brain, particularly since initial reports of its nonuniform distribution and evidence of its concentration in specific areas such as the hippocampus and the cerebral cortex (Frederickson, 1989; McLardy, 1962). Because each of these regions has been associated with emotion, learning, and memory (Baxter and Murray, 2002), it is presumed that zinc may play a prominent role in these processes and, critically, may play a prominent role in neuropathologies associated with these regions.

Dietary zinc deficiency has been shown to induce severe malformations of the CNS, and more recently has been shown to adversely affect learning, memory, and other behaviors in animals and humans (Golub et al., 1995; Sandstead et al., 1998). There is also evidence that abnormal zinc metabolism may play an integral role in the development and pathogenesis of neurodegenerative diseases such as Alzheimer's and ALS (Estevez and Jordan, 2002; Bush, 2000).

The many and varied roles of zinc and the numerous pathologies associated with zinc in the brain point to its importance as an essential mineral critical to brain function. However, our fundamental understanding of the biochemical basis of zinc-related pathologies in the brain is just beginning. As a result, the use of zinc as a therapeutic strategy in treating neuropathologies must be approached with caution, despite the promise of exciting new developments that undoubtedly lie in the near future.

REFERENCES

Adolphs, R., Tranel, D., Damasio, H. and Damasio, A. (1994) Impaired recognition of emotion in facial expressions following bilateral damage to the human amygdala. *Nature,* 372, 669–672.

Aiken, S. P., Horn, N. M. and Saunders, N. R. (1992) Effects of histidine on tissue zinc distribution in rats. *Biometals,* 5, 235–243.

Aksenov, M. Y. and Markesbery, W. R. (2001) Changes in thiol content and expression of glutathione redox system genes in the hippocampus and cerebellum in Alzheimer's disease. *Neurosci Lett,* 302, 141–145.

Assaf, S. Y. and Chung, S. H. (1984) Release of endogenous Zn^{2+} from brain tissue during activity. *Nature,* 308, 734–746.

Atwood, C. S., Moir, R. D., Huang, X., Scarpa, R. C., Bacarra, N. M., Romano, D. M., Hartshorn, M. A., Tanzi, R. E. and Bush, A. I. (1998) Dramatic aggregation of Alzheimer A beta by Cu(II) is induced by conditions representing physiological acidosis. *J Biol Chem,* 273, 12817–12826.

Bagchi, D., Vuchetich, P. J., Bagchi, M., Tran, M. X., Krohn, R. L., Ray, S. D. and Stohs, S. J. (1998) Protective effects of zinc salts on TPA-induced hepatic and brain lipid peroxidation, glutathione depletion, DNA damage and peritoneal macrophage activation in mice. *Gen Pharmacol,* 30, 43–50.

Bales, C. W., DiSilvestro, R. A., Currie, K. L., Plaisted, C. S., Joung, H., Galanos, A. N. and Lin, P. H. (1994) Marginal zinc deficiency in older adults: responsiveness of zinc status indicators. *J Am Coll Nutr,* 13, 455–462.

Baxter, M. G. and Murray, E. A. (2002) The amygdala and reward. *Nat Rev Neurosci,* 3, 563–573.

Beckman, J. S., Estevez, A. G., Crow, J. P. and Barbeito, L. (2001) Superoxide dismutase and the death of motoneurons in ALS. *Trends Neurosci,* 24, S15–S20.

Bentley, M. E., Caulfield, L. E., Ram, M., Santizo, M. C., Hurtado, E., Rivera, J. A., Ruel, M. T. and Brown, K. H. (1997) Zinc supplementation affects the activity patterns of rural Guatemalan infants. *J Nutr,* 127, 1333–1338.

Bergmann, K. E., Makosch, G. and Tews, K. H. (1980) Abnormalities of hair zinc concentration in mothers of newborn infants with spina bifida. *Am J Clin Nutr,* 33, 2145–2150.

Bettger, W. J. and O'Dell, B. L. (1981) A critical physiological role of zinc in the structure and function of biomembranes. *Life Sci,* 28, 1425–1438.

Bray, T. M. and Bettger, W. J. (1990) The physiological role of zinc as an antioxidant. *Free Radic Biol Med,* 8, 281–291.

Brown, R. H., Jr. (1995) Amyotrophic lateral sclerosis: recent insights from genetics and transgenic mice. *Cell,* 80, 687–692.

Brown, R. H., Jr. (1996) Superoxide dismutase and familial amyotrophic lateral sclerosis: new insights into mechanisms and treatments. *Ann Neurol,* 39, 145–146.

Burnet, F. M. (1981) A possible role of zinc in the pathology of dementia. *Lancet,* 1, 186–188.

Bush, A. I. (2000) Metals and neuroscience. *Curr Opin Chem Biol,* 4, 184–191.

Bush, A. I., Pettingell, W. H., Multhaup, G., d Paradis, M., Vonsattel, J. P., Gusella, J. F., Beyreuther, K., Masters, C. L. and Tanzi, R. E. (1994) Rapid induction of Alzheimer A beta amyloid formation by zinc. *Science,* 265, 1464–1467.

Bush, A. I., Whyte, S., Thomas, L. D., Williamson, T. G., Van Tiggelen, C. J., Currie, J., Small, D. H., Moir, R. D., Li, Q. X., Rumble, B. et al. (1992) An abnormality of plasma amyloid protein precursor in Alzheimer's disease. *Ann Neurol,* 32, 57–65.

Cavdar, A. O., Babacan, E., Asik, S., Arcasoy, A., Ertem, U., Himmetoglu, O., Baycu, T. and Akar, N. (1983) Zinc levels of serum, plasma, erythrocytes and hair in Turkish women with anencephalic babies. *Prog Clin Biol Res,* 129, 99–106.

Cherny, R. A., Atwood, C. S., Xilinas, M. E., Gray, D. N., Jones, W. D., McLean, C. A., Barnham, K. J., Volitakis, I., Fraser, F. W., Kim, Y., Huang, X., Goldstein, L. E., Moir, R. D., Lim, J. T., Beyreuther, K., Zheng, H., Tanzi, R. E., Masters, C. L. and Bush, A. I. (2001) Treatment with a copper-zinc chelator markedly and rapidly inhibits beta-amyloid accumulation in Alzheimer's disease transgenic mice. *Neuron,* 30, 665–676.

Cherny, R. A., Legg, J. T., McLean, C. A., Fairlie, D. P., Huang, X., Atwood, C. S., Beyreuther, K., Tanzi, R. E., Masters, C. L. and Bush, A. I. (1999) Aqueous dissolution of Alzheimer's disease A beta amyloid deposits by biometal depletion. *J Biol Chem,* 274, 23223–23228.

Cole, T. B., Wenzel, H. J., Kafer, K. E., Schwartzkroin, P. A. and Palmiter, R. D. (1999) Elimination of zinc from synaptic vesicles in the intact mouse brain by disruption of the ZnT3 gene. *Proc Natl Acad Sci USA*, 96, 1716–1721.

Connor, J. R., Snyder, B. S., Beard, J. L., Fine, R. E. and Mufson, E. J. (1992) Regional distribution of iron and iron-regulatory proteins in the brain in aging and Alzheimer's disease. *J Neurosci Res*, 31, 327–335.

Cousins, R. J. (1994) Metal elements and gene expression. *Annu Rev Nutr*, 14, 449–469.

Cousins, R. J. (1996) In *Present Knowledge in Nutrition* (Ziegler, E. E. and Filer, L. J., Eds.), pp. 293–306. ILSI Press, Washington, D.C.

Cousins, R. J. and Lee-Ambrose, L. M. (1992) Nuclear zinc uptake and interactions and metallothionein gene expression are influenced by dietary zinc in rats. *J Nutr*, 122, 56–64.

Coyle, J. T. and Puttfarcken, P. (1993) Oxidative stress, glutamate, and neurodegenerative disorders. *Science*, 262, 689–695.

Crow, J. P., Sampson, J. B., Zhuang, Y., Thompson, J. A. and Beckman, J. S. (1997) Decreased zinc affinity of amyotrophic lateral sclerosis-associated superoxide dismutase mutants leads to enhanced catalysis of tyrosine nitration by peroxynitrite. *J Neurochem*, 69, 1936–1944.

Cuajungco, M. P., Goldstein, L. E., Nunomura, A., Smith, M. A., Lim, J. T., Atwood, C. S., Huang, X., Farrag, Y. W., Perry, G. and Bush, A. I. (2000) Evidence that the beta-amyloid plaques of Alzheimer's disease represent the redox-silencing and entombment of abeta by zinc. *J Biol Chem*, 275, 19439–19442.

Cuajungco, M. P. and Lees, G. J. (1997) Zinc metabolism in the brain: relevance to human neurodegenerative disorders. *Neurobiol Dis*, 4, 137–169.

Davis, S. R. and Cousins, R. J. (2000) Metallothionein expression in animals: a physiological perspective on function. *J Nutr*, 130, 1085–1088.

Deibel, M. A., Ehmann, W. D. and Markesbery, W. R. (1996) Copper, iron, and zinc imbalances in severely degenerated brain regions in Alzheimer's disease: possible relation to oxidative stress. *J Neurol Sci*, 143, 137–142.

Deng, H. X., Hentati, A., Tainer, J. A., Iqbal, Z., Cayabyab, A., Hung, W. Y., Getzoff, E. D., Hu, P., Herzfeldt, B., Roos, R. P. et al. (1993) Amyotrophic lateral sclerosis and structural defects in Cu,Zn superoxide dismutase. *Science*, 261, 1047–1051.

Dvergsten, C. L., Fosmire, G. J., Ollerich, D. A. and Sandstead, H. H. (1983) Alterations in the postnatal development of the cerebellar cortex due to zinc deficiency. I. Impaired acquisition of granule cells. *Brain Res*, 271, 217–226.

Estevez, A. G., Crow, J. P., Sampson, J. B., Reiter, C., Zhuang, Y., Richardson, G. J., Tarpey, M. M., Barbeito, L. and Beckman, J. S. (1999) Induction of nitric oxide-dependent apoptosis in motor neurons by zinc-deficient superoxide dismutase. *Science*, 286, 2498–2500.

Estevez, A. G. and Jordan, J. (2002) Nitric oxide and superoxide, a deadly cocktail. *Ann N Y Acad Sci*, 962, 207–211.

Falchuk, K. H. (1998) The molecular basis for the role of zinc in developmental biology. *Mol Cell Biochem*, 188, 41–48.

Frederickson, C. J. (1989) Neurobiology of zinc and zinc-containing neurons. *Int Rev Neurobiol*, 31, 145–238.

Frederickson, C. J. and Danscher, G. (1990) Zinc-containing neurons in hippocampus and related CNS structures. *Prog Brain Res*, 83, 71–84.

Gaither, L. A. and Eide, D. J. (2001) Eukaryotic zinc transporters and their regulation. *Biometals*, 14, 251–270.

Gibbs, P. N., Gore, M. G. and Jordan, P. M. (1985) Investigation of the effect of metal ions on the reactivity of thiol groups in human 5-aminolaevulinate dehydratase. *Biochem J*, 225, 573–580.

Golub, M. S., Keen, C. L., Gershwin, M. E. and Hendrickx, A. G. (1995) Developmental zinc deficiency and behavior. *J Nutr*, 125, 2263S–2271S.

Graziani, L. J., Mitchell, D. G., Kornhauser, M., Pidcock, F. S., Merton, D. A., Stanley, C. and McKee, L. (1992) Neurodevelopment of preterm infants: neonatal neurosonographic and serum bilirubin studies. *Pediatrics*, 89, 229–234.

Guerinot, M. L. (2000) The ZIP family of metal transporters. *Biochim Biophys Acta*, 1465, 190–198.

Gunshin, H., Mackenzie, B., Berger, U. V., Gunshin, Y., Romero, M. F., Boron, W. F., Nussberger, S., Gollan, J. L. and Hediger, M. A. (1997) Cloning and characterization of a mammalian proton-coupled metal-ion transporter. *Nature*, 388, 482–488.

Gurney, M. E., Pu, H., Chiu, A. Y., Dal Canto, M. C., Polchow, C. Y., Alexander, D. D., Caliendo, J., Hentati, A., Kwon, Y. W., Deng, H. X. et al. (1994) Motor neuron degeneration in mice that express a human Cu,Zn superoxide dismutase mutation. *Science*, 264, 1772–1775.

Halas, E. S., Hunt, C. D. and Eberhardt, M. J. (1986) Learning and memory disabilities in young adult rats from mildly zinc deficient dams. *Physiol Behav*, 37, 451–458.

Hambidge, K. M., Neldner, K. H. and Walravens, P. A. (1975) Zinc, acrodermatitis enteropathica, and congenital malformations (Letter). *Lancet*, 1, 577–578.

Harris, E. D. (2002) Cellular transporters for zinc. *Nutr Rev*, 60, 121–124.

Hempe, J. M. and Cousins, R. J. (1991) Cysteine-rich intestinal protein binds zinc during transmucosal zinc transport. *Proc Natl Acad Sci USA*, 88, 9671–9614.

Hensley, K., Carney, J. M., Mattson, M. P., Aksenova, M., Harris, M., Wu, J. F., Floyd, R. A. and Butterfield, D. A. (1994) A model for beta-amyloid aggregation and neurotoxicity based on free radical generation by the peptide: relevance to Alzheimer disease. *Proc Natl Acad Sci USA*, 91, 3270–3274.

Hidalgo, J., Aschner, M., Zatta, P. and Vasak, M. (2001) Roles of the metallothionein family of proteins in the central nervous system. *Brain Res Bull*, 55, 133–145.

Hoadley, J. E., Leinart, A. S. and Cousins, R. J. (1988) Relationship of 65Zn absorption kinetics to intestinal metallothionein in rats: effects of zinc depletion and fasting. *J Nutr*, 118, 497–502.

Howell, G. A., Welch, M. G. and Frederickson, C. J. (1984) Stimulation-induced uptake and release of zinc in hippocampal slices. *Nature*, 308, 736–738.

Huang, L. and Gitschier, J. (1997) A novel gene involved in zinc transport is deficient in the lethal milk mouse. *Nat Genet*, 17, 292–297.

Huang, L., Kirschke, C. P. and Gitschier, J. (2002) Functional characterization of a novel mammalian zinc transporter, ZnT6. *J Biol Chem*, 277, 26389–26395.

Hurley, L. S. and Swenerton, H. (1966) Congenital malformations resulting from zinc deficiency in rats. *Proc Soc Exp Biol Med*, 123, 692–696.

Julien, J. P. (2001) Amyotrophic lateral sclerosis: unfolding the toxicity of the misfolded. *Cell*, 104, 581–591.

Kambe, T., Narita, H., Yamaguchi-Iwai, Y., Hirose, J., Amano, T., Sugiura, N., Sasaki, R., Mori, K., Iwanaga, T. and Nagao, M. (2002) Cloning and characterization of a novel mammalian zinc transporter, zinc transporter 5, abundantly expressed in pancreatic beta cells. *J Biol Chem*, 277, 19049–19055.

Kasarskis, E. J. (1984) Zinc metabolism in normal and zinc-deficient rat brain. *Exp Neurol*, 85, 114–127.

Keller, K. A., Chu, Y., Grider, A. and Coffield, J. A. (2000) Supplementation with L-histidine during dietary zinc repletion improves short-term memory in zinc-restricted young adult male rats. *J Nutr*, 130, 1633–1640.

Klug, A. and Schwabe, J. W. (1995) Protein motifs 5. Zinc fingers. *FASEB J*, 9, 597–604.

Lanningham-Foster, L., Green, C. L., Langkamp-Henken, B., Davis, B. A., Nguyen, K. T., Bender, B. S. and Cousins, R. J. (2002) Overexpression of CRIP in transgenic mice alters cytokine patterns and the immune response. *Am J Physiol Endocrinol Metab*, 282, E1197–E203.

Lee, J. Y., Cole, T. B., Palmiter, R. D., Suh, S. W. and Koh, J. Y. (2002) Contribution by synaptic zinc to the gender-disparate plaque formation in human Swedish mutant APP transgenic mice. *Proc Natl Acad Sci USA*, 99, 7705–7710.

Levenson, C. W., Shay, N. F., Lee-Ambrose, L. M. and Cousins, R. J. (1993) Regulation of cysteine-rich intestinal protein by dexamethasone in the neonatal rat. *Proc Natl Acad Sci USA*, 90, 712–715.

Magneson, G. R., Puvathingal, J. M. and Ray, W. J., Jr. (1987) The concentrations of free Mg^{2+} and free Zn^{2+} in equine blood plasma. *J Biol Chem*, 262, 11140–11148.

Markesbery, W. R., Ehmann, W. D., Alauddin, M. and Hossain, T. I. (1984) Brain trace element concentrations in aging. *Neurobiol Aging*, 5, 19–28.

McLardy, T. (1962) Zinc enzymes in the hippocampal mossy fibre system. *Nature*, 194, 300–302.

McMahon, R. J. and Cousins, R. J. (1998) Regulation of the zinc transporter ZnT-1 by dietary zinc. *Proc Natl Acad Sci USA*, 95, 4841–4846.

Miura, T., Suzuki, K., Kohata, N. and Takeuchi, H. (2000) Metal binding modes of Alzheimer's amyloid beta-peptide in insoluble aggregates and soluble complexes. *Biochemistry*, 39, 7024–7031.

Molinari, M., Filippini, V. and Leggio, M. G. (2002) Neuronal plasticity of interrelated cerebellar and cortical networks. *Neuroscience*, 111, 863–870.

Nitzan, Y., Sekler, I., Hershfinkel, M., Moran, A. and Silverman, W. (2002) Postnatal regulation of ZnT-1 expression in the mouse brain. *Brain Res Dev Brain Res*, 137, 149.

Noseworthy, M. D. and Bray, T. M. (1998) Effect of oxidative stress on brain damage detected by MRI and *in vivo* ^{31}P-NMR. *Free Radic Biol Med*, 24, 942–951.

Noseworthy, M. D. and Bray, T. M. (2000) Zinc deficiency exacerbates loss in blood-brain barrier integrity induced by hyperoxia measured by dynamic MRI. *Proc Soc Exp Biol Med*, 223, 175–182.

O'Dell, B. L., Becker, J. K., Emery, M. P. and Browning, J. D. (1989) Production and reversal of the neuromuscular pathology and related signs of zinc deficiency in guinea pigs. *J Nutr*, 119, 196–201.

Ohlsson, A. (1981) Acrodermatitis enteropathica: reversibility of cerebral atrophy with zinc therapy. *Acta Paediatr Scand*, 70, 269–273.

Olanow, C. W. (1993) A radical hypothesis for neurodegeneration. *Trends Neurosci*, 16, 439–444.

Oteiza, P. I., Olin, K. L., Fraga, C. G. and Keen, C. L. (1995) Zinc deficiency causes oxidative damage to proteins, lipids and DNA in rat testes. *J Nutr*, 125, 823–829.

Palmiter, R. D. (1995) Constitutive expression of metallothionein-III (MT-III), but not MT-I, inhibits growth when cells become zinc deficient. *Toxicol Appl Pharmacol*, 135, 139–146.

Palmiter, R. D. and Findley, S. D. (1995) Cloning and functional characterization of a mammalian zinc transporter that confers resistance to zinc. *EMBO J*, 14, 639–649.

Powell, S. R. (2000) The antioxidant properties of zinc. *J Nutr*, 130, 1447S–1454S.

Prasad, A., Mial, A., Farid, Z., Schulert, A., Sandstead, H. H. (1963) Zinc metabolism in patients with the syndrome of iron deficiency anemia, hypogonadism and dwarfism. *J Lab Clin Med*, 61, 537.

Prasad, A. S. (1988) Zinc in growth and development and spectrum of human zinc deficiency. *J Am Coll Nutr*, 7, 377–384.

Price, D. L., Whitehouse, P. J., Struble, R. G., Coyle, J. T., Clark, A. W., Delong, M. R., Cork, L. C. and Hedreen, J. C. (1982) Alzheimer's disease and Down's syndrome. *Ann N Y Acad Sci*, 396, 145–164.

Ratovitski, T., Corson, L. B., Strain, J., Wong, P., Cleveland, D. W., Culotta, V. C. and Borchelt, D. R. (1999) Variation in the biochemical/biophysical properties of mutant superoxide dismutase 1 enzymes and the rate of disease progression in familial amyotrophic lateral sclerosis kindreds. *Hum Mol Genet*, 8, 1451–1460.

Richards, M. P. and Cousins, R. J. (1977) Isolation of an intestinal metallothionein induced by parenteral zinc. *Biochem Biophys Res Commn*, 75, 286–294.

Rodriguez, J. A., Valentine, J. S., Eggers, D. K., Roe, J. A., Tiwari, A., Brown, R. H., Jr. and Hayward, L. J. (2002) Familial amyotrophic lateral sclerosis-associated mutations decrease the thermal stability of distinctly metallated species of human copper/zinc superoxide dismutase. *J Biol Chem*, 277, 15932–15937.

Sandstead, H. H., Fosmire, G. J., McKenzie, J. M. and Halas, E. S. (1975) Zinc deficiency and brain development in the rat. *Fed Proc*, 34, 86–88.

Sandstead, H. H., Frederickson, C. J. and Penland, J. G. (2000) History of zinc as related to brain function. *J Nutr*, 130, 496S–502S.

Sandstead, H. H., Penland, J. G., Alcock, N. W., Dayal, H. H., Chen, X. C., Li, J. S., Zhao, F. and Yang, J. J. (1998) Effects of repletion with zinc and other micronutrients on neuropsychologic performance and growth of Chinese children. *Am J Clin Nutr*, 68, 470S–475S.

Sandstead, H. H., Strobel, D. A., Logan, G. M., Jr., Marks, E. O. and Jacob, R. A. (1978) Zinc deficiency in pregnant rhesus monkeys: effects on behavior of infants. *Am J Clin Nutr*, 31, 844–849.

Sandstrom, B. (1989) In *Zinc in Human Biology* (Mills, C., Ed.), pp. 57–78. Springer-Verlag, New York.

Sawashita, J., Takeda, A. and Okada, S. (1997) Change of zinc distribution in rat brain with increasing age. *Brain Res Dev Brain Res*, 102, 295–298.

Sayre, L. M., Perry, G. and Smith, M. A. (1999) Redox metals and neurodegenerative disease. *Curr Opin Chem Biol*, 3, 220–225.

Sazawal, S., Bentley, M., Black, R. E., Dhingra, P., George, S. and Bhan, M. K. (1996) Effect of zinc supplementation on observed activity in low socioeconomic Indian preschool children. *Pediatrics*, 98, 1132–1137.

Seubert, P., Vigo-Pelfrey, C., Esch, F., Lee, M., Dovey, H., Davis, D., Sinha, S., Schlossmacher, M., Whaley, J., Swindlehurst, C. et al. (1992) Isolation and quantification of soluble Alzheimer's beta-peptide from biological fluids. *Nature*, 359, 325–327.

Shah, D. and Sachdev, H. P. (2001) Effect of gestational zinc deficiency on pregnancy outcomes: summary of observation studies and zinc supplementation trials. *Br J Nutr*, 85(Suppl. 2), S101–S108.

Swenerton, H., Shrader, R. and Hurley, L. S. (1969) Zinc-deficient embryos: reduced thymidine incorporation. *Science,* 166, 1014–1015.

Takeda, A., Kawai, M. and Okada, S. (1997) Zinc distribution in the brain of Nagase analbuminemic rat and enlargement of the ventricular system. *Brain Res,* 769, 193–195.

Takeda, A., Minami, A., Takefuta, S., Tochigi, M. and Oku, N. (2001) Zinc homeostasis in the brain of adult rats fed zinc-deficient diet. *J Neurosci Res,* 63, 447–452.

Taylor, C. M., Bacon, J. R., Aggett, P. J. and Bremner, I. (1991) Homeostatic regulation of zinc absorption and endogenous losses in zinc-deprived men. *Am J Clin Nutr,* 53, 755–763.

Tully, C. L., Snowdon, D. A. and Markesbery, W. R. (1995) Serum zinc, senile plaques, and neurofibrillary tangles: findings from the Nun Study. *Neuroreport,* 6, 2105–2018.

Valente, T., Auladell, C. and Perez-Clausell, J. (2002) Postnatal development of zinc-rich terminal fields in the brain of the rat. *Exp Neurol,* 174, 215–229.

Vallee, B. L. and Falchuk, K. H. (1993) The biochemical basis of zinc physiology. *Physiol Rev,* 73, 79–118.

Vallee, B. L. and Galdes, A. (1984) The metallobiochemistry of zinc enzymes. *Adv Enzymol Relat Areas Mol Biol,* 56, 283–430.

Volkl, A., Berlet, H. and Ule, G. (1974) Trace elements (Cu, Fe, Mg, Zn) of the brain during childhood. *Neuropadiatrie,* 5, 236–242.

Volpe, J. J. (1990) Brain injury in the premature infant: is it preventable? *Pediatr Res,* 27, S28–S33.

Wallwork, J. C., Milne, D. B., Sims, R. L. and Sandstead, H. H. (1983) Severe zinc deficiency: effects on the distribution of nine elements (potassium, phosphorus, sodium, magnesium, calcium, iron, zinc, copper and manganese) in regions of the rat brain. *J Nutr,* 113, 1895–1905.

Wang, K., Zhou, B., Kuo, Y. M., Zemansky, J. and Gitschier, J. (2002) A novel member of a zinc transporter family is defective in acrodermatitis enteropathica. *Am J Hum Genet,* 71, 66–73.

Washabaugh, M. W. and Collins, K. D. (1986) Dihydroorotase from *Escherichia coli.* Sulfhydryl group-metal ion interactions. *J Biol Chem,* 261, 5920–5929.

Widdowson, E. M., Dauncey, J. and Shaw, J.C. (1974) Trace elements in early fetal and postnatal development. *Proc Nutr Soc,* 33, 272.

Willis, W. D. (1998) In *Physiology* (Koeppen, B. M., Ed.), pp. 81–249. Mosby, Chicago.

Xu, Z., Squires, E. J. and Bray, T. M. (1994) Effects of dietary zinc deficiency on the hepatic microsomal cytochrome P450 2B in rats. *Can J Physiol Pharmacol,* 72, 211–216.

Yamashita, T., Shimada, S., Guo, W., Sato, K., Kohmura, E., Hayakawa, T., Takagi, T. and Tohyama, M. (1997) Cloning and functional expression of a brain peptide/histidine transporter. *J Biol Chem,* 272, 10205–10211.

Yoshiike, Y., Tanemura, K., Murayama, O., Akagi, T., Murayama, M., Sato, S., Sun, X., Tanaka, N. and Takashima, A. (2001) New insights on how metals disrupt amyloid beta-aggregation and their effects on amyloid-beta cytotoxicity. *J Biol Chem,* 276, 32293–32299.

Zola-Morgan, S., Squire, L. R. and Amaral, D. G. (1986) Human amnesia and the medial temporal region: enduring memory impairment following a bilateral lesion limited to field CA1 of the hippocampus. *J Neurosci,* 6, 2950–2967.

17 Copper and Brain Function

W. Thomas Johnson

CONTENTS

17.1 INTRODUCTION

Copper derived from dietary sources is present in low concentrations in the central nervous system (CNS) and is essential for normal brain function. Early recognition of the essentiality of copper for brain function occurred when Bennetts and Chapman (1937) discovered that low copper levels in the pastures of western Australia were associated with enzootic ataxia in lambs and subnormal copper concentrations in the blood and tissues of ewes and affected lambs. These initial findings were extended by subsequent research showing that maternal copper deficiency in laboratory animals leads to a variety of pathophysiological outcomes in the CNS of neonates. Examples of the neonatal pathologies associated with maternal copper deficiency include hypomyelination in rats and mice (DiPaolo et al., 1974; Prohaska and Smith, 1982), hypomyelination and cerebellar malformation in guinea pigs (Everson et al., 1968), abnormal hippocampal formation (Hunt and Idso, 1995), and impaired auditory response (Prohaska and Hoffman, 1996). Although the neuropathology in the offspring of copper-deficient mothers varies somewhat, depending on the species, experimental findings clearly indicate that copper is required for brain development and that the neurological effects of copper deficiency depend on the biological functions of this metal during developmental stages. After the brain has matured and neurons and glia are fully differentiated, dietary copper deficiency has little effect on brain function.

In addition to the neurological effects observed in neonates when copper deficiency is initiated during gestational development, mutations that cause copper malabsorption and abnormal cellular copper transport during developmental stages in humans and mice also confirm the essential role of copper in brain development and function. The brindled allele in the mottled mouse series of X-chromosomal mutants is associated with poor absorption of dietary copper. As a result, copper concentrations in brains of brindled mice are very low. These mice live only for ca. 2 weeks and die from neurological defects related to neuronal degeneration in the cerebral cortex and thalamic nuclei (Yajima and Suzuki, 1979). Menkes disease is also an X-chromosomal mutation that causes malabsorption of copper and extreme copper deficiency in humans (Danks et al., 1972a, 1972b). As a result of the defects in copper metabolism, brain copper concentrations are abnormally low in patients with Menkes disease (Danks, 1987). Thus, mental retardation, seizures, and severe neurodegeneration in the cerebellum and cerebrum associated with Menkes disease (Menkes et al.,

1962; Iwata et al., 1979) are clinical and pathological consequences resulting from copper deprivation in the brain.

The importance of copper for normal brain function and development has been unequivocally demonstrated in animal models showing that maternal dietary copper deficiency produces neurological defects in neonates and in laboratory animals and humans with inherited defects in copper transport that cause neuropathology by depriving the brain of copper. In this chapter, knowledge of the mechanisms underlying copper biochemistry is placed in perspective with brain function and the development of neurodegenerative diseases.

17.2 DISTRIBUTION OF COPPER IN THE BRAIN

Although there is interspecies variation, brain copper concentrations in mammals generally increase following birth (Smith, 1983). In humans, adult concentrations are achieved at ca. 11 years. Rats experience a gradual rise in brain copper concentration during early life until the concentration stabilizes at adult levels 3 to 4 weeks following birth when brain weight also becomes stabilized (Terao and Owen, 1977; Prohaska and Wells, 1974). Rates at which copper accumulates postnatally in the rat brain depend on the region of the brain. The medulla, pons, and hypothalmus accumulate copper most rapidly early in postnatal life, 5 to 10 days following birth, whereas the cerebellum, cortex, and midbrain accumulate copper most rapidly at 10 to 20 days following birth (Lai et al., 1985).

Mean adult concentrations of copper in brain are 24 µg/g dry tissue in humans and 14 µg/g dry tissue in rats (Smith, 1983). In the human brain, copper concentrations vary considerably between regions and are higher in the gray matter than in the white matter. Although the highest copper concentrations are in the substantia nigra and locus ceruleus, the cerebellar cortex, pallidum, and putamen also contain substantial concentrations (Table 17.1). Copper concentrations are less variable between regions in the rat brain than in the human brain. Among regions of the rat brain, the hypothalamus stands out as having the highest copper concentration. However, the apparent lack of differences between copper concentrations in different regions of the rat brain may result more from imprecise dissection of the brain into high-copper and low-copper regions than from actual similarity in copper content.

TABLE 17.1
Regional Distribution of Copper in Rat and Human Brain

Rat		Human	
Region	µg Cu/g Dry Tissue	Region	µg Cu/g Dry Tissue
Cerebellum	14.3	Centrum semiovale	13.7
Cerebral cortex	15.0	Cerebral cortex (frontal)	24.7
Midbrain	14.5	Cerebellar cortex	33.1
Medulla	18.3	Cerebellar white matter	13.5
Hippocampus	11.1	Corpus callosum	9.8
Hypothalamus	14.6	Hippocampus	21.0
Striatum	14.9	Locus ceruleus	201
		Palladium	30.3
		Substantia nigra	59.9
		Putamen	32.9

Source: Values for copper concentrations from Prohaska, J.R. (1987) *Physiol. Rev.* 67, 858–901. With permission.

The accrual of copper in the rat brain following birth occurs during a time of brain growth, neuronal and glial maturation, and synaptogenesis. Depriving the brain of copper during postnatal brain development, through either dietary copper deficiency or inherited defects in copper absorption, likely produces neuropatholgy and defects in brain maturation by impairing copper-dependent processes that regulate maturation, cellular differentiation, and synaptogenesis. Copper deficiency in rats and mice during pregnancy and lactation greatly attenuates copper accrual in the neonatal brain and results in major reductions in brain copper concentrations (Prohaska and Wells, 1974, Prohaska and Smith, 1982; Prohaska and Bailey, 1993; Johnson et al., 2000). No region of the brain is spared from the reduction in copper accrual resulting from copper deficiency during neonatal copper deficiency. Copper concentrations in the cerebrum, midbrain, hypothalamus, striatum, cerebellum, and medulla are reduced by 45% to more than 80% in 4-week-old neonates of copper-deficient mice and rats compared with 4-week-old controls (Prohaska and Bailey, 1993, 1994). Furthermore, reductions in brain copper concentrations that occur during postnatal development in mice and rats are irreversible once brain maturation is complete. In mice fed adequate copper after being deprived of copper during postnatal brain development, copper concentrations remained at 40 to 60% of control values in the cerebrum, midbrain, cerebellum, and medulla (Prohaska and Bailey, 1993). Similar results have been obtained with rats. Copper concentrations in the cerebrum, hypothalamus, midbrain, striatum, medulla, and cerebellum of rats that are copper deficient for 4 weeks following birth remain at abnormally low levels after 1, 2, 3, and even 5 months of copper repletion (Prohaska and Bailey, 1995a, 1995b; Prohaska and Hoffman, 1996). However, the hypothalamus and medulla are exceptions in that copper concentrations in these regions are more readily restored by copper repletion. Also, the auditory startle response in rats who are copper deficient as neonates cannot be fully restored by copper repletion and remains diminished after they consume adequate dietary copper for 1, 3, or 5 months. Thus, copper deprivation during the terminal stages of brain development irreversibly lowers copper concentrations in most regions of the brain if sufficient dietary copper is not provided before brain maturation is complete. Furthermore, brain maturation in itself may be affected by copper deprivation in a manner that has irreversible, detrimental consequences on neurological function and behavior.

17.3 CUPROENZYMES IN THE BRAIN

The influence of copper on biological function is mediated by a fairly large number of cuproenzymes that depend on copper for their catalytic activity. The most common cuproenzymes found in vertebrates are listed in Table 17.2. Molecular configurations of the copper sites in these enzymes

TABLE 17.2
Copper-Dependent Enzymes and Their Functions in Animals

Enzyme	Function
Cytochrome c oxidase	Mitochondrial electron transfer
CuZn superoxide dismutase	Dismutation of superoxide ($O_2^{\cdot-}$)
Dopamine-β-monooxygenase	Synthesis of norepinephrine
Lysyl oxidase	Cross-linking during elastin and collagen synthesis
Peptidylglycine α-amidating monooxygenase	C-terminus amidation and activation of glycine-extended peptides
Ceruloplasmin	Cu transport and Fe^{2+} oxidation
Amine oxidase	Oxidative deamination
Diamine oxidase	Oxidative deamination

lead to high positive reducing potentials that are ideal for the catalysis of redox reactions in which cuproenzymes participate as electron receptors. As a consequence of the redox properties of cuproenzymes, they serve exclusively as oxidoreductases in a variety of biochemical pathways. Some cuproenzymes are widely distributed in a variety of organs and types of cells, whereas others are confined to cells having specialized functions in which they catalyze specific reactions. Cuproenzymes having important metabolic roles in the CNS and whose activities are affected by development and copper deficiency during development are cytochrome c oxidase, CuZn superoxide dismutase, dopamine-β-monoxygenase, and peptidyl α-amidating monooxygenase.

Cytochrome c oxidase (CCO, EC 1.9.3.1) is the terminal oxidase of the mitochondrial electron transport chain. This enzyme is compartmentalized exclusively to the inner mitochondrial membrane, where it catalyzes the reduction of molecular oxygen to water through the oxidation of cytochrome c as shown in Equation 17.1:

$$4 \text{ Cytochrome } c(Fe^{+2}) + O_2 + 4 \text{ H}^+ \rightarrow 4 \text{ Cytochrome } c(Fe^{+3}) + 2 \text{ H}_2O \qquad (17.1)$$

In mammals, CCO is a multimer having at least 13 subunits with a total molecular weight of approximately 200,000 Da (Capaldi, 1990).

The three largest subunits (I, II, and III) are encoded by the mitochondrial DNA and the rest of the subunits are encoded by nuclear DNA and imported into the mitochondria. This complex protein contains two heme centers, cytochrome a and cytochrome a_3, and two copper atoms, Cu_A and Cu_B, localized to subunits I and II. The locus of Cu_A most likely resides in subunit II. The copper sites Cu_A and Cu_B are associated with cytochromes a and a_3, respectively, and participate in electron transport by cycling between reduced and oxidized states. CCO is a major cuproenzyme in the brain and contains ca. 20% of the total brain copper (Prohaska, 1987).

The activity of CCO in the rat brain increases several fold over a 4-week period following birth and parallels the increase in copper concentration that occurs in the brain of neonates (Prohaska and Wells, 1974). Neuronal development in rat brain continues postnatally for 3 to 4 weeks and the increase in CCO activity during this period may reflect the creation of mitochondria in newly formed brain structures.

Postnatal copper deficiency caused by either low dietary intakes or genetic defects in transport impairs the expression of CCO activity during neonatal brain development. When copper deficiency is initiated in rats during gestation and continued through lactation, CCO in the brain at postnatal day 21 is reduced by 75 to 80% (Prohaska and Wells, 1974, 1975). Deprivation of copper in brain through genetic defects in copper transport affects CCO activity in a manner similar to that for dietary deprivation. Prohaska (1983) found that CCO activity was reduced by 70% in brains of mottled mice (Mo$^{br/y}$) compared with controls (Mo$^{+/y}$). Although copper is required for catalytic activity, inhibition of CCO by copper deficiency may be the consequence of a bimodal effect. Copper is required for normal heme synthesis, and it has been shown that copper deficiency causes a fourfold reduction in the cytochrome a + a_3 content in mitochondria isolated from brains of copper-deficient suckling rats (Prohaska and Wells, 1975). Thus, the inhibitory effect of copper deficiency on CCO activity in the developing brain is likely a consequence of both impaired synthesis of heme a and the loss of copper from the active centers of the enzyme.

The biological consequences of impaired CCO activity caused by copper deficiency during brain development are not clear. Although CCO activity is severely reduced, energy metabolites ATP, ADP, and creatine-P remain unchanged in the brains of copper-deficient rats (Prohaska and Wells, 1975). However, even though energy metabolism in the brain mitochondria is not grossly affected by lower CCO activity, the lactate/pyruvate and α-glycerol-P/dihydroxyacetone-P ratios are increased, indicating that the brain is in a more reduced state during copper deficiency. This indicates that impaired CCO activity associated with copper deficiency can damage brain mitochondria and affect their ability to oxidize electron donors such as NADH, placing the brain in a higher reduced state without greatly affecting ATP content.

Activation of the mitochondrial pathway for apoptosis is another potential biological consequence related to the inhibitory effect of copper deficiency on CCO activity in the brain. A high
number of apoptotic cells are present in the hippocampus and neocortex of mottled mice (Mo$^{br/y}$).
Apoptosis in these regions is associated with a decrease in the expression of Bcl-2, an antiapoptotic
protein, and an increase in the release of cytochrome c from mitochondria. Reduced brain copper
concentration and CCO activity accompany hippocampal and neocortical apoptosis in mottled mice
(Rossi et al., 2001), indicating that mitochondrial damage resulting from the inhibitory effect of
copper deprivation on CCO activity may cause neurodegeneration by triggering the mitochondrial
apoptotic pathway.

CuZn superoxide dismutase (EC 1.15.1.1), also known as SOD1, is another important cuproenzyme that is found in a variety of organs, including the brain. SOD1 is a homodimer with a molecular
weight of 32,000 Da. Each subunit contains one atom of copper and one atom of zinc (Fridovich,
1974). This enzyme is a major cuproenzyme of the brain and constitutes ca. 25% of total brain
copper (Prohaska, 1987). SOD1 is found primarily in the cytosol, where it catalyzes the formation
of hydrogen peroxide and oxygen through the dismutation of superoxide ($O_2^{·-}$) as shown in Equation
17.2:

$$O_2^{·-} + O_2^{·-} + 2H^+ + H_2O_2 + O_2 \qquad (17.2)$$

SOD1 is found in all regions of the brain, but activity is highest in the brain stem and hypothalamus
(Thomas et al., 1976).

The activity of SOD1 is low at birth and increases during postnatal brain development in the
rat. However, activity tends to decline in the cerebellum 3 weeks following birth (Prohaska and
Wells, 1974) and generally declines in the whole brain with age (Ledig et al., 1982). Copper
deficiency does not affect SOD1 activity during the early stages of neonatal brain development,
but significantly suppresses activity beyond the third week following birth (Prohaska and Wells,
1974, 1975). SOD1 activity also is reduced by 20% in the brains of mottled mice (Mo$^{br/y}$) compared
with controls (Mo$^{+/y}$; Prohaska, 1983). These findings indicate that depriving the brain of copper
during development either through dietary deficiency or genetic defects in transport impairs the
expression of SOD1 activity.

Biological consequences resulting from the suppression of SOD1 activity by copper deficiency
during brain development are likely to occur from increased concentration of $O_2^{·-}$. Superoxide
serves as a reactant for the formation of hydroxyl radical ($·OH$), which can oxidize numerous
biomolecules, including polyunsaturated lipids, proteins, and DNA. During brain development,
SOD1 activity increases in all brain regions, but the maximal increase occurs in the medulla-pons,
a region having high catecholamine content (Ledig et al., 1982). Catecholamines are susceptible
to oxidation by $O_2^{·-}$ and the presence of SOD1 in brain regions having high catecholamine content
indicates that an important biological role for this enzyme centers around its ability to destroy $O_2^{·-}$.
Although $O_2^{·-}$ is not highly reactive, its destruction by SOD1 protects catecholamines and other
biomolecules in the brain that are susceptible to oxidation by reactive oxygen species formed from
$O_2^{·-}$. However, copper deficiency during brain development does not increase lipid peroxidation in
brains of 28-day-old rats (Prohaska and Wells, 1975), suggesting that copper deficiency does
not suppress SOD1 activity sufficiently in developing rat brain to impair its function through
increased lipid peroxidation. Oxidation of proteins, nuclear DNA, and mitochondrial DNA has not
been examined in brains of copper-deficient animals, and it is possible that increased oxidation of
these biomolecules as a result of decreased SOD1 activity contributes to impaired brain development
and function.

Dopamine β-monooxygenase (DBMO, EC 1.14.17.1) is a cuproenzyme found in low abundance
in the brain and constitutes less than 0.1% of total brain copper (Prohaska, 1987). Although a minor
cuproenzyme in terms of abundance, DBMO catalyzes an important reaction: the formation of
norepinephrine from dopamine (Equation 17.3).

$$3,4\text{-Dihydroxyphenylethylamine (dopamine)} + O_2 + 2\text{ Ascorbate} + NAD^+ \rightarrow$$

$$\text{Norepinephrine} + NADH + 2\text{ Semidehydroascorbate} + H_2O \qquad (17.3)$$

This enzyme is a glycoprotein tetramer with a molecular weight of 290,000 Da. Each subunit requires two atoms of Cu for maximal activity (Stewart and Klinman, 1988). The low abundance of DBMO relative to other cuproenzymes in the brain is related to its specific biological role in the synthesis of norepinephrine. This specialization limits the distribution of the enzyme to the locus ceruleus, brain stem, and posterior hypothalamus, which are regions of the brain enriched with cell bodies of noradrenergic neurons (Coyle and Axelrod, 1972; Cimarusti et al., 1979).

As expected from the dependence of DBMO on copper for the catalytic conversion of dopamine to norepinephrine, copper deprivation decreases norepinephrine concentrations in several regions of the brain. When copper deficiency is initiated during late gestation and maintained for 4 weeks following birth, norepinephrine concentrations decrease in most regions of the brain in male and female offspring, except for the hypothalamus in females, whereas dopamine concentrations increase in the cerebellum and medulla (Prohaska and Bailey, 1993). Similarly, initiation of copper deficiency during pregnancy in rats greatly decreases the ratio of norepinephrine to dopamine in the cerebellum and medulla and modestly decreases the ratio in the hypothalamus of the offspring (Prohaska and Bailey, 1994). These findings indicate that copper deficiency causes DBMO to become rate limiting for the synthesis of norepinephrine in brain regions innervated by noradrenergic neurons.

Although DBMO activity depends on bound Cu and measurements of catecholamine concentrations are consistent with decreased DBMO activity, copper deficiency actually increases the measurable activity of DBMO. In mice subjected to dietary copper deficiency during neonatal brain development, DBMO activity in brain is elevated by 25% and mottled mice ($Mo^{br/y}$) also exhibit a 50% increase in brain DBMO activity. DBMO activity is also increased in the midbrain and medulla of rats subjected to dietary copper deficiency during brain development (Prohaska and Bailey, 1993). The discrepancy between the effects of copper deficiency on catecholamine concentrations and DBMO activity in the brain may in itself reflect the suppressive effect of copper deficiency on the synthesis of norepinephrine. The depletion of norepinephrine may induce the synthesis of DBMO, which, because of copper deprivation, remains inactive as an apoprotein. Although the increase in DBMO production in the brain does not help alleviate the depletion of norepinephrine concentrations caused by copper deficiency because the enzyme is inactive *in vivo*, the higher content becomes apparent when the enzyme is maximally activated *in vitro* by trace amounts of Cu used in sample preparation or assaying activity (Prohaska and Smith, 1982).

Peptidylglycine α-amidating monooxygenase (PAM, EC 1.14.17.3) has two independent enzymatic domains, a monooxygenase domain and a lyase domain. These domains catalyze two sequential reactions for the amidation of peptides containing glycine at their carboxy terminus as shown in Equation 17.4:

$$\text{Peptide-glycine} + O_2 + 2\text{ Ascorbate } \alpha \rightarrow \text{Peptide-}\alpha\text{-hydroxyglycine}$$

$$+ 2\text{ Semidehydroascorbate} + H_2O \rightarrow \text{Peptide-NH}_2 + \text{Glyoxylate} \qquad (17.4)$$

The monooxygenase domain is ascorbate dependent and contains two Cu atoms (Prigge et al., 1997). PAM activity varies within brain regions in rats. The activity is highest in the hypothalamus and midbrain and lowest in the cerebellum, with intermediate activities in the cerebrum, striatum, and medulla of male and female rats. PAM activity is lowered in all brain regions of 4-week-old male and female rats subjected to copper deficiency during postnatal brain development. However, reductions are greatest in the midbrain, cortex, and cerebellum and least in the hypothalamus. After 1 month of copper repletion, PAM activity was restored to normal in all brain regions except the cerebrum of male and female rats, the cerebellum of male rats, and the medulla of female rats (Prohaska and Bailey, 1995b). This suggests that regional differences in the recovery of PAM activity following copper deficiency may exist.

Many peptide hormones and neuropeptides exist in glycine-extended forms that require amidation at the carboxy terminus to exhibit full activity. These peptides include gastrin, oxytocin, neuropeptide Y, substance P, choleocystokinin, and calcitonin. More research is required to determine whether copper deficiency lowers PAM activity sufficiently to reduce the activation of glycine-extended peptides and impair their physiological functions. However, reducing PAM activity *in vivo* by treating animals with copper chelators decreases the content of amidated gastrin and increases the content of glycine-extended progastrin in the antrum. This leads to increased basal and gastrin-stimulated gastric acid secretion, possibly as a result of upregulation of the gastrin receptor (Dickenson et al., 1990). Other experiments have shown that treatment of mice with a copper-specific chelator increases their pain threshold. These mice also exhibit changes in the contents of substance P and unamidated glycine-extended forms of substance P in their CNS (Marchand et al., 1990). Although these studies employed copper chelators, they suggest that dietary copper deficiency, particularly during postnatal development, has the potential to affect brain function by limiting the catalytic activity of PAM and altering the activation of glycine-extended neuropeptides.

17.4 COPPER-TRAFFICKING PROTEINS IN THE BRAIN

Research has clearly shown that copper deficiency resulting from low dietary intake or from genetic defects in copper transport, such as that found in Menkes disease and mottled mice, leads to abnormal brain development and function. The neurological defects associated with copper deficiency arise from the impaired function of metabolically essential copper-dependent enzymes and proteins. However, copper can be toxic if the cellular concentration exceeds the metabolic requirement. An example of copper toxicity that produces neurological abnormalities is Wilson's disease. This is an autosomal recessive disorder that causes pathological accumulations of copper primarily in the liver, brain, cornea, and kidney. Patients with Wilson's disease experience chronic liver disease with cirrhosis, renal dysfunction, abnormal corneal pigmentation, neurological symptoms such as behavior and movement disorders, and dysarthria. To counteract the potential toxicity of copper, homeostatic mechanisms are present in organ systems, including the nervous system, that provide the critical balance needed for supplying copper to essential proteins while preventing the accumulation of copper to toxic levels. However, the redox chemistry of copper provides a serious complication in the delivery of this metal to proteins that require it. Copper ions participate in reactions that produce reactive oxygen species that are capable of oxidizing lipids, proteins, DNA, and RNA, and these biomolecules must be protected during intercellular copper transport and delivery. Such protection is provided by a variety of trafficking proteins and lower-molecular-weight chaperones that regulate copper homeostasis by carefully orchestrating copper uptake and distribution within cells and its removal while maintaining copper in a bound state, which prevents its participation in undesirable redox reactions.

A profound leap in knowledge regarding intercellular copper trafficking occurred with the discovery of the genes responsible for Menkes and Wilson's diseases. The basic defect in Menkes disease is failure in the cellular export of copper. This results in copper deprivation in the brain and other organs, because adsorbed copper cannot be mobilized from the intestinal epithelial cells. The gene responsible for Menkes disease was identified in 1993 by three different laboratories as a gene that codes for a membrane-associated P-type ATPase (Chelly et al., 1993; Mercer et al., 1993; Vulpe et al., 1993). The Menkes protein (MNK) contains six successive repeats of a highly conserved copper-binding motif (Gly-Met-Thr-Cys-X-X-Cys) within its amino-terminal region. Mutations in the Menkes gene (*ATP7A*) and the murine homolog of *ATP7A* lead to the defects in copper metabolism that occur in Menkes disease and in mottled mouse mutants that serve as models for Menkes disease (Mercer, 1998).

Evidence for the role of MNK in copper efflux has come from experiments with Chinese hamster ovary cells. When these cells overexpress MNK as a result of *ATP7A* amplification, copper efflux increases in a manner that depends on the extent of MNK expression. The underlying mechanism for the involvement of MNK in copper efflux has been elucidated by immunofluoresence studies showing that MNK is located intracellularly within the Golgi apparatus of Chinese hamster ovary cells (Petris et al., 1996) and HeLa cells (Suzuki and Gilin, 1999). In both these cell types, increasing copper concentrations in the culture medium causes a redistribution of MNK from the trans-Golgi to a vesicular distribution throughout the cytoplasm and to the plasma membrane. Furthermore, the redistribution of MNK is reversible when copper concentrations are reduced in the medium. These findings indicate that MNK senses and incorporates intracellular copper for its removal through the secretory pathway. A unique feature of MNK is that its sequential array of copper-binding motifs is ideal for sensing copper and allowing copper to regulate the intercellular trafficking of its transport protein.

Expression of *ATP7A* occurs in the brain, heart, lung, muscle, kidney, pancreas, and placenta, with little or no expression in the liver. Northern blot analysis has shown that the murine homolog of *ATP7A* is expressed strongly in the choroid plexus, Ammon's horn, dentate gyrus, Purkinje cells, and the granular layer of the cerebellum in both normal and mottled mice. Furthermore, the levels of expression in these brain regions are similar in normal and mutant mice (Murata et al. 1997). Also, the mRNA for MNK is the same in size (8.3 kb) and extent of expression in brains of normal mice and mottled mutants following birth. The location and expression of *ATP7A* has also been determined by *in situ* hybridization. Hybridization signals are present in the hippocampal CA1 and CA3 regions, dentate gyrus, olfactory bulb nuclei, cerebellar granular cell layer and Purkinje cells, choroid plexus, and ependyma (Iwase, 1996). The location and expression of *ATP7A* determined by *in situ* hybridization also do not differ in normal and mottled mutant mice on postnatal days 4, 10, and 13. These findings indicate that no obvious differences in *ATP7A* expression occur between the brains of normal mice and mottled mutants during postnatal brain development. However, the distribution of *ATP7A* in the mottled mouse brain correlates well with regions that experience neuronal necrosis. This indicates that neuronal degeneration in specific regions of the brain and the neurological manifestations that characterize mottled mutants are functional outcomes of perturbations in copper metabolism resulting directly from mutations in MNK.

The *ATP7A* gene and MNK protein have also been detected in cultured C6 rat glioma cells and PC12 neuron-like cells (Qian et al., 1997). This has led to a proposed model in which MNK mediates not only the efflux of copper but also the transfer of copper from glial cells to neurons and from neurons to glial cells when neuronal copper is in excess.

MNK may also participate in the trafficking of copper to intercellular compartments for specific metabolic purposes. Pituitary adenylate cyclase activating polypeptide (PACAP) stimulates neuronal proliferation and survival of olfactory neurons. PACAP is a glycine-extended polypeptide that requires α-amidation for the full expression of its activity. Thus, olfactory epithelium expresses not only PACAP but also copper-dependent PAM, which is required for the amidation and activation of PACAP. It has been shown that mottled mice experience a loss in olfactory receptor neurons as a direct result of impaired production of amidated PACAP (Hansel et al., 2001). This indicates that the mutated form of MNK in mottled mice is ineffective in delivering copper to PAM for full maintenance of its α-amidating activity. The loss of PAM activity in this particular instance has a direct negative effect on the amidation and activation of PACAP that results in impaired olfactory neurogenesis. This finding indicates that MNK has a direct role in delivering copper to cuproenzymes in the secretory pathway of neuronal cells.

The gene responsible for Wilson's disease is known as *ATP7B* and codes for the protein WND, which is also a P-type ATPase with 60% similarity to MNK (Bull et al., 1993; Tanzi et al., 1993; Yamaguchi et al., 1993). The structure of *ATP7B* is almost identical to that of the Menkes gene, and WND contains six copper-binding motifs within the amino-terminal region. In HepG2 cells, WND is located primarily in the trans-Golgi, but redistributes to cytosolic vesicles in the presence of excess copper in the culture medium. The redistribution of WND from the Golgi to cytosolic vesicles is reversed on removal of excess copper from the medium (Suzuki and Gilin, 1999). Thus, WNK is similar to MNK in its ability to sense copper concentrations and use the cellular secretory pathway to remove copper from the cell. However, the primary function of WNK in the Golgi is to incorporate copper into ceruloplasmin, the major extracellular copper transport protein, whereas MNK regulates copper release at the outer cellular membrane.

WND and its mRNA are detectable in neuronal cells of the hippocampus, olfactory bulbs, cerebellum, cerebral cortex, and the pontine and lateral reticular nuclei of the brain stem in rats (Saito et al., 1999). These regions are also the regions containing high amounts of detectable copper and the cuproenzymes DBMO and SOD1. Mutations of *ATP7B* cause abnormal catecholamine metabolism and lower DBMO content in the cerebral cortex in a rat model of Wilson's disease (Saito et al., 1996). Thus, WND may help regulate DBMO by controlling copper homeostasis in brain regions innervated by noradrenergic neurons. Mutations in WND may lead to the neurological manifestations of Wilson's disease by causing copper to accumulate in critical brain regions and disrupting catecholamine metabolism by impairing DBMO activity.

The Menkes and Wilson's disease proteins are important copper-binding proteins that are confined to the Golgi and secretory vesicles, where they regulate the trafficking of copper through the secretory pathway. A group of smaller peptides known as copper chaperones are present in the cytosolic compartment and plasma membrane, where they form complexes with copper and deliver it to specific molecules or subcellular organelles. Four copper chaperones, CCS, Atx1, Ctr1, and Cox17, have been identified in the brain.

The mammalian copper chaperone CCS shares 28% amino acid identity with the yeast chaperone LYS7. CCS contains 274 amino acids and has a highly conserved copper-binding motif, Met-X-Cys-X-X-Cys, within the amino-terminal region (Culotta et al., 1997). This chaperone is expressed in many tissues and delivers copper specifically to SOD1. In the brain and spinal cord, CCS is confined primarily to neurons. The highest amounts of CCS are found in the Purkinje cells, deep cerebellar nuclei, pyramidal cortical neurons, and motor neurons in the spinal cord. In conjunction with its specificity for copper delivery, expression of CCS generally parallels that of SOD1 in the brain and spinal cord (Rothstein et al., 1999).

Another copper chaperone originally found in yeast is Ctr1. Ctr1 is a high-affinity copper transport protein responsible for transporting copper across the plasma membrane. Human Ctr1 contains 190 amino acids and has 92% identity to mouse Ctr1, which has 188 amino acids. Both the human and mouse peptides have three transmembrane domains and two methionine-rich and two histidine-rich regions within the extracellular amino-terminal domain, which may serve as extracellular copper-binding sites for high-affinity copper uptake (Lee et al., 2000).

The gene for Ctr1 is expressed in a variety of human and mouse tissues, including the brain (Zhou and Gitschier, 1997; Lee et al., 2000). During embryogenesis in the mouse, Ctr1 is expressed in the forebrain at embryonic day 14.5 and in the choroid plexus at embryonic day 16.5. In adult mouse brain, Ctr1 expression is highest in the epithelia of the choroid plexus. The important role of Ctr in copper transport in brain has been demonstrated in mice having a null mutation in the *Ctr1* gene (Kuo et al., 2001; Lee et al., 2001). Although this is a lethal mutation in homozygotes that causes severe embryonic malformation with no survival beyond embryonic day 9.5, heterozygote embryos are phenotypically indistinguishable from wild type. Heterozygous adults are also indistinguishable from wild type, except that brain copper concentration is reduced by 50%. Reduction in copper concentrations is not significant in the intestine, kidney, or liver of heterozygous adults. Furthermore, reductions in brain copper concentrations in heterozygous adults are accompanied by significant reductions in the activities of CCO and SOD1 (Lee et al., 2001). These findings indicate that Crt1 is essential for embryonic development and has an important role in copper transport in the brain. The choroid plexus is the site for the synthesis of cerebral spinal fluid, and the high degree of Crt1 expression in this region indicates that Crt1 may transport copper through the choroid plexus epithelia to be taken up and used by copper-dependent enzymes and proteins in neurons and glia. However, expression of Crt1 in the brain is likely not affected by cellular copper availability because mRNA for Crt1 is not lower in the hypothalamus of copper-deficient rats even though cerebellar copper is reduced by 84% (Lee et al., 2000).

The copper chaperone Atx1, referred to as HAH1 in humans, was also first discovered in yeast. This peptide is responsible for high-affinity transport of intracellular copper to copper-requiring proteins in the secretory pathway (Lin et al., 1997; Pufahl et al., 1997). Atx1 has a single copper-binding motif, Met-Thr-Cys-X-X-Cys, which is similar to the motif found in the Menkes and Wilson's disease proteins and may directly transfer copper to these proteins in the secretory pathway.

A wide variety of human tissues express mRNA for HAH1. These tissues include several brain regions: the amygdala, caudate, corpus callosum, hippocampus, hypothalamus, substantia nigra, subthalamic nucleus, and thalamus (Klomp et al., 1997). Northern blot analysis has also shown that Atx1 is expressed in the cerebellum, brain stem, hippocampus, cortex, thalamus, and striatum of the rat brain. Histochemically, Atx1 is expressed primarily in the pyramidal neurons of the cerebral cortex and hippocampus and in neurons of the locus coeruleus (Naeve et al., 1999). The areas of highest Atx1 expression are targets of noradrenergic fibers from the locus coeruleus and are also brain regions that sequester high levels of copper. The correlation of high Atx1 expression and high copper content in the adrenergic system of the brain suggests that Atx1 has a role in directing copper to cells that have a special need for this metal to maintain the activity of DBMO and other cuproenzymes in these regions.

In yeast, the assembly of functional CCO depends on the chaperone Cox17 to deliver copper to the mitochondria for its insertion into CCO. Homology between yeast and mammalian Cox17 is high and both yeast and mammalian peptides contain copper-binding motifs with the amino acid sequence Lys-Pro-Cys-Cys-X-Cys. A comprehensive examination of Cox17 in rodent tissues and cells (Kako et al., 2000) has shown that Cox17 is highly expressed in the mouse pituitary; moderately expressed in the cerebral cortex, cerebellum, and brain stem; and poorly expressed in the hippocampus and hypothalamus. Expression of Cox17 in the rat brain varies with postnatal age, temporarily declining until postnatal day 8 before increasing to adult levels. The period of lowest expression coincides with a period of peak natural apoptosis in CNS neurons in rats before progressive formation of the neural network occurs. Thus, the temporal changes in Cox17 expression in brain following birth suggest that this chaperone may has a role in regulating copper trafficking during different periods of brain development.

The discovery of copper transport proteins and chaperones in the CNS has led to a general picture of copper trafficking in neurons as shown in Figure 17.1. Copper uptake in the nervous system is likely to be mediated by the copper transporter Ctr1 in the plasma membrane. From Ctr1, copper is transferred to CCS, Cox17, and Atox1, which direct copper specifically to SOD1, mitochondria, and the trans- Golgi, respectively. Copper directed to the mitochondria is incorporated into CCO and copper directed to the Golgi becomes bound to MNK and WND. Both MNK and WND maintain copper homeostasis by carrying the copper to the secretory pathway, where they release copper to secretory vesicles or to extracellular transport proteins such as ceruloplasmin before returning to the trans-Golgi. MNK also has the additional roles of transferring copper to cuproenzymes such as PAM in the secretory pathway and reversibly transporting copper between glial cells and neurons.

17.5 COPPER AND NEURODEGENERATIVE DISEASE

As discussed previously, copper homeostasis in the brain is maintained by a number of chaperones and binding proteins that regulate copper trafficking to cellular compartments containing proteins that either require copper for activity or have a role in copper transport through the secretory pathway. However, the delivery of copper to some neural proteins may be directly involved in the

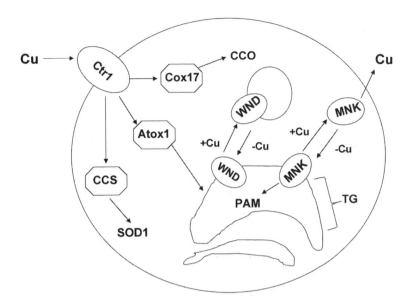

FIGURE 17.1 Schematic representation of copper (Cu) absorption and trafficking in neuronal cells. Arrows trace the movement of Cu through the plasma membrane transporter Ctr1 to specific chaperones (CCS, Atox1, and Cox17) that deliver intracellular copper to cuproenzymes or to copper transport proteins (MNK and WND) in the trans-Golgi network (TG). Copper bound to MNK in the TG is either trasferred to the plasma membrane for efflux or delivered to peptidylglycine α-amidating monooxygenase (PAM) and other secretory proteins. Both MNK and WND regulate copper homeostasis by sensing intracellular Cu levels, trafficking Cu to either the plasma membrane for efflux or cytosolic vesicles.

pathogenesis of neural degenerative diseases, particularly familial amyotrophic lateral sclerosis and Alzheimer's disease.

Mutations in SOD1 are known to be associated with the development of familial amyotrophic lateral sclerosis (FALS). Approximately 10 to 20% of FALS can be linked to 90 different autosomal-dominant mutations in the gene encoding SOD1 (Rowland and Schneider, 2001). Curiously, the mutations in SOD1 do not significantly affect enzymatic activity or the ability of the enzyme to scavenge $O_2^{\cdot-}$ (Winterborn et al., 1995; Marklund et al., 1997). Thus, the deleterious effects of the mutations on motor neurons that ultimately culminates in FALS result from a gain in function of SOD1 and are not directly related to impaired $O_2^{\cdot-}$ metabolism resulting from loss of SOD1 activity.

The gain in function associated with the mutant forms of SOD1 apparently results from changes in the affinity of the enzyme for zinc. Relative to the wild-type enzyme, zinc affinity of several FALS-associated SOD1 mutants is decreased by 5- to 50-fold whereas copper affinity is similar to the wild-type enzyme (Lyons et al., 1996; Crow et al., 1997). Low affinity of the SOD1 mutants for zinc may result in the loss of zinc and the formation of neurotoxic forms of the enzyme that lead to the selective degeneration of motor neurons. When either wild-type SOD1 or several FALS mutants are depleted of zinc, they all initiate apoptosis in motor neurons cultured in the presence of the brain-derived neurotrophic factor. However, neither the wild-type nor the FALS mutants can initiate apoptosis if they are replete with both copper and zinc (Estevez et al., 1999). Thus, neurotoxicity of SOD1 requires copper but not zinc to be bound to the protein. The loss of zinc from SOD1 decreases its ability to react with $O_2^{\cdot-}$ and allows nitric oxide (NO) to effectively compete with SOD1 for $O_2^{\cdot-}$ and form the highly reactive oxidant peroxynitrite ($ONOO^-$). Neurotoxicity leading to the degeneration of motor neurons may then result from increased $ONOO^-$-mediated tyrosine nitrosylation of proteins having essential metabolic roles in neuronal function (Estevez et al., 1999; Crow et al., 1997). The FALS-linked mutations in SOD1 may also increase inflammation in motor neurons. Genes related to the inflammatory process and apoptosis are upregulated during neurodegeneration in transgenic mice expressing a FALS mutation (G93A; Yoshihara et al., 2002). This indicates that neuronal death and the development of FALS resulting from mutations in SOD1 may occur through mechanisms involving inflammation and apoptosis.

Whether the gain in function of SOD1 resulting from the FALS mutation causes neurotoxicity through inflammation and apoptosis, oxidative modifications involving nitric oxide, or a combination of these events is not entirely clear. However, the neurotoxicity of SOD1 in FALS depends on the presence of copper in the enzyme and several FALS-linked SOD1 mutants are known to acquire copper *in vivo* through specific action of the copper chaperone CCS (Corson et al., 1998). This implies that CCS has a role in FALS by delivering copper to mutated forms of SOD1, which become neurotoxic following acquisition of copper. However, ablation of the gene encoding CCS does not modify the onset or progression of motor neuron disease in mice having an FALS-linked mutation in SOD1 (Subramaniam et al., 2002). Thus, motor neuron disease characteristic of FALS can occur without copper loading of mutated SOD1 by CCS. Although the presence of copper in mutated forms of SOD1 lacking zinc is sufficient to produce neurotoxicity *in vitro*, the delivery of copper to SOD *in vivo* by CCS apparently is not necessary to produce FALS. The role of copper in FALS-linked SOD1 remains controversial, and more research is needed to determine whether the delivery of copper to mutated forms of SOD1 by CCS is an important step in producing motor neuron disease.

Alzheimer's disease is characterized by selective loss of neurons in hippocampal and frontal cortices and the presence of neurofibrillary tangles, neutrophil threads, and β-amyloid peptide (Aβ)-rich senile plaques. Although the mechanisms leading to neuronal degeneration in Alzheimer's disease are not entirely clear, transition metals, including copper, may be involved in the biochemical events that occur in this disease. Aβ is derived from specific endoproteolytic cleavage of the membrane-bound amyloid precursor protein (APP) both during and after APP is trafficked through the secretory pathway. Both Aβ and APP can bind and reduce Cu(II) (Kontush, 2001; Waggoner et al., 1999). Aβ, particularly the more amyloidogenic form containing 42 amino acids, has very high affinity for Cu(II), with stability constants of ca. 10^{10} to 10^{17} M^{-1}. Under mildly acidic conditions, such as those that may be encountered during inflammation, the binding of copper to Aβ induces a conformational change that leads to a higher degree of β-sheet configuration within the secondary structure of the protein. Aggregation of Aβ then occurs as a result of cross-linking between the β-sheets and copper atoms. The interaction of copper with Aβ and the resultant conformational change leading to aggregation of Aβ may be important biochemical steps in the pathology of Alzheimer's disease because Aβ must be aggregated to fibrils in order to be neurotoxic.

Although the formation of Aβ aggregates is an essential biochemical event in the development of Alzheimer's disease, the binding of Cu(II) to APP and its reduction to Cu(I) also may be important. Once APP binds and reduces copper, the reduced copper remains bound as APP and is delivered from the cell body to the axonal cell surface and dendritic plasma membrane. Although this suggests that APP may have a physiological role in trafficking copper through the endoplasmic reticulum, Golgi, and secretory pathways of neurons, the binding and reduction of copper by APP may also be conducive for the formation of Aβ. Copper bound to APP can be rapidly reoxidized by hydrogen peroxide, and once reoxidation occurs smaller peptides are formed from APP through site-specific fragmentation (Waggoner et al., 1999). One of the peptides formed from the fragmentation is the amino terminus of Aβ. This suggests that metabolic perturbations leading to increased intercellular hydrogen peroxide concentrations could provide a mechanism for copper to mediate the formation and aggregation of Aβ directly from APP as bound copper is reoxidized.

Inhibition of CCO with azide or the uncoupling of mitochondrial respiration increases proteolysis of APP (Gabuzda et al., 1994). This finding suggests that impaired mitochondrial function contributes to the development of Alzheimer's disease and possibly to other neurodegenerative disorders as well. Partial inhibition of CCO increases the reduction state of upstream mitochondrial respiratory complexes. This higher reduction state increases mitochondrial superoxide production through a single-electron transfer to molecular oxygen, particularly at respiratory complex I.

Superoxide produced by mitochondria is converted to hydrogen peroxide through the action of mitochondrial MnSOD, and because mitochondrial membranes are permeable to hydrogen peroxide the net effect of CCO inhibition may be increased intercellular hydrogen peroxide concentrations. Brain CCO activity normally decreases with age and this may produce an intercellular redox environment that is favorable for the oxidation of copper bound to APP. Copper deficiency during brain development may accelerate neurodegenerative processes related to mitochondrial dysfunction. As noted earlier, copper deficiency during gestation and postnatal development irreversibly lowers CCO activity in the brain. Copper deficiency during brain development could place adult CCO activity at such a level that further declines during aging prematurely decrease CCO activity below a threshold level wherein mitochondrial hydrogen peroxide production becomes sufficiently elevated to cause APP fragmentation through the oxidation of bound copper. This could accelerate the formation and aggregation of Aβ and the formation of amyloid plaques.

17.6 SUMMARY

Evidence for the essentiality of copper for normal development and function of the CNS has been obtained from numerous investigations with copper-deficient animals and from animals and humans having genetic defects in copper metabolism. Neuropathologies associated with copper deprivation include ataxia, hypomyelination, abnormal structure in some brain regions, neuronal degeneration, seizures, mental retardation, and impaired sensory responses. Although the biochemical mechanisms underlying the various pathological outcomes associated with copper deprivation in the brain are not well understood in all instances, reductions in the activities of cuproenzymes and resultant changes in metabolic pathways containing those enzymes are likely involved. The effects of copper deprivation are most pronounced if it occurs during the later stages of fetal or neonatal brain development. Copper deprivation during this period of brain development reduces copper accrual, which leads to impaired developmental expression of several cuproenzyme activities. Once the glia and neurons have fully differentiated, brain copper content becomes stabilized and some biochemical defects in the brain caused by copper deprivation persist even when adequate dietary copper is supplied. Thus, copper deprivation during brain development has not only immediate effects on brain copper and cuproenzyme activities but also lasting effects on brain function.

High intercellular copper concentrations and free copper ions can be toxic. However, several copper transport proteins and chaperones have evolved to assure that cellular copper requirements are met without allowing copper to reach toxic levels. The chaperones protect the cell from free copper ions by keeping copper ions chelated during delivery to copper-dependent proteins. The discovery that mutations in two genes, *ATP7A* and *ATP7B,* are responsible for the aberrations in copper metabolism associated with Menkes and Wilson's diseases, respectively, was an important step for understanding the intercellular mechanisms for processing copper. Both *ATP7A* and *ATP7B* are present in the CNS, where their products are transport proteins responsible for carrying copper to the secretory pathway in neuronal cells, where it can be released to secretory vesicles or copper-dependent proteins within the secretory pathway. Defective copper metabolism resulting from mutations in *ATP7A* and *ATP7B* is a major factor in producing the neuropathology found in Menkes and Wilson's diseases. In addition to copper transport proteins, several chaperones are also present in the CNS that deliver copper to specific intercellular sites for incorporation into proteins that require copper.

Copper also participates directly in the neurodegeneration of FALS and may have a role in Alzheimer's disease. Mutations in SOD1 associated with FALS do not alter the enzyme's ability to scavenge superoxide, but cause a gain in function that increases its neurotoxicity. The neurotoxicity of the FALS-linked SOD1 depends on the presence of copper in the enzyme. Copper may also facilitate the formation of senile plaques characteristic of Alzheimer's disease by increasing the formation of Aβ from APP and causing configurational changes in Aβ that promote its aggregation.

REFERENCES

Bennetts, H.W. and Chapman, F.E. (1937) Copper deficiency in sheep in Western Australia: a preliminary account of the etiology of enzooic ataxia of lambs and an anemia of yews. *Aust. Vet. J.* 13, 138–149.

Bull, P.C., Thomas, G.R., Rommens, J.M., Forbe, J.R. and Cox, D.W. (1993) The Wilson disease gene is a putative copper transporting P-type ATPase similar to the Menkes gene. *Nat. Genet.* 5, 327–337.

Capaldi, R.A. (1990) Structure and function of cytochrome c oxidase. *Annu. Rev. Biochem.* 59, 569–596.

Chelly, J., Tumer, Z., Tonnesen, T., Petterson, A., Ishikawa-Brush, Y., Tommerup, N., Horn, N. and Monaco, A.P. (1993) Isolation of a candidate gene for Menkes disease that encodes a potential heavy metal binding protein. *Nat. Genet.* 3, 14–19.

Cimarusti, D.L., Saito, K., Vaughn, J.E., Barber, R., Roberts, E. and Thomas, P.E. (1979) Immunochemical localization of dopamine-beta-hydroxylase in rat coeruleus and hypothalamus. *Brain Res.* 16, 66–67.

Corson, L.B., Srain, J.J., Culotta, V.C. and Cleveland, D.W. (1998) Chaperone-facilitated copper binding is a property common to several classes of familial amytrophic lateral sclerosis-linked superoxide dismutase mutants. *Proc. Natl. Acad. Sci. USA* 95, 6361–6366.

Coyle, J.T. and Axelrod, J. (1972) Dopamine-β-hydroxylase in the rat brain: developmental characteristics. *J. Neurochem.* 19, 449–459.

Crow, J.P., Sampson, J.B., Zhuang, Y., Thompson, J.A. and Beckman, J.S. (1997) Decreased zinc affinity of amyotrophic lateral sclerosis-associated superoxide dismutase leads to enhanced catalysis of tyrosine nitration by peroxynitrite. *J. Neurochem.* 69, 1936–1944.

Culotta, V.C., Klomp, L.W.J., Strain, J., Casareno, R.L.B., Krems, B. and Gitlin, J.D. (1997) The copper chaperone for superoxide dismutase. *J. Biol. Chem.* 272, 23469–23472.

Danks, D.M. (1983) Copper deficiency in infants with particular reference to Menkes' disease. In *Copper in Animals and Man,* Vol. II (Howell, J.M. and Gawthorne, J.M., Eds.), pp. 29–51. CRC Press, Boca Raton, FL.

Danks, D.M., Stevens, B.J., Campbell, P.E., Gillespie, J.M., Walker-Smith, J., Blomfield, J. and Turner, B. (1972a) Menkes' kinky-hair syndrome. *Lancet* 1, 1100–1102.

Danks, D.M., Campbell, P.E., Stevens, B.J., Mayne, V. and Cartwright, E. (1972b) Menke's kinky hair syndrome: an inherited defect in copper absorption with widespread effects. *Pediatrics* 50, 188–201.

Dickenson, C.J., Marino, L. and Yamada, T. (1990) Inhibition of the alpha-amidation of gastrin: effects on gastric acid secretion. *Am. J. Physiol.* 258, G810–G814.

DiPaolo, R.V., Kanfer, J.N. and Newberne, P.M. (1974) Copper deficiency and the central nervous system. *J. Neuropath. Exp. Neurol.* 33, 226–236.

Estevez, A.G., Crow, J.P., Sampson, J.B., Reiter, C., Zhuang, Y., Richardson, G.J., Tarpey, M.M., Barbeito, L. and Beckman, J.S. (1999) Induction of nitric oxide-dependent apoptosis in motor neurons by zinc-deficient superoxide dismutase. *Science* 286, 2498–2500.

Everson, G.J., Shrader, R.E. and Wang, T. (1968) Chemical and morphological changes in the brains of copper-deficient guinea pigs. *J. Nutr.* 96, 115–125.

Fridovich, I. (1974) Superoxide dismutases. *Adv. Enzymol.* 41, 36–97.

Gabuzda, D., Busciglio, J., Chem, L.B., Matsudaira, P. and Yankner, B.A. (1994) Inhibition of energy metabolism alters the processing of amyloid precursor protein and induces a potentially amyloidogenic derivative. *J. Biol. Chem.* 269, 13623–13628.

Hansel, D.E., May, V., Eipper, B.A. and Ronnett, G.V. (2001) Pituitary adenyl cyclase-activating peptides and α-amidation in olfactory neurogenesis and neuronal survival *in vitro. J. Neurosci.* 21, 4625–4636.

Hunt, C.D. and Idso, J.P. (1995) Moderate copper deprivation during gestation and lactation affects dentate gyrus and hippocampal maturation in immature male rats. *J. Nutr.* 125, 2700–2710.

Iwase, T., Nishimura, M., Sugimura, H., Igarashi, H., Ozawa, F., Shinmura, K., Suzuki, M., Tanaka, M. and Kino, I. (1996) Localization of Menkes gene expression in the mouse brain: its association with neurological manifestations in Menkes model mice. *Acta Neuropathol. (Berl.)* 91, 482–488.

Iwata, M., Hirano, A. and French, J.H. (1979) Degeneration of the cerebellar system in X-chromosome-linked copper malabsorption. *Ann. Neurol.* 5, 542–549.

Johnson, W.T., Thomas, A.C. and Lozano, A.A. (2000) Maternal copper deficiency impairs the developmental expression of protein kinase C a, b, and g isoforms in neonatal rat brain. *Nutr. Neurosci* 3, 113–122.

Kako, K., Tsumori, K., Ohmasa, Y., Takahashi, Y. and Eisuke, M. (2000) The expression of Cox17p in rodent tissues and cells. *Eur. J. Biochem.* 267, 6699–6707.

Klomp, L.W.J., Lin, S.J., Yuan, D.S., Klausner, R.D., Culotta, V.C. and Gitlin, J.D. (1997) Identification and functional expression of HAH1, a novel human gene involved in copper homeostasis. *J. Biol. Chem.* 272, 9221–9226.

Kontush, A. (2001) Amyloid-β: an antioxidant that becomes a pro-oxidant and critically contributes to Alzheimer's disease. *Free Radic. Biol. Med.* 31, 1120–1131.

Kuo, Y.M.K., Zhou, B., Cosco, D. and Gitschier, J. (2001) The copper transporter CTR1 provides an essential function in mammalian embryonic development. *Proc. Natl. Acad. Sci. USA* 98, 6836–6841.

Lai, J.C.K., Chan, A.W.K., Minski, M.J. and Lim, L. (1985) Roles of metal ions in brain development and aging. In *Metal Ions in Neurology and Psychiatry* (Gabay, S., Harris, J. and Ho, B.T., Eds.), pp. 49–67. Alan R. Liss, New York.

Ledig, M., Fried, R., Ziessel, M. and Mandel, P. (1982) Regional distribution of superoxide dismutase in rat brain during postnatal development. *Brain Res.* 256, 333–337.

Lee, J., Prohaska, J.R., Dagenais, S.L., Glover, T.W. and Thiele, D.J. (2000) Isolation of a murine copper transporter gene, tissue specific expression and functional complementation of a yeast copper transport mutant. *Gene* 254, 87–96.

Lee, J., Prohaska, J.R. and Thiele, D.J. (2001) Essential role for mammalian copper transporter Ctr1 in copper homeostasis and embryonic development. *Proc. Natl. Acad. Sci. USA* 98, 6842–6847.

Lin, S.J., Pufahl, R.A., Dancis, A., O'Halloran, T.V. and Culotta, V.C. (1997) A role for the *Saccharomyces cerevisiae ATX1* gene in copper trafficking and iron transport. *J. Biol. Chem.* 272, 9215–9220.

Lyons, T.J., Liu, H., Goto, J.J., Nerissian, A., Roe, J.A., Graden, J.A., Café, C., Ellerby, L.M., Bredesen, D.E., Gralla, E.B. and Valentine, J.S. (1996) Mutations in copper-zinc superoxide dismutase that cause amyotrophic lateral sclerosis alter the zinc binding site and the redox behavior of the protein. *Proc. Natl. Acad. Sci. USA* 93, 12240–12244.

Marchand, J.E., Hershman, K., Kumar, M.S., Thompson, M.L. and Kream, R.M. (1990) Disulfiram administration affects substance P-like immunoreactive and monoaminergic neural systems in rodent brain. *J. Biol. Chem.* 265, 264–273.

Marklund, S.L., Anderson, P.M., Forsgren, L., Nilsson, P., Ohlsson, P., Wikander, G. and Öberg, A. (1997) Normal binding and reactivity of copper in mutant superoxide dismutase isolated from amyotrophic lateral sclerosis patients. *J. Neurochem.* 69, 675–681.

Menkes, J.H., Alter, M., Steigleder, G., Weakley, D.R. and Sung, J.H. (1962) A sex-linked recessive disorder with retardation of growth and peculiar hair, and focal cerebral and cerebellar degeneration. *Pediatrics* 29, 764–779.

Mercer, J.F., Livingston, J., Hall, B., Paynter, J.A., Begy, C., Chandrasekharappa S., Lockhart, P., Grimes, A., Bhave, M., Siemieniak, D. and Glover, T.W. (1993) Isolation of a partial candidate gene for Menkes disease by positional cloning. *Nat. Genet.* 3, 20–25.

Mercer, J.F.B. (1998) Menkes syndrome and animal models. *Am. J. Clin. Nutr.* 67, 1022S–1028S.

Murata, Y., Kodama, H., Abe, T., Ishida, N., Nishimura, M., Levinson, B., Gitschier, J. and Packman, S. (1997) Mutation analysis and expression of the mottled gene in the macular mouse model of Menkes disease. *Pediatr. Res.* 42, 436–442.

Naeve, G.S., Vana, A.M., Eggold, J.R., Kelner, G.S., Maki, R., Desouza, E.B. and Foster, A.C. (1999) Expression profile of the copper homeostasis gene, *rAtox1*, in the rat brain. *Neuroscience* 93, 1179–1187.

Petris, M.J., Mercer, J.F., Culvenor, J.G., Lockhart, P., Gleeson, P.A. and Camakaris, J. (1996) Ligand-regulated transport of the Menkes copper P-type ATPase efflux from the Golgi apparatus to the plasma membrane: a novel mechanism of regulated trafficking. *EMBO J.* 15, 6084–6095.

Prigge, S.T., Kolhekar, A.S., Eipper, B.A., Mains, R.E. and Amzel, L.M. (1997) Amidation of bioactive peptides: the structure of peptidyl α-hydroxylating monooxygenase. *Science* 278, 1300–1305.

Prohaska, J.R. (1983) Changes in tissue growth, concentrations of copper, iron, cytochrome oxidase and superoxide dismutase subsequent to dietary or genetic copper deficiency in mice. *J. Nutr.* 113, 2148–2158.

Prohaska, J.R. (1987) Functions of trace elements in brain metabolism. *Physiol. Rev.* 67, 858–901.

Prohaska, J.R. (1990) Biochemical changes in copper deficiency. *J. Nutr. Biochem.* 1, 452–461.

Prohaska, J.R. and Bailey, W.R. (1993) Persistent regional changes in brain copper, cuproenzymes and catecholamines following perinatal copper deficiency in mice. *J. Nutr.* 123, 1226–1234.

Prohaska, J.R. and Bailey, W.R. (1994) Regional specificity in alterations of rat brain copper and catecholamines following perinatal copper deficiency. *J. Neurochem.* 63, 1551–1557.

Prohaska, J.R. and Bailey, W.R. (1995a) Persistent neurochemical changes following perinatal copper deficiency in rats. *J. Nutr. Biochem.* 6, 275–280.

Prohaska, J.R. and Bailey, W.R. (1995b) Alterations of rat brain peptidylglycine α-amidating monooxygenase and other cuproenzyme activities following perinatal copper deficiency. *Proc. Soc. Exp. Biol. Med.* 210, 107–116.

Prohaska, J.R. and Hoffman, R.G. (1996) Auditory startle response is diminished in rats after recovery from perinatal copper deficiency. *J. Nutr.* 126, 618–627.

Prohaska, J.R. and Smith, T.L. (1982) Effect of dietary or genetic copper deficiency on brain catacholamines, trace metals and enzymes in mice and rats. *J. Nutr.* 112, 1706–1717.

Prohaska, J.R. and Wells, W.W. (1974) Copper deficiency in the developing rat brain: a possible model for Menkes' steely-hair disease. *J. Neurochem.* 23, 91–98.

Prohaska, J.R. and Wells, W.W. (1975) Copper deficiency in the developing rat brain: evidence for abnormal mitochondria. *J. Neurochem.* 25, 221–228.

Pufahl, R.A., Singer, C.P., Peariso, K.L., Lin, S.J., Schmidt, P.J., Fahrni, C.J., Culotta, V.C., Penner-Hahn, J.E. and O'Halloran, T.V. (1997) Metal ion chaperone function of the soluble Cu(I) receptor Atx1. *Science* 278, 853–856.

Qian, Y., Tiffinay-Castiglioni, E. and Harris, E.D. (1997) A Menkes P-type ATPase involved in copper homeostasis in the central nervous system of the rat. *Mol. Brain Res.* 48, 60–66.

Rossi, L., DeMartino, A., Marchese, E., Piccirilli, S., Rotilio, G. and Ciriolo, M.R. (2001) Neurodegeneration in the animal model of Menkes' disease involves Bcl-2-linked apoptosis. *Neuroscience* 103, 181–188.

Rothstein, J.D., Dykes-Hoberg, M., Corson, L.B., Becker, M., Clevland, D.W., Price, D.L., Culotta, V.C. and Wong, P.C. (1999) The copper chaperone CCS is abundant in neurons and astrocytes in human and rodent brain. *J. Neurochem.* 72, 422–429.

Rowland, L.P. and Schneider, N.A. (2001) Amyotrophic lateral sclerosis. *N. Engl. J. Med.* 344, 1688–1700.

Saito, T., Nagao, T., Okabe, M. and Saito, K. (1996) Neurochemical and histochemical evidence for an abnormal catecholamine metabolism in the cerebral cortex of the Long–Evans Cinnamon rat before excessive copper accumulation in the brain. *Neurosci. Lett.* 216, 195–198.

Saito, T., Okabe, M., Hosokawa, T., Kurasaki, M., Hata, A., Endo, F., Nagano, K., Matsuda, I., Urakami, K. and Saito, K. (1999) Immunohistochemical determination of the Wilson copper-transporting P-type ATPase in the brain tissues of the rat. *Neurosci. Lett.* 266, 13–16.

Smith, R.M. (1983) Copper in the developing brain. In *Neurobiology of the Trace Elements,* Vol. 1 (Dreosti, I.E. and Smith, R.M., Eds.), pp. 1–40, Humana Press, Clifton, NJ.

Stewart, L.C. and Klinman, J.P. (1988) Dopamine beta-hydroxylase of adrenal chromaffin granules: structure and function. *Annu. Rev. Biochem.* 57, 551–592.

Subramanian, J.R., Lyons, W.E., Liu, J., Bartnikas, T.B., Rothstein, J., Price, D.L., Cleveland, D.W., Gitlin, J.D. and Wong, P.C. (2002) Mutant SOD1 causes motor neuron disease independent of copper chaperone-mediated copper loading. *Nat. Neurosci.* 5, 301–307.

Suzuki, M. and Gilin, J.D. (1999) Intracellular localization of the Menkes and Wilson's disease proteins and their role in intracellular copper transport. *Pediatr. Int.* 41, 436–442.

Tanzi, R.E., Petrukhin, K., Chernov, I., Pellequer, J.L., Wasco, W., Ross, B., Ramano, D.M., Parano, E., Pavone, L., Brzustowicz, L.M., Devoto, M., Peppercorn, J., Bush, A.I., Sternlieb, I., Pirastu, M., Gusella, J.F., Evgrafov, O., Penchaszadeh, G.K., Honig, B., Edelman, I.S., Soares, M.B., Scheinberg, I.H. and Gilliam, T.C. (1993) The Wilson disease gene is a copper transporting ATPase with homology to the Menkes disease gene. *Nat. Genet.* 5, 344–350.

Terao, T. and Owen, C.A. Jr. (1977) Copper metabolism in pregnant and postpartum rat and pups. *Am. J. Physiol.* 232, E172–E179.

Thomas, T.N., Priest, D.G. and Zemp, J.W. (1976) Distribution of superoxide dismutase in rat brain. *J. Neurochem.* 27, 309–310.

Vulpe, C., Levinson, B., Whitney, S., Packman, S. and Gitschier, J. (1993) Isolation of a candidate gene for Menkes disease and evidence that it encodes a copper-transporting ATPase. *Nat. Genet.* 3, 7–13.

Waggoner, D.J., Bartnikas, T.B. and Gitlin, J.D. (1999) The role of copper in neurodegenerative disease. *Neurobiol. Dis.* 6, 221–230.

Winterbourn, C.C., Domigan, N.M. and Broom, J.K. (1995) Decreased thermal stability of red blood cell glu^{100}→gly superoxide dismutase from a family with amyotrophic lateral sclerosis. *FEBS Lett.* 368, 449–451.

Yajima, K. and Suzuki, K. (1979) Neuronal degeneration in the brain of the brindled mouse. *Acta Neuropathol.* 45, 17–25.

Yamaguchi, Y., Heiny, M.E. and Gitlin, J.D. (1993) Isolation and characterization of a human cDNA as a candidate for Wilson disease. *Biochem. Biophys. Res. Commn.* 197, 271–277.

Yoshihara, T., Ishigaki, S., Yamamoto, M., Liang, Y., Niwa, J., Takeuchi, H., Doyu, M. and Sobue, G. (2002) Differential expression of inflammation- and apoptosis-related genes in spinal cords of a mutant SOD1 transgenic mouse model of familial amyotrophic lateral sclerosis. *J. Neurochem.* 80, 158–167.

Zhou, B. and Gitschier, J. (1997) hCTR1: a human gene for copper uptake identified by complementation in yeast. *Proc. Natl. Acad. Sci. USA* 94, 7481–7486.

18 Roles of Selenium in Function of the Brain

Chiho Watanabe

CONTENTS

18.1 SELENIUM (SE) AS AN ESSENTIAL TRACE ELEMENT

Selenium (Se) is among the essential trace elements for mammals, including humans. Its essentiality was first found in 1957 as a factor that prevents liver necrosis induced by vitamin E deficiency in rats. Biochemical basis of its essentiality was established in 1973 by identification of the enzyme glutathione peroxidase (GPx), which has Se in its catalytic center. Chambers et al. (1986) revealed a unique property of GPx: the gene coding for GPx contains a stop codon, TGA, which actually codes for selenocysteine (SeCys), a selenium-containing amino acid. Thus, unlike other trace elements such as iron, cupper, zinc, and manganese, Se is incorporated into the amino acid sequence cotranslationally as the twenty-first amino acid. This finding stimulated molecular studies of Se-containing proteins and their biosynthesis mechanisms, eventually leading to the identification of many Se-containing proteins besides GPx.

In the mammalian body, most Se is associated with proteins, and it is thought that physiological functions of Se are exerted through the functions of such Se-containing proteins (Allan et al., 1999; Behne and Kyriakopoulos, 2001; Himeno and Imura, 2002). In these proteins, Se exists either as SeCys or as selenomethionine (SeMet). Substitution of SeCys with cysteine usually results in loss of catalytic property, whereas substitution of SeMet by methionine does not change the property of the original protein. This is because mammalian metabolism does not distinguish SeMet from methionine and it handles the former as the latter. Therefore, SeMet is usually found in the position of methionine in ordinal proteins, and only proteins that contain Se as SeCys may be important in terms of Se functions, and the term *selenoproteins* often refers to these proteins.

The most illustrating finding regarding the essentiality of Se for mammalian life was reported by Bosl et al. (1997), using a knockout of the gene coding for SeCys tRNA, an indispensable component for selenoprotein biosynthesis. It was found that this knockout, which would disrupt the biosynthesis of all selenoproteins, ended in a 100% mortality of the embryo, indicating that one or more of the selenoproteins are essential for survival of the developing mouse embryo. Despite this finding, the question regarding which protein or combination of proteins is most critical for survival remains unanswered. Although more than 20 selenoproteins have been reported so far, physiological functions of individual selenoproteins are largely unknown. The number of seleno-proteins is likely to keep growing with the aid of computer-assisted analyses of base sequence, and so are the number of proteins with no known functions.

This chapter reviews existing findings regarding brain-specific or brain-related functions of Se. Such functions can be achieved through existence of a brain-specific selenoprotein, brain-specific function of a ubiquitously expressed selenoprotein, or a brain-specific manifestation of deficiency or excess of Se. In the last two cases, specificity may arise from the specific structure, biochemistry, and physiology of the brain tissue rather than from those of the selenoprotein. Thus, the focus of this chapter is on nutritional or physiological aspects of Se. Toxicological and pharmacological aspects of Se are referred to, but to a minimum extent. Note that exogenously administered inorganic Se can induce pharmacological effects on the nervous system. These include lowered temperature set point after administration of selenite in mice (Watanabe and Suzuki, 1986) and altered wakefulness in rats after intracerebral infusion of a tetravalent selenium compound (Matsumura et al., 1991). Prevention of cancer with prophylactic intake of a "supranutritional" level of (nonproteinous) Se, becoming an important research field, is not covered.

The following sections discuss experimental findings followed by findings from human studies. First, characteristics of Se in the brain compared with those in other organs are described, then established and putative functions of the each selenoprotein detected in the brain are discussed, followed by a section describing neurochemical and behavioral consequences of Se deficiency. Then a brief summary of human nutrition is presented, followed by a discussion on Se deficiency in humans and findings from the analyses of autopsy brains.

18.2 CHARACTERISTICS OF SE IN THE BRAIN

The concentrations of Se vary among organs in the mammalian body (Table 18.1). The kidney, liver, and testes are among the organs that have higher concentrations, whereas other tissues such as the muscle, heart, and brain have low concentrations. Despite these lower concentrations compared with tissues such as the liver, brain Se is hardly depleted when dietary intake of Se is restricted. For example, Se concentrations in the liver of rodents can be reduced to less than one tenth of the

TABLE 18.1
Tissue Selenium Levels in Mammals

	Human Autopsy				Rodents			
	Japan	Germany	Poland	U.S. Tube Fed	Mouse	Deficient Mouse[a]	Rat	Deficient Rat[b]
n, sex, age	23-40, M and F, 43	13–19, M, 39	46, M and F, 45	2, M, 69,70				
Kidney	1.08/1.13[c]	0.77	0.47	2.75, 1.73			1.07	0.22
Liver	0.73	0.29	0.39	0.51, 0.52	1.00	0.04	0.72	0.22
Testis		0.27					1.03	0.59
Muscle		0.11	0.05	0.13, 0.13			0.10	0.02
Heart	0.33	0.16	0.09	0.44, 0.28				
Brain	0.22/0.26[d]	0.11	0.09	0.26, 0.44	0.13	0.07	0.17	0.11
Ref.	Yoshinaga et al. (1990)	Oster et al. (1988)	Zachara et al. (2001)	Feller et al. (1987)	Watanabe and Satoh (1994)	Watanabe and Satoh (1994)	Whanger and Butler (1988)	Whanger and Butler (1988)

Note: Values are given in µg/g wet weight except for the U.S. tube-fed data, which are given on a dry weight basis; blank cells: not reported.

[a] Born to mothers fed low-Se diet before and during gestation and lactation.

[b] Fed low-Se diet for 9 weeks from weaning.

[c] Values for cortex (left)/medulla (right).

[d] Values for gray matter (left)/white matter (right) of cerebrum.

normal with several weeks of dietary restriction, whereas those in the brain retain more than half of the normal value. When Se supplementation is resumed for these depleted animals, the brain is among the organs that show the most rapid uptake. Recently, it has been shown that uptake of a stable isotope of Se by the brain is linear up to 3 weeks. This is in contrast to other tissues such as the liver, kidney, and testis, which show slowing down of uptake rate during the same period, suggesting that uptake of Se by nonproliferating cells is greater in the brain than in other organs (Shiobara et al., 2000). Taken together, these observations imply the importance of the element in the brain.

In addition to such organ-specific metabolism of Se, a hierarchy exists among the selenoproteins (Behne et al., 1988) in terms of metabolism; that is, some selenoproteins are preferentially preserved when Se supply is limited. In most organs, including the brain, classical or cellular GPx (cGPx) is the selenoprotein depleted most rapidly and most extensively when Se intake is restricted. On the other hand, selenoproteins such as phospholipid hydroperoxide GPx (phGPx) and selenoprotein W (Se-W) are well preserved even under a limited supply of Se. Such a hierarchic regulation is accomplished at least partly by varying the structure of the so-called SECIS (selenocystein insertion sequence) region, a specific sequence that enables the stop codon to be read as a codon for SeCys in the untranslated area of the mRNA for each selenoprotein (Berry and Larsen, 1992). Thus, the mRNA of each selenoprotein contains an SECIS region of slightly difference structure. The efficacy of incorporating SeCys into the polypeptide chain differs among these slightly different SECISs, especially when the supply of Se is limited.

Little is known about the mode of Se entry into the brain. How Se is transported in the bloodstream is also not known, though this point is discussed later. Little knowledge is available on the possible kinetics and functions of Se-containing compounds other than selenoproteins (Allan et al., 1999). One should always bear in mind the lack of these basic components in Se physiology.

18.3 SELENOPROTEINS IN THE BRAIN AND THEIR POSSIBLE FUNCTIONS

18.3.1 OVERVIEW

The functions of Se in the mammalian body are thought to be exerted primarily as functions of selenoproteins. Many selenoproteins are distributed ubiquitously, whereas some show organ specificity. For example, cGPx is expressed in almost all organs, but gastrointestinal GPx is expressed in the gastrointestinal (GI) tract only and the extracellular glutathione peroxidase (eGPx) is synthesized in the kidney and liver and secreted into the bloodstream. In terms of subcellular distribution, phGPx is located in cytosolic and membrane-bound fractions, whereas cGPx and GI-GPx are cytosolic proteins. About 10 selenoproteins have been so far detected in the brain in terms of either mRNA expression or protein per se (Table 18.2). These include two of five GPxs (cGPx and phGPx), two of three known iodothyronine deiodinases (DI2 or type II and DI3 or type III), two of three known thioredoxin reductases (TRxRs), selenoprotein P (Se-P), Se-W, the 15-kDa selenoprotein, and selenophosphate synthetase 2. Although some selenoproteins have not been detected in the brain (e.g., GI-GPx), no brain-specific proteins, which should be expressed in the brain only, have been identified except for the mRNA signal for a Se-P-like bovine protein, which may be confined to the brain (Saijoh et al., 1995). Although most selenoproteins appear to be involved in redox reactions, knowledge about the precise *in vivo* functions of these proteins is very limited.

18.3.2 CLASSICAL GLUTATHIONE PEROXIDASES (cGPx)

cGPx was the first selenoprotein identified in mammalian species. Although the chemical reaction catalyzed by cGPx has been known for a long time, the physiological importance of cGPx has not been completely understood. In other words, that an enzyme catalyzes a certain chemical reaction

---

TABLE 18.2
Major Selenoproteins Identified in the Mammalian Brain

Selenoprotein	(Putative) Functions, Characteristics	Tissue Distribution (Other Than Brain)
Classical glutathione peroxidase (cGPx, GPx-1)	Tetramer of 23-kDa subunits; reduces peroxides; the first protein identified as selenoprotein; knockout mice of this protein apparently normal	Ubiquitous
Phopholipid hydroperoxide glutathione peroxidase (phGPx, GPx-3)	19.7 kDa; reduces peroxides, including phospholipid hydroperoxides	Ubiquitous
Type 2 deiodothyronine deiodinase	30 kDa, 5-monodeiodination of thyroid hormones; converts T4 to T3 or reverse T3 to T2	Brown adipose tissue, pituitary, placenta
Type 3 deiodothyronine deiodinase	32 kDa, 5-monodeiodination of thyroid hormones; converts T4 to reverse T3, or T3 to T2	Placenta, skin
Thioredoxin reductase 1	Homodimer of 56-kDa subunits; cytosolic enzyme; reduces oxidized thioredoxin	Ubiquitous
Thioredoxin reductase 2	53- or 56-kDa mitochondrial enzyme; reduces oxidized thioredoxin	Ubiquitous
Selenoprotein P	Transporting selenium? antioxidant activity? synthesized and secreted by glial cells	Liver (plasma)
Selenoprotein W	Antioxidant?	Ubiquitous
Selenophosphate synthetase 2	Biosynthesis of selenophosphate	Ubiquitous

in test tube does not necessarily mean that the reaction also takes place *in vivo,* being catalyzed by the same enzyme. cGPx reduces peroxides at the expense of (reduced) glutathione, a tripeptide existing at millimolar levels in most organs. Although it is expected that suppression of the enzyme results in accumulation of peroxides in cells, which would eventually lead to malfunctions of the cells, *in vivo* evidence for this has not been obtained.

The findings with cGPx knockout (GPKO) mice further cast a doubt on the physiological importance of this enzyme. It has been shown that GPKO is not lethal, reproduces normally, and is resistant to hyperoxygen exposure to the same extent as the wild type is. In addition, measurement of lipid peroxidation, oxidized proteins, and other antioxidant systems such as catalase or superoxide dismutases does not provide any evidence for the enhanced oxidative stress in GPKO mice (Ho et al., 1997). These observations do not support the physiological importance of cGPx, but rather favor the proposed role of this selenoenzyme as a pool of Se.

More recently, however, it was shown that the GPKO does differ from the wild type in the responses against paraquat treatment, which has been known to increase the oxidative stress through production of superoxide (Cheng et al., 1998). The role of cGPx depends on the dose of paraquat given. When the paraquat dose is high enough to kill the mice, survival time of the mice correlates well with tissue GPx activity, whereas with a moderate and sublethal dose the drug-induced lesion is independent of GPx activity. It has been speculated that cGPx is less important in protecting animals against a sublethal dose of paraquat. GPKO mice (as well as Se-depleted mice) are also different from wild-type mice when challenged with an avirulent strain of the Coxackie virus (Beck et al., 1998). The avirulent strain acquires virulence in GPKO mice as well as in Se-depleted or vitamin E-deficient mice but not in control mice. These observations suggest that the viral conversion is associated with enhanced oxidative stress, which in turn is enhanced in GPKO. Overall, these observations indicate that the importance of GPKO will be revealed only when the oxidative stress is moderately enhanced for a prolonged period.

Interestingly, transgenic mice overexpressing human cGPx are more sensitive to high temperature as judged from the survival time. In these transgenic mice, brain cGPx activity was four times higher than that of normal mice. Several responses to heat (hyperthermia) were examined and compared among the transgenic as well as control mice. In transgenic mice, both induction of HSP70 and production of peroxides in the brain (as well as lung and muscle) were suppressed and synthesis of prostaglandin E2 in the brain differed (Mirochnitchenko et al., 1995). These results suggest that the level of peroxide production in the hyperthermic condition depends on cGPx activity.

Compared with organs such as liver and kidneys, the brain exhibits much lower activity of cGPx. The distribution of this enzyme among different brain regions, however, suggests the possible importance of cGPx in this organ. Higher cGPx activities are found in the caudate putamen and substantia nigra (SN), systems containing the dopaminergic (DAergic) cells or projections (Brannan et al., 1980). The DAergic pathways are vulnerable to oxidative stress because oxidation of dopamine (DA) by monoamine oxidase or by autooxidation would generate hydrogen peroxide (Olanow, 1993). cGPx may contribute to the sequestration of these locally generated reactive oxygen species (ROS). Trepanier et al. (1996) examined the distribution of cGPx proteins in the mouse brain by using a polyclonal antibody. The reticular thalamic and red nuclei, cerebrum cortex, CA1 region of the hippocampus, dentate gyrus, and pontine nucleus exhibited higher contents of cGPx, followed by the caudate putamen, hippocampus, thalamus, and hypothalamus. Interestingly, no signals were detected from the cholinergic neurons in the septic nuclei and DAergic neurons in the SN, the target cells in Parkinson's disease.

There might be a localization of cGPx according to cell type. Savolainen (1978) showed that cGPx activity was much higher in glia than in neurons, whereas Cafe et al. (1995) reported similar levels in these two cell types. Among studies that used immunological detection, two detected cGPx signals from both the cell types in mice (Trepanier et al., 1996) and humans (Takizawa et al., 1994), whereas Damier et al. (1993) detected them exclusively from glia in the human brain. Thus, although the regional and histological distributions of cGPx have been examined, the pictures obtained so far are not consistent, and the reasons for the apparent discrepancies among these reports need to be clarified. Further elucidation of the regional distribution of cGPx is warranted.

18.3.3 Phospholipid Hydroperoxide Glutathione Peroxidases (phGPx)

phGPx can catalyze the degradation of phospholipid hydroperoxides, thus serving an important role in protecting membranes from oxidative damages. This is especially important for the brain, which has a higher lipid content. However, direct evidence for the importance of this selenoenzyme in brain functions has not been obtained. The difficulty may arise partly because phGPx in the brain is hardly depleted by dietary manipulations of Se intake. For example, brain cGPx activity of newborn rats born to dams raised on a low-Se diet for 8 weeks was reduced to half as compared with those raised on a normal diet, whereas phGPx activity was not different from control levels (Mitchell et al., 1997). Such sparing of the activity was achieved, at least partly, by the increase of mRNA levels; in the cerebrum of the Se-depleted newborn rats, mRNA of phGPx increased by 130 to 170% of levels in control (Se-adequate) rats.

18.3.4 Iodothyronine Deiodinases (DIs)

Iodothyronine deiodinases (DIs) are enzymes controlling the level of active thyroid hormones in various tissues. Three isozymes of DIs, DI1, DI2, and DI3 (also referred to as type I, II, and III enzymes, respectively), are known, all of which have been identified as selenoproteins (although there have been arguments against this for DI2). All share the reaction in which an iodine is removed from iodothyronines: DI1 and DI2 primarily catalyze 5'- or outer-ring deiodination, whereas DI3 catalyzes the 5- or inner-ring deiodination. When thyroxine (T4) is the substrate, 5'-deiodination produces triiodothyronine (T3), the most active thyroid hormone, whereas 5-deiodination produces reverse T3 (rT3), an inactivated hormone. In this way, DIs regulate the levels of active hormones in tissues as well as in circulating blood. DI1 is ubiquitously expressed, having both 5'- and 5-catalytic properties, although the emphasis is on the first one. The DI2 catalyzes 5'-deiodination and is expressed in only the brain, thyroid, and brown adipose tissue, whereas DI3 catalyzes 5-deiodination and is expressed in the placenta as well as fetal organs, including the brain. The 5-deiodination activity is high in the fetal brain but low in the adult brain.

It has been known that thyroid hormone plays a crucial role in differentiation of the developing brain. Both deficiency and excess of the hormone lead to abnormal development of the brain. In the fetal brain, T3 is locally produced through deiodination of T4 derived from fetal circulation. The origin of circulating T4 depends on the stage of development, and in rodents fetal thyroid becomes functional to synthesize and secrete thyroid hormones only in the later phase of fetal life. In the earlier phase of development, circulating T4 is derived from maternal circulation. The placenta is relatively permeable to T4 but not to T3; an increase in DI3 activity in the placenta reduces the T4 supply from the mother to the fetus. In the fetal brain, the level of T3 is controlled by the balance between DI3 and DI2 activities. Thus, the level of T3 in the fetal brain depends on DI activities both in the placenta and in the fetal brain.

The distribution of DI2 activity in the brain has been reported. In adult rats, in decreasing order, the cerebellum, cerebrum, and midbrain exhibit the activity, the ratio of distribution of activity in the cerebellum to that in the midbrain being less than twofold. In contrast, brain regions of preweaning rats show considerable regional difference in activity: cerebellum has the highest whereas cerebrum has the lowest activity, with brain stem and midbrain being in between (Kohrle, 1996). The significance of such regional differences in the brain needs to be clarified.

The effect of Se deficiency on the levels of these enzymes has been examined by several researchers. It was consistently found that in adult animals, dietary Se deficiency hardly changes the levels of these enzymes. On the contrary, in the brain of rat neonates (on the 11th day of postnatal life; PND11) born to Se-depleted mothers (fed a low-Se diet from 8 weeks before gestation until the end of the experiment), DI1 activity did not differ from that of Se-adequate controls but DI2 activity was reduced to less than half that of the control (Mitchell et al., 1998). In a three-generation deficiency experiment in rats (Campos-Barros et al., 1997), the activity of DI2 in the brains of third-generation rats (at the age of ca. 3 months) decreased to one fourth of the control levels. These rats had brain T4 levels similar to those of control rats, but T3 levels were significantly reduced in the hippocampus, hypothalamus, and striatum while they were not reduced in the cerebrum, cerebellum, and brain stem. In contrast, brains of fetal and neonatal rats born to the Se-depleted second-generation mothers had normal levels of DI2 and DI3 (although DI1 activity was slightly reduced) (Bates et al., 2000). Another experiment conducted under similar conditions also reported unchanged thyroid hormone levels in maternal as well as fetal circulations (Chanoine et al., 1993). The reasons for these inconsistent findings have not been well understood, although the timing of evaluation as well as duration of Se deficiency might play a significant role in determining the results. A longitudinal observation of the responses of the maternal–fetal complex against Se deficiency appears to be crucial in clarifying the importance of Se nutrition in the developmental process.

Mitchell et al. (1998) showed that mRNA expression of the brain-derived neurotrophic factor (BDNF) was significantly reduced in Se-depleted offspring on PND 4 and 11, but expression of myelin basic protein did not show any effects. Because thyroid hormones are reported to be involved in the regulation of BDNF expression (Koibuchi et al., 1999), the reduction in BDNF might be a direct consequence of the compromised DI regulation. Although the importance of thyroid hormone for normal development of the brain has been well established, the quantitative relationship has been hardly examined.

The changes in DI activities induced by Se depletion are different from those induced by iodine depletion. A clear but notable consequence of this is that possible brain dysfunctions induced by Se depletion should be different from those caused by iodine depletion. The Se-depletion-induced decrease of BDNF mRNA is not observed in the iodine-depleted group, which instead shows an increase (Mitchell et al., 1998).

We now discuss findings with the human population. The most severe manifestation of the dysfunction of thyroid hormones during perinatal life is cretinism, characterized by retardation in both physical growth and intellectual development. In northern Zaire, myxedematous cretinism, a minor subtype of endemic cretinism, is observed among children. The schoolchildren of this region, regardless of the disease status, show, other than the expected low urinary iodine excretion, low plasma Se levels. Because poor Se nutritional status could compromise the thyroid hormone metabolism, Se supplementation (before iodine supplementation) was introduced as an intervention trial. The result of this intervention was unexpected. After 2 months of supplementation with selenomethionine, the already-impaired thyroid indices, for example, low plasma T4 and high plasma TSH, in cretin children worsened, although the Se nutrition indices, plasma Se, and plasma GPx activity were improved. It was speculated that Se deficiency, through suppressed deiodination, maintains T4 supply to the fetus, providing protection against iodine deficiency and brain damage to the fetus, whereas through decreased GPx activity and the resultant oxidative stress, it exaggerates the damage to the thyroid gland and facilitates development of the myxedematous type of cretinism peculiar to this geographical area (Contempre et al., 1991). Interaction of Se status with iodine deficiency was also observed in Polish children with goiter.

18.3.5 SELENOPROTEIN P (Se-P)

Se-P was first recognized as the major form of Se-containing proteins in the plasma of humans and rats. In the human plasma, Se is associated with three proteins, Se-P, eGPx, and albumin, of which Se-P binds the largest portion (ca. 60%) of plasma Se. The glycoprotein, synthesized in many organs and secreted into the bloodstream, has a molecular weight of ca. 57 kDa and contains 7 to 8 SeCys per molecule (which consists of a single subunit) in the plasma.

In an immunostaining experiment using a polyclonal antibody, Se-P was found to be associated with the endothelium of liver as well as with capillary endothelial cells of the kidney glomeruli. In addition, the signal was also recognized in the brain, where the signal was confined to vascular endothelial cells. It was thought that Se-P attached to the surface of endothelial cells by its heparin-binding domain (Burk et al., 1997). Later, it was shown that cultured rat astrocytes express mRNA for Se-P and secrete Se-P protein into the culture medium. Therefore, Se-P can exist not only in the vascular space but also in the brain side of the blood-brain barrier (Yang et al., 2000).

The physiological function of Se-P was not known for a long time. Because of its localization to plasma and its rapid appearance in the plasma after injection of labeled selenite, a transporting function has been suggested. Although the large number of Se atoms per protein is advantageous for a transporter, this putative transporting process would require energy-consuming processes such as biosynthesis, secretion, and degradation of the protein. The brain of Se-depleted rats takes up injected Se-P more efficiently than that of control rats. Such an effect is not observed in the other tissues or with GPx (Burk et al., 1991). These observations suggest that handling of Se-P by the brain is different from that by other tissues. Recently, two laboratories reported successful generation of Se-P knockout mice (Hill et al., 2003; Schomburg et al., 2003). Schomburg et al. (2003) reported that Se-P KO had abnormally high mortality (9/29 by 5 wks of age), markedly reduced body weight gain, loss of motor coordination as confirmed by the RotorRod test, decreased Se concentrations in brain, testis, kidney, and plasma, and increased Se in liver. While this study used only one dietary level of Se (0.29 ppm), Hill et al. (2003) fed the mice diets with various Se levels (<0.02 to 2 mg/kg). Most of those characteristics of the Se-P KO mice observed by Schomburg et al. were confirmed but it was so only when the Se-P KO mice were fed a Se-deficient diet (less than 0.1 ppm). At the higher dietary Se levels, tissue Se concentrations were not modified and the mortality as well as motor functions appeared to be normal. Thus, the abnormal mortality and motor function of Se-P KO appeared to be associated with altered Se distribution. Although this suggests that the primary function of Se-P is transporting Se, it cannot be excluded that locally produced Se-P (e.g., Yang et al., 2000) could also contribute to the observed phenotypes. It should be noted that based on the fact that Se-P KO could respond to the change in dietary level of Se, Hill et al. (2003) suggested existence of transporting form of Se other than Se-P.

On the other hand, several observations suggest that Se-P acts as an extracellular antioxidant. First, Se-P derived from human plasma has phGPx-like activity, although its relevance *in vivo* is unknown (Saito et al., 1999); GSH concentration in the brain extracellular space is estimated to be as low as several micromolars, which appears to be too low to support GPx-like activity in the brain. Second, it was proposed that Se-P might scavenge peroxynitrite, a highly reactive molecule responsible for the inflammation process (Burk et al., 1997; Behne and Kyriakopoulos, 2001).

Se-P may have beneficial effects on neurons. A protein purified from bovine serum, cross-reactive with an antibody against Se-P, supported the survival of cultured neurons when added to the culture medium. Although this result suggested that the protein acts as an Se transporter to the neuron, that Se-P-like protein was six times (based on the number of Se atoms) more efficient in the survival-promoting effect than selenite suggested that the survival-promoting effect might at least involve some processes other than transport (Yan and Barrett, 1998). Although it is not known whether the neuron could uptake Se-P derived from circulation or that synthesized and secreted by glial cells, the function of Se-P in this culture system should be further evaluated. The expressions of proteolipid proteins, myelin basic protein, and myelin-associated protein were reduced in a glial cell culture devoid of Se in the medium. Because these proteins are involved in the maturing of oligodendrocytes, the observation suggests that Se may play a role in the maturation of glial cells (Gu et al., 1997). In this study, Se was added as sodium selenite, and the selenoprotein responsible for this effect was not determined.

18.3.6 SELENOPROTEIN W (Se-W)

Se-W is a low-molecular-weight (9.5-kDa) protein whose function has not been elucidated. It is expressed in many tissues, including the brain, muscle, heart, spleen, liver, kidney, and testis. When rats or sheep were kept on a low-Se diet for up to 14 weeks, Se-W was decreased in all the tissues except the brain (Sun et al., 2001).

The putative function of Se-W was examined by expressing the protein derived from mouse brain in CHO cells and H1299, a human cancer cell line. The cells overexpressing Se-W showed enhanced tolerance to H_2O_2-induced cytotoxicity by suppressing the production of peroxide. The enhanced tolerance depended on the presence of intracellular GSH, and a SeCys and a Cys residue were required for the antioxidant activity *in vivo* (Jeong et al., 2002). Thus, Se-W might serve as an antioxidant in the brain.

18.3.7 THIOREDOXIN REDUCTASES (TRxRs)

Three isozymes of thioredoxin reductase (TRxR) have been identified as selenoproteins. TRxR catalyzes the reduction of oxidized thioredoxin (TRx) which, in its reduced form, serves as a reductant for many redox-dependent systems relevant to such processes as DNA synthesis and transcription regulation. In addition to TRx, TRxR can accept hydroperoxides, dehydroascorbate, and various enzymes as substrate, and the physiological functions of TRxR appear to be multifold, including the reduction of ascorbic acid and cell growth regulation. Disruption of the TRx gene is embryonic lethal.

The activity of TRxR in the brain is well preserved when animals are kept on a low-Se diet. In rats fed on low-Se diets for 14 weeks after weaning, TRxR activity in the liver and kidney decreased to 4.5 and 11% of that of control rats, respectively, whereas the activity in the brain did not change (116% of the control). In contrast, the brain GPx activity in these rats decreased to 70% of the activity in the control.

18.3.8 OTHER SELENOPROTEINS FOUND IN THE BRAIN

The 15-kDa selenoprotein was first identified in human T cells, but is actually expressed in many tissues, including the brains of mice and humans. Although the function of this protein has not been elucidated, from the expression pattern in normal and in malignant tissues it was speculated that this protein has a role in cancer etiology (Kumaraswamy et al., 2000). Selenoprotein M is another newly found protein, with a 3-kb gene devoid of the conventional SECIS element, which is invariably found in the untranslated regions of all other mammalian selenoproteins found so far. This protein is also expressed in many mammalian tissues, but among the 14 tissues examined in mice the highest expression is in the brain (Korotkow et al., 2002). The 56-kDa Se-binding protein, originally discovered by Bansal et al. (1990), differs from other selenoproteins in that it does not possess UGA coding for SeCys. The chemical form of Se in this protein has not been elucidated. The protein is found to be an intra-Golgi protein transporter, although the significance of Se for the protein's function has not been established.

18.4 EXPERIMENTAL SE DEFICIENCY AND BRAIN-ASSOCIATED FUNCTIONS

Experimental animals raised on low-Se diets exhibit several changes in the brain and behavior. Although the mechanism of these alterations, including the identity of the responsive selenoprotein, has not been elucidated, these observations suggest the possible roles of Se in brain functions and are thus worth considering.

18.4.1 Neurochemical Aspects: Effects on the Dopaminergic System

Although some contradictive findings exist, Se and cGPx distribute preferentially along the DAergic transmission pathways in the brain, suggesting a linkage between Se and DAergic systems. This possible linkage has been explored by examining the effects of dietary-induced Se deficiency on the DAergic pathway. In adult rats, a depletion of Se for as little as 2 weeks brought about many changes in the metabolism of DA in the nigrostriatal pathway and the SN. Both the concentration and the turnover of DA were elevated by the low-Se diet (Castano et al., 1993, 1995). Similar observations were obtained in the striatum and hippocampus, although the effects were not as distinctive as those in the SN. In the SN as well as the hippocampus, the activity of tyrosine hydroxylase, the rate-limiting enzyme of DA biosynthesis, was increased with the enhanced expression of mRNA. In addition, mRNA of the DA transporter was increased in the SN and DA uptake by striatal synaptosomes was enhanced. All these observations are consistent with the accelerated DAergic activity in these pathways, which the authors postulated. The authors speculated that enhanced oxidative stress due to Se deficiency led to accelerated turnover of DA, which in turn facilitated DA-driven oxidative stress, thus creating a feed-forward loop. Thus, enhanced oxidative stress ensues, which leads to DAergic system pathology. Although such interpretation is intriguing, the activity of cGPx was reduced by only 15 to 20%, which is unlikely to lead to enhanced oxidative stress considering the results of GPKO mice. Alternatively, the enhancement of oxidative stress may be a local event, which takes place in the proximity of DA neurons or terminals and can only be detected by examining local oxidative status.

The function of DAergic pathway in the Se-depleted brain of mice has been examined (Watanabe et al., 1997). In mice fed with a low-Se diet for 11 to 13 weeks after weaning, striatal extracellular outflow of released DA was monitored by microdialysis. Although the baseline outflow in the Se-depleted group was not different from that in the control group, the outflow under the depolarization stimulus (for 15 minutes) by a high-K^+ perfusing solution was significantly elevated in the Se-depleted group. This observation suggests that DA available for vesicular release from the terminal was increased in the Se-depleted group, which is consistent with the neurochemical changes. In this case, brain Se concentration and GPx activity were reduced to 71% and 60% of those in the control brain, respectively. The change was not observed in the group that was Se-depleted for 4 weeks, in which Se concentrations and GPx activity were 81% and 75% of those of the control, respectively, suggesting that for neurophysiological change to occur a substantial and prolonged reduction is required.

18.4.2 Behavioral Effects of Se Deficiency

Behavior in its broader sense is the outcome of brain functions, and wide variety of molecular, neurochemical, or neurophysiological alterations of the brain can be detected as a deviation of certain behavior from the normal behavior. If a depletion of dietary Se induces behavioral alteration, the alteration might be due to change in a certain brain function.

In a two-generation Se-deficiency experiment, female mice were kept on a low-Se diet (<0.02 mg Se/kg diet) before and during pregnancy as well as the lactation period. Offspring were also kept on the same diet after weaning. In these offspring mice, several behavioral anomalies were found: compared with Se-adequate controls, development of walking behavior was significantly delayed, open-field activity was decreased, performance in the Morris water maze was poorer, and these mice exhibited a preference for a warmer temperature. In the Se-deficient offspring mice, brain Se concentration was ca. 60% of that in control mice (Watanabe and Satoh, 1994, 1995). Because these offspring mice weighed less than control mice, a part of the observed behavioral effects might not be specific to Se deficiency but arise from general nutritional constraints. In another attempt that used three-generation Se deficiency in rats (Bastug et al., 1998), though no behavioral anomalies were recognized, plasma Se concentration of the third generation of deficient rats was as high as half the third-generation controls, suggesting that the diet regimen used was not severe enough to induce a deficient status.

Because some neurochemical changes were observed in the brain of rodents kept on low-Se diet after weaning, the possible behavioral changes caused by the same manipulation were explored in mice. Female ICR mice were kept on a low-Se diet for 4 to 14 weeks and their behavior was examined (unpublished results). In Se-deficient mice, contrary to what was observed for the perinatally induced low-Se diet, open-field activity was enhanced in Se-deficient mice. The difference may be reasonable, because among the selenoproteins exist the DIs, which are crucial in controlling the level of thyroid hormone in the developing brain, which in turn is crucial for normal development and construction of the brain. This enhanced open-field activity was dependent on mice strain; among the five strains evaluated, C57BL/6J was the most sensitive. This strain was the only strain in which Se-depleted mice showed significantly poorer performance than control mice on the rotarod, with which sensory–motor coordination was evaluated by measuring the ability of mice to keep walking on a rotating shaft. Because the weight of Se-deficient mice was equal to or heavier than control mice, the observed effects could not be due to general malnutrition. The cause of such strain difference might be related to strain difference in the sensitivity to oxidative stress.

18.5 EFFECTS OF SE ON THE NEUROTOXICITY OF CHEMICALS

Se has been known for its ability to interact with many chemicals, such as heavy metals and synthetic drugs, resulting in most of the cases in alleviating the toxicity. Among chemicals that interact with Se are those that either generate ROS or affect DAergic systems. This section focuses on reports in which altered dietary intake of Se affects the toxicity of chemicals, as they should be more relevant to the physiological functions of Se. The effects of acutely and exogenously administered Se are not covered here.

18.5.1 METHYLMERCURY

Methylmercury has been known to be the causative agent for the Minamata disease, a severe and irreversible, sometime fatal, neurological disease with motor and sensory dysfunctions. Although severe intoxications such as the Minamata disease have become rare, the possible neurotoxicity of this compound after *in utero* exposure (through mother's fish consumption) has become a public health issue because of the possibility that very low levels of this compound may affect the developing brain. Unlike the interaction between inorganic mercury (Hg^{2+}) and Se, in which a direct chemical reaction would lead to the formation of inert Hg–Se complex and resultant detoxification, the interaction between methylmercury and nutritional levels of Se has been much less explored. In general, neurotoxicity and fetotoxicity of methylmercury *in utero* are exaggerated when Se supply to the animals is reduced. On the other hand, toxicities are suppressed when excess amounts (supranutritional level) of Se are added to the diet (Watanabe, 2002).

The mode of interaction remains largely unknown. First, direct interaction between methylmercury and selenite was reported to form a complex bis(methylmercuric) selenide. The complex, however, was so unstable that it rapidly disappeared from tissues. Second, in a long-term exposure of monkeys, brain Se was increased along with the accumulation of inorganic mercury. This inorganic mercury must have been derived from demethylation of the administered methylmercury, suggesting that there was a direct chemical interaction between the inorganic mercury (Hg^{2+}) and selenide. Although intriguing and consistent with previous observations on the autopsy brains of mercury miners (who had been exposed to high doses of elementary Hg vapor), no similar observation has been obtained in human populations chronically exposed to low levels of methylmercury through fish consumption. Third, methylmercury may affect the metabolism of Se; in the brain of fetal mice exposed to methylmercury for 3 consecutive days during gestation, 5-DI activity was decreased and 5′-DI activity was increased. Because T4 in the fetal circulation was normal, the activity changes would favor excess production of T3 in the fetal brain, which could eventually lead to abnormal development of the brain. Although the Se level unchanged, GPx activity was suppressed in the fetal brain (Watanabe et al., 1999).

18.5.2 PARAQUAT AND DIQUAT

These bipyridil-derivative pesticides exert their toxicity by generating ROS in cells. Se-deficient rats are more vulnerable to diquat toxicity than are Se-adequate controls, which coincides with the different levels of Se-P but not those of GPx (Behne and Kyriakopoulos, 2001). As discussed earlier, GPKO mice showed poorer resistance to the lethal dose of paraquat but showed similar resistance to the moderate dose of paraquat (Cheng et al., 1998). Thus, the relative importance of Se-P and cGPx (and perhaps other selenoproteins) as antioxidants may also vary depending on the dose.

Recently, an *in vitro* study was conducted in which hepatocytes isolated from GPKO mice and those from wild-type mice were compared in terms of the cytotoxicity and oxidative stress induced by diquat and peroxynitrite (Fu et al., 2001). Although GPKO cells were less resistant to the apoptosis induced by diquat, they were more resistant to the apoptosis induced by peroxynitrite. The role of GPx in the protection against oxidative stress might not be straightforward as has been expected.

The primary target organ of these herbicides is the lung, and the oxidative stress in this tissue resulting in lung fibrosis has been regarded as the cause of death in human suicide cases. Recently, it was shown that repeated exposure to paraquat induced loss of DA neurons in the SN, accompanied by reduced ambulatory activity, both of which were also caused by repeated exposure to MPTP, a DAergic toxicant (see next section). It would be intriguing to study how the nutritional status of Se affects paraquat-induced brain pathology.

18.5.3 MPTP AND METHAMPHETAMINE (MA)

Both MPTP (1-methyl-1,4-phenyl-1,2,3,6-tetrahydropyridine) and methamphetamine (MA) have well-known DA toxicity and thus have been used as study tools for DAergic systems. Se supplementation to drinking water (3 mg Se/l) 1 week both before and after injections of MA to mice was shown to attenuate the formation of 3-nitrotyrosine in the striatum, a marker of peroxynitrite production. Together with the *in vitro* observation that the formation of 3-nitrotyrosine was associated with the depletion of DA and its metabolites, this observation led the authors to conclude that Se attenuates DAergic toxicity of MA by suppressing the production of peroxynitrite (Imam and Ali, 2000).

On the other hand, mice fed on a low-Se diet (<0.01 ppm Se) for 90 days showed enhanced MA-induced DAergic toxicity in the SN, evaluated by the loss of tyrosine hydroxylase-like immunoreactivity and decrease of DA and its metabolites (Kim et al., 2000). The mechanism that links the depletion of Se and this DAergic toxicity remains to be elucidated.

In an earlier study, the effect of a low-Se diet on DAergic toxicity was evaluated in mice (Sutphin and Buckman, 1991). Although neither the low-Se diet (6–8 weeks) nor MPTP (15–30 mg/kg, IV) affected brain oxidative stress (evaluated by the level of malondialdehyde), MPTP treatment induced enhanced oxidative stress in the Se-deficient brain.

Responses of rats and mice to 3,4-methylenedioxymethamphetamine (MDMA), which causes relatively selective long-term neurotoxic damage, were compared (Sanchez et al., 2003). Although the depletion of monoamine by MDMA was comparable between Se-deficient and adequate groups in rats, it was much more exaggerated in the Se-deficient group than in the Se-adequate group in mice. Without MDMA, a comparative degree of brain GPx suppression by Se deficiency was observed in rats and mice, while increased lipid peroxidation by Se deficiency was observed only in mice. These results were interpreted as mice being more susceptible to Se deficiency than rats.

18.5.4 GLUTAMATE

While glutamate (Glu) is an excitatory neurotransmitter, excess Glu can exert neurotoxicity via a mechanism known as excitotoxicity. Using a hippocampal cell line HT22, it was shown that addition of Se to the medium as sodium selenite or selenate protected the cells from excitotoxicity of excess Glu. While the protective action of Se does not involve glutathione, it requires *de novo* protein synthesis. Activation of NF-kB and AP-1 as well as generation of peroxides, all induced by excess Glu, are suppressed by Se, and protection by Se may depend on protein-dependent anti-apoptotic pathway (Savaskan et al., 2003). In addition, rats maintained on a Se-deficient diet were much more susceptible to the convulsive effect of kainate and subsequent cellular toxicity. The Se content of the brain of these Se-deficient rats was decreased but only by 10%, which appears too modest to account for the enhanced susceptibility to kainate neurotoxicity. As the authors mentioned, the effects could involve some non-selenoprotein(s) that could be influenced by Se deficiency; a number of genes are activated under low Se status at least in peripheral tissues (Rao et al., 2001).

18.6 SE DEFICIENCY AND EXCESS IN HUMAN POPULATIONS

18.6.1 SE INTAKE IN VARIOUS POPULATIONS

The body of the "reference man," with 70 kg of body weight, contains 13 mg of Se (ICRP, 1975). In terms of the amount in the body, Se is comparable to such trace elements as vanadium, manganese, or iodine, but is less abundant than elements such as iron, zinc, or copper.

Most of the Se in the body comes from the diet. Se content of food items varies widely, and the relative contribution of various foodstuffs to total Se intake depends on the population. Se concentrations in plant food depend on the soil content of Se, which varies widely among different geographical regions. If the livestock depends on the locally grown feed, then the Se content of meat or dairy products also depends on soil Se. Thus, in China, where both Se-deficient and Se-rich (excess) regions exist, the Se concentration of plant and animal foods reflected the regional soil content of Se (Combs, 2001). After Se-deficient soils of Finland were fortified with Se on a nationwide scale, Se concentrations of wheat, meat, dairy products, and the residents' plasma showed appreciable increases. Meat and cereals serve as the major sources of Se in most Western countries. Fish contains high concentrations of Se and could serve as a major contributor for Se intake in such countries as Japan and Norway, although the bioavailability of Se in some fish species may not be very high. Vegetables usually contribute little for Se intake (Combs, 2001).

The average daily intake of Se per person ranges from ca. 10 μg/day in Se-deficient areas in China to several hundreds μg/day in seleniferous areas in China or Venezuela. Although such countries as Japan, Canada, and the U.S. have relatively rich intakes of Se (moe than 100 μg/day), other countries, including New Zealand and most European countries (e.g., U.K., France, Sweden, and Slovakia) have intakes of less than 50 μg/day (Rayman, 2000; Combs, 2001).

Based on the epidemiological data obtained from China and New Zealand (Rayman, 2000; Combs, 2001), the recommended daily allowance by the U.S. National Academy of Science (2000) is set at 55 μg/day for both adult men and women (refer to www.preventivehealthtoday.com/nutrition/ for details), whereas the dietary reference intake (the upper safety limit of intake) is set at 400 μg/day by the U.S. and some European countries.

18.6.2 DEFICIENCY OF Se IN THE HUMAN POPULATION

Because of different Se intakes in various regions, some regions or countries are at a risk of Se deficiency and, though rare, Se excess. The most well documented pathology associated with Se deficiency is the Keshan disease, a type of cardiomyopathy found in the northeastern to southcentral area of China, which could be fatal in severe cases (Combs, 2001). Because supplementation of Se (in the form of sodium selenite) drastically reduces the incidence, the Keshan disease is thought to be a manifestation of Se deficiency in humans. The epidemiology of the disease, however, reveals seasonal fluctuation and sex, age, and area dependence of the incidence, suggesting that other contributing factors are involved in the etiology. An intriguing finding is that a nonvirulent strain of the Coxackie virus, a putative contributing factor for the disease, acquires virulence in the body of Se-deficient mice, presumably through the differential selection of the virus having virulent mutations in its RNA sequence. Kashin–Beck disease, a deforming arthritis, is another disease to which deficiency of Se may contribute as an etiological factor. Like the Keshan disease, however, the disease is thought to have a multifactorial etiology.

People under chronic total parenteral nutrition (TPN) may suffer from deficiency of various trace elements, including Se. Many reports have been published on Se deficiency symptoms developed in TPN patients, such as decolorized hair, deformed fingernails, cardiomyopathy, muscle pain in legs, and macrocytic anemia. These symptoms rapidly disappear on supplementation of Se to the TPN solution.

Patients of Keshan disease, Kashin-Back disease, or those receiving chronic TPN have not been reported to develop symptoms that suggest clear involvement of the brain or nervous system. In other words, no known neurological symptoms in humans are primarily ascribable to Se deficiency.

18.6.3 Se Nutrition and Nervous System-Related Conditions

Because many selenoproteins have antioxidant property, a link between inadequate Se nutrition and increased risk for diseases associated with oxidative stress has been suggested. For example, inverse relationships have been reported between Se intake and increased risk for some cardiovascular diseases and for various type of cancer, whereas supranutritional prophylactic intake of Se decreases the risk for cancer.

Several studies have examined the Se level of biological media (RBC, plasma, urine, or hair) obtained from patients with some neurological disease. For example, plasma and RBC Se and GPx activity levels were evaluated in multiple sclerosis (MS) patients and compared with those in control subjects. One study observed no difference between the two groups (Smith et al., 1989), whereas the other one found lowered plasma and RBC Se and GPx in patients. In this type of study, use of such media as plasma, urine, or hair might be questionable because Se or GPx activity in such media may not reflect the nutritional status of Se in the brain. In countries such as the U.S. that have a rich intake of Se, the individual variation of plasma Se may simply reflect the amount of albumin-bound Se and not the amount of functional selenoproteins (Se-P and eGPx). In autopsy samples of apparently healthy people ($n = 46$, mean age 42.5 years), none of the correlations between hair Se and Se in the cerebrum or cerebellum was significant, although correlations among the tissues, including the brain, were highly significant. In this sample population, brain Se varied considerably, the maximum:minimum ratio being 2.5 (cerebrum white matter) or 4 (cerebellum). Thus, hair Se was not sensitive enough to reflect such large interindividual variation of brain Se as found in apparently healthy subjects (Yoshinaga et al., 1990).

Apart from the credibility of these media as surrogates of the brain, cross-sectional relationship does not provide information on causal relationship. A prospective study or nutritional intervention (in case of reversible effects) will be much more informative and valuable. A successful intervention trial on four children with intractable seizures was documented. Two of them showed low RBC GPx activities, high plasma GPx activity, and normal whole-blood Se. In the remaining two children, all parameters showed a decrease. One patient died later, and the autopsied brain showed reduced weight and gliosis. A supplementation with Se resulted in clinical improvement, although no details of the supplementation were given in the article (Weber et al., 1991).

18.6.4 STUDIES USING AUTOPSY BRAIN

Analyses of autopsy brain can provide important information in this respect. In a group of German males (mean age 39 years), autopsy brain contained 0.11 mg/kg of Se, whereas in a Japanese group consisting of 31 males and 14 females of various ages, the cerebellum and cerebrum contained 0.25 and 0.22 mg/kg of Se, respectively. (Unless otherwise noted, figures in this section are given on a wet weight basis.) Within-country individual variation more or less equals between-country differences in this case.

Distribution of Se among various brain structures has also been reported. According to these studies, differences in the Se concentrations among brain regions are as high as threefold but not more than that; thus, such structures as the caudate putamen, SN, and thalamus contain high concentrations of Se, whereas the corpus callosum has low concentrations (Larsen et al., 1979). The distribution roughly approximates the distribution of the activity of cGPx in rat brain, which shows the highest activity in the caudate putamen and SN and the lowest activity in the cerebral cortex and corpus callosum (Brannan et al., 1980).

Several reports have compared the levels of Se and GPx in the normal and the diseased brain. In Larsen et al. (1979), concentrations of trace elements, including Se, were determined by neutron activation analysis in the normal brain as well as in brains of patients with Parkinson's disease and amyotrophic lateral sclerosis. There were no differences between the normal and pathological brains in terms of Se as well as of manganese, although the sample size was very small (not exceeding five per group). In another study, autopsy brains from the confirmed cases of Parkinson's disease were examined in terms of localization of the immunoreactive cGPx protein (Damier et al., 1993). In midbrains of both control and Parkinson's disease, immunopositive signals were detected exclusively in glial cells. In control brains, the signal was obscured in the zona compacta of SN, the primary pathological site of Parkinson's disease, whereas intense signals were detected in the A8 DAergic neurons as well as in the central gray, which are relatively spared in the disease. In patients with Parkinson's disease, the signals were also detected from the area surrounding the survived DAergic neurons. Based on these observations, it was speculated that DAergic neurons located in the area with weak GPx activity would be vulnerable to toxic insult.

Alzheimer's disease is another type of neurodegenerative disease responsible for a considerable portion of dementia cases. The temporal lobes in the autopsy brains of patients with Alzheimer's disease were examined in terms of 13 trace element (and heavy metals) concentrations, including Se (Wenstrup et al., 1990). Compared with the control brain ($n = 12$), the brains of patients with Alzheimer's disease ($n = 10$) had higher Hg concentration and lower (by 5 to 25%) Se concentration, which was most prominent in the microsome fraction. These authors speculated that the elevated Hg was related to the pathology and that the lowered Se was used for Hg detoxification. More recently, various brain sites of patients with Alzheimer's disease ($n = 10$) and control subjects ($n = 10$) were compared in terms of TRxR activity as well as the amount of TRx (Lovell et al., 2000). Brains of patients with Alzheimer's disease had significantly lower amounts of thioredoxin and significantly higher activity of TRxR. Because a supplemental *in vitro* experiment showed a protective effect of both thioredoxin and TRxR against neurotoxicity of amyloid-β peptide, the authors speculated that in the brains of patients with Alzheimer's disease, the decrease of thioredoxin contributed to enhanced oxidative stress, which could not be alleviated by the increased TRxR activity.

In the autopsy brains of sporadic motor neuron disease patients, Se concentration and GPx activity were higher than those in control brains by 40% and 30%, respectively (Ince et al., 1994). However, the "control" group in this study included the patients of Parkinson's or Alzheimer's diseases, making interpretation of the results difficult. In brains of patients who died from cerebral infarction or cerebral hemorrhage, the localization was examined by using anti-human GPx monoclonal antibodies. In brains without infarction ($n = 2$), immunopositive signals were detected from both the neurons and glia. Signals were detected from macrophages in those who died within 5 days of an ischamic attack, whereas they were detected from the cytosol of glial cells surrounding the infracted area in brains of those who died at least 6 days after the attack. These observations suggested that glial GPx would prevent or alleviate the oxidative lesions induced by the ischaemic-reperfusion process (Takizawa et al., 1994).

Little is known about the extent to which Se is depleted in the human brain by low Se intake. In two male TPN patients who died in their 70s after receiving tube feeding for at least 4 months, brain Se concentrations were reported to be 0.26 and 0.44 µg/g dry tissue (Feller et al., 1987). If we apply a tentative conversion factor of 5 to estimate the wet weight-based concentration, these values do not suggest low Se status. They were lower, however, than the dry-wet-based "normal" values of 0.6 to 0.7 µg/g dry weight reported in another independent study (Malmqvist et al., 1988).

Being a cross-sectional evaluation, the value of the studies that have used autopsy brains is limited in elucidating the ultimate cause of the disease, but at the same time it is suggestive in many respect for further researches. Thus, it appears that any measurements of Se of selenoproteins at the level of whole brain are of very limited value because of (1) relatively large individual variation as well as brain region-associated variation (in the concentration of Se or expression of selenoproteins), (2) the multiplicity of selenoproteins, and (3) unresponsiveness of brain Se to dietary Se intake. Instead, as typically illustrated by the study of GPx localization after the infarction (Takizawa et al., 1994), it is likely the progress of certain pathological changes would be chrono-logically associated with the local change in the spatial distribution or expression and metabolism of Se or selenoproteins. These factors should be taken into account in designing future studies both in humans and experimental animals.

18.6.5 MOOD AND SE

If Se deficiency has any consequences on the human brain, the manifestation must be a subtle and subclinical one, presumably associated with minor biochemical changes. Based on such assump-tions, the effect of Se nutritional status on mood has been examined in the U.K. as well as in the U.S.

An intervention trial with 50 male and female volunteers aged 14 to 74 was conducted in a cross-over design, in which one group was assigned to Se supplementation and then to placebo treatment and the other group went vice versa (Benton and Cook, 1991). The intervention included 5 weeks of Se supplementation, during which 100 µg/day of yeast-derived Se was given to subjects. Subjects were asked to report subjective feelings (mood), using the Profile of Mood Status. The Se-supplemented group showed an improvement in the mood status. The effects were more dis-tinctive in the subgroup with lower dietary intake of Se (28 to 63 µg Se/day).

In the U.S., a similar intervention trial was conducted for a much longer period (approximately 3 months) but with fewer subjects (11 males) (Hawkes and Hornbostel, 1996). Of the 11 subjects, 5 were given a high-Se diet containing 356 µg Se/day, and the rest of the subjects a low-Se diet containing 13 µg Se/day for 99 days. Although the authors did not see any effects of Se supple-mentation in terms of change in mood status, those with low RBC Se levels showed an exaggerated worsening by a low-Se diet. Another intervention trial conducted in the U.S. showed that 15 weeks of a high-Se diet (226.5 µg Se/day) or a low-Se diet (32.6 µgSe/day) resulted in improvement or worsening of mood status, respectively.

A major drawback of these studies is the lack of appropriate assessment of Se nutrition status. In the U.K. study, although dietary Se intake was estimated, no measurements were made by using biological media (plasma or urine). In the U.S. study, although the RBC Se was measured at the beginning of the experimental period, no measurement was made after the dietary regimen. There-fore, the impact of the dietary regimen used in these studies on the Se nutritional status of subjects remained unknown. In addition, the sample sizes were relatively small, which limits the power of these studies. Therefore, although suggestive, results of these studies must be deliberately inter-preted. At present, another intervention trial with much larger sample size is being undertaken in the U.K., which includes the measurement of plasma Se both before and after dietary intervention.

18.6.6 EXCESS SE

Although known as an essential trace element, an excess amount of Se gives rise to toxicity. Actually, the element has been known as a toxicant rather than a nutrient for a long time. The toxicity has been mostly observed in domestic livestock, and human poisoning rarely occurs except for occupational exposures and some accidental exposure through consumption of wrongly manufactured Se tablets (as supplementary diet). The only known large-scale endemic intoxication was reported in a geographically Se-rich area of China in the 1960s, where the common symptoms included loss of hair and nails, and skin and nervous system involvement was also described in severely affected subpopulations. Symptoms associated with the nervous system were likely due to polyneuritis, including hyperreflexia of the tendon, convulsions, paralysis, motor disturbance, and hemiplegia. In this area, daily Se intake was estimated to be ca. 5 mg/day, although the detailed methods for this estimation were not provided (Yang et al., 1983).

18.7 CONCLUSIONS AND RESEARCH NEED

From human and epidemiological literatures, it is apparent that deficiency of Se, even to the most severe degree as experienced by patients with Keshan disease or long-term TPN patients, does not have serious neurological or behavioral manifestations. This fact indicates that Se does little for normal brain functions and that there is little neurological, neurobehavioral risk from low Se intake. These inferences need to be carefully examined, however.

As shown in the experimental literatures, brain Se is hard to deplete and there are virtually no data regarding the Se status in the human brain under the most severe deficiency. The hierarchic regulation of selenoprotein metabolism (Behne et al., 1988), though not demonstrated, is likely to operate in humans, which would preserve the most important selenoproteins, especially in the brain. Thus, in a human population, Se deficiency may never reach an extent severe enough to cause depletion of Se in the brain or depletion of some crucial selenoproteins in the brain.

Instead of resulting in severe neurological symptoms, Se deficiency may have mild or modest manifestations. The experimental findings on DAergic pathways and open-field behavior, as well as the human intervention resulting in the change of mood status, might be examples of such mild or modest manifestations, although all these effects should be validated by further data. To establish these effects as the consequence of Se deficiency, a mechanistic evaluation involving biochemical and neurochemical parameters should be included.

In addition, Se deficiency may predispose the organism susceptible to other environmental insults, as illustrated by the interaction between the neurotoxic metal and pesticides. The example of GPKO mice is also in line with this notion; although GPKO mice are apparently normal, challenge with radical-generating chemicals as well as an RNA virus revealed the vulnerability of the Se-deficient (GPx-deficient) status. Such possibilities are important in terms of health risk.

In view of the recent discovery of multiple selenoproteins in mammalian tissues, studies on the possible roles of Se in brain functions need to include molecular understanding of the event. In this respect, the most well documented brain-specific function of Se in experimental literatures is that exerted through DIs. The physiological functions of DIs have been well characterized, and it is clear that these enzymes, especially DI2 and DI3, play critical roles for development of the brain. However, in human populations, there have been no known cases of pure Se deficiency resulting in disturbed thyroid function and deranged brain development such as cretinism. That is, contribution of Se nutrition to thyroid-related pathology has been recognized only when Se deficiency combines with iodine deficiency; no thyroid complication has been reported in the Se-deficient area of China or from long-term TPN patients. Such a discrepancy again supports the notion that the extent of Se deficiency never reaches the extent that is achieved in multigeneration animal experiments.

Several observations suggest possible roles of other selenoproteins in the brain. Among them, the brain may have specific uptake mechanism for Se-P and this protein is synthesized and secreted by glial cells *in situ*. Se-P probably promotes the survival of cultured neurons. Taken together, these observations strongly imply the importance of this protein in the brain. On the contrary, *in vivo* evidence of the cGPx's role, using GPKO mice, was obtained only when the animals were under certain physiological load. As pointed out by Behne and Kyriakopoulos (2001), there exists some similarity among the putative roles of multiple selenoproteins, which may make the demonstration of physiological roles of a single protein difficult. Therefore, to bridge the gap between the apparently healthy GPKO and embryo-lethal knockout of SeCys-tRNA, double or multiple knock-outs may be among the approaches of choice in the future.

Finally, as apparent from the preceding discussions, to reveal the possible role of Se in the brain, one has to consider not only the multiplicity of selenoproteins (as opposed to Se as a whole) but also the heterogeneity within the brain. In making any kind of measurement, the brain region, cell type (glia or neuron), cell environment (affected by, e.g., the neurotransmitters used by the cell and its neighboring cells), and location in a cell (e.g., soma or synapse) need to be addressed.

REFERENCES

Allan, C., G. Lacourciere and T. Stadtman (1999). Responsiveness of selenoproteins to dietary selenium. *Annu. Rev. Nutr.* 19: 1–16.

Bansal, M., T. Mukhopadhyay, J. Scott, R. Cook, R. Mukhopadhyay and D. Medina (1990). DNA sequencing of a mouse liver protein that binds selenium: implications for selenium's mechanism of action in cancer prevention. *Carcinogenesis* 11: 2071–2073.

Bastug, M., S. Ayhan and B. Turan (1998). The effect of altered selenium and vitamin E nutritional status on learning and memory of third-generation rats. *Biol. Trace Elem. Res.* 64(1–3): 151–160.

Bates, J. M., V. L. Spate, J. S. Morris, D. L. St. Germain and V. A. Galton (2000). Effects of selenium deficiency on tissue selenium content, deiodinase activity, and thyroid hormone economy in the rat during development. *Endocrinology* 141(7): 2490–2500.

Beck, M., R. Esworthy, Y.-S. Ho and F. Chu (1998). Glutathione peroxidase protects mice from viral-induced myocarditis. *FASEB J.* 12: 1143–1149.

Behne, D., H. Hilmert, S. Scheid, H. Gessner and W. Elger (1988). Evidence for specific selenium target tissues and new biologically important selenoproteins. *Biochim. Biophys. Acta* 966: 12–21.

Behne, D. and A. Kyriakopoulos (2001). Mammalian selenium-containing proteins. *Annu. Rev. Nutr.* 21: 453–473.

Benton, D. and R. Cook (1991). The impact of selenium supplementation on mood. *Biol. Psychiatry* 29: 1092–1098.

Berry, M. and R. Larsen (1992). The role of selenium in thyroid hormone action. *Endocrinol. Rev.* 13: 207–219.

Bosl, M. R., K. Takaku, M. Oshima, S. Nishimura and M. M. Taketo (1997). Early embryonic lethality caused by targeted disruption of the mouse selenocysteine tRNA gene (Trsp). *Proc. Natl. Acad. Sci. USA* 94(11): 5531–5534.

Brannan, T., H. Maker, C. Weiss and G. Cohen (1980). Regional distribution of glutathione peroxidase in the adult rat brain. *J. Neurochem.* 35: 1013–1014.

Burk, F., K. Hill, M. Boeglin, F. Ebner and H. Chittum (1997). Selenoprotein P associates with endothelial cells in rat tissue. *Histochem. Cell Biol.* 108: 11–15.

Burk, R., K. Hill, R. Read and T. Bellew (1991). Response of rat selenoprotein P to selenium administration and fate of its selenium. *Am. J. Physiol.* 261: E26–E30.

Cafe, C., C. Torri, L. Bertorell, F. Tartara, F. Tancioni, P. Gaetani, R. Rodriguez y Baena and F. Marzatico (1995). Oxidative events in neuronal and glial cell-enriched fractions of rat cerebral cortex. *Free Rad. Biol. Med.* 19: 853–857.

Campos-Barros, A., H. Meinfold, B. Walzog and D. Behne (1997). Effects of selnium and iodine deficiency on thyroid hormone concentrations in the central nervous system of the rat. *Eur. J. Endocrinol.* 136: 316–323.

Castano, A., A. Ayala, J. Rodriguez-Gomez, C. de la Cruz, R. Cano and A. Machado (1995). Increase in dopamine turnover and tyrosine hydroxylase enzyme in hippocampus of rats fed on low selenium diet. *J. Neurosci. Res.* 42: 684–691.

Castano, A., J. Cano and A. Machado (1993). Low selenium diet affects monoamine turnover differentially in substantia nigra and striatum. *J. Neurochem.* 61: 1302–1307.

Chambers, I., J. Frampton, P. Colgfarb, N. Affara, W. MacBain and P. Harrison (1986). The structure of mouse glutathione peroxidase gene: the selenocysteine in the active site is encoded by the 'termination' codon, TGA. *EMBO J.* 5: 1221–1227.

Chanoine, J.-P., S. Alex, S. Stone, S. Fang, I. Veronikis, J. Leonard and L. Braverman (1993). Placental 5-deiodinase activity and fetal thyroid hormone economy are unaffected by selenium deficiency in the rat. *Pediatr. Res.* 34: 288–292.

Cheng, W.-H., Y.-S. Ho, B.A. Valentine, D.A. Ross, G.F. Combs and X.G. Lei (1998). Cellular glutathione peroxidase is the mediator of body selenium to protect against paraquat lethality in transgenic mice. *J. Nutr.* 128: 1070–1076.

Combs, G. (2001). Selenium in global food systems. *Br. J. Nutr.* 85: 517–547.

Contempre, B., J. Dumont, B. Ngo, C. Thilly, A. Diplock and J. Vanderpas (1991). Effect of selenium supplementation in hypothyroid subjects of an iodine and selenium deficient area: the possible danger of indiscriminate supplementation of iodine-deficient subjects with selenium. *J. Clin. Endocrinol. Metab.* 73: 213–215.

Damier, P., E. Hirsch, P. Zhang and F. Javoy-Agid (1993). Glutathione peroxidase, glial cells and Parkinson's disease. *Neuroscience* 52: 1–6.

Feller, A., D. Rudman, P. Erve, R. Johnson, J. Boswell, D. Jackson and D. Mattson (1987). Subnormal concentrations of serum selenium and plasma carnitine in chronically tube-fed patients. *Am. J. Clin. Nutr.* 45: 475–483.

Fu, Y., H. Sies, and X.G. Lei (2001). Opposite roles of selenium-dependent glutathione peroxidase-1 in superoxide generator diquat- and peroxynitrite-induced apoptosis and signaling. *J. Biol. Chem.* 276, 43004–43009.

Gu, J., J. E. Royland, R. C. Wiggins and G. W. Konat (1997). Selenium is required for normal upregulation of myelin genes in differentiating oligodendrocytes. *J. Neurosci. Res.* 47(6): 626–635.

Hawkes, W. and L. Hornbostel (1996). Effects of dietary selenium on mood in healthy men living in a metabolic research unit. *Biol. Psychiatry* 39: 121–128.

Hill K. E., J. Zhou, W. J. McMahan, A. K. Motley, J. F. Atkins, R. F. Gesteland and R. F. Burk (2003). Deletion of selenoprotein P alters distribution of selenium in the mouse. *J. Biol. Chem.* 278(16):13640–13646.

Himeno, S. and N. Imura (2002). Selenium in nutrition and biology. In *Heavy Metals in the Environment* (B. Sarkar, Ed.), pp. 587–629. Marcel Dekker, New York.

Ho, Y.-S., J.-L. Magnenat, R. Bronson, J. Cao, M. Gargano, M. Sugawara and C. Funk (1997). Mice deficient in cellular glutathione peroxidase develop normally and show no increased sensitivity to hyperoxia. *J. Biol. Chem.* 272: 16644–16651.

Imam, S. Z. and S. F. Ali (2000). Selenium, an antioxidant, attenuates methamphetamine-induced dopaminergic toxicity and peroxynitrite generation. *Brain Res.* 855(1): 186–191.

Ince, P., P. Shaw, J. Candy, D. Mantle, L. Tandon, W. Ehmann and W. Markesbery (1994). Iron, selenium, and glutathione peroxidase activity are elevated in sporadic motor neuron disease. *Neurosci. Lett.* 182: 87–90.

Jeong, D., T. Kim, Y. Chung, B. Lee and I. Kim (2002). Selenoprotein W is a glutathione-dependent antioxidant *in vivo*. *FEBS Lett.* 517: 225–228.

Kim, H., W. Jhoo, E. Shin and G. Bing (2000). Selenium deficiency potentiates methamphetamine-induced nigral neuronal loss; comparison with MPTP model. *Brain Res.* 862(1–2): 247–252.

Kohrle, J. (1996). Thyroid hormone deiodinases: a selenoenzyme family acting as gate keepers to thyroid hormone action. *Acta Med. Austriaca* 23: 17–30.

Koibuchi, N., H. Fukuda and H. Chin (1999). Promotor-specific regulation of the brain-derived neurotropic factor gene by thyroid hormone in the developing rat cerebellum. *Endocrinology* 140: 3955–3961.

Korotkow, K., S. Novoselov, D. Hatfield and V. Gladyshev (2002). Mammalian selenoprotein in which selenocysteine (Sec) incorporation is supported by a new form of Sec insertion sequence element. *Mol. Cell. Biol.* 22: 1402–1411.

Kumaraswamy, E., A. Malykh, K. V. Korotkov, S. Kozyavkin, Y. Hu, S. Y. Kwon, M. E. Moustafa, B. A. Carlson, M. J. Berry, B. J. Lee, D. L. Hatfield, A. M. Diamond and V. N. Gladyshev (2000). Structure-expression relationships of the 15-kDa selenoprotein gene. Possible role of the protein in cancer etiology. *J. Biol. Chem.* 275(45): 35540–35547.

Larsen, N., H. Pakkenberg, E. Damsgaard and K. Heydom (1979). Topographical distribution of arsenic, manganese, and selenium in the normal human brain. *J. Neurol. Sci.* 42: 407–416.

Lovell, M., X. Chengsong, S. Gabbita and W. Markesbery (2000). Decreased thioredoxin and increased thioredoxin reductase levels in Alzheimer's disease brain. *Free Rad. Biol. Med.* 28: 418–427.

Malmqvist, K., A. Brun, K. Inamura, E. Martins, L. Salford, B. Siesjo, U. Tapper and K. Themner (1988). Proton microprobe and paticle induced X-ray emission (PIXE) analysis for studies of pathological brain tissue. *Scan Microsc.* 2: 1685–1693.

Matsumura, H., R. Takahara and O. Hayaishi (1991). Inhibition of sleep in rats by inorganic selenium compounds, inhibitors of prostaglandin D synthase. *Proc. Natl. Acad. Sci. USA* 88: 9046–9050.

Mirochnitchenko, O., U. Palnitkar, M. Philbert and M. Inouye (1995). Thermosensitive phenotype of trangenic mice overproducing human glutathione peroxidases. *Proc. Natl. Acad. Sci. USA* 92: 8120–8124.

Mitchell, J., F. Nicol, G. Beckett and J. Arthur (1997). Selenium and iodine deficiencies: effects on brain and brown adipose tissue selenoenzyme activity and expression. *J. Endocrinol.* 155: 255–263.

Mitchell, J., F. Nicol, G. Beckett and J. Arthur (1998). Selenoprotein expression and brain development in preweanling selenium- and iodine-deficient rats. *J. Mol. Endocrinol.* 20: 203–210.

Olanow, C. (1993). A radical hypothesis for neurodegeneration. *Trends Neurosci.* 16: 439–444.

Rao, L., B. Puschner and T. A. Prolla (2001). Gene expression profiling of low selenium status in the mouse intestine: transcriptional activation of genes linked to DNA damage, cell cycle control and oxidative stress. *J. Nutr.* 131: 3175–3181.

Rayman, M.P. (2000). The importance of selenium to human health (In Process Citation). *Lancet* 356(9225): 233–241.

Saijoh, K., N. Saito, M. Lee, M. Fujii, T. Kobayashi and K. Sumino (1995). Molecular cloning of cDNA encoding a bovine selenoprotein P-like protein containing 12 selenocysteines and a (His-Pro) rich domain insertion, and its regional expression. *Mol. Brain Res.* 30: 301–311.

Saito, Y., T. Hayashi, A. Tanaka, Y. Watanabe, M. Suzuki, E. Saito and K. Takahashi (1999). Selenoprotein P in human plasma as an extracellular phospholipid hydroperoxide glutathione peroxidase: isolation and enzymatic characterization of human selenoprotein p. *J. Biol. Chem.* 274(5): 2866–2871.

Sanchez, V., J. Camarero, E. O'Shea and A. R. Green (2003). Differential effect of dietary selenium on the long-term neurotoxicity induced by MDMA in mice and rats. *Neuropharmacology* 44(4): 449–461.

Savaskan, N. E., A. U. Brauer, M. Kuhbacher, I. Y. Eyupoglu, A. Kyriakopoulos, O. Ninnemann, D. Behne and R. Nitsch (2003). Selenium deficiency increases susceptibility to glutamate-induced excitotoxicity. *FASEB J.* 17(1): 112–114.

Savolainen, H. (1978). Superoxide dismutase and glutathione peroxidase activities in rat brain. *Res. Commn. Chem. Pathol. Pharmacol.* 21: 173–176.

Schomburg, L., U. Schweizer, B. Holtman, L. Flohe, M. Sendtner and J. Kohrle (2003). Gene disruption discloses role of selenoprotein P in selenium delivery to target tissues. *Biochem. J.* 370: 397–402.

Shiobara, Y., Y. Ogra and K. Suzuki (2000). Exchange of endogenous selenium for dietary selenium as 82Se-enriched selenite in brain, liver, kidneys and testes. *Life Sci.* 67: 3041–3049.

Smith, D., E. Feldman and D. Feldman (1989). Trace element status in multiple sclerosis. *Am. J. Clin. Nutr.* 50: 136–140.

Sun, Y., Q. P. Gu and P. D. Whanger (2001). Selenoprotein W in overexpressed and underexpressed rat glial cells in culture. *J. Inorg. Biochem.* 84(1–2): 151–156.

Sutphin, M. and T. Buckman (1991). Effects of low selenium diets on antioxidant status and MPTP toxicity in mice. *Neurochem. Res.* 16: 1257–1263.

Takizawa, S., K. Matsushima, Y. Shinohara, S. Ogawa, N. Komatsu, H. Utsunomiya and K. Watanabe (1994). Immunohistochemical localization of glutathione peroxidase in infarcted human brain. *J. Neurol. Sci.* 122: 66–73.

Trepanier, G., D. Furling, J. Puymirat and M. E. Mirault (1996). Immunocytochemical localization of seleno-glutathione peroxidase in the adult mouse brain. *Neuroscience* 75(1): 231–243.

Watanabe, C. (2002). Modification of mercury toxicity by selenium: pratical importance? *Tohoku J. Exp. Med.* 196: 71–77.

Watanabe, C. and H. Satoh (1994). Brain selenium status and behavioral development in selenium-deficient preweanling mice. *Physiol. Behav.* 56(5): 927–932.

Watanabe, C. and H. Satoh (1995). Effects of prolonged selenium deficiency on open field behavior and Morris water maze performance in mice. *Pharmacol. Biochem. Behav.* 51(4): 747–752.

Watanabe, C. and T. Suzuki (1986). Sodium selenite-induced hypothermia in mice: indirect evidence for a neural effects. *Toxicol. Appl. Pharmacol.* 86: 372–379.

Watanabe, C., Y. Kasanuma, and H. Satoh (1997). Deficiency of selenium enhances the K+-induced release of dopamine in the striatum of mice. *Neurosci. Lett.* 236, 49–52.

Watanabe, C., K. Yoshida, Y. Kasanuma and H. Satoh (1999). In utero exposure to methylmercury exposure differentially affects the activities of selenoenzymes in the fetal mouse brain. *Environ. Res.* 80: 208–214.

Weber, G., P. Maertens, X. Meng and C. Pippenger (1991). Glutathione peroxidase deficiency and childhood seizures. *Lancet* 337: 1443–1444.

Wenstrup, D., W. Ehmann and W. Markesbery (1990). Trace element imbalances in isolated subcellular fractions of Alzheimer's disease brains. *Brain Res.* 533: 125–131.

Yan, J. and J. Barrett (1998). Purification from bovine serum of a survival-promoting factor for cultured central neurons and its identification as selenoprotein-P. *J. Neurosci.* 18: 8682–8691.

Yang, G., S. Wang, R. Zhou and S. Sun (1983). Endemic selenium intoxication of humans in China. *Am. J. Clin. Nutr.* 37: 872–881.

Yang, X., K. E. Hill, M. J. Maguire and R. F. Burk (2000). Synthesis and secretion of selenoprotein P by cultured rat astrocytes. *Biochim. Biophys. Acta* 1474(3): 390–396.

Yoshinaga, J., H. Imai, M. Nakazawa and T. Suzuki (1990). Lack of significantly positive correlations between elemental concentrations in hair and in organs. *Sci. Total Environ.* 99: 125–135.

Part IV

Foods and Supplements that Modulate Brain Function

19 Food-Derived Neuroactive Cyclic Dipeptides

Chandan Prasad

CONTENTS

19.1 INTRODUCTION

Cyclic dipeptides [also known as 2,5 dioxopiperazines; 2,5 diketopiperazines; cyclo(dipeptides); or dipeptide anhydrides] are relatively simple compounds and therefore are among the most common peptide derivatives found in nature. Curtius and Gloebel (1888) synthesized the first cyclic dipeptide, cyclo(Gly-Gly); however, the existence of cyclic dipeptides as a special group of compounds in nature was not recognized until early in the 20th century (Fischer, 1906; Fischer and Raske, 1906; Abderhalden and Komm, 1924; Abderhalden and Haas, 1926). Simple cyclic dipeptides give rise to very characteristic mass spectral fragmentation patterns in which the parent is generally prominent and is followed by fragmentation products that include loss of CO or CHO, amine fragmentation (a diagnostic tool for determining structure of cyclic dipeptide), and elimination of HNCO. In addition, optical rotatory dispersion (ORD) spectra of cyclic dipeptides show Cotton effects, the position of which is solvent dependent (Greenfield and Fasman, 1969; Balasubramanian and Wetlaufer, 1966), attributed to conformational changes in different media; such spectroscopic properties are of interest as models for more complex peptides (Kopple and Maar, 1967). Also the *cis*-amide bonds of cyclic dipeptides exhibit modified infrared absorption properties from those associated with *trans*-amide bonds. Such unique properties of cyclic peptides aroused interest of chemists and, therefore, between the late 1800s and mid 1900s, many simple diketopiperazines such as cyclo(Gly-Gly) were synthesized for the sole purpose of examining their interesting physicochemical properties (Curtius and Goebel, 1888; Fischer, 1906; Fischer and Raske, 1906; Abderhalden and Komm, 1924; Abderhalden and Haas, 1926). In later years, a variety of dipeptide diketopiperazines were shown to exist in protein and polypeptide hydrolysates as well as fermentation broths and cultures of yeast, lichens, and fungi (Sammes, 1975; Johne and Groger, 1977; Prasad, 1995). Some of these diketopiperazines were thought to result from nonenzymatic cyclization of dipeptides and their amides, inasmuch as they are often formed during chemical and thermal manipulations as well as during storage of peptides and proteins (Tamura et al., 1964; Sammes, 1975; Johne and Groger, 1977; Prasad, 1989; Nagayama et al., 1990; Vitt et al., 1990; Kertscher et al., 1993).

19.2 FOOD-DERIVED CYCLIC DIPEPTIDES WITH KNOWN BIOLOGICAL ACTIVITY

19.2.1 Cyclo(His-Pro)

Cyclo(His-Pro) is the first cyclic dipeptide shown to be biologically active and also to be endogenous to animal tissues and body fluids. Perry et al. (1965) first reported the presence of cyclo(His-Pro) in human urine. These authors also found an increased amount of cyclo(His-Pro) in the urines of phenylketonurics receiving Lofenalac® as part of their diets as well as in the Lophenalac formula itself (Perry et al., 1965). Because there was no known biological activity associated with cyclo(His-Pro), no further attention was paid to the report by Perry et al. (1965). More than a decade later, while studying the metabolism of thyrotropin-releasing hormone (TRH) metabolism, we observed that *in vitro* incubation of [3H-Pro]-TRH (pGlu-His-ProNH$_2$) with hamster hypothalamic extracts or intraventricular administration of [3H-Pro]-TRH into rat brain led to the formation of a new metabolite characterized as CHP (Prasad and Peterkofsky, 1976; Prasad et al., 1977). The CHP formation first required the cleavage of the amino-terminal pyroglutamic acid from TRH by pyroglutamyl aminopeptidase, followed by cyclization of His-ProNH$_2$. The latter appears to be a nonenzymatic process with a maximal velocity ($t = 140$ min) at pH 6 to 7 and 37°C (Prasad et al., 1977; Moss and Bundgaard, 1990). This observation was followed by development of a sensitive radioimmunoassay for CHP and demonstration of CHP-like immunoreactivity in a variety of animal tissues and body fluids (Prasad, 1985, 1989). The structural similarities between TRH and CHP led us to examine whether some of the biological activities previously attributed to TRH could be due to CHP. In 1977, we compared CHP and TRH for their potencies in attenuating ethanol narcosis; the results of this study showed CHP to be several times more potent than TRH (Prasad et al., 1977). This initial report led to a flurry of activities investigating the biological activities of CHP (Prasad, 1985, 1989).

19.2.1.1 Origin and Distribution in Food

A variety of cyclic dipeptides exist in protein and polypeptide hydrolysates, as well as cultures of yeast, lichens, and fungi (Table 19.1). Some of these diketopiperazines were thought to result from nonenzymatic cyclization of dipeptides and their amides during their chemical and thermal manipulations. A recent study designed to evaluate species differences in urinary excretion of CHP-LI uncovered high levels of peptide in urine of the leopard (a carnivore; 35 ± 3 µg/dl), followed by humans (omnivores, 27.5 ± 2.5 µg/dl) and the hippopotamus (a herbivore, 18.5 ± 1.0 µg/dl) (Prasad et al., 1991). These data led us to explore whether at least part of the endogenous CHP may be derived from dietary protein. To this end, rats were fed a 95% protein–5% fat or a 95% carbohydrate–5% fat diet for a 7-day period, and the urinary and plasma levels of CHP were measured. The data showed no significant difference in the levels of CHP, suggesting that dietary proteins do not contribute significantly toward plasma or urine levels of CHP in rats. In contrast, a decrease

TABLE 19.1
Examples of Cyclic Dipeptides that Occur in Food

Cyclic Dipeptide	Food Source	Other Sources	Biological Activity	Ref.
Cyclo(Pro-Ile)	Roasted coffee			Ginz and Engelhardt, 2000
Cyclo(Pro-Leu)	Roasted coffee	Fungi		Ginz and Engelhardt, 2000; Johnson et al., 1951; Chen, 1960
Cyclo(Pro-Phe)	Roasted coffee	Fungi	Antibacterial, attenuates physical dependence on morphine, induces differentiation in HT-29 cells *in vitro*	Ginz and Engelhardt, 2000; Graz et al., 1999; Milne et al., 1998; Chen, 1960; Walter et al., 1979
Cyclo(Pro-Pro)	Roasted coffee			Ginz and Engelhardt, 2000
Cyclo(Pro-Val)	Roasted coffee	Fungi		Ginz and Engelhardt, 2000; Kodaira, 1961; Chen, 1960
Cyclo(Pro-Tyr)	Commercial yeast extract		Blocks delayed-rectifier potassium channels, induces differentiation in HT-29 cells *in vitro*	Graz et al., 1999; Milne et al., 1998; Tamura et al., 1964
Cyclo(Pro-Trp)		Fungi	Antifungal, antibacterial, blocks calcium channels, blocks delayed-rectifier potassium channels	Graz et al., 1999; Milne et al., 1998; Plieninger and Herzog, 1967
Cyclo(Pro-Gly)	Commercial yeast extract	Rat brain	Facilitates memory consolidation	Gudasheva et al., 1996, 1997; Tamura et al. 1964
Cyclo(Leu-Gly)			Facilitates memory consolidation	Walter et al., 1979
Cyclo(Trp-Trp)			Antifungal, blocks calcium channels, induces differentiation in HT-29 cells *in vitro*	Graz et al., 1999; Milne et al., 1998
Cyclo(His-Pro)	Many protein-rich processed foods	A varity of animal tissues and body fluids	Multiple	Prasad, 1989, 1995

in the availability of calories from protein during the neonatal period led to an increase and decrease in the content of CHP in the cerebellum and the remainder of the brain, respectively (Mori et al., 1983). We then searched for CHP-LI in protein-derived processed food and found that many food items contained a CHP-LI that was chromatographically and immunologically identical to authentic CHP. These food items included nutritional supplements (e.g., Ensure plus, 300 ng/ml; Two Cal® HN, 4763 ng/ml), potted meat, and cold cuts (Hilton et al., 1990, 1991). Furthermore, ingestion of 250 ml Ensure (112.5 μg CHP) by eight human volunteers led to a time-dependent rise in the plasma level of CHP, reaching a maximum 80% increase from the baseline within 60 min (Prasad et al., 1991).

In conclusion, these data suggest that ingestion of a diet rich in protein is unlikely to raise the plasma level of CHP to any significant level. However, consumption of a diet containing CHP may appreciably increase circulating CHP levels.

19.2.1.2 Absorption from Food and Biological (Pharmacological) Activity

Cyclic dipeptides are often formed from proteins and peptides during manipulations such as treatments with bases, heating, or pyrolysis (Tamura et al., 1964; Sammes, 1975; Johne and Groger, 1977; Prasad, 1989; Nagayama et al., 1990; Vitt et al., 1990; Kertscher et al., 1993). Once formed, these cyclic dipeptides generally behave as relatively stable molecules. Little attention has been paid to their metabolic fate, largely because these molecules have been considered until very recently to be inert. In a study conducted on eight human volunteers, we observed that consumption of 250 ml Ensure (113 ± 16 μg CHP-like immunoreactivity) resulted in a steady time-dependent increase in plasma CHP levels, reaching to a maximum of 180% of the basal at 60 min (Prasad et al., 1991b). This increase in plasma CHP-LI was not due to increase in the release of endogenous CHP because after 50 g oral glucose the plasma CHP at 60 min was not different from the basal value (Prasad et al., 1991b). In a follow-up study conducted on 28 volunteers, we again observed a similar rise in plasma CHP (Mizuma et al., 1996). In this study, we also evaluated the effect of increased CHP on plasma glucose, insulin, and C-peptide; the results show that CHP in nutritional supplements may be absorbed when ingested orally but it does not change circulating insulin, C-peptide, or glucose. This is not to be interpreted as the rise in CHP having no biological significance. It simply means that such a rise in CHP may affect some but not all biological activities regulated by CHP. For example, alcohol-preferring C57BL mice, receiving drinking water containing CHP that raised serum CHP to about 8.5 nM, exhibited a marked decrease in voluntary alcohol preference (Prasad, 2001). This serum level of CHP (ca. 10nM) was similar to that obtained in humans receiving 250 ml of Ensure (Mizuma et al., 1996). A recent report shows that cyclic dipeptides are metabolically stable in the gut and absorbed from the intestine by the oligopeptide transporter, and therefore the active cyclic dipeptides can be developed into orally active therapeutic agents (Mizuma et al., 1998).

19.2.2 Cyclo(Pro-Tyr)

Cyclo(Pro-Tyr) comes from two sources: commercial yeast extracts (Tamura et al., 1964) and the fungus *Alternaria alternata* (Stierle et al., 1988). Cyclo(Pro-Tyr), commonly known as maculosin, is a host-specific phytotoxin produced by *Alternaria alternata*, a pathogen for spotted knapweed (*Centaurea maculosa* L.; Stierle et al., 1988). Of the many cyclic dipeptides produced by this fungus, only cyclo(Pro-Tyr) causes black necrotic lesions on the leaves of spotted knapweed (Stierle et al., 1988). A recent study designed to investigate the mechanism of action of this peptide led to the identification of a membrane-associated and a soluble cytoplasmic receptor for cyclo(Pro-Tyr) in knapweed leaves (Park and Strobel, 1994). On further characterization by Sepharose 6B column chromatography, the soluble receptor protein eluted as an aggregated mass of ~600,000 Da; it appeared to be made up of four subunits, three of which have a molecular weight ~60,000 Da each. The fourth, a minor component, is of ~14,000 Da (Park and Strobel, 1994). One of the larger subunits and the small subunit of the cyclo(Pro-Tyr) receptor exhibit immunoidentity with the enzyme ribulose-1,5-diphosphate carboxylase (Park and Strobel, 1994); furthermore, cyclo(Pro-Tyr) inhibits ribulose-1,5-diphosphate carboxylase *in vitro* (Chu and Passhan, 1972). Therefore, the mechanism underlying the phytotoxic action of cyclo(Pro-Tyr) may reside in its ability to inhibit ribulose-1,5-diphosphate carboxylase, an enzyme catalyzing the CO_2-fixing reaction in photosynthesis (Miziorko and Lorimer, 1983).

Pharmacological studies conducted with synthetic cyclic dipeptides have shown to facilitate induction of differentiation in HT-29 cells *in vitro* and to block delayed-rectifier potassium channels (Milne et al., 1998; Graz et al., 1999).

19.2.3 CYCLO(PRO-PHE)

Cyclo(Pro-Phe) exists in roasted coffee (Ginz and Engelhardt, 2000) and a variety of fungi, including *Rosellinia nectrix* (Tamura et al., 1964). Exogenous cyclo(Pro-Phe) causes a potent diminution in the development of physical dependence on morphine (Walter et al., 1979). Its potency equals that of cyclo(Leu-Gly), an oxytocin-related cyclic dipeptide (Walter et al., 1975). Cyclo(Pro-Phe) exhibits broad-spectrum antibacterial activity when tested against *Streptococcus pneumoniae*, *Staphylococcus aureus*, and *Bacillus subtilis* (Ginz and Engelhardt, 2000; Graz et al., 1999; Milne et al., 1998). In addition, it facilitates induction of differentiation in HT-29 cells *in vitro* (Ginz and Engelhardt, 2000; Graz et al., 1999; Milne et al., 1998).

19.2.4 CYCLO(ASP-PHE)

Of all the cyclic dipeptides, cyclo(Asp-Phe) has received special attention during the past several years (Anonymous, 1980; Ishii, 1981; Ishii et al., 1981; Lipton et al., 1991; Geha et al., 1993; Prodolliet and Bruelhart, 1993). Cyclo(Asp-Phe) is a degradation product of the artificial sweetener aspartame (N-L-α-Asp-L-Phe-1-methyl ester). Although aspartame is relatively stable in dry powder form, when exposed to high temperatures and extremes of pH, as is during food preparation, it is converted into a variety of products, including cyclo(Asp-Phe) (Lipton et al., 1991; Prodolliet and Bruelhart, 1993). Interest in this cyclic dipeptide emerged after a suggestion that aspartame administered in large dosage to rats for 2 years may have caused brain tumors (Anonymous, 1980). However, a detailed follow-up study of the toxicology of cyclo(Asp-Phe) did not reveal an increased incidence of cancer with the use of this peptide (Ishii, 1981; Ishii et al., 1981; Geha et al., 1993).

19.2.5 CYCLO(PRO-GLY)

Cyclo(Pro-Gly) exists in the commercial yeast extract (Tamura et al., 1964) and rat brain (Gudasheva et al., 1996). By using gas chromatography, the level of cyclo(Pro-Gly) in the rat brain was estimated to be 2.8 ± 0.3 nmol/g wet brain or roughly 5.6 nmol/rat brain (Gudasheva et al., 1996). Intraperitoneal administration (100 μg/kg or roughly 21 μg or 90 nmol/rat) of synthetic cyclo(Pro-Gly) to rat elicits an antiamnestic response in the passive avoidance test (Gudasheva et al., 1996).

19.3 FOOD-DERIVED CYCLIC DIPEPTIDES OF UNKNOWN BIOLOGICAL ACTIVITY

Cyclo(Pro-Ile), cyclo(Pro-Leu), and cyclo(Pro-Pro) exist in roasted coffee. However, to our knowledge, there has been no attempt to investigate whether these cyclic dipeptides may possess any biological activity. A host of other cyclic dipeptides are reportedly endogenous to different living organisms or exist as either by-products of manufacturing processes or simply degradation products of larger proteins and peptides. Although there is no doubt that many of these cyclic dipeptides may possess interesting and possibly economically beneficial biological activities, evaluations have just begun.

19.4 OTHER BIOLOGICALLY ACTIVE CYCLIC DIPEPTDES

19.4.1 CYCLO(LEU-GLY)

The cyclic dipeptide cyclo(Leu-Gly) is structurally related to oxytocin (Cys-Tyr-Ile Glu-Asn-Cys-Pro-Leu-GlyNH$_2$) and melanocyte-stimulating hormone release inhibiting factor (MIF, Pro-Leu-GlyNH$_2$). However, there is no evidence to suggest the presence or absence of an enzymatic system capable of generating cyclo(Leu-Gly) from these two peptides. Furthermore, it is not known whether this cyclic dipeptide is endogenous to the central nervous system (CNS). The answer must wait until the development of an assay for cyclo(Leu-Gly). Walter et al. (1975), while studying the role of neurohypophyseal hormones in memory, found cyclo(Leu-Gly), a cyclized carboxy-terminal portion of oxytocin, to be a potent blocker of puromycin-induced amnesia. This observation led to a series of pharmacological studies, the results of which showed an attenuation or blockade of (1) the development of physical dependence on morphine (Walter et al., 1979); (2) the development of tolerance to the pharmacological effects (analgesia, hypothermia, and catalepsy) of β-endorphin (Bhargawa, 1981b); (3) the development of tolerance to haloperidol-induced catalepsy and hypothermia (Bhargawa, 1981a); (4) dopaminergic supersensitivity after chronic morphine (Lee et al., 1986); and (5) a slight but significant increase in response latency measured in the rat yeast paw test following oral administration of the peptide (Chipkin and Latranyi, 1984). Little is known, however, about the molecular and neurochemical mechanisms through which cyclo(Leu-Gly) acts in the brain. The nature of its interaction with opioids and haloperidol has led to suggestions that it may act through a dopaminergic mechanism. Studies to test this suggestion show supersensitivity to the behavioral effects of apomorphine after chronic cyclo(Leu-Gly) (Lee et al., 1984), with no change in the regulation of striatal acetylcholine or dopamine metabolism (LeDourin et al., 1984). We lack conclusive studies on the interaction (acute or chronic) between cyclo(Leu-Gly) and dopaminergic (D1 and D2) receptors in the striatum.

19.4.2 CYCLO(TRP-TRP)

To our knowledge, the existence of cyclo(Trp-Trp), like cyclo(Leu-Gly), has not been examined. This is true for many cyclic dipeptides. The main reason for this is an absence of an easy method to screen for their presence and a lack of interest due to a perceived notion that cyclic dipeptides are generally biologically inactive. Pharmacological studies with synthetic cyclo(Trp-Trp) have yielded some very interesting results showing that the peptide blocks calcium channels, induces differentiation in HT-29 cells *in vitro*, and exhibits broad-spectum antifungal properties (Graz et al., 1999; Milne et al., 1998).

19.4.3 CYCLO(TYR-ARG)

Cyclo(Tyr-Arg) is a synthetic analog of kyotorphin (Tyr-Arg). Takagi et al. (1979) isolated Tyr-Arg from bovine brain, reporting that this peptide was more potent than met-enkephalin in its antinociceptive activity. Both cyclo(Tyr-Arg) and its *N*-methyl tyrosine derivatives are many times more potent than kyotorphin or met-enkephalin (Sakurada et al., 1982). Tyr-Arg is clearly distributed throughout the rat brain (Ueda et al., 1980), but no attempts have been made to demonstrate the presence of cyclo(Tyr-Arg) or cyclo(*N*-methyl Tyr-Arg) in the CNS.

19.4.4 CYCLO(ASP-PRO)

Enterostatin (Val-Pro-Asp-Pro-Arg, or VPDPR) is a pentapeptide generated during tryptic activation of procolipase to colipase (Borgstrom and Erlanson-Albertsson, 1984). Exogenous VPDPR administered to rats decreases caloric intake (Erlanson-Albertsson and Larsson, 1988); furthermore, the decrease has been associated with a selective decrement in dietary fat intake (Okada et al., 1991).

Results of a recent study designed to determine the minimal structure of enterostatin required to inhibit caloric intake and dietary fat preference show cyclo(Asp-Pro) or Pro-Asp-Pro to be as effective as enterostatin (Lin et al., 1994).

19.5 CYCLIC DIPEPTIDES AND BRAIN DISEASES

Cyclo(His-Pro) is the most investigated of all cyclic dipeptides. The role of CHP in human disease or in an animal model of human disease, if any, is unknown because of a scarcity of such investigations. Because of ethical considerations, studies into the pharmacology of CHP in humans are lacking. A few investigators have examined the relationship between endogenous levels of CHP and different human diseases (Jackson et al., 1976; Kurahasi et al., 1986; Steiner et al., 1989; Prasad et al., 1991a; Wisniewski et al., 1994). CHP levels in the CSF of a group of patients diagnosed with various neurological or neuropsychiatric disorders were found to be elevated (Wisniewski et al., 1994); however, the number of patients in any specific disease group was not large enough to allow an examination of possible relationships between CHP and the specific disease state. In a later study, we found an elevated level of CHP in the CSF of never-medicated schizophrenics compared with controls or medicated schizophrenics (Prasad et al., 1991a). In contrast, there was a significant increase in spinal cord but not CSF levels of CHP in amyotrophic lateral sclerosis patients (Jackson et al., 1976). Because CHP is known to modulate caloric intake in rodents (Kow and Pfaff, 1991a, 1991b), we examined whether changes in blood CHP levels may play a role in two eating disorders — anorexia nervosa and bulimia (Steiner et al., 1989). Results of this study show that CHP levels correlate significantly with weight in restrictor ($r = -0.449$, $p < 0.05$) and bulimic anorexics ($r = +0.489$, $p < 0.01$).

It is of concern, however, that measurement of CHP in the blood does not tell us anything about the origin (peripheral, central, or exogenous) of the change in the peptide; therefore, it may be difficult to understand the nature of disease association, if any. Second, not only is stress a common component of many disorders, but the spinal tap procedure for CSF sampling may be rather stressful as well; under such circumstances it may be difficult to determine whether observed changes in CHP are related to the disease state or to the stress. A recent study by Wisniewski et al. (1994) reported a progressive increase in serum CHP levels, with an increase in clinical stage of HIV infection in AIDS patients. The changes in CHP obtained in this study (Wisniewski et al., 1994) are unlikely to be secondary to stress because the levels of cortisol, a marker for stress, were unaffected. In another study in which 3-acetyl pyridine was used to produce ataxia through selective degeneration of the inferior olivary nucleus in the cerebellum, an area rich in CHP (Mori et al., 1982), administration of CHP significantly improved ataxic gate during the early but not the late stage of the disease (Kurahasi et al., 1986).

REFERENCES

Abderhalden, E. and Haas, R. (1926) Further studies on the structure of proteins: studies on the physical and chemical properties of 2,5-diketopiperazines. *Z. Physiological Chemistry,* 151, 114–119.

Abderhalden, E. and Komm, E. (1924) The formation of diketopiperazines from polypeptides under various conditions. *Z. Physiological Chemistry,* 139, 147–152.

Anonymous. (1980) Aspartame: availability of board of inquiry decision. *Federal Register,* 45, 69558.

Balasubramanian, D. and Wetlaufer, D.B. (1966) Optical rotatory properties of diketopiperazines. *Journal of the American Chemical Society,* 88, 3449–3453.

Bhargava, H.N. (1981a) The effects of a hypothalamic peptide factor, MIF, and its cyclic analog on tolerance to haloperidol in the rat. *Life Sciences,* 29, 45–51.

Bhargava, H.N. (1981b) Inhibition of tolerance to the pharmacologic effects of human beta endorphin by prolyl-leucyl-glycinamide and cyclo(Leu-Gly) in the rat. *Journal of Pharmacology and Experimental Therapeutics,* 218, 404–408.

Borgstrom, B. and Erlanson-Albertsson, C. (1984) Pancreatic colipase. In *Lipases* (Borgstrom, B. and Brockman, H.L.), pp. 151–184. Elsevier, Amsterdam.

Chen, Y.-S. (1960) Studies on the metabolic products of *Roselinia necatrix*. I. Isolation and characterization of several physiologically active, neutral substances. *Bulletin of the Agricultural Chemistry Society of Japan,* 24, 372–377.

Chipkin, R.E. and Latranyi, M.B. (1984) Effect of cyclo(Leu-Gly) in the rat yeast-paw test. *European Journal of Pharmacology,* 100, 239–243.

Chu, D.K. and Passhan, J.A. (1972) Inhibition of ribulose 1,5-diphosphate carboxylase by 6 phosphogluconate. *Plant Physiology,* 50, 224–227.

Curtius, T. and Goebel, F. (1988) Uber Glycollather J. Prakt. *Chemistry,* 37, 50–181.

Erlanson-Albertsson, C. and Larsson, A. (1988) The activation peptide of pancreatic procolipase decreases food intake in rats. *Regulatory Peptides,* 22, 325–331.

Fischer, E. (1906) Untersuchungen uber aminosauren, polypeptide und proteine. *Bericht,* 39, 530–536.

Fischer, E. and Raske, K. (1906) Beitrag zur stereochemie der 2,5-diketopiperazine. *Bericht,* 39, 3981–3984.

Geha, R., Buckley, C.E., Greenberg, P., Patterson, R., Polmar, S., Saxon, A., Rohr, A., Yang, W. and Drouin, M. (1993) Aspartame is no more likely than placebo to cause urticaria/angioedema: results of a multi-center, randomized, double-blind, placebo controlled, crossover study. *Journal of Allergy and Clinical Immunology,* 92, 513–520.

Ginz, M. and Engelhardt, U.H. (2000) Identification of proline-based diketopiperazines in roasted coffee. *Journal of Agricultural and Food Chemistry,* 48, 3528–3532.

Graz, M., Hunt, A., Jamie, H., Grant, G. and Milne, P. (1999) Antimicrobial activity of selected cyclic dipeptides. *Pharmazie,* 54, 772–775.

Greenfield, N.J. and Fasman, G.D. (1969) Optical activity of simple cyclic amides in solution. *Biopolymers,* 7, 595–598.

Gudasheva, T.A., Boyko, S.S., Akparov, V.K., Ostrovskaya, R.U., Skoldinov, S.P., Rozantsev, G.G., Voronina, T.A., Zherdev, V.P. and Seredenin, S.B. (1996) Identification of a novel endogenous memory facilitating cyclic dipeptide cyclo-prolylglycine in rat brain. *FEBS Letters,* 391, 149–152.

Gudasheva, T.A., Boyko, S.S., Ostrovskaya, R.U., Voronina, T.A., Akparov, V.K., Trofimov, S.S., Rozantsev, G.G., Skoldinov, A.P., Zherdev, V.P. and Seredenin, S.B. (1997) The major metabolite of dipeptide piracetam analogue GVS-111 in rat brain and its similarity to endogenous neuropeptide cyclo-L-prolylglycine. *European Journal of Drug Metabolism and Pharmacokinetics,* 22, 245–252.

Hilton, C.W., Prasad, C., Svec, F., Vo, P. and Reddy, S. (1990) Cyclo(His-Pro) in nutritional supplements. *Lancet,* 336, 1455.

Hilton, C.W., Prasad, C., Vo, P. and Mouton, C. (1991) Food contains the bioactive peptide, cyclo(His-Pro). *Journal Clinical Endocrinology and Metabolism,* 75, 375–378.

Ishii, H. (1981) Incidence of brain tumor in rats fed aspartame. *Toxicology Letters,* 7, 433–437.

Ishii, H., Koshimizu, T., Usami, S. and Fujimoto, T. (1981) Toxicity of aspartame and its diketopiperazine for Wistar rats by dietary administration for 104 weeks. *Toxicology,* 21, 91–94.

Jackson, I.M.D., Adelman, L.D., Munsat, T.L., Forte, S. and Lechan, R.M. (1986) Amyotrophic lateral sclerosis: TRH and cyclo(His-Pro) in the spinal cord and cerebrospinal fluid. *Neurology,* 36, 1218–1223.

Johne, S. and Groger, D. (1977) Naturstoffe mit diketopiperazin-struktur. *Pharmazie,* 32, 1–14.

Johnson, J.L., Jackson, W. and Eble, T.E. (1951) Isolation of L-leucyl-L-proline anhydride from microbiological formulations. *Journal of the American Chemical Society,* 73, 2947–2952.

Kertscher, U., Bienert, M., Krause, E., Sepetov, N.F. and Mehlis, B. (1993) Spontaneous chemical degradation of substance P in the solid phase and in solution. *International Journal of Peptide and Protein Research,* 41, 207–211.

Kodaira, Y. (1961) Toxic substances to insects produced by *Aspergillus achraceus* and *Oospora destructor*. *Agricultural Biological Chemistry (Tokyo),* 25, 261–270.

Kopple, K.D. and Maar, D.H. (1967) Conformations of cyclic dipeptides: the folding of cyclic dipeptides containing aromatic side chain. *Journal of the American Chemical Society,* 89, 6193–6198.

Kow, L.M. and Pfaff, D.W. (1991a) Cyclo(His-Pro) potentiates the reduction of food intake induced by amphetamine, fenfluramine, or serotonin. *Pharmacology Biochemistry and Behavior,* 38, 365–369.

Kow, L.M. and Pfaff, D. (1991b) The effects of the TRH metabolite cyclo(His-Pro) and its analogs on feeding. *Pharmacology Biochemistry and Behavior,* 38, 359–364.

Kurahasi, K., Kannari, K., Kimura, K., Matsunaga, M. and Takebe, K. (1986) Histidyl-proline diketopiperazine (HPD), a metabolite of thyrotropin-releasing hormone (TRH) improves the ataxic gait in 3-acetylpyridine treated rats. *No To Shinkei,* 38, 893–898.

LeDourin, C., Fage, D. and Scatton, B. (1984) Effect of cyclo(Leu-Gly) on neurochemical indices of striatal dopaminergic supersensitivity induced by prolonged haloperidol treatment. *Life Sciences,* 34, 393–399.

Lee, J.M., DeLeon-Jones, F., Fields, J.Z. and Ritzman, R.F. (1986) Cyclo(Leu-Gly) attenuates the striatal dopaminergic supersensitivity induced by chronic morphine. *Alcohol and Drug Research,* 7, 1–10.

Lee, J.M., Ritzman, R.F. and Fields, J.Z. (1984) Cyclo(Leu-Gly) has opposite effects on D2 dopamine receptors in different brain areas. *Peptides,* 5, 7–10.

Lin, L., Okada, S., York, D.A. and Bray, G.A. (1994) Structural requirements for the biological activity of enterostatin. *Peptides,* 15, 849–854.

Lipton, W.E., Li, Y.N., Younoszai, M.K. and Steginck, L.D. (1991) Intestinal absorption of aspartame decomposition products in adult rats. *Metabolism,* 40, 1337–1345.

Milne, P.J., Hunt, A.L., Rostoll, K., Van Der Walt, J.J. and Graz, C.J. (1998) The biological activity of selected cyclic dipeptides. *Journal of Pharmacy and Pharmacology,* 50, 331–337.

Miziorko, H.M.and Lorimer, G.H. (1983) Ribulose-1,5-bisphosphate carboxylase-oxygenase. *Annual Review of Biochemistry,* 52, 507–535.

Mizuma, H., Legardeur, B.Y., Prasad, C. and Hilton, C.W. (1996) The bioactive peptide cyclo(His-Pro) may be absorbed following ingestion of nutritional supplements that contain it. *Journal of the American College of Nutrition,* 15, 175–179.

Mizuma, T., Masubuchi, S. and Awazu, S. (1998) Intestinal absorption of stable cyclic dopeptides by the oligopeptide transporter in rat. *Journal of Pharmacy and Pharmacology,* 50, 167– 172.

Mori, M., Jayaraman, A., Prasad, C., Pegues, J., and Wilber, J.F. (1982) Distribution of histidyl proline diketopiperazine cyclo-(His-Pro) and thyrotropin-releasing hormone (TRH) in the primate central nervous system. *Brain Research,* 245, 183–186.

Mori, M., Prasad, C., Wilber, J.F. and Nakamoto, T. (1983) Protein-energy malnutrition alters brain thyrotropin-releasing hormone and cyclo (His-Pro) in the neonatal rat. *Neuroscience Letters,* 43, 241–244.

Moss, J. and Bundgaard, H. (1990) Kinetics and mechanism of the facile cyclization of histidyl prolieamide to cyclo(His-Pro) in aqueous solution and the competitive influence of human plasma. *Journal of Pharmacy and Pharmacology,* 42, 7–12.

Nagayama, M., Takaoka, O., Iomata, K. and Yamagata, Y. (1990) Diketopiperazine-mediated peptide formation in aqueous solution. *Origin of Life Evolutionary Biosphere,* 20, 249–257.

Okada, S., York, D., Bray, G.A. and Erlanson-Albertsson, C. (1991) Enterostatin (Val-Pro-Asp- ro Arg), the activation peptide of procolipase, selectively reduces fat intake. *Physiology and Behavior,* 49, 1185–1189.

Park, S.H. and Strobel, G.A. (1994) Cellular protein receptors of maculosin, a host specific phytotoxin of spotted knapweed (*Centaurea maculosa* L.). *Biochimica Biophysica Acta,* 1199, 13–19.

Perry, T.L., Richardson, K.S.C.. Hansen, S. and Friesen, A.J.D (1965). Identification of the diketopiperazine of histidyl-proline in human urine. *Journal of Biological Chemistry,* 240, 4540–4542.

Plieninger, H. and Herzog, H. (1967) Synthesis of O- and C-alkylated indoxyl derivatives. Preliminary work for the synthesis of echinulin. *Monatsh Chem,* 98, 807–811.

Prasad, C. (1989) Neurobiology of cyclo(His-Pro). *Annals of the New York Academy of Sciences,* 553, 232–251.

Prasad, C. (1995) Bioactive cyclic dipeptides. *Peptides,* 16, 151–164.

Prasad, C. (2001) Role of endogenous cyclo(His-Pro) in voluntary alcohol consumption by alcohol-preferring C57Bl mice. *Peptides,* 22(12), 2113–2117.

Prasad, C., Hilton, C.W., Lohr, J.B. and Robertson, H.J.R. (1991a) Increased cerebrospinal fluid cyclo(His-Pro) content in schizophrenia. *Neuropeptides,* 20, 187–190.

Prasad, C., Hilton, C.W., Svec, F., Onaivi, E.S. and Vo, P. (1991b) Could dietary proteins serve as cyclo(His-Pro) precursors? *Neuropeptides,* 19, 17–22.

Prasad, C., Matsui, T. and Peterkofsky, A. (1977) Antagonism of ethanol narcosis by histidyl-proline diketopiperazine. *Nature (London),* 268, 142–144.

Prasad, C. and Peterkofsky, A. (1976) Demonstration of pyroglutamyl peptidase and amidase activities toward thyrotropin-releasing hormone in hamster hypothalamic extracts. *Journal of Biological Chemistry,* 251, 3229–3234.

Prodolliet, J. and Bruelhart, M. (1993) Determination of aspartame and its major decomposition products in foods. *Journal of the AOAC International*, 76, 275–282.

Sakurada, S., Sakurada, T., Jin, H., Sato, T., Kisara, K., Sasaki, Y. and Suzuki, K. (1982) Antinociceptive activities of synthetic dipeptides in mice. *Journal of Pharmacy and Pharmacology*, 34, 750–751.

Sammes, P.G. (1975) Naturally occurring 2,5-dioxopiperazines and related compounds. *Chemistry Org. Naturstoffe (Wien)*, 32, 51–118.

Steiner, H., Wilber, J.F., Prasad, C., Rogers, D. and Rosenkranz, R.R. (1989) Histidyl-proline diketopiperazine [cyclo(His-Pro)] in eating disorders. *Neuropeptides*, 14, 185–189.

Stierle, A.C., Cardellina, J.H., III and Stroebel, G.A. (1988) Maculosin, a host-specific phytotoxin for spotted knapweed from *Alternaria alternata*. *Proceedings of the National Academy of Sciences of the United States of America*, 85, 8008–8011.

Takagi, H., Shiomi, H., Ueda, H. and Amano, H. (1979) A novel analgesic dipeptide from bovine brain is a possible met-enkephalin releaser. *Nature*, 282, 410–412.

Tamura, S., Suzuki, A., Aoka, A. and Otaki, N. (1964) Isolation of several dioxopiperazines from peptone. *Agricultural Biological Chemistry (Japan)*, 28, 650–654.

Ueda, H., Shiomi, H. and Takagi, H. (1980) Regional distribution of a novel analgesic dipeptide kyotorphin (Tyr-Arg) in the rat brain and spinal cord. *Brain Research*, 198, 460–464.

Vitt, S.V., Paskonova, E.A., Saporovskaia, M.B. and Belikov, V.M. (1990) The composition of amino acid-peptide mixtures obtained during hydrolysis of proteins. *Prikl Biokhim Mikrobiol*, 26, 279–282.

Walter, R., Hoffman, P.L., Flexner, J.B. and Flexner, L.B. (1975) Neurohypophyseal hormones, analogs and fragments: their effect on puromycin-induced amnesia. *Proceedings of the National Academy of Sciences of the United States of America*, 72, 4180–4184.

Walter, R., Ritzmann, R.F., Bhargava, H.N. and Flexner, L.B. (1979) Prolyl-leucyl-glycinamide, cyclo(Leu-Gly) and derivatives block development of physical dependence on morphine in mice. *Proceedings of the National Academy of Sciences of the United States of America*, 76, 518–520.

Wisniewski, T.L., Mendel, E., Morse, E.V., Hilton, C.W. and Svec, F. (1994) Relationship between serum cyclo(His-Pro) concentrations and the nutritional status of HIV-infected patients. *Southern Medical Journal*, 87, 348–351.

20 Caffeine

Andrew Smith

CONTENTS

20.1 INTRODUCTION

Caffeine is the most widely consumed behaviorally active substance in the world. The behavioral effects of caffeine have been reviewed many times (e.g., Lieberman, 1992; Fredholm et al., 1999; Smith, 2002) and this chapter overviews the main findings and provides an update of recent research. Caffeine produces its behavioral effects through adenosine receptor antagonism and subsequent changes in many neurotransmitter systems. This results in increased alertness, and caffeine may be especially beneficial in low arousal situations (e.g., working at night, prolonged work, or sleep deprivation). It improves performance on tasks that are impaired when alertness is low (vigilance and sustained response). Such effects largely reflect increased turnover of central noradrenaline. In addition, it increases the speed of encoding and response to new stimuli, an effect that probably reflects changes in cholinergic functioning. More complex cognitive tasks show less consistent effects of caffeine, which is again consistent with its arousal-increasing effect. Beneficial effects have been observed in simulations of real-life activities such as driving and in interpolated tasks used in the field. Similarly, there is some evidence that regular consumption of caffeine has benefits, although this may depend on age. In terms of negative effects, there is little convincing evidence that caffeine produces major health problems. Sleep may be impaired if large doses are ingested late at night. Mental health problems (e.g., anxiety) have also been observed when very large doses are consumed or caffeine is given to those with existing psychopathology. Caffeine is not a typical drug of dependence and many of the withdrawal effects appear to be weak or transient. Overall, evidence shows that levels of caffeine consumed by most people have largely positive effects on behavior. The next section briefly describes sources of caffeine, metabolism of caffeine, results from animal studies, and our knowledge of underlying CNS mechanisms.

20.2 SOURCES OF CAFFEINE

Caffeine occurs naturally in dietary sources (food and drinks), is added to a number of products, and is present in certain medications. The major sources of caffeine vary from country to country and across different age groups. For example, in many Asian countries tea is the caffeine-containing beverage of choice, whereas in North America and many European countries coffee is the major source of caffeine. Cola beverages and energy drinks also contain caffeine and are popular worldwide and are often the major source of caffeine in younger groups. Caffeine is also added to many analgesics, and caffeine pills are sold in pharmacies as stimulants. Small amounts are also present in chocolate products (both drinks and chocolate bars). When caffeine is added to a product, the amount that is present is constant. However, where caffeine occurs in natural products, such as coffee or tea, the growing conditions, plant variety, processing and storage, and the method of preparation increase variability in the caffeine concentration of the final beverage. For example, coffee prepared by the drip method may contain more than 100 mg caffeine per cup, whereas instant (soluble) coffee contains ca. 60 mg per cup. Tea contains less caffeine than coffee (ca. 40 mg of caffeine) and most colas have 30 to 40 mg caffeine.

20.3 PATTERNS OF CONSUMPTION

Caffeine consumption from all sources has been estimated to be 70–75 mg/person/day worldwide (Gilbert, 1984), but reaches 210–238 mg/day in the U.S. and Canada and more than 400 mg/day in Scandinavia, where 80–100% of caffeine intake comes from coffee alone (Debry, 1994; Barone and Roberts, 1996). A recent study in the U.K. (Brice and Smith, 2002a) reported an average caffeine intake of about 220 mg/day, with about two thirds coming from coffee. It is also important to emphasize the large individual variation in consumption, with some individuals never drinking caffeinated beverages and a small minority ingesting more than 1000 mg/day.

20.4 CAFFEINE ABSORPTION, DISTRIBUTION, AND PHARMACOKINETICS

Caffeine (1,3,7-trimethylxanthine) is one member of a class of naturally occurring substances termed methylxanthines. Absorption from the gastrointestinal tract is rapid and reaches 99% in humans ca. 45 min after ingestion. The hydrophobic properties of caffeine allow its passage through all biological membranes and there is no blood-brain barrier to caffeine. The time for peak plasma concentration is variable (15 to 120 min) and caffeine half-lives range from 2.5 to 4.5 h. Caffeine half-life is reduced by 30 to 50% in smokers and is approximately doubled in those taking oral contraceptives.

20.5 CENTRAL NERVOUS SYSTEM (CNS) MECHANISMS

CNS mechanisms have been reviewed in detail by Fredholm et al. (1999). Most of the data suggest that caffeine, in the doses that are commonly consumed, acts primarily by blocking adenosine A_1 and A_{2a} receptors. Even though the primary action of caffeine may be to block adenosine receptors, this leads to very important secondary effects on many classes of neurotransmitters, including noradrenaline, acetylcholine, dopamine, serotonin, glutamate, and GABA (Daly, 1993). Such effects show that caffeine has the ability to increase alertness, a possible reason why people consume caffeine-containing beverages. There are other effects of caffeine on the CNS (e.g., direct release of intracellular calcium and effects on alkaline phosphatase), but many of these only occur at doses well above the range of human consumption.

20.6 ANIMAL STUDIES

Much of our knowledge of the CNS mechanisms of caffeine comes from animal studies. It is difficult to extrapolate these findings to humans because of the issue of comparing doses in different species. This topic is discussed in detail in Fredholm et al. (1999), and it is generally assumed that 10 mg/kg in a rat represents ca. 250 mg of caffeine in a 70-kg human (3.5 mg/kg).

The effect of caffeine on locomotor behavior of animals has been widely studied. The threshold for such effects is 1 to 3 mg/kg and the peak effect between 10 and 40 mg/kg. Above 50 mg/kg there is evidence of reduced responding (see Fredholm et al. 1999). Caffeine has also been shown to increase cortical and hippocampal activity, which provides a plausible basis for examining cognitive effects of caffeine. Indeed, evidence from animal studies suggests that caffeine improves maze learning and visual discrimination (Daly, 1993). It has been suggested that blockade of the A_{2a} receptors may underlie the effects of caffeine on locomotion whereas blockade of adenosine A_1 receptors may be responsible for the effects of caffeine on cognitive tasks.

20.7 STUDIES IN HUMAN VOLUNTEERS

20.7.1 CAFFEINE AND MENTAL PERFORMANCE

This section is subdivided into two parts: research before 1990 [reviewed in detail by Lieberman (1992)] and research before 2000 [reviewed by Smith (2002)]. Recent developments in the topic are covered at the end of the chapter.

20.7.1.1 Research before 1990

When caffeine is consumed in moderate amounts, it is often regarded as a mild stimulant, a view suggested more than 400 years ago by Pietro della Vale [cited in Tannahill (1989)]. Reviews written in the 1980s (e.g., Dews, 1984) suggested that the effects were highly variable, which probably reflects the numerous problems associated with early studies of caffeine (e.g., use of insensitive tests and designs). Later reviews (e.g., Lieberman, 1992) argued that effects are task or situation specific and the following section considers results from studies using different types of task.

20.7.1.1.1 Sensory Functions

Lieberman (1992) stated, "there is no evidence to suggest that moderate doses of caffeine have direct effects on sensory function, although well controlled studies using state-of-the-art methods have not been conducted."

20.7.1.1.2   Reaction Time

A number of studies have shown beneficial effects of caffeine on simple reaction time (e.g., Clubley et al., 1979) and choice reaction time (e.g., Lieberman et al., 1987). Other research has demonstrated such effects in some groups but not others (e.g., the elderly but not the young — Swift and Tiplady, 1988) and with some doses but not others (e.g., Roache and Griffiths, 1987).

20.7.1.1.3   Sustained Attention

Several studies have shown that caffeine improves sustained attention (e.g., Clubley et al., 1979). Other researchers (e.g., Loke and Meliska, 1984) have failed to demonstrate significant effects of caffeine on vigilance, but there is little evidence to suggest that it may actually lead to impairments of performance of tasks requiring sustained attention.

20.7.1.1.4   Memory and Cognition

Effects of caffeine on memory are less evident, with most studies finding that it has no effect (e.g., Fine et al., 1982; Loke, 1988). This may, at least in part, reflect the limited number of studies of the topic and the restricted tasks investigated.

20.7.1.1.5   Simulation of Real-Life Tasks

Regina et al. (1974) examined the effects of caffeine on a simulated driving task. The results showed beneficial effects of caffeine and confirmed findings by using laboratory vigilance tasks. Studies conducted by the military [cited by Lieberman (1992)] have also shown that caffeine can improve a critical military task, namely, sentry duty.

20.7.1.1.6   Caffeine and Reduced Alertness

A major research issue has been whether caffeine can remove impairments produced by fatigue or drugs. A number of studies from the late 1980s and early 1990s show that caffeine removes performance impairments produced by sleep loss, fatigue, working at night, or sedative drugs (File et al., 1982; Nicholson et al., 1990; Johnson et al., 1990, 1991; Zwyghuizen-Doorenbos et al., 1990; Roache and Griffiths, 1987; Rogers et al., 1989). These findings have important implications for safety-critical jobs and for maintaining operational efficiency when arousal is reduced.

20.7.1.1.7   Caffeine, Personality, and Time of Day

Revelle and his colleagues [see Revelle et al. (1987) for a review] showed that caffeine improved the performance of high impulsive individuals and impaired the performance of low impulsive individuals doing complex cognitive tests in the morning. In the evening, the opposite pattern of results was observed. This has been interpreted in terms of relationships between optimum levels of arousal and complex task performance. Such effects do not appear with simple tasks, in which even high levels of alertness usually facilitate performance.

20.7.1.1.8   Adverse Effects of Caffeine on Performance

> *Fine motor performance:* Anecdotally, it has been suggested that the increased arousal induced by consumption of caffeine impairs hand steadiness. However, early studies failed to demonstrate such effects (e.g., Lieberman et al., 1987).
> *Caffeine withdrawal:* Lieberman (1992) discusses the effects of caffeine withdrawal on headache and mood but cites no evidence to suggest that it influences performance.

20.7.1.1.9   Cost–Benefit Analysis of Early Studies on the Behavioral Effects of Caffeine

Lieberman (1992) reaches the following conclusions about the beneficial and adverse behavioral effects of caffeine. "When caffeine is consumed in the range of doses found in many foods, it

improves the ability of individuals to perform tasks requiring sustained attention, including simulated automobile driving. In addition, when administered in the same dose range, caffeine increases
self-reported alertness and decreases sleepiness. Adverse behavioral effects occur when caffeine is
consumed in excessive doses or by individuals who are overly sensitive to the substance."

20.7.1.2   Research after 1990

20.7.1.2.1   Confirmation of Earlier Findings
More recent studies of effects of caffeine on performance have confirmed many of the earlier
results. For example, the beneficial effects of caffeine on psychomotor speed and vigilance have
been replicated (e.g., Fine et al., 1994; Frewer and Lader, 1991). Similarly, the absence of effects
in episodic memory tasks has also been confirmed (e.g., Loke, 1990; Smith et al., 1997a).

20.7.1.2.2   Consideration of Other Aspects of Memory
The effects of caffeine on other aspects of memory have also been investigated. For example,
components of Baddeley's working memory model have been examined and the results show no
effects of caffeine on the articulatory loop (Smith et al., 1999) or the visuospatial sketchpad
(Warburton, 1995) but improved central executive function as shown by improved speed and
accuracy of performing a logical reasoning task (Smith et al., 1992, 1994). Semantic memory has
also been studied and results show that caffeine improves the speed of retrieval of semantic
information. This effect appears to be very consistent with the majority of studies showing improved
performance after caffeine (Smith et al., 1992, 1994, 1999).

An alternative research strategy has been to consider memory processes (speed of retrieval,
recognition vs. recall, levels of processing, and implicit memory). Speed of retrieval has been
studied by using the memory scanning paradigm and the effects of caffeine have been inconsistent
(Kerr, 1991; Hindmarch et al., 1998; Hogervorst et al., 1998). Recognition memory, whether
immediate or delayed, also shows few effects of caffeine except in cases in which word lists are
long (e.g., Anderson and Revelle, 1994; Bowyer et al., 1983), a situation wherein caffeine might
reduce attentional lapses during encoding. Implicit memory has received little attention and the
one study on the effects of caffeine showed no effect (Turner, 1993), although this might reflect
the methods used. Research on effects of caffeine on memory following different levels of encoding
has shown consistent interactions among caffeine, impulsivity, and level of processing (Gupta,
1991, 1993). These results have been interpreted in terms of arousal level, with high arousal
facilitating the processing of physical features of verbal stimuli and low arousal improving memory
following processing of semantic information. Interactions between caffeine and impulsivity are
less consistent in other types of tasks. About half the studies investigating this topic have failed to
obtain significant interactions and only one (Anderson and Revelle, 1994) has verified the predictions made by the model of Humphreys and Revelle (1984).

20.7.1.2.3   Stages of Processing
Research has continued to study the effects of caffeine on attention tasks, with one aim being to
identify mechanisms underlying the effects of caffeine. For example, Lorist and Snel (1997) have
shown that target detection and response preparation are enhanced by caffeine, and Ruijter et al.
(1999) have demonstrated that the quantity of information processed is higher after caffeine. Smith
et al. (1999) showed that caffeine increases the speed of processing new stimuli, confirming results
reported by Streufert et al. (1997). In contrast to its effects on encoding and sustained attention,
caffeine not been shown to reduce resistance to distraction (Kenemans and Verbaten, 1998).
Similarly, caffeine appears to have little effect on output processes (e.g., movement time — Lorist,
1998), although occasional reports of caffeine-induced impairments on hand steadiness can be
found (e.g., Bovim et al., 1995).

20.7.1.2.4   Caffeine and Low Levels of Alertness

Research has shown that the decreased alertness produced by consumption of lunch can be eliminated by consuming caffeinated coffee (Figure 20.1; Smith et al., 1991; Smith and Phillips, 1993). Furthermore, alertness is often reduced by minor illnesses such as the common cold, and caffeine can remove the impaired performance and negative mood associated with these illnesses (Smith et al., 1997a). The ability of caffeine to counteract the effects of fatigue has been confirmed by using simulations of driving (Horne and Reyner, 1996; Reyner and Horne, 1997). A study of simulated assembly-line work (Muehlbach and Walsh, 1995) also demonstrated significant improvements after caffeine on five consecutive nights and showed no decrements when caffeine was withdrawn.

Some of these studies allow one to assess the magnitude of the effects of caffeine. For example, Smith et al. (1993a) found that consumption of caffeine at night maintained individuals at the levels seen in the day. Another approach has been to compare the effects of caffeine with other methods aimed at counteracting sleepiness. Bonnet and Arand (1994a, 1994b) report that the combination of a prophylactic nap and caffeine was more effective in maintaining nocturnal alertness than was the nap alone. Other studies have continued to demonstrate that caffeine can remove impairments produced by sedative drugs (e.g., alcohol, Hasenfratz et al., 1993; scopolamine, Riedel et al., 1995; lorazepam, Rush et al., 1994a; triazolam, Rush et al., 1994b).

One issue is whether positive effects of caffeine are largely restricted to low alertness situations. Battig and Buzzi (1986) argued that caffeine can improve performance beyond a mere restoration of fatigue. Other studies have shown that fatigued subjects show larger performance changes after caffeine than do well-rested volunteers (Lorist et al., 1994a, 1994b). Another issue is whether caffeine exacerbates negative effects produced by stressful conditions (e.g., electrical shocks, Hasenfratz and Battig, 1992; noise, Smith et al., 1997b) and results suggest that it does not.

20.7.1.2.5   Different Doses of Caffeine

A number of studies (e.g., Lieberman et al., 1987; Durlach, 1998; Smith et al., 1999) have shown that beneficial effects of doses of caffeine typically found in commercial products can now be demonstrated in both measures of mood and performance. A linear dose–response curve has also been shown in a number of studies (Amendola et al., 1998; Smith, 1999) although, like the animal literature, beneficial effects often disappear at very high doses.

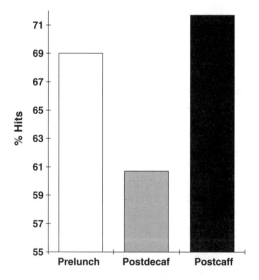

FIGURE 20.1 Effects of caffeine on the postlunch dip in sustained attention. Pre-lunch, performance before lunch; postdecaf, performance after lunch and decaffeinated coffee; postcaf, performance after lunch and caffeinated coffee. (After Smith et al., *Neuropsychobiology*, 23, 160–163, 1991.)

20.7.1.2.6 Expectancy Effects

Fillmore and colleagues (e.g., Fillmore, 1994, 1999) have demonstrated that effects of caffeine depend on a person's expectations. These expectations can generalize to placebo conditions if individuals are led to expect that they are consuming a caffeinated beverage. In many experiments, the role of expectations has not been assessed and these could account for at least some of the conflicting results in the caffeine literature.

20.7.1.2.7 Habitual Consumption

There has been far less research on the effects of regular caffeine consumption than on acute effects. However, a number of papers suggest that high consumers demonstrate better performance (e.g., Loke, 1988, 1989). The strongest evidence for beneficial effects of regular caffeine consumption comes from a study by Jarvis (1993). He examined the relationship between habitual coffee and tea consumption and cognitive performance by using data from a cross-sectional survey of a representative sample of more than 9000 British adults. Subjects completed tests of simple reaction time, choice reaction time, incidental verbal memory, and visuospatial reasoning, in addition to providing self-reports of usual coffee and tea intake. After controlling extensively for potential confounding variables, a dose–response trend to improved performance with higher levels of coffee consumption (best performance associated with ca. 400 mg caffeine/day) was found for all tests. Estimated overall caffeine consumption showed a dose–response relationship to improved cognitive performance that was strongest in those who had consumed high levels for the longest time period (the 55 years plus age group). Studies by Hogervorst et al. (1998) and Rogers and Dernoncourt (1998) failed to replicate these effects by using acute caffeine challenges, which suggests that these effects reflect regular consumption patterns rather than recent intake of caffeine.

20.7.1.2.8 Beneficial Effects of Caffeine or Removal of Negative Effects of Withdrawal

Overall, studies discussed in the preceding sections confirm that the effects of caffeine on performance are largely beneficial. However, this view has been questioned by James (1994), who argues that the beneficial effects of caffeine are really only removal of negative effects produced by caffeine withdrawal. Smith (1995) argued against this general view of caffeine effects on a number of grounds. First, it cannot account for the behavioral effects seen in animals or nonconsumers (see later), in which withdrawal cannot occur. Second, caffeine withdrawal cannot account for behavioral changes following caffeine consumption after a short period of abstinence (Warburton, 1995; Smith et al., 1994) or the greater effects of caffeine when arousal is low. Finally, claims about the negative effects of caffeine withdrawal require closer examination as they can often be interpreted in ways other than caffeine dependence (e.g., expectancy, Smith, 1996; Rubin and Smith, 1999). In most studies that have demonstrated increases in negative effects following caffeine withdrawal, the volunteers were not blind but were told or even instructed to abstain from caffeine. This is clearly very different from the double-blind methodology typically used to study effects of caffeine challenge.

The view that beneficial effects of caffeine reflect degraded performance in caffeine-free conditions (James, 1994) crucially depends on the strength of the evidence for withdrawal effects. James (1994) states, "there is an extensive literature showing that caffeine withdrawal has significant adverse effects on human performance." If one examines the details of the studies cited to support this view, one finds that some of them do not even examine performance, and when they do, any effects are selective, not very pronounced, and largely unrelated to the beneficial effects of caffeine reported in the literature.

Rogers et al. (1995a) reviewed a number of studies of caffeine withdrawal and performance. They concluded, "in a review of recent studies we find no unequivocal evidence of impaired psychomotor performance associated with caffeine withdrawal." They found that caffeine improved performance in both deprived volunteers and nonconsumers (Richardson et al., 1994). Furthermore,

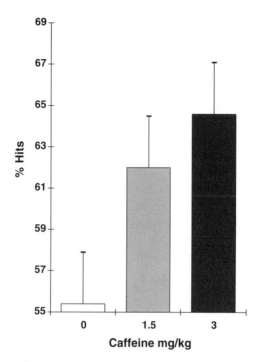

FIGURE 20.2 Effects of caffeine in volunteers deprived of caffeine for 7 days: hits in cognitive vigilance task. (After Smith, A.P., in *Caffeine and Behavior*, CRC Press, Boca Raton, FL, 1999).

other studies which suggest that withdrawal may impair performance (e.g., Bruce et al., 1991; Rizzo et al., 1988) can be interpreted in other ways than deprivation (e.g., changes in state).

The effects of caffeine withdrawal are still controversial. James (1998) showed that caffeine withdrawal impaired short-term memory performance but caffeine ingestion had no effect. In contrast, Smith (1999) reported that caffeine improved attention in both those who had been deprived of caffeine for a short period and those who had had no caffeine for 7 days (Figure 20.2).

Other studies (e.g., Comer et al., 1997) suggest that effects of withdrawal are restricted to mood and that performance is unaltered. Like many areas of caffeine research, some of the effects that have been attributed to withdrawal are open to other interpretations. For example, Lane (1997), Phillips-Bute and Lane (1997), and Lane and Phillips-Bute (1998) compared days when mid-morning coffee was either caffeinated or decaffeinated. Caffeine consumption was associated with better performance and mood. The authors interpret this as a negative effect of caffeine withdrawal, whereas one could interpret it as a positive effect of caffeine. Other studies of caffeine withdrawal effects have methodological problems such as the lack of predrink baselines (e.g., James, 1998; Robelin and Rogers, 1998) or failure to consider possible asymmetric transfer when using within-subject designs (e.g., James, 1998). This topic is dealt with again when very recent research is considered.

Literature suggests that caffeine often has alerting effects. This may be beneficial in many circumstances, but it can be an unwanted effect when the person is trying to sleep. This issue is covered in the next section.

20.7.2 EFFECTS OF CAFFEINE ON SLEEP

Much of the research on caffeine and sleep has been concerned with removing unwanted sleepiness either when persons are working at night or when they are sleep deprived. The fact that caffeine can remove sleepiness means that it can interfere with normal sleep. However, patterns of consumption suggest that individuals usually control their caffeine intake to prevent interference with sleep. If large amounts of caffeine are consumed shortly before trying to sleep, then it will undoubtedly disturb sleep. The experimental evidence for such effects is well established and is briefly summarized next.

A number of studies have shown that caffeine increases sleep latency (e.g., Zwyghuizen-Doorenbos et al., 1990) and reduces sleep duration (Hicks et al., 1983). Caffeine often produces its effects by increasing latencies in the first half of the night (Bonnet and Webb, 1979), which is different from the insomnia seen in hypnotic withdrawal (Brezinova et al., 1975). It would not appear, therefore, that caffeine-induced insomnia acts as a good general model of insomnia, as suggested by some researchers (Alford et al., 1996).

There are large individual differences in the effects of caffeine on sleep. For example, one study showed that caffeine given even in the early morning can influence the subsequent night's sleep (Landolt et al., 1995), whereas other individuals report that they can consume caffeine-containing beverages before bedtime with no adverse impact on their sleep (Colton et al., 1967; Levy and Zylber-Katz, 1983). There are probably many reasons for these differences, but it appears to be established that high consumers appear less likely to report sleep disturbance than those who consume caffeine only infrequently (Snyder and Sklar, 1984). Indeed, other results suggest that tolerance develops to effects of caffeine on sleep (Zwyghuizen-Doorenbos et al., 1990), but there are no withdrawal effects on sleep when caffeine is no longer given (Searle, 1994). It is also unclear whether the sleep disturbance produced by caffeine has an impact on behavior the next day, with one study showing no changes in mood and performance following caffeine-disturbed sleep (Smith et al., 1993b).

It is quite easy to demonstrate effects of late-night caffeine on sleep, but it is much harder to find evidence that high levels of consumption per se will affect sleep. Hicks et al. (1983) conducted a survey to examine the associations between daily caffeine consumption, habitual sleep duration, and sleep satisfaction. The results showed an inverse relationship between level of daily consumption of caffeinated drinks and habitual sleep duration, but no significant association between caffeine consumption and sleep satisfaction. Dekker et al. (1993) examined the impact of caffeine consumption on the sleep of locomotive engineers and their spouses. For the engineers only, caffeine consumption was correlated with longer sleep latency. The effect was not apparent in their spouses.

Other surveys have found little evidence of associations between caffeine consumption and sleep. For example, Lee (1992) examined data from 760 nurses. The results showed that age and family factors contributed to differences in sleep much more than caffeine did. Similarly, Greenwood et al. (1995) found no effect of caffeine consumption on the sleep of 72 rotating-shift workers. Finally, a study of sleep in elderly women found no differences in level of caffeine consumption in good and poor sleepers (Bliwise, 1992).

Overall, the research on the effects of caffeine on sleep leads to three main conclusions. First, large amounts of caffeine (e.g., more than 3 mg/kg in a single beverage) consumed in the late evening prevent individuals from going to sleep and reduce sleep duration. Effects of smaller doses show large individual variation, with high consumers being more resistant to effects of caffeine on sleep. Second, the impact of caffeine-induced changes in sleep on behavior the next day and long-term health is not known. Finally, high levels of caffeine consumption do not appear to be strongly related to sleep parameters. This again suggests that consumption is usually controlled to avoid any potential adverse effects on sleep.

20.7.3 EFFECTS OF CAFFEINE ON MOOD

20.7.3.1 Increases in Alertness and Hedonic Tone

Many studies have shown that consumption of caffeine leads to increased alertness (or reduced fatigue) that may or may not be accompanied by an increase in hedonic tone (feeling happier or more sociable). These effects have often been demonstrated by using paradigms involving low alertness situations, but beneficial effects of caffeine have also been demonstrated in individuals in an alert state (e.g., Leathwood and Pollet, 1982; Rusted, 1994, 1999; Smith et al., 1994; and Warburton, 1995). Many of these studies have used quite high doses of caffeine that would not be typically consumed in a single drink in real-life situations. However, other studies have demonstrated similar effects with realistic doses (e.g., Leathwood and Pollet, 1982; Warburton, 1995). Many of the studies have administered caffeine in coffee and it is unclear whether it is the caffeine alone or caffeine in combination with other compounds in the coffee that underlies the behavioral effects. Recent research (Smith et al., 1999) has shown that it is the caffeine rather than a combination of the caffeine and the type of drink in which it is presented that is important. Similar results have also been demonstrated with caffeine given as a capsule and in a drink.

One must now consider why some studies (e.g., Svensson et al., 1980; Swift and Tiplady, 1988) have failed to find effects of caffeine on alertness. This lack of effect could possibly, in some studies, reflect sample size or other details of the methodology. Lieberman (1992) suggests that beneficial effects of caffeine on alertness are most easily demonstrated when circadian alertness is low and mood is measured in the context of doing demanding performance tasks. Rusted (1999) also suggests that mood effects occur after changes in performance and this may account for the absence of effects in certain studies. Smith et al. (1997a) demonstrated that caffeine reduced the drop in alertness seen over the course of performing a battery of tests. Another possible explanation of the failure to find positive mood changes in certain studies is that they are masked by increases in negative mood. A number of results suggest that caffeine may increase anxiety and these are reviewed next.

20.7.3.2 Increases in Anxiety

Anecdotal evidence suggests that when individuals have an excessive amount of caffeine they may become anxious. Similarly, some psychiatric patients attribute their problems to consumption of caffeine, which has led to a diagnosis of "caffeinism." Other patients, especially those with anxiety disorders, report that caffeine may exacerbate their problems. The validity of these statements is assessed by considering the literature on these topics.

Lieberman (1992) stated, "It appears that caffeine can increase anxiety when administered in single bolus doses of 300 mg or higher, which is many times greater than the amount present in a single serving of a typical caffeine-containing beverage. However, in lower doses it appears to have little effect on this mood state or, under certain circumstances, it may even reduce anxiety levels. It has also been observed that caffeine reduces self-rated depression when administered in moderate doses (Lieberman, 1988)."

The literature supports Lieberman's view, because only a small proportion of the studies reviewed show increases in anxiety following administration of caffeine. Loke (1988) found that caffeine reduced fatigue but also led to increased tension and nervousness. Loke et al. (1985) reported increased anxiety following caffeine when the doses were high (either 3 or 6 mg/kg). Green and Suls (1996) also found that caffeine increased anxiety and again the volunteers had consumed very high amounts (125 mg caffeine per cup of coffee over the day). Similarly, Sicard et al. (1996) found increased anxiety following consumption of 600 mg of caffeine. Stern et al. (1989) found that individuals who chose a high dose of caffeine reported positive mood changes whereas nonchoosers reported anxiety and dysphoria. Overall, these results suggest that increases in anxiety following caffeine are often only found following consumption of amounts that would rarely be ingested by the majority of people.

It is important to assess whether caffeine leads to mood problems when the person ingesting it already has a high level of anxiety. It has been claimed that some people abstain from caffeinated drinks because of the accompanying jitteriness and nervousness (Goldstein et al., 1969). Other authors have even gone as far as to suggest that caffeine acts as a "fairly convincing model of generalised anxiety" (Lader and Bruce, 1986). Caffeinism refers to a constellation of symptoms associated with very high caffeine intake that are virtually indistinguishable from severe chronic anxiety (Greden, 1974). Caffeinism is usually associated with daily intakes of between 1000 and 1500 mg. However, it appears to be a rather specific condition and there is little evidence for correlations between caffeine intake and anxiety in either nonclinical volunteers (Lynn, 1973; Hire, 1978) or psychiatric outpatients (Eaton and Mcleod, 1984). Other research has investigated whether caffeine is capable of increasing the anxiety induced by other stressors. Shanahan and Hughes (1986) found that 400 mg of caffeine increased anxiety when paired with a stressful task. However, other research (e.g., Hasenfratz and Battig, 1992; Smith et al., 1997b) has not been able to provide any evidence of interactive effects of caffeine and stress. The next section considers another area in which caffeine is claimed to be associated with adverse effects, namely, when it is withdrawn.

20.7.3.3   Caffeine Withdrawal and Mood

Caffeine withdrawal has been widely studied because it is meant to provide crucial evidence on whether caffeine is addictive or leads to some kind of dependence. The most frequent outcome measure has been reporting of headache, but mood has been examined in other studies. Ratcliff-Crain et al. (1989) reported that caffeine deprivation led to increased reporting of stress by heavy coffee drinkers. This has been confirmed by Schuh and Griffiths (1997), who found that caffeine withdrawal was associated with feelings of fatigue and decreased feelings of alertness. Silverman et al. (1992) found that ca. 10% of volunteers with a moderate daily intake (235 mg/day) reported increased depression and anxiety when caffeine was withdrawn.

Richardson et al. (1995) examined the effects of varying time periods of caffeine deprivation (90 min, overnight, and 7 days) on mood. They report that overnight caffeine deprivation produced dysphoric symptoms and these mood effects were reduced but still present after longer-term abstinence. However, close examination of the results does not support this conclusion, with only 1 of the 17 mood scales showing a significant effect.

20.7.4   REINFORCING EFFECTS OF CAFFEINE

Caffeine acts as a reinforcer in several animal species but is unable to maintain self-administration behavior, which contrasts with other psychostimulant drugs (Fredholm et al., 1999). The reinforcing effects of caffeine in human volunteers appear to vary with dose, with high doses sometimes associated with aversion (Garrett and Griffiths, 1998). The relationship between the reinforcing properties of caffeine and preexposure to caffeine has also been studied. Rogers et al. (1995) investigated caffeine reinforcement by assessing changes in preference for a novel drink consumed with or without caffeine. Caffeine had no significant effects on drink preferences of low habitual caffeine consumers, but the high-caffeine consumers developed a relative dislike for drinks lacking caffeine. However, another study (Brauer et al., 1994) found that ratings of the pleasantness of a coffee taste were not significantly altered by caffeine deprivation. In studies assessing preference by choice of caffeinated or placebo beverages, only 10 to 50% of participants reliably chose caffeine over placebo (Silverman et al., 1994). Other studies have assessed the reinforcing effects of caffeine by determining how much work would be performed or money spent in order to get access to caffeine. Griffiths et al. (1989) found that decaffeinated coffee was as valuable to volunteers as caffeinated coffee and that it was it was only placebo capsules that were not deemed worthy of any work at all. Situational factors also have a substantial effect on caffeine reinforcement. Silverman et al. (1994) found that volunteers chose caffeine before doing a vigilance task, whereas placebo was chosen before relaxation. This confirms the view that the great majority of consumers drink caffeinated beverages in a controlled manner, although a small minority may use caffeine compulsively, which could lead to problems when intake is stopped.

20.7.5   Recent Developments

20.7.5.1   Methodological Issues

20.7.5.1.1   Consumption Regime

Most studies of the effects of caffeine have administered a single large dose, often equivalent to the person's total daily consumption level. Caffeine is usually ingested in a number of smaller doses and it is unclear whether effects observed after a single large dose are the same as those produced by an identical level produced by consuming several caffeine containing drinks over a longer time period. Brice and Smith (2002b) examined this issue and found that the improved mood and enhanced performance found after a single dose of 200 mg were also observed following four doses of 65 mg given at hourly intervals (which resulted in a final level identical to the single 200-mg dose).

20.7.5.1.2   Effects of Single Doses Typically Found in Commercial Beverages

Smit and Rogers (2000) examined the effects of 0, 12.5, 25, 50, and 100 mg of caffeine on cognitive performance and mood. They concluded that all doses of caffeine affected cognitive performance and that the dose–response relationships were rather flat. The effects were also more marked in individuals with higher levels of habitual caffeine consumption. Unfortunately, the order of the caffeine conditions was not included in the analyses and the similar effects of different caffeine doses may reflect transfer effects across conditions. Subjective alertness was only significant in the 100-mg condition and the benefits of this dose became greater over the test session.

20.7.5.1.3   Metabolism of Caffeine and Behavior

Most of the beneficial effects of caffeine show a linear dose–response relationship up to ca. 300 mg, which is then followed by either a flattening of the curve or, sometimes, impaired performance at higher doses. Brice and Smith (2001a) examined the relationship between metabolism of a fixed dose of caffeine (as indicated by saliva levels) and mood and performance changes and found that there was no strong association between the two. This is not too surprising in that it is not caffeine levels in the periphery per se but secondary CNS mechanisms that produce the behavioral changes. The individual differences in the metabolism of the caffeine may be very different from the individual differences in the CNS mechanisms, which plausibly accounts for the lack of a strong association between plasma or saliva levels and behavioral changes.

20.7.5.2   Caffeine Withdrawal

Recent research in this area has been concerned with two main topics: (1) what underlies the increase in symptoms following caffeine withdrawal and (2) whether the effects of caffeine reflect removal of negative effects of withdrawal. Dews et al. (2002) considered factors underlying caffeine withdrawal and concluded, "non pharmacological factors related to knowledge and expectation are the prime determinants of symptoms and their reported prevalence on withdrawal of caffeine after regular consumption."

In contrast, some researchers still suggest that caffeine only has beneficial effects on performance when the person has been caffeine withdrawn. Yeomans et al. (2002) reported that caffeine improved performance on a sustained attention task and increased rated alertness when volunteers had been caffeine deprived but had no such effects when they were no longer deprived. However, the results showed an effect of order of treatments, with those who received caffeine first continuing to show better performance even when subsequently given placebo.

Smith et al. (submitted a) examined effects of caffeine in the evening after a day of normal caffeine consumption. Caffeine improved performance (Figure 20.3), which casts doubt on the view that reversal of caffeine withdrawal is a major component underlying effects on performance.

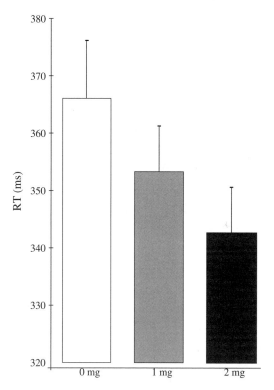

FIGURE 20.3 Effects of caffeine on simple reaction time of fatigued volunteers following a day of normal caffeine consumption. (After Smith, Christopher and Sutherland, *Psychopharmacology*, in press.)

Further evidence against the caffeine withdrawal explanation comes from recent studies of nonconsumers (Smith et al., 2001). These studies not only detected few negative effects of withdrawal (Figure 20.4) but also showed that caffeine improved the performance of both withdrawn consumers and nonconsumers (Figure 20.5), a finding that argues strongly against the withdrawal reversal explanation.

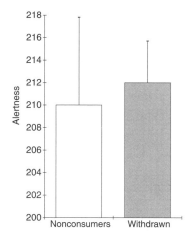

FIGURE 20.4 Alertness ratings given before caffeine ingestion by nonconsumers of caffeine and withdrawn consumers. High scores = greater alertness; the two groups do not differ. (After Smith et al., in *Ninteenth International Scientific Colloquium on Coffee*, ASIC, p. 500, 2001.)

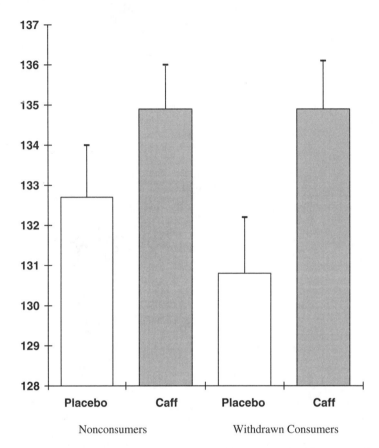

FIGURE 20.5 Effects of caffeine (4 mg/kg) on the number of items completed in a semantic processing task by nonconsumers and withdrawn consumers. Both groups show improved performance after caffeine. (After Smith et al., in *Nineteenth International Scientific Colloquium on Coffee*, ASIC, p. 500, 2001.)

20.7.5.3   Real-Life Performance

Recent research has shown that caffeine can have beneficial effects on performance in real-life situations. Lieberman et al. (2002) investigated whether caffeine would reduce the adverse effects of sleep deprivation and exposure to severe environmental and operational stress. They studied U.S. Navy Sea-Air-Land trainees and found that even in the most adverse circumstances moderate doses of caffeine improved vigilance, learning, memory and mood state. A dose of 200 mg appeared to be optimal under such conditions. Lieberman et al. (2002) concluded, "when cognitive performance is critical and must be maintained during exposure to severe stress, administration of caffeine may provide a significant advantage." Such beneficial effects of caffeine have been reported in many real-life activities (Weinberg and Beale, 2002) and a recent study suggests that performance at work may be improved (Brice and Smith, 2001a).

20.7.5.4   Underlying CNS Mechanisms

Animal studies of the CNS effects of caffeine show that it can potentially influence behavior through a number of mechanisms. In contrast to this, research with human volunteers is often based on the assumption that all the observed changes can be accounted for by a single mechanism. Evidence for distinct effects of caffeine comes from pharmacological challenge studies. Low states of alertness can be induced by reducing the turnover of central noradrenaline by giving clonidine. In a recent

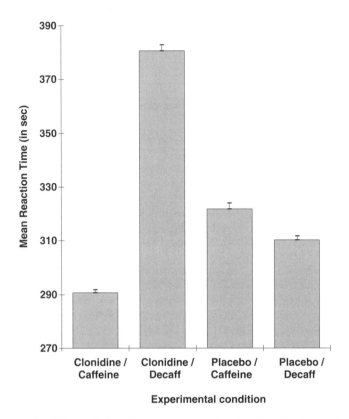

FIGURE 20.6 Effects of caffeine and clonidine on mean reaction time in a simple reaction time task. (After Smith et al., *Journal of Psychopharmacology*, 17, 283–293, 2003.)

study (Smith et al., 2003b), we have shown that caffeine can reverse the effect of clonidine (Figure 20.6).

However, certain types of tasks (e.g., a cognitive vigilance task) were not impaired by clonidine yet showed significant improvements following ingestion of caffeine. These tasks are known to be sensitive to cholinergic challenges and prior research has shown that caffeine can reverse these (Riedel et al., 1995). These cholinergic effects reflect an increase in the speed of encoding of information and a reduction in variability in performance (Warburton et al., 2001) and are not restricted to low-alertness situations. This dual mechanism model is clearly an oversimplification of the effects of caffeine, but it represents a move toward mapping the behavioral effects with the underlying neurotransmitter changes.

20.8 CONCLUSIONS

This chapter has reviewed the effects of caffeine on mood, mental performance, and sleep. In all areas it is apparent that there is a big difference between the effects of amounts of caffeine that are normally consumed and those observed when excessive amounts are ingested or when very sensitive individuals are studied. Most of the research has examined acute effects of single doses, and further studies are needed to produce a more detailed profile of effects of regular levels of consumption. However, the general picture to emerge is that when caffeine is consumed in moderation by the majority of the population there are unlikely to be many negative effects. The positive effects may be important in maintaining efficiency and safety at both the workplace and other environments. Excessive consumption of caffeine produces problems, and appropriate information should be given to minimize effects in psychiatric patients and other sensitive groups. It is important to balance this with information on the benefits of caffeine, for most consumers can usually control their intake to maximize the beneficial effects and reduce or prevent adverse effects due to over-consumption or consumption at inappropriate times. The behavioral effects of caffeine may reflect a variety of different neurotransmitter changes, and further research is needed to identify the mechanisms underlying specific effects. Future studies must use appropriate designs and tasks so that development of the topic is not slowed by the methodological problems frequently seen in the existing literature.

REFERENCES

Alford, C., Bhatti, J., Leith, T., Jamieson, A. and Hindmarch, I. (1996) Caffeine-induced sleep disruption: effects on waking the following day and its reversal with an hypnotic. *Human Psychopharmacology Clinical and Experimental* 11, 185–198.

Amendola, C.A., Gabrieli, J.D.E and Lieberman, H.R. (1998) Caffeine's effects on performance and mood are independent of age and gender. *Nutritional Neuroscience* 1, 269–280.

Anderson, K.J. and Revelle, W. (1994) Impulsivity and time of day: is rate of change in arousal a function of impulsivity? *Journal of Personality and Social Psychology* 67, 334–344.

Barone, J.J. and Roberts, H.R. (1996) Caffeine consumption. *Food Chemical Toxicology* 34, 119–129.

Battig, K. and Buzzi, R. (1986) Effect of coffee on the speed of subject-paced information processing. *Neuropsychobiology* 16, 126–130.

Bliwise, N.G. (1992) Factors related to sleep quality in healthy elderly women. *Psychology and Aging* 7, 83–88.

Bonnet, M.H. and Arand, D.L. (1994a) The use of prophylactic naps and caffeine to maintain performance during a continuous operation. *Ergonomics* 37, 1009–1020.

Bonnet, M.H. and Arand, D.L. (1994b) Impact of naps and caffeine on extended nocturnal performance. *Physiology and Behavior* 56, 103–109.

Bonnet, M.H. and Webb, W.B. (1979) The return to sleep. *Biological Psychology* 8, 225–233.

Bovim, G., Naess, P., Helle, J. and Sand, T. (1995) Caffeine influence on the motor steadiness battery in neuropsychological tests. *Journal of Clinical and Experimental Neuropsychology* 17, 472–476.

Bowyer, P., Humphreys, M.S., and Revelle, W. (1983) Arousal and recognition memory: the effects of impulsivity, caffeine and time on task. *Personality and Individual Differences* 4, 41–49.

Brauer, L.H., Buican, B. and de Wit, H. (1994) Effects of caffeine deprivation on taste and mood. *Behavioural Pharmacology* 5, 111–118.

Brezinova, V., Oswald, I. and Loudon, J. (1975) Two types of insomnia: too much waking or not enough sleep. *British Journal of Psychiatry* 126, 439–445.

Brice, C.F. and Smith, A.P. (2001a) The effects of caffeine on simulated driving, subjective alertness and sustained attention. *Human Psychopharmacology* 16, 523–531.

Brice, C.F. and Smith, A.P. (2001b) Caffeine levels in saliva: associations with psychosocial factors and behavioural effects. *Human Psychopharmacology* 16, 507–521.

Brice, C.F. and Smith, A.P. (2002a) Factors associated with caffeine consumption. *International Journal of Food Sciences and Nutrition* 53, 55–64.

Brice, C.F. and Smith, A.P. (2002b) Effects of caffeine on mood and performance: a study of realistic consumption. *Psychopharmacology* 164, 188–192.

Bruce, M., Scott, N., Shine, P. and Lader, M. (1991) Caffeine withdrawal: a contrast of withdrawal symptoms in normal subjects who have abstained from caffeine for 24 hours and for 7 days. *Journal of Psychopharmacology* 5, 129–134.

Clubley, M., Bye, C.E., Henson, T.A., Peck, A.W. and Riddington, C.J. (1979) Effects of caffeine and cyclizine alone and in combination on human performance, subjective effects and EEG activity. *British Journal of Clinical Pharmacology* 7, 157–163.

Colton, T., Gosselin, R.E. and Smith, R.P. (1967) The tolerance of coffee drinkers to caffeine. *Clinical Pharmacology and Therapeutics* 9, 31–39.

Comer, S.D., Haney, M., Foltin, R.W. and Fischman, M.W. (1997) Effects of caffeine withdrawal in humans living in a residential laboratory. *Experimental and Clinical Psychopharmacology* 5, 399–403.

Daly, J.W. (1993) Mechanism of action of caffeine. In *Caffeine, Coffee and Health* (Garattini, S., Ed), pp. 97–150. Raven Press, New York.

Debry, G. (1994) *Coffee and Health*. John Libbey, Paris.

Dekker, D.K., Pakey, M.J., Popkin, S.M. and Tepas, D.I. (1993) Locomotive engineers and their spouses: coffee consumption, mood, and sleep reports. Special issue: Night and shiftwork. *Ergonomics* 36, 233–238.

Dews, P.B. (1984) Behavioral effects of caffeine. In *Caffeine* (Dews, P.B., Ed.). Springer, New York.

Dews, P.B., O'Brien, C.P. and Bergman, J. (2002) Caffeine: behavioural effects of withdrawal and related issues. *Food and Chemical Toxicology* 40, 1257–1261.

Durlach, P.J. (1998) The effects of a low dose of caffeine on cognitive performance. *Psychopharmacology* 140, 116–119.

Eaton, W.W. and McLeod, J. (1984) Consumption of coffee or tea and symptoms of anxiety. *American Journal of Public Health* 74, 66–68.

File, S.A., Bond, A.J. and Lister, R.G. (1982) Interaction between effects of caffeine and lorazepam in performance tests and self-ratings. *Journal of Clinical Psychopharmacology* 2, 102–106.

Fillmore, M.T. (1994) Investigating the behavioral effects of caffeine: the contribution of drug-related expectancies. Special Issue: caffeine research. *Pharmacopsychoecologia* 7, 63–73.

Fillmore, M.T. (1999) Behavioral effects of caffeine: the role of drug-related expectancies. In *Caffeine and Behavior: Current Views and Research Trends* (Gupta, B.S. and Gupta, B., Eds), pp. 207–219. CRC Press, Boca Raton, FL.

Fine, B.J., Kobrick, J.L., Lieberman, H.R., Marlowe, B., Riley, R.H. and Tharion, W.J. (1994) Effects of caffeine or diphenhydramine on visual vigilance. *Psychopharmacology* 114, 233–238.

Fredholm, B.B., Battig, K., Holmen, J., Nehlig, A. and Zvartau, E.E. (1999) Actions of caffeine in the brain with special reference to factors that contribute to its widespread use. *Pharmacological Reviews* 91, 83–133.

Frewer, L. J. and Lader, M. (1991) The effects of caffeine on two computerized tests of attention and vigilance. *Human Psychopharmacology Clinical and Experimental* 6, 119–128.

Garrett, B.E. and Griffiths, R.R. (1998) Physical dependence increases the relative reinforcing effects of caffeine versus placebo. *Psychopharmacology* 139, 195–202.

Gilbert, R.M. (1984) Caffeine consumption. In *The Methylzanthine Beverages and Foods: Chemistry, Consumption, and Health Effects* (Spiller G.A., Ed), pp. 185–213. Alan R Liss, New York.

Goldstein, A., Kaizer, S. and Whitby, O. (1969) Psychotropic effects of caffeine in man: IV. Quantitative and qualitative differences associated with habituation to coffee. *Clinical Pharmacology and Therapeutics* 10, 489–497.

Greden, J.F. (1974) Anxiety or caffeinism: a diagnostic dilemma. *American Journal of Psychiatry* 131, 1089–1092.

Green, P.J. and Suls, J. (1996) The effects of caffeine on ambulatory blood pressure, heart rate, and mood in coffee drinkers. *Journal of Behavioral Medicine* 19, 111–128.

Greenwood, K.M., Rich, W.J. and James, J.E. (1995) Sleep hygiene practices and sleep duration in rotating-shift shiftworkers. *Work and Stress* 9, 262–271.

Griffiths, R.R., Bigelow, G.E. and Liebson, I.A. (1989) Reinforcing effects of caffeine in coffee and capsules. *Journal of the Experimental Analysis of Behaviour* 52, 127–140.

Gupta, U. (1991) Differential effects of caffeine on free recall after semantic and rhyming tasks in high and low impulsives. *Psychopharmacology* 105, 137–140.

Gupta, U. (1993) Effects of caffeine on recognition. *Pharmacology, Biochemistry and Behavior* 44, 393–396.

Hasenfratz, M. and Battig, K. (1992) No psychophysiological interactions between caffeine and stress? *Psychopharmacology* 109, 283–290.

Hasenfratz, M. and Battig, K. (1994) Acute dose-effect relationships of caffeine and mental performance, EEG, cardiovascular and subjective parameters. *Psychopharmacology* 114, 281–287.

Hasenfratz, M., Bunge, A., Dal Pra, D. and Battig, K. (1993) Antagonistic effects of caffeine and alcohol on mental performance parameters. *Pharmacology, Biochemistry and Behavior* 46, 463–465.

Hicks, R.A., Hicks, G.J., Reyes, J.R. and Cheers, Y. (1983) Daily caffeine use and the sleep of college students. *Bulletin of the Psychonomic Society* 21, 24–25.

Hindmarsh, I., Quinlan, P.T., Moore, K.L. and Parkin, C. (1998) The effects of black tea and other beverages on aspects of cognition and psychomotor performance. *Psychopharmacology*, 230–238.

Hire, J.N. (1978) Anxiety and caffeine. *Psychological Reports* 42, 833–834.

Horne, J.A. and Reyner, L.A. (1996) Counteracting driver sleepiness: effects of napping, caffeine, and placebo. *Psychophysiology* 33, 306–309.

Hogervorst, E., Riedel, W.J., Scmitt, J.A.J. and Jolles, J. (1998) Caffeine improves memory performance during distraction in middle-aged but not in young or old subjects. *Human Psychopharmacology* 13, 277–284.

Humphreys, M.S. and Revelle, W. (1984) Personality, motivation and performance: a theory of the relationship between individual differences and information processing. *Psychological Review* 91, 153–184.

James, J.E. (1994) Does caffeine enhance or merely restore degraded psychomotor performance? *Neuropsychobiology* 30, 124–125.

James, J.E. (1998) Acute and chronic effects of caffeine on performance, mood, headache and sleep. *Neuropsychobiology* 38, 32–41.

Jarvis, M.J. (1993) Does caffeine intake enhance absolute levels of cognitive performance? *Psychopharmacology* 110, 45–52.

Johnson, L.C., Freeman, C.R., Spinweber, C.L. and Gomez, S.A. (1991) Subjective and objective measures of sleepiness: effect of benzodiazepine and caffeine on their relationship. *Psychophysiology* 28, 65–71.

Johnson, L.C., Spinweber, G.L. and Gomez, S.Z. (1990) Benzodiazepines and caffeine: effect on daytime sleepiness, performance, and mood. *Psychopharmacology* 101, 160–167.

Kenemans, J.L. and Verbaten, M. (1998) Caffeine and visuo-spatial attention. *Psychopharmacology* 135, 353–360.

Kerr, J. S., Sherwood, N. and Hindmarch, I. (1991) Separate and combined effects of the social drugs on psychomotor performance. *Psychopharmacology* 104, 113–119.

Lader, M. and Bruce, M. (1986) States of anxiety and their induction by drugs. *British Journal of Clinical Pharmacology* 22, 252–261.

Landolt, H. P., Werth, E., Borbely, A.A. and Dijk, D.J. (1995) Caffeine intake (200 mg) in the morning affects human sleep and EEG power spectra at night. *Brain Research* 675, 67–74.

Lane, J.D. (1997) Effects of brief caffeinated-beverage deprivation on mood, symptoms and psychomotor performance. *Pharmacology, Biochemistry and Behavior* 58, 203–208.

Lane, J.D. and Phillips-Bute, B.G. (1998) Caffeine deprivation affects vigilance performance and mood. *Physiology and Behavior* 65, 171–175.

Leathwood, P.D. and Pollet, P. (1982) Diet-induced mood changes in normal populations. *Journal of Psychiatric Research* 17, 147–154.

Lee, K.A. (1992) Self-reported sleep disturbances in employed women. *Sleep* 15, 493-498.

Levy, M. and Zylber-Katz, E. (1983) Caffeine metabolism and coffee-attributed sleep disturbances. *Clinical Pharmacology and Therapeutics* 33, 770–775.

Lieberman, H.R. (1988) Beneficial effects of caffeine. In *Twelfth International Scientific Colloquium on Coffee*. pp. 102–107. Association Scientifique Internationale du Café (ASIC), Paris.

Lieberman, H.R. (1992) Caffeine. In *Handbook of Human Performance* (Smith, A.P. and Jones, D.M., Eds.), Vol. 2, pp. 49–72. Academic Press, London.

Lieberman, H.R., Tharion,W.J., Shukitt-Hale, B., Speckman, K.L. and Tulley, R. (2002) Effects of caffeine, sleep loss, and stress on cognitive performance and mood during U.S. Navy SEAL training. *Psychopharmacology* 164, 250–261.

Lieberman, H.R., Wurtman, R.J., Emde, G.G., Roberts, C. and Covielle, I.L.G. (1987) The effects of low doses of caffeine on human performance and mood. *Psychopharmacology* 92, 308–312.

Loke, W.H. (1988) Effects of caffeine on mood and memory. *Physiology and Behavior* 44, 367–372.

Loke, W.H. (1989) Effects of caffeine on task difficulty. *Psychologica-Belgica* 29, 51–62.

Loke, W.H. (1990) Effects of repeated caffeine administration on cognition and mood. *Human Psychopharmacology Clinical and Experimental* 5, 339–348.

Loke, W.H., Hinrichs, J.V. and Ghoneim, M.M. (1985) Caffeine and diazepam: separate and combined effects on mood, memory, and psychomotor performance. *Psychopharmacology* 87, 344–350.

Loke, W.H. and Meliska, C.J. (1984) Effects of caffeine use and indigestion on a protracted visual vigilance task. *Psychopharmacology* 84, 54–57.

Lorist, M.M. (1998) Caffeine and information processing in man. In *Nicotine, Caffeine and Social Drinking: Behavior and Brain Function* (Snel, J. and Lorist, M.M., Eds.), pp. 185–200. Academic Press, Amsterdam.

Lorist, M.M. and Snel, J. (1997) Caffeine effects on perceptual and motor processes. *Electroencephalography and Clinical Neurophysiology* 102, 401–413.

Lorist, M.M., Snel, J. and Kok, A. (1994a) Influence of caffeine on information processing stages in well rested and fatigued subjects. *Psychopharmacology* 113, 411–421.

Lorist, M.M., Snel, J., Kok, A. and Mulder, G. (1994b) Influence of caffeine on selective attention in well-rested and fatigued subjects. *Psychophysiology* 31, 525–534.

Lynn, R. (1973) National differences in anxiety and the consumption of coffee. *British Journal of Social Clinical Psychology* 12, 92–93.

Muehlbach, M.J. and Walsh, J.K. (1995) The effects of caffeine on simulated night-shift work and subsequent daytime sleep. *Sleep* 18, 22–29.

Nicholson, A.N., Pascoe, P.A. and Stone, B.M. (1990) The sleep-wakefulness continuum: interactions with drugs which increase wakefulness and enhance alertness. *Alcohol, Drugs and Driving* 5, 287–301.

Phillips-Bute, B.G. and Lane, J.D. (1997) Caffeine withdrawal symptoms following brief caffeine deprivation. *Physiology and Behavior* 63, 35–39.

Ratliff-Crain, J., O'Keeffe, M.K. and Baum, A. (1989) Cardiovascular reactivity, mood, and task performance in deprived and nondeprived coffee drinkers. *Health Psychology* 8, 427–447.

Regina, E.G., Smith, G.M., Keiper, C.G. and McKelvey, R.K. (1974) Effects of caffeine on alertness in simulated automobile driving. *Journal of Applied Psychology* 59, 483–489.

Revelle, W., Anderson, J.K. and Humphries, M.S. (1987) Empirical tests and theoretical extensions of arousal-based theories of personality. In *Personality Dimensions and Arousal* (Strelau, J. and Eysenck, H.J., Eds.), Plenum Press, London.

Reyner, L.A. and Horne, J.A. (1997) Suppression of sleepiness in drivers: combination of caffeine with a short nap. *Psychophysiology* 34, 721–725.

Richardson, N.J., Rogers, P.J., Elliman, N.A. and O'Dell, R.J. (1994) Mood and performance effects of caffeine in relation to acute and chronic caffeine deprivation. *Pharmacology, Biochemistry and Behavior* 52, 313–320.

Riedel, W., Hogervorst, E., Leboux, R. and Verhey, F. (1995) Caffeine attenuates scopolamine-induced memory impairment in humans. *Psychopharmacology* 122, 158–168.

Rizzo, A.A., Stamps, L.E. and Lawrence, A.F. (1988) Effects of caffeine withdrawal on motor performance and heart rate changes. *International Journal Psychophysiology* 6, 9–14.

Roache, J.O. and Griffiths, R.R. (1987) Interactions of diazepam and caffeine: behavioral and subjective dose effects in humans. *Pharmacology, Biochemistry and Behavior* 26, 801–812.

Robelin, M. and Rogers, P.J. (1998) Mood and psychomotor performance effects of the first but not subsequent, cup-of-coffee equivalent doses of caffeine consumed after overnight caffeine abstinence. *Behavioral Pharmacology* 9, 611–618.

Rogers, A.S., Spencer, M.B., Stone, B.M. and Nicholson, A.N. (1989) The influence of a 1-hr nap on performance overnight. *Ergonomics* 32, 1193–1205.

Rogers, P.J. and Dernoncourt, C. (1998) Regular caffeine consumption: a balance of adverse and beneficial effects for mood and psychomotor performance. *Pharmacology, Biochemistry and Behavior* 59, 1039–1045.

Rogers, P.J., Richardson, N.J. and Dernoncourt, C. (1995a) Caffeine use: is there a net benefit for mood and psychomotor performance? *Neuropsychobiology* 31, 195–199.

Rogers, P.J., Richardson, N.J. and Elliman, N.A. (1995b) Overnight caffeine abstinence and negative reinforcement of preference for caffeine containing drinks. *Psychopharmacology* 120, 457–462.

Rubin, G.J. and Smith, A.P. (1999) Caffeine withdrawal and headaches. *Nutritional Neuroscience* 2, 123–126.

Ruijter, J., Lorist, M.M. and Snel, J. (1999) The influence of different doses of caffeine on visual task performance. *Journal of Psychophysiology* 13, 37–48.

Rush, C.R., Higgins, S.T., Bickel, W.K. and Hughes, J.R. (1994a) Acute behavioral effects of lorazepam and caffeine, alone and in combination, in humans. *Behavioral Pharmacology* 5, 245–254.

Rush, C.R., Higgins, S.T., Hughes, J.R. and Bickel, W.K. (1994b) Acute behavioral effects of triazolam and caffeine, alone and in combination, in humans. *Experimental and Clinical Psychopharmacology* 2, 211–222.

Rusted, J. (1994) Caffeine and cognitive performance: effects on mood or mental processing? *Pharmacopsychoecologia* 7, 49–54.

Rusted, J. (1999) Caffeine and cognitive performance: effects on mood or mental processing? In *Caffeine and Behavior: Current Views and Research Trends* (Gupta, B.S. and Gupta, U., Eds), pp. 221–230. CRC Press, Boca Raton, FL.

Schuh, K.J. and Griffiths, R.R. (1997) Caffeine reinforcement: the role of withdrawal. *Psychopharmacology* 130, 320–326.

Searle, G.F. (1994) The effect of dietary caffeine manipulation on blood caffeine, sleep and disturbed behavior. *Journal of Intellectual Disability Research* 38, 383–391.

Shanahan, M.P. and Hughes, R.N. (1986) Potentiation of performance-induced anxiety by caffeine in coffee. *Psychological Reports* 59, 83–86.

Sicard, B.A., Perault, M.C., Enslen, M., Chauffard, F., Vandel, B. and Tachon, P. (1996) The effects of 600 mg of slow release caffeine on mood and alertness. *Aviation, Space, and Environmental Medicine* 67, 859–862.

Silverman, K., Evans, S.M., Strain, E.C. and Griffiths, R.R. (1992) Withdrawal syndrome after the double-blind cessation of caffeine consumption. *New England Journal of Medicine* 327, 1109–1114.

Silverman, K., Mumford, G.K. and Griffiths, R.R. (1994) Enhancing caffeine reinforcement by behavioral requirements following drug ingestion. *Psychopharmacology* 114, 424–432.

Smit, H.J. and Rogers, P.J. (2000) Effects of low doses of caffeine on cognitive performance, mood and thirst in low and high caffeine consumers. *Psychopharmacology* 152, 167–173.

Smith, A.P. (1995) Caffeine, caffeine withdrawal and psychomotor performance: a reply to James. *Neuropsychobiology* 31, 200–201.

Smith, A.P. (1996) Caffeine dependence: an alternative view. *Nature Medicine* 2, 494.

Smith, A.P. (1999) Caffeine, caffeine withdrawal and performance efficiency. In *Caffeine and Behavior: Current Views and Research Trends* (Gupta, B.S. and Gupta, U., Eds), pp.161–178. CRC Press, Boca Raton, FL.

Smith, A.P. (2002) Effects of caffeine on human behaviour. *Food and Chemical Toxicology* 40, 1243–1255.

Smith, A.P., Brice, C.F., Nash, J., Rich, N. and Nutt, D.J. (2003) Caffeine and central noradrenaline: effects on mood and cognitive performance. *Journal of Psychopharmacology,* 17, 283–293.

Smith, A.P., Brice, C.F. and Nguyen-van-Tam, D. (2001) Beneficial effects of caffeinated coffee and effects of withdrawal. In *Nineteenth International Scientific Colloquium on Coffee,* p. 500. ASIC, Trieste.

Smith, A.P., Brockman, B., Flynn, R., Maben, A. and Thomas, M. (1993a) Investigation of the effects of coffee on alertness and performance during the day and night. *Neuropsychobiology* 27, 217–223.

Smith, A.P., Christopher, G. and Sutherland, D. (in press) Effects of repeated doses of caffeine on mood and performance of alert and fatigued volunteers. *Psychopharmacology.*

Smith, A.P., Clark, R. and Gallagher, J. (1999) Breakfast cereal and caffeinated coffee: effects on working memory, attention, mood and cardiovascular function. *Physiology and Behavior* 67, 9–17.

Smith, A.P., Kendrick, A.M. and Maben, A.L. (1992) Effects of breakfast and caffeine on performance and mood in the late morning and after lunch. *Neuropsychobiology* 26, 198–204.

Smith, A.P., Kendrick, A., Maben, A. and Salmon, J. (1994) Effects of breakfast and caffeine on cognitive performance, mood and cardiovascular functioning. *Appetite* 22, 39–55.

Smith, A.P., Maben, A. and Brockman, P. (1993) The effects of caffeine and evening meals on sleep and performance, mood and cardiovascular functioning the following day. *Journal of Psychopharmacology* 7, 203–206.

Smith, A.P., Maben, A., and Brockman, P. (1994b) Effects of evening meals and caffeine on cognitive performance, mood and cardiovascular functioning. *Appetite* 22, 57–65.

Smith, A.P. and Phillips, W. (1993) Effects of low doses of caffeine in coffee on human performance and mood. In *Fifteenth International Scientific Colloquim on Coffee,* Vol. 2, pp. 461–469. ASIC, Paris.

Smith, A.P., Rusted, J.M., Eaton-Williams, P., Savory, M. and Leathwood, P. (1991) Effects of caffeine given before and after lunch on sustained attention. *Neuropsychobiology* 23, 160–163.

Smith, A.P., Sturgess, W. and Gallagher, J. (1999) Effects of a low dose of caffeine given in different drinks on mood and performance. *Human Psychopharmacology* 14, 473–482.

Smith, A.P., Thomas, M., Perry, K. and Whitney, H. (1997a) Caffeine and the common cold. *Journal of Psychopharmacology* 11, 319–324.

Smith, A.P., Whitney, H., Thomas, M., Perry, K. and Brockman, P. (1997b) Effects of caffeine and noise on mood, performance and cardiovascular functioning. *Human Psychopharmacology* 12, 27–33.

Snyder, S.H. and Sklar, P. (1984) Behavioral and molecular actions of caffeine: focus on adenosine. *Journal of Psychiatric Reseach* 18, 91–106.

Stern, K.N., Chait, L.D. and Johanson, C.E. (1989) Reinforcing and subjective effects of caffeine in normal human volunteers. *Psychopharmacology* 98, 81–88.

Streufert, S., Satish, U., Pogash, R., Gingrich, D., Landis, R., Roache, J. and Severs, W. (1997) Excess coffee consumption in simulated complex work settings: detriment or facilitation of performance? *Journal of Applied Psychology* 82, 774–782.

Svensson, E., Persson, L.O. and Sjoberg, L. (1980) Mood effects of diazepam and caffeine. *Psychopharmacology* 67, 73–80.

Swift, C.G. and Tiplady, B. (1988) The effects of age on the response to caffeine. *Psychopharmacology* 94, 29–31.

Tannahill, R. (1989) *Food in History.* Crown Publishers, New York.

Turner, J. (1993) Incidental information processing: effects of mood, sex and caffeine. *International Journal of Neuroscience* 72, 1–14.

Warburton, D.M. (1995) Effects of caffeine on cognition and mood without caffeine abstinence. *Psychopharmacology* 119, 66–70.

Warburton, D.M., Bersellini, E. and Sweeney, E. (2001) An evaluation of a caffeinated taurine drink on mood, memory and information processing in healthy volunteers without caffeine abstinence. *Psychopharmacology* 158, 322–328.

Weinberg, B.A. and Bealer, B.K. (2002) *The Caffeine Advantage.* The Free Press, New York.

Yeomans, M.R., Ripley, T., Davies, L.H., Rusted, J.M. and Rogers, P.J. (2002) Effects of caffeine on performance and mood depend on the level of caffeine abstinence. *Psychopharmacology* 164, 241–249.

Zwyghuizen-Doorenbos, A., Roehrs, T.A., Lipschutz, L., Timms, V. and Roth, T. (1990) Effects of caffeine on alertness. *Psychopharmacology* 100, 36–39

21 Tyrosine

Jan Berend Deijen

CONTENTS

21.1 INTRODUCTION

21.1.1 BACKGROUND

Tyrosine is a large neutral amino acid (LNAA) and one of the nonessential amino acids. Nonessential means that the body can synthesize tyrosine in adequate amounts for normal function. It does not mean that this amino acid is not an essential constituent of the proteins, but rather it is not essential to include in the diet because tissues can make their own supply from the essential amino acid phenylalanine, the precursor of tyrosine. Phenylalanine and tyrosine together lead to the formation of thyroxin and epinephrine (Krause and Mahan, 1984). In addition, tyrosine is the precursor from which the pigment of skin and hair is made (Brown, 2001). Thus, most phenylalanine is converted to tyrosine. Excess phenylalanine is normally eliminated from the body by hydroxylation to tyrosine. Formation of L-tyrosine from L-phenylalanine occurs in the liver by the action of the enzyme phenylalanine hydroxylase and, to a very limited extent, in the brain by a secondary action of tyrosine hydroxylase on L-phenylalanine. As the enzyme phenylalanine hydroxylase is essential for this reaction, inactivity of this enzyme leads to an accumulation of phenylalanine in the blood and from there the excess is excreted in urine. This inborn error of metabolism, characterized by a virtual absence of phenylalanine hydroxylase activity and an elevation of plasma phenylalanine, is called phenylketonuria (PKU). PKU frequently results in mental retardation. The exact cause of the mental retardation is not known, but it is the consequence of the biochemical defect. Regarding tyrosine, anomalies of tyrosine metabolism are called tyrosinemia, in which the enzyme tyrosine aminotransferase is deficient and may, like PKU, lead to mental retardation and eventually keratitis and dermatitis. Treatment consists of controlled phenylalanine and tyrosine intake.

Essential amino acids must be supplied to the body through the diet. As tyrosine can be a metabolite of phenylalanine, daily requirements have been estimated for these two aromatic amino acids together. The adult minimum daily requirement for phenylalanine and tyrosine is 14 mg/kg body weight and an absolute minimum is 1.10 g (Krause and Mahan, 1984; Bender and Bender, 1997). This amount is readily obtained from four slices of bread and one pint of milk. The daily requirements in infancy and childhood (3 months to 12 years) range from 22 to 125 mg/kg. The reference range for plasma tyrosine is 64 ± 19 mol/l (mean ± SD). The solubility of L-tyrosine is low (0.453g/l in water at 25°C; Bender and Bender, 1997). Various tyrosine-containing dipeptides, however, are more water soluble than tyrosine itself and are capable of raising serum tyrosine concentration. These dipeptides are L-tyrosyl-L-proline (TYR-PRO), L-tyrosyl-L-alalanine (TYR-ALA), L-alanyl-L-tyrosine (ALA-TYR), and L-tyrosyl-L-tyrosine (TYR-TYR). If given intravenously, they all cause significant increases in serum tyrosine.

The past decade has witnessed a resurgence of interest in the aromatic amino acid requirements. Although tyrosine is indispensable in neonates, it is a dispensable amino acid in healthy adults, because it can be synthesized though hepatic phenylalanine hydroxylation when sufficient phenylalanine is provided in the diet. However, in conditions such as hepatic and renal disease, where aromatic amino acid metabolism is impaired, a preformed balanced source of the amino acid may be required to minimize aromatic amino acid excess and catabolism. Tyrosine requirements have been estimated in adults at fixed and adequate, but not excessive, phenylalanine intakes. The mean daily tyrosine requirement of healthy men receiving an average intake of dietary phenylalanine was estimated to be 6 mg/kg. The safe population estimate of daily intake of tyrosine is 7 mg/kg (Roberts et al., 2001). This is half of the Food and Agricultural Organization/World Health Organization/United Nations University (FAO/WHO/UNU) upper requirements of 14 mg/kg for total aromatic amino acids tyrosine and phenylalanine (Basile-Filho et al., 1998).

21.1.2 Tyrosine-Dependent Control of Brain Catecholamine Synthesis

Along with leucine, isoleucine, methionine, valine, phenylalanine, and tryptophan, L-tyrosine is an LNAA. As a consequence, tyrosine shares a sodium-dependent active transport system in the gut with these other neutral amino acids. Before intracellular metabolism, amino acids must be transported from the interstitial space across the cell membrane. This transport requires the presence of a carrier system in the cell membrane and is mostly an active process because the amino acid concentration gradient between the cell's interior and the bloodstream is often unfavorable. The transport process is usually associated with the operation of a sodium ion pump. At least seven different carriers — A, ASCP, L, Ly, dicarboxylate, N, and β (beta) systems (Skeie et al., 1990) — having overlapping specificity for the different amino acids exist. For uptake into the brain, L-tyrosine shares carrier system L with the LNAAs leucine, isoleucine, valine, phenylalanine, tryptophan, and methionine. The affinity of L-tyrosine for the transport system is higher than the affinities of some of the other LNAAs. As a consequence of this shared carrier system, the concentration of L-tyrosine in relation to the concentration of all the LNAAs determines the uptake into the brain. Thus, the tyrosine:LNAAs plasma ratio is predictive of brain tyrosine concentration. In mice, brain tyrosine reaches its maximum concentration 1 h after oral ingestion and returns to the baseline level after 4 h (Topall and Laborit, 1989).

At levels up to 15 g/day, the amount of protein in the diet has been shown to influence the levels of tyrosine in plasma. High protein and high carbohydrate meals also influence the plasma ratios among LNAAs. For example, the tyrosine:LNAAs ratio is low following a meal high in protein, because plasma levels of the neutral branch-chain amino acids (BCAAs) leucine, isoleucine, and valine are increased with high protein intake. The BCAAs are transported into the brain by the same carrier that transports the aromatic amino acids (AAAs). Consequently, competition for entry into the brain between BCAAs and AAAs may influence the rate of synthesis and the level of neurotransmitters. Therefore, although the initial metabolism of BCAAs occurs primarily in the skeletal muscle, BCAAs can influence behavior (Skeie et al., 1990). Thus, following a meal high in protein, tyrosine is at a comparative disadvantage for uptake into the brain. Conversely, after a meal high in carbohydrates, the tyrosine:LNAAs ratio is high because the insulin secreted in response to the carbohydrates also promotes entry of the BCAAs into peripheral tissues. As a result, less competition for tyrosine uptake into the brain from LNAAs exists after a meal high in carbohydrate.

L-Tyrosine thus is a precursor for several biologically active substances, including brain catecholamine neurotransmitters (norepineperine and dopamine). Urinary excretion products of the catecholamines include methoxyhydroxyphenylethyleneglycol (MOPEG) and homovanillic acid (HVA).

Wurtman (1992) formulated a sequence of biochemical processes necessary for any nutrient precursor to affect the synthesis and release of its neurotransmitter product.

1. Plasma levels of the precursor must be allowed to increase after its administration or after its consumption as a constituent of foods. Indeed, plasma levels of tyrosine vary severalfold after the consumption of normal foods.
2. Brain levels of the precursor must be dependent on its plasma level; that is, there must not be an absolute blood-brain barrier for circulating tyrosine. Such a barrier does not exist. Rather, facilitated diffusion mechanisms operate that allow tyrosine to enter the brain at rates that depend on its plasma level.
3. The rate-limiting enzyme within presynaptic nerve terminals that initiates the conversion of tyrosine to its neurotransmitter product must be unsaturated with its substrate so that it can accelerate synthesis of the neurotransmitter when presented with more tyrosine.
4. The activity of tyrosine hydroxylase cannot be subject to local end-product inhibition, that is, the products of tyrosine hydroxylation and sodium. DA itself may not suppress tyrosine hydroxylation activity. However, tyrosine hydroxylase activity is probably subject to some end-product inhibition when the enzyme protein is in its nonphosphorylated state. But once the enzyme is phosphorylated, when the nerve cells containing it become active it is apparently freed from this constraint.

At present, L-tyrosine or tyrosine-containing diets are known to increase the plasma tyrosine:LNAAs ratio and brain tyrosine levels in animals and humans. This occurrence may be followed by an increase in brain noradrenaline (NA) or dopamine (DA), or both. The ability of tyrosine supplementation to enhance catecholamine (NA and DA) synthesis has been established in a variety of animal and human studies.

21.2 ANIMAL STUDIES

Interest in the amino acid tyrosine centers on its role as a precursor to neurotransmitters, in particular NE and DA. The question is whether elevating tyrosine concentration in brain NE and DA neurons can stimulate neurotransmitter synthesis. The first question is whether plasma levels of tyrosine rise after its administration and whether the level of tyrosine in the brain is dependent on its plasma level. Studies on these issues have been mainly conducted on animals, because techniques for determining the concentration of brain neurotransmitters cannot be applied to humans. The usual procedure involves decapitating the animal after stress induction, removing the brain, freezing, and storing until dissection. Brain tissue can than be taken from different regions. At present, an extensive amount of animal data is available on the effects of tyrosine administration on neurotransmitter synthesis, neuroendocrine and autonomic variables, and behavior.

21.2.1 PROTEIN-RICH MEALS AND PLASMA TYROSINE LEVELS

In rat studies it has been shown that the acute ingestion of a single meal varying in protein content from 6 to 18 to 40% casein results in elevations of the serum tyrosine level from 73 to 121 to 344 nmol/ml, respectively. However, although the *ad libitum* consumption of meals varying in protein content for 3 days caused tyrosine serum levels to rise from 100 to 200 nmol/ml as protein intake increased from 6 to 24% casein, the serum level declined after intake of 40% casein. The increase in tyrosine from 6 to 24% casein was larger than the rise in levels of the competing LNAAs, and the serum tyrosine ratio also rose. In rats fed high-protein (40% casein) diets, the serum tyrosine level and ratio decreased slightly with ingestion of casein diets. All parameters of retinal DA synthesis and release, that is, retinal DA, DOPAC, and HVA levels, reflected the changes in precursor level, peaking at 24% casein intake. The increase in the amount of dietary protein to the normal diet increased retinal tyrosine levels, which in turn increased retinal DA synthesis and release (Gibson, 1988). The finding that after a single meal the serum tyrosine level directly varies with dietary protein only to decline with a 40% casein diet can be explained by adaptation in the liver tyrosine-metabolizing enzyme. Thus, serum tyrosine levels in rats consuming 40% casein diets seem to be related to an increase in the liver enzyme tyrosine aminotransferase (TAT), a major tyrosine-metabolizing enzyme. Increase in the activity of TAT prevents the continued rise in serum tyrosine levels in the case of chronic ingestion of high-protein diets. An additional cause may be the influx of other amino acids, for example, leucine or phenylalanine, which may inhibit or compete for the synthetic enzyme tyrosine hydroxylase.

Several years later, groups of young adult male rats were studied that ingested *ad libitum* diets containing 2, 5, 10, or 20% protein for 2 weeks (Fernstrom and Fernstrom, 1995a). Serum tyrosine levels and serum tyrosine:LNAAs ratio rose as dietary protein content increased from 2 to 10%. In addition, tyrosine levels in the retina, hypothalamus, and cerebral cortex were lowest in rats consuming 2% protein and rose progressively to a plateau in rats ingesting 10% protein. No consistent increase occurred between 10 and 20% protein intake. Tyrosine hydroxylation rate in retina and hypothalamus, but not prefrontal cortex, rose in parallel with the increments in tyrosine serum level between 2 and 10% protein, showing no further increase between 10 and 20% protein. These results indicate that the differences in tyrosine levels in the retina and hypothalamus may influence directly the rate of tyrosine hydroxylation, and thus perhaps the overall rate of catecholamine synthesis. This relationship might provide the hypothalamus, a region that is important in food intake control, with a signal for monitoring and ultimately modulating the chronic level of protein intake.

21.2.2 NEUROTRANSMITTER AND BEHAVIORAL STUDIES IN ANIMALS

Tyrosine may be especially useful in counteracting any stress-related NE depletion and associated behavioral disturbances. The rationale behind this hypothesis is that stress induces brain depletion of catecholamines, especially NE, in animals. As a consequence of the decrease of brain NE levels, behavior deficits may become visible in these stressed animals. Supplementation with tyrosine may increase depleted brain NE levels and thereby counteract stress-induced behavioral impairment. Thus, most animal studies on tyrosine, brain NE, and behavior explore the potentially beneficial effects of tyrosine under stressful conditions. Exposure of mice or rats to an acute, uncontrollable stressor such as cold-swim or tail shock can increase brain NE turnover and decrease locomotor and exploratory behavior. Cold-swim stress was used to determine the effects of dietary supplements of L-tyrosine on brain neurotransmitters as well as on aggressive behavior and locomotor activity in young and aged mice presented with or without this type of stress (Brady et al., 1980). Forty-eight minutes before observing aggressive behavior, half the animals were placed in a tank filled with cold water (2 to 6°C) for 3 min. Nonstressed animals were placed in a mouse cage for only 3 min. In the absence of stress exposure, aggressive behavior in the young mice increased after tyrosine supplementation (a diet consisting of 21% casein supplemented with 4% L-tyrosine). However, no increase in aggressive behavior was seen in the aged mice. The induction of cold-swim stress decreased the aggressiveness of both young and aged mice. Tyrosine prevented the stress-induced decrease of aggressiveness in both groups. In stressed older mice, tyrosine prevented decreases in locomotion. In addition to the behavioral effects, tyrosine was found to increase brain tyrosine and DA. The authors concluded that tyrosine is effective in reducing behavioral deficits of stressed animals. This observed effect may be related to the prevention of NE depletion. The authors explain the increase of aggressive behavior in young, unstressed mice by tyrosine with a reciprocal relationship between DA and (NE + 5-HT) for the facilitation of aggressive behavior. Thus, aggressive behavior may be related to a lower brain NE and 5-HT relative to DA. An enhanced metabolism of 5-HT, indicated by increases in 5-HIAA, was observed in the aging mice. In addition, the DA:(NE+ 5-HIAA) ratios were highly correlated with mean aggression scores in young and aged animals. As the aged mice had the highest levels of NE and 5-HIAA, tyrosine could not facilitate aggression in these older animals but it could in young animals with lower brain NE and 5-HT. This finding suggests that tyrosine increases aggression if DA levels are relatively higher than NE and 5-HT levels.

The other type of stress induction, tail shock, has also been studied in male Sprague-Dawley rats. Animals were subjected to a 60-min period of tail shock or no shock after pretreatment with tyrosine (200 mg/kg, IP). Effects were compared with those in animals pretreated with saline. Norepinephrine and 3-methoxy-4-hydroxy-phenylethylene-glycol sulfate (MHPG-SO$_4$) levels within the locus coeruleus, hippocampus, and hypothalamus were altered after exposure to shock. Brain NE appeared to be depleted and its turnover increased. Behavioral deficits were observed by using measures of locomotion, standing on hind legs, and hole poking in an open-field apparatus. Animals given tyrosine before shock displayed neither shock-induced NE depletion nor the deficits in locomotion and hole poking. Also, brain MHPG-SO$_4$ levels tended to be higher than after shock alone. Thus, after tail-shock stress NE is released more rapidly from some neurons and can be restored by synthesis or reuptake. The consequence is that noradrenergic transmission and NE-dependent exploratory behaviors are impaired. Tyrosine prevents both the neurochemical and behavioral deficits in tail-shocked animals, presumably by enhancing NE synthesis (Reinstein et al., 1984).

A similar study was performed in rats given a diet enriched with tyrosine and exposed to tail-shock stress. Animals receiving the tyrosine-enriched diet displayed neither the stress-induced depletion of NE nor behavioral deficits like reduction of exploratory motor activity, of hole-poking, or of the frequency of standing on their hind legs. As tyrosine did not change NE turnover in animals not subjected to stress, one can conclude that catecholaminergic neurons seem to respond to the precursor amino acid only when they are physiologically active (Lehnert et al., 1984).

These studies clearly indicate that stress may deplete NE brain levels and thereby impair behavior. In addition, these effects may be prevented or reversed by tyrosine. It has been proposed that the technique of measuring striatal DA metabolism in untreated animals is too insensitive to demonstrate an increase in DA release, despite it being present. Thus, a more sensitive intracerebral dialysis method that directly assesses DA levels within the intrastriatal extracellular fluid was used to determine whether tyrosine may increase CA brain levels when neurons are not physiologically active, that is, if no stress is induced to accelerate the firing of CA neurons. The effect of the intraperitoneal administration of 50 to 200 mg/kg tyrosine on DA release *in vivo* in anesthetized rats on responses of DA neurons in the corpus striatum and the nucleus accumbens has been investigated in some studies. DA release was assessed using brain microdialysis to monitor extracellular fluid from rat striatum and nucleus accumbens. DA concentration from both the striatum and nucleus accumbens increased in response to tyrosine. A larger tyrosine effect was seen in the nucleus accumbens than in the corpus striatum. Tyrosine administration at 50 to 200 mg/kg IP causes a dose-related increase in extracellular fluid DA levels. However, the rise in DA was brief, suggesting that a receptor-mediated feedback mechanism responded to the increased DA release by diminishing neuronal firing or sensitivity to tyrosine. The levels of DOPC and HVA, its major metabolites, were slightly elevated, and this increase was not dose related. Thus, if the measurement of changes in striatal DOPAC and HVA is negative, it need not rule out increases in nigrostriatal DA release. From these findings it can be concluded that tyrosine is able to increase DA release even when no additional treatment is given to accelerate the firing of DA neurons (Acworth et al., 1988; During et al., 1988).

Although tyrosine may enhance DA release in nonstressed animals, there is no indication that the same is true for NE release. However, if the concentration of NE and DA is measured in animals after stress induction, no changes in the DA concentration are found (Weiss, 1991). Therefore, more severe stress seems to be required to alter DA levels in the brain than NE levels. The release of DA may be more related to coping behaviors than to the uncontrollability of the stressor, which appears to be the crucial determinant of the NE response (Lehnert et al., 1984). These data suggest that NE is more easily released than DA in nontreated animals.

21.2.3 REGULATION OF BLOOD PRESSURE IN ANIMALS

In animals, tyrosine not only reduces blood pressure in hypertensive rats but also elevates blood pressure in animals with hemorrhagic shock. Thus, tyrosine may normalize blood pressure. In one study, rats were injected with 25, 50, or 100 mg/kg tyrosine after having been bled until 25 to 50% of blood volume was lost (about 45 min). The 50- and 100-mg/kg doses produced significant increases in systolic blood pressure. The 50-mg/kg dose produced an increase after 5 min, the 100-mg/kg dose after 5, 15, and 30 min. Therefore, a higher dose continued to increase blood pressure 30 min after tyrosine administration (Conlay et al., 1981). With respect to hypertension, a 100-mg/kg injection of 15 µg L-tyrosine resulted in a significant reduction in blood pressure in the spontaneously hypertensive rat (SHR). At the same time, an increase in the turnover of NA as indicated by MOPEG-SO$_4$ was observed. Blood pressure depression was maximal at 2 h following injection, with pressure returning to the preinjection value by 4 h (Yamori et al., 1980). Unfortunately, the use of IV tyrosine to enhance catecholamine release in hemorrhagic shock and other cardiovascular diseases, or as a constituent of nutrient mixtures used for total parenteral nutrition, is limited by the unusually poor solubility of the amino acid in water. Fortunately, as mentioned in the introduction, various tyrosine-containing dipeptides are much more water soluble than tyrosine itself. Within 5 min of administering an IV of these dipeptides (12.5 to 25 mg/kg) to hemorrhaged, hypotensive rats, significant increases in serum tyrosine and significant elevations in systolic blood pressure were observed. The extent of blood pressure increase after 2 h ranged from 50 to 88%. In addition, in SHRs all these dipeptides given intraperitoneally at 100 mg/kg lowered systolic blood pressure. The dipeptide TYR-TYR also reduced blood pressure at 50 mg/kg. These observations suggest that tyrosine-containing dipeptides may be useful in maintaining or elevating blood TYR levels (Maher et al., 1990). These studies indicate that tyrosine may accelerate catecholamine synthesis, depending on the firing frequency, which may be high in hypertension and low in hypotension, of catecholamine-producing cells in the brain stem.

Tyrosine then raises blood pressure by enhancing the peripheral synthesis and release of catecholamines. The data available indicate that the effect of tyrosine on blood pressure is in the range of several hours, peaking at ca. 2 h after administration. As the use of parenteral tyrosine is limited by its poor water solubility, it would be interesting to examine the possibility to use tyrosine-containing dipetides pharmacologically. If there were more knowledge of the efficacy of these dipeptides to use as tyrosine substitutes in parenteral solutions, it may be of clinical use, particularly for normalization of blood pressure.

21.2.4 BEHAVIORAL STUDIES IN ANIMALS

Exposure to cold stress or other stressors such as tail shock or (simulated) altitude can impair working memory. This may be attributed to a reduced synthesis of brain catecholamines. Some studies have been performed to determine whether tyrosine would protect rats from the adverse effects of cold stress or hypobaric hypoxia (simulated altitude) on working memory. In the cold stress condition, working memory was assessed by a delayed matching-to-sample (DMTS) task, which was administered in a standard operant chamber. There were two response levers on the front wall and a third lever at the rear wall. A light above one of the two response levers was illuminated and the rat had to press the lever beneath it. The light was extinguished and a cue light was turned on above the rear response lever, which started a delay interval of 1, 2, 4, 8 or 16 sec. The first response on the rear wall lever after the delay interval illuminated both lights above the other two response levers on the front wall. The rat then had to push the cued lever previously responded to and a food pellet was delivered. By this procedure, the retention interval for remembering the right lever was varied from 1 to 16 sec. In the hypoxia study, memory was assessed by the Morris water maze with a hidden escape platform. Reference memory was established by allowing each rat 120 sec to escape onto the platform. If the rat failed, it was guided to the platform. After having stayed for 10 sec on the platform, working memory was assessed by determining the latency of reaching the platform on a second trial.

Cold stress was induced by providing an ambient temperature of either 2 or 22°C. Hypobaric stress was induced at 2 and 6 h of an 8-h exposure to a simulated altitude of 5950 m. During cold exposure, doses of 100 and 200 mg/kg tyrosine significantly improved overall matching accuracy but did not completely reverse the effects of cold exposure. An IP dose of 400 mg/kg of tyrosine, administered 30 min before testing, reversed the memory decrement that was induced by altitude exposure. Thus, supplemental tyrosine is partially effective in ameliorating the effects of stress on working memory performance, possibly by preventing a stress-induced reduction in brain catecholamine levels (Shukitt-Hale et al., 1996; Shurtleff et al., 1993).

21.2.5 EFFECTS OF TYROSINE ON THE COGNITIVE AND BIOCHEMICAL CONSEQUENCES OF WEIGHT LOSS

Animal studies have provided some indications that tyrosine might be a potential therapy for cognitive and mood problems related to maintaining a reduced body weight. It can probably be used to treat eating disorders such as anorexia nervosa and obesity. Chronic voluntary and involuntary weight loss may lead to changes in brain neurotransmission, which may be counteracted by administering tyrosine. Work in rats has shown that catecholamine pathways in the hypothalamus are involved in feeding behavior. Stimulation of α-adrenergic receptors in the area of the paraventricular nucleus of the hypothalamus induces feeding and stimulation of β-adrenergic receptors inhibits feeding behavior. Animals depleted of DA may stop eating and starve to death. This observation appears to be similar to patients with anorexia nervosa. In these patients, cerebrospinal fluid (CSF) levels of tyrosine, NE, and metabolites of NE are decreased. After weight loss, the inability to eat and restore body weight may be related to neurochemical dysfunctions (Avraham et al., 1996). Thus, as a consequence of insufficient NE or DA, appetite in these patients may be suppressed. If weight is returned to normal, NE levels can remain abnormal for a long time after recovery. Studying the effects of malnutrition on catecholamine regulation and cognitive function in animals can shed more light on the pathophysiology of eating disorders in humans and may lead to more successful treatments.

One method to decrease body weight in mice is to place them in a cage with plexiglass partitions. The mice are separated from each other by plexiglass and can smell and see each other without having any physical contact. This separation stress decreases body weight. Within 14 days, a reduction of 35% in body weight can be observed. Separation-induced behavioral deficits were assessed by measuring the number of spontaneous alternations in the T-maze. Spontaneous alternation is a two-trial phenomenon in which an animal is said to alternate if its choice on the second trial of testing is opposite to that of the first trial. Alternation behavior in the T-maze is considered to be a test of hippocampal function, related to catecholaminergic and cholinergic pathways. A reduction in the number of spontaneous alternations is seen as impaired behavior. Separation reduced the ability of mice to spontaneously alternate in the T-maze when compared with control groups. This reduction was related to depletion of NE and DA. Increasing tyrosine availability by injections (dose of 100 mg/kg) 1 h before testing restored performance to control levels and repleted DA. In addition, tyrosine injections increased MHPG and the MHPG:NE ratio, suggesting an increase of NE. However, tyrosine did not affect body weight (Hao et al., 2001). Another way to decrease weight is by chronically restricting the diet and then studying the resulting cognitive and neurochemical alterations. Young female rats were fed 100, 60, and 40% of the calculated daily nutritional requirements for up to 18 days. For the mice on a 60% diet restriction weight loss was ca. 15%, whereas the mice on the 40% diet restriction experienced a 25% weight loss. Cognitive function was evaluated by an eight-arm maze with water as a reward. In animals on the 40% diet restriction, maze performance was impaired. Tyrosine injections of 100 mg/kg/day restored performance. This was associated with changes in α and β receptor density and increased NE and MHPG. Regarding adrenergic receptors, diet restriction to 40 and 60% increased the number of α-receptors and decreased gradually the number of β receptors in the hypothalamus. Tyrosine treatment reversed these alterations, reducing the number of α-receptors in both 40 and 60% diet-restricted mice and increasing β receptors in the 40% diet-restricted mice. There were no changes in body weight (Avraham et al., 1996).

Thus, at present there is evidence from animal studies that tyrosine restores the neurobiological disturbances caused by diet restriction without causing an increase in body weight. These insights may be directed toward improving the treatment of anorexia nervosa and related eating disorders in humans. Unfortunately, studies on tyrosine administration in human subjects with eating disorders are nonexistent.

21.3 HUMAN STUDIES

Behavioral effects of stress in humans are well documented. The main cognitive effects of stress are (1) attentional narrowing, (2) increased speed in information processing paired with less accuracy, and (3) decrease in the capacity of short-term memory. Whether these behavioral stress effects are associated with NE depletion is still unclear. The physiological effects of stress on the peripheral nervous system can be summarized as being increased heart rate, blood pressure, muscle tonus, skin conductance, and respiration rate. In addition, biochemical changes take place, such as an increase in the concentration of catecholamines (Deijen and Orlebeke, 1994). In humans, NE depletion causes insuppressible eye movements during smooth pursuit and visual search. The finding that catecholamine depletion increases the amplitude and frequency of saccadic intrusions during fixation and pursuit implies that brainstem neurons use catecholamines to suppress saccades (Tychsen and Sitaram, 1989). The well-known association between anxiety and the frequency of eye movements may be circumstantial evidence that stressful events cause NE depletion in human brain.

21.3.1 TYROSINE/CATECHOLAMINE DEPLETION

Low levels of catecholamines may predispose humans to lowered mood and psychomotor retardation. Acute depletion of the catecholamine precursors phenylalanine and tyrosine can reduce catecholamine synthesis.

An amino acid cocktail has been devised and tested in rats to determine whether it reduces central nervous system (CNS) tyrosine levels. In addition, the effect of this mixture on *in vivo* tyrosine hydroxylation rate was examined. The amino acid cocktail consisted of 10 amino acids lacking tyrosine and phenylalanine. Serum TYR levels, the serum ratio of TYR to the sum of its transport competitors, and CNS TYR concentrations fell within 1 h of intubation and remained low for at least 3 h. *In vivo* tyrosine hydroxylation rate, determined in the hypothalamus and retina 2 h after amino acid administration, also declined. Thus, such a mixture can be used to acutely reduce TYR levels and hydroxylation rate in the CNS. This paradigm has been proved to be a useful tool for studying catecholamine involvement in normal and impaired behavior in humans (Fernstrom and Fernstrom, 1995b). The use of this acute phenylalanine/tyrosine depletion (APTD) paradigm in humans indicated that participants ingest a mixture of essential amino acids that is deficient in the catecholamine precursors phenylalanine and tyrosine. This induces protein synthesis, which diminishes the body's stores of phenylalanine and tyrosine. Because tyrosine hydroxylase, the rate-limiting enzyme in catecholamine synthesis, is normally not fully saturated with tyrosine, the reduced availability of tyrosine will likely reduce catecholamine synthesis. Effects of APTD on mood and anxiety have been examined in several human studies. In one study, 5 h after ingesting an amino acid mixture lacking tyrosine and phenylalanine, healthy female participants aged 19 to 39 years underwent a mild threatening psychological challenge, a social stress test. This test consists of preparing and presenting a 5-minute speech followed by 5 min of solving arithmetic problems. Mood scales were administered before ingestion of the amino acid mixture and again before and immediately after the social stress test, which took place 5 h after ingestion. APTD lowered mood and energy and increased irritability scores. Effects were only significant after the psychological challenge. APTD did not attenuate the anxiety caused by the psychological challenge. This study shows that lowered catecholamine synthesis impairs mood, but only after subjects have been exposed to unpleasant events (Leyton et al., 2000). The same paradigm of inducing acute tyrosine depletion has been studied in male and female volunteers who received an amino acid drink lacking tyrosine and phenylalanine (TYR-free) on one occasion and on the other received a balanced amino acid drink (BAL). Plasma prolactin, amino acid levels, mood, and cognitive functions were assessed at different timepoints within a 6-h period following the drinks. As plasma prolactin levels rose 5 to 6 h following the TYR-free drink relative to the BAL-drink, decreased DA neurotransmission within the hypothalamus was inferred. The subjects reported that they felt worse after the TYR-free drink than after the BAL drink. Regarding cognitive functions, ca. 6 h following the TYR-free drink, spatial recognition memory and spatial working memory were impaired. Both types of memory functions are sensitive to frontostriatal dysfunction, particularly dopaminergic pathways. Thus, tyrosine depletion in healthy volunteers may impair executive functions, such as planning and set shifting, involving neuronal networks connected with or situated within the prefrontal cortex (Harmer et al., 2001). It has been suggested that DA plays a major role in executive functions as DA neurons appear to be more vulnerable to TYR depletion than do NA neurons. Administration of a TYR-free amino acid load also reduced DOPA accumulation, especially in the striatum (–44%) and nucleus accumbens (–34%), areas with a predominant dopaminergic innervation. Smaller decreases (–20 to 24%) were detected in the cortex, hippocampus, and hypothalamus. The effect on DOPA was prevented by supplementing the mixture with tyrosine/phenylalanine. However, the TYR-free amino acid mixture did not alter the basal or amphetamine-evoked release of NA in the hippocampus. Therefore, administration of a TYR-free amino acid mixture to rats depletes brain TYR to cause a substantial decrease in brain DA synthesis and release (McTavish et al., 1999).

21.3.2 ENDOCRINE/BIOCHEMICAL STUDIES

At present there is compelling evidence that the synthesis and release of a number of brain neurotransmitters (e.g., acetylcholine, serotonin, and the catecholamines) depends on brain levels of their precursor nutrients, that is, choline, tryptophan, and tyrosine. Wurtman (1981) introduced this idea.

The effects of tyrosine on plasma tyrosine and catecholamine synthesis and release have been examined in many human studies. For instance, an oral load of 100 mg/kg tyrosine significantly increases the plasma tyrosine level. In fasting recumbent normal subjects, the plasma concentration was nearly doubled 1 h after tyrosine ingestion and doubled 2 h after ingestion (Cuche et al., 1985).

In addition to the effects of L-tyrosine alone, the combined effect of tyrosine and concurrent food consumption on plasma tyrosine levels and the plasma tyrosine ratio has been examined. On the first day, 11 subjects consumed three equal portions of a diet containing 113 g of protein at 8 a.m., 12 p.m., and 5 p.m. On the other day, they took 100 mg/kg of L-tyrosine in three equally divided doses before the same meals. Plasma tyrosine levels rose during the day when subjects consumed the diet alone. The concentrations were highest between 1 p.m. and 9 p.m. and lowest at 5 p.m. The consumption of the protein-rich meals caused daytime postprandial plasma amino acid levels to rise. Ingestion of L-tyrosine 1 h before each meal elevated plasma tyrosine levels much more. Tyrosine administration did not change plasma concentrations of the other neutral amino acids that compete with tyrosine for entry into the brain. Peak tyrosine plasma levels of Day 2 (178 nmol/ml) were more than double the peak levels of Day 1 (79 nmol/ml). The plasma tyrosine ratio increased from 0.13 (only diet) to 0.21 on the day subjects received the combination of the diet and tyrosine. Because the amount of tyrosine that enters the brain depends on the plasma tyrosine ratio, the authors concluded that tyrosine administration might thus increase brain tyrosine levels and may accelerate catecholamine synthesis in humans with diseases in which catecholamine synthesis or release is deficient (Melamed et al., 1980).

Another step then is to determine the relationship between the plasma tyrosine level and the CSF concentration of tyrosine. The biochemical relationship between CSF tyrosine and plasma tyrosine was determined in 52 normal individuals. Males showed higher concentrations of plasma tyrosine than did females. In both females and males, CSF tyrosine was significantly related to plasma tyrosine and the plasma ratio of tyrosine to LNAAs. Age does not appear to be related to these biochemical parameters in males or females (Möller et al., 1996). Effects of oral amino acid supplementation on physical and neuroendocrine variables were investigated in male subjects who cycled in four trials until exhaustion. In one of the four conditions, 20 g of tyrosine was administered. Plasma TYR/BCAAs was augmented. Plasma prolactin (PRL) was increased after 30 min of exercise. Tyrosine administration did not alter physical performance (exercise time to physical exhaustion). However, the observed rise in plasma PRL concentration indicates that tyrosine administration affected transport characteristics of the L-carrier and central neurotransmitters systems. Augmented PRL secretion is likely to have resulted from an increased activity of the dopaminergic system induced by supplemental plasma tyrosine. This study provides further evidence that tyrosine enhances DA release, but only from activated neurons (Strüder et al., 1998). From these studies it may be concluded that ingestion of L-tyrosine leads to higher plasma tyrosine levels and is associated with higher CSF tyrosine levels, which in turn may increase levels of brain catecholamines.

21.3.3 TREATMENT OF DEPRESSION

Only two studies on the possible use of tyrosine as an antidepressant have been performed. The first study was a single case and provides evidence for an antidepressant effect of tyrosine. In a double-blind, placebo-controlled crossover trial, a depressed woman was administered L-tyrosine 100 mg/kg/day by mouth in three daily doses for 5 weeks. After 2 weeks of tyrosine administration the depression decreased. The patient felt an improvement in mood, self-esteem, sleep, and energy level, and she also experienced less anxiety and somatic complaints. Her clinical global impression went from "moderately ill" to "not ill at all." Within 1 week of placebo substitution, the depressive symptoms returned. No unwanted effects of tyrosine ingestion were observed. Ten years later, the same author reported negative findings in a large sample of patients. In this double-blind study, 65 outpatients with major depression were treated with 100 mg/kg/day oral L-tyrosine, imipramine, or placebo for 4 weeks. MHPG excretion rose significantly after tyrosine administration, suggesting an increase in catecholamine turnover. However, HAM-D scores did not show evidence of tyrosine's antidepressant activity (Gelenberg et al., 1980, 1990). In spite of these negative results it may still be possible that a subgroup of depressed patients with low pretreatment tyrosine plasma levels is responsive to tyrosine. However, no data concerning such depressed patients are available.

21.3.4 TREATMENT OF ATTENTION DEFICIT DISORDER

Tyrosine supplementation has also been used to treat attention deficit disorder (ADD) in children and adults. In seven children with ADD, 100mg/kg/day of tyrosine was given for 3 weeks. The treatment effects were evaluated by the parents on the 10-item Connors rating scale and by their teachers on the Connors Teaching Rating Scale. Furthermore, the children were given a vigilance task, the Children's Checking Task. There were no treatment effects (Eisenberg et al., 1988). For the treatment of adults, the daily dose of L-tyrosine given to treat ADD was 150 mg/kg in all studies. The treatment periods in adults ranged from 8 to 16 weeks. Although in some studies treatment effects were seen after ca. 4 weeks, at 6 weeks all patients who responded to L-tyrosine became tolerant to its therapeutic effects (Wood et al., 1985; Reimherr et al., 1987; Eisenberg et al., 1988). It is clear from these findings on ADD that tyrosine has no clinical utility in the treatment of this disorder.

Another way to use tyrosine for treating ADD is to combine it with methylphenidate. Studies of rats have examined whether exogenous tyrosine could potentiate the methylphenidate-induced increase in extracellular DA. Male rats were implanted with microdialysis probes in the right nucleus accumbens. Samples were collected from awake animals beginning 22 h after surgery for 3 consecutive days. On a given day, an animal was infused with methylphenidate, tyrosine, or methylphenidate plus tyrosine. Methylphenidate plus tyrosine significantly increased extracellular levels of DA in comparison to the drug alone. This effect was long lasting and peaked 40 min after the peak induced by methylphenidate alone. Tyrosine alone induced a small but significant increase in extracellular DA in the absence of any treatment to accelerate the firing of DA cells (Woods and Meyer, 1991). Tyrosine or a tyrosine-rich diet given to attention deficit hyperactivity disorder (ADHD) children may thus increase the supply of brain DA. In addition, if supplemental tyrosine does sustain the methylphenidate-induced increase in extracellular DA in humans by preventing the depletion of tyrosine, exogenous tyrosine along with methylphenidate may prolong the effect of such drugs as Ritalin®. Thus, administration of tyrosine in addition to methylphenidate may reduce the side effects of methylphenidate by reducing the frequency of administration. Although these considerations may have implications for treating ADHD, there are no other studies on this topic. It would be worthwhile to carry out clinical studies in patients to evaluate the effects of tyrosine combined with methylphenidate.

21.3.5 TYROSINE AND PARKINSON'S DISEASE (PD)

To determine whether L-tyrosine administration can enhance DA synthesis in humans, the levels of tyrosine and the major DA metabolite HVA were measured in lumbar spinal fluids of 23 patients with Parkinson's disease (PD) before and following ingestion of 100 mg/kg/day of tyrosine (Growdon et al., 1982). Nine patients received 100 mg/kg probenecid, which blocks the transport of HVA across the CSF-blood barrier and allows it to accumulate in the CSF. L-Tyrosine increased both

CSF tyrosine levels in all patients as well as HVA levels in patients pretreated with probenecid. Thus, L-tyrosine can increase DA turnover in Parkinson's patients in whom dopaminergic neurotransmission should be enhanced. The question then remains whether L-tyrosine has a therapeutic effect in Parkinson's patients. Daily doses of 4.7, 4.2, and 1.8 g of L-m-tyrosine in three Parkinson's patients in addition to a decarboxylase inhibitor did not result in any improvement in tremor, rigidity, or hypokinesia. The treatment duration, however, is not reported (Cotzias et al., 1973). In addition to this pilot-like study, there is one long-term study on tyrosine and PD with positive results. Ten PD patients were tested in an open study. Five patients with ongoing and unsatisfactory L-dopa or DA agonist treatment, or both, shifted to L-tyrosine treatment, and five patients were initially treated with L-tyrosine. Patients were administered a mean daily dose of 2.24 g L-tyrosine for 5 to 36 months (mean treatment duration 14.8 months). A gradual improvement, as assessed by a scale of 0 to 4 concerning rigidity, tremor, akinesia, and gait, was observed over several months, reaching a maximum within ca. 6 months. Tyrosine administration was not associated with side effects and on–off episodes did not occur. The authors conclude that after 3 years of treatment, L-tyrosine treatment resulted in positive clinical results, negligible side effects, and a possible neurons-sparing capability (Lemoine et al., 1989). However, as no comparisons were made with a reference group and no statistical analyses were carried out, the claims of the authors should be confirmed in other placebo-controlled studies. At present, there is no experimental evidence on the positive effects of L-tyrosine in the treatment of PD.

21.3.6 TYROSINE LEVELS IN AGING AND ALZHEIMER'S DISEASE

One study to date provides data of an absence of a relationship among tyrosine, aging, and Alzheimer's disease (AD). This study was directed not on tyrosine but on the 3-nitrotyrosine concentration, which was determined in the CSF of neurologically normal controls and patients with AD. However, at the same time, the CSF concentration of tyrosine was determined. The 3-nitrotyrosine concentration and the 3-nitrotyrosine:tyrosine ratios significantly increased with advancing age. The tyrosine concentration did not change with increasing age. In patients with AD, the 3-nitrotyrosine concentration and the 3-nitrotyrosine:tyrosine ratio was significantly higher than those of controls of similar age and increased with decreasing cognitive functions. Regarding tyrosine, it appeared that the CSF concentration in patients with AD was not different from that of age-matched controls. Thus, an activation of tyrosine nitration, increase in nitrated tyrosine-containing proteins, and its degradation may be involved in brain aging and play an important role in the pathogenesis of AD (Toghi et al., 1999). The tyrosine concentration itself is not related to age and does not seem to play a role in the development of AD.

21.3.7 TYROSINE AND STRESS-INDUCED BEHAVIORAL DEFICITS

Tyrosine may be useful in counteracting performance decrement and mood impairments that are caused by stress. The hypothesis advanced is that various stressors may induce brain depletion of catecholamines, especially NE, in animals. Lower brain NE levels are associated with a decreased performance in animals. The administration of tyrosine may reduce or even reverse stress-induced performance decrement and mood deterioration by increasing depleted brain NE levels. In animals, these relationships are supported by the results of many studies. However, in humans it is more difficult to find support for these claims as there is no direct evidence that stress depletes NE in the human brain. However, there is some indirect evidence. For instance, it is generally known that humans under stress are not able to keep their eyes stable and fixed. Under stress, humans are characterized by an increase in involuntary saccadic eye movements. These eye movements seem to be associated with catecholaminergic pathways. There is one study in which ocular motor pathways in humans were studied after catecholamines were experimentally depleted. Subjects received metyrosine (α-methylparatyrosine), a drug that temporarily depletes DA and NE, as measured by a reduction in the metabolite 3-methoxy-4-hydroxy-phenylethyleneglycol (MHPG). Metyrosine induced an increase in the amplitude and frequency of saccadic intrusions during fixation and pursuit. The increase in the number of saccadic intrusions can be explained by a modulatory role of catecholamines of the activity of a subpopulation of suppressor motor neurons in the human brain stem (Tychsen and Sitaram, 1989). As stress induces an increase in eye movements and eye movements are related to NE levels, it can be inferred that stress depletes brain catecholamines, especially NE, in humans. The next step is to provide evidence that NE depletion is related to impaired attention. It has been found that central NE is important in maintaining attention. Brain levels of NE can be lowered by clonidine, an α-2-adrenoceptor agonist. In low doses, clonidine acts presynaptically to decrease noradrenergic cell firing and NE release. The reduced NE release can be reversed by idazoxan, a selective α-2-adrenoceptor antagonist. To study the relationship between attention processes and NE levels, subjects were assigned to groups receiving placebo, clonidine, or idazoxan + clonidine. They had to perform, in both a quiet and noisy environment, a two-choice RT task, which was adapted to record lapses of attention (RT > 1.5 msec). It was found that the number of lapses of attention increased after clonidine adminis-tration. This effect was reversed, that is, attention was normalized, by coadministration of idazoxan (Smith and Nutt, 1996).

The relation between attention and the inhibition of NE release by clonidine suggests an interrelationship between NE and cognitive processes. Although insight into the mechanisms explaining the relationship between stress, NE, and performance is lacking in humans compared with animals, catecholamines may be a mediating factor between stress and performance in humans. Based on the evidence available, the usefulness of tyrosine in military sustained operations has been proposed. Sustained operations consist of continuous work periods exceeding 12 h. The resulting sleep loss and fatigue can lead to stress, anxiety, deterioration of mood, and performance decrements (Owasoyo et al., 1992). Techniques for coping with combat stress to prevent emotional breakdown mostly rely on such stress-reducing approaches as effective leadership, rest, relaxation, and being in good physical condition. As these techniques are not adequate to deal with combat stress, the military needs more innovative techniques (Salter, 1989). As a consequence, some studies have been performed to study the possible usefulness of tyrosine to reverse performance decrements and mood deterioration in humans. These studies have been largely directed on learning and memory functions, including mood states. One of the first studies on tyrosine and behavior included 16 healthy men, aged 18 to 45, who were administered tryptophan (50 mg/kg), tyrosine (100 mg/kg), and a placebo in a single dose, using a double-blind crossover design. Before ingestion, the subject fasted for 12 h. Four tests of sensorimotor performance and two self-reported mood questionnaires were administered 2 h after ingestion of the substances. In this study, no effects of tyrosine were observed (Lieberman et al., 1982). However, subjects in this study did not experience experimental stressors. In later studies, it was recognized that tyrosine would be expected to have positive effects on behavior only under stressful conditions.

The first study in which the behavioral effects of tyrosine were examined in humans subjected to acute stress was conducted in young men exposed to cold and hypoxia. Subjects who were most vulnerable to the stressors exhibited fewer stress symptoms, such as headache, tension, and fatigue, and showed less psychomotor impairment after being supplemented with 100 mg/kg tyrosine (Banderet and Lieberman, 1989).

Since then, these findings have been replicated in healthy volunteers and extended. Tyrosine in oral dose ranges from 85 to 170 mg/kg given 1 to 2 h after stress induction, such as cold stress (4°C) or noise (90 dB), restored memory performance and mood particularly in individuals who were most affected by the stressors. Tyrosine did not seem to have dose-dependent effects (Deijen and Orlebeke, 1994; Shurtleff et al., 1994). In one of these studies, tyrosine decreased blood pressure in humans. About 15 min after ingestion, diastolic blood pressure was decreased and 1 h after ingestion diastolic blood pressure returned to baseline level (Deijen and Orlebeke, 1994). These studies suggest that tyrosine is only beneficial in counteracting cognitive impairments under stress.

However, positive effects of tyrosine on neuropsychological test performance have been found without overt exposure to stress. One hour after the administration of 150 mg/kg of L-tyrosine or placebo, healthy volunteers performed a number of tests of a multiple-test battery, assessing working memory, arithmetic skills, and visual and auditory monitoring simultaneously. A comparison was made with the performance on a simple battery, measuring working memory and visual monitoring. Tyrosine was found to prevent decrements of working memory only in the multiple-task condition. Thus, tyrosine may sustain working memory performance when there are competing requirements by other tasks (Thomas et al., 1999). Although no stress was overtly induced in this study, it may well be true that performing more tasks at the same time is a stressful requirement that is comparable with the induction of cold or noise.

In addition to these laboratory-like studies, tyrosine has been tested in real-life stress settings. Behavioral effects of tyrosine were examined during an episode of continuous nighttime work involving one night's sleep loss. Subjects, male U.S. Marines aged 21 to 27 years, performed nine iterations of a battery of performance tasks and mood scales for ca. 13 h, beginning at 19:30 and ending at 08:20. They remained awake throughout the day on which the experiment began and were awake until testing was completed (totaling 24 h). Six hours after the experiment had begun, subjects received 150 mg/kg tyrosine or placebo. Tyrosine reduced the performance decline on a psychomotor task and lapse probability on a vigilance task. These improvements lasted for 3 h (Neri et al., 1995).

In another study, the effects of a protein-rich drink containing 2 g tyrosine on cognitive task performance, mood, blood pressure, and the norepinephrine metabolite MHPG were determined in a group of cadets after a 5-day demanding military combat training course. Subjects received the tyrosine-rich drink for 5 days. Effects were compared with subjects who received a carbohydrate-rich drink with the same amount of calories (255 kcal). Assessments were made both immediately before the combat course and on the sixth day of the course. The group treated with the tyrosine-rich drink performed better on a memory and a tracking task than the group given the carbohydrate-rich drink. In addition, supplementation with tyrosine decreased systolic blood pressure (Deijen et al., 1999). These findings suggest that supplementation with tyrosine or even a tyrosine-rich drink may, under operational circumstances characterized by psychosocial and physical stress, reduce the negative effects of stress and fatigue on cognitive task performance.

21.3.8 BLOOD PRESSURE AND CARDIOVASCULAR STRESS

Tyrosine decreased diastolic blood pressure in normotensive young subjects 15 min after ingestion, whereas blood pressure returned to baseline value 1 h after ingestion (Deijen and Orlebeke, 1994). Similar effects were found in normotensive cadets receiving a tyrosine-rich drink during a demanding military combat training course. The group supplemented with the tyrosine-rich drink showed a decrease in systolic blood pressure (Deijen et al., 1999). Thus, tyrosine can reduce blood pressure in healthy, normotensive persons during or after stressful situations. Data from animal studies suggest that tyrosine has a unique regulatory function, lowering high blood pressure while increasing low blood pressure. Decreased blood pressure can be induced experimentally by cardiovascular stress, that is, subjecting them to lower-body negative pressure (LBNP; Dollins et al., 1995). LBNP is a technique used to simulate gravitational stress (orthostasis) by exposing the lower body to subatmospheric pressures. Subjects exposed to LBNP initially respond with decreased blood pressure and increased heart rate. In the course of this cardiovascular stress exposure, consciousness is lost if LBNP is not terminated. After oral administration of 100 mg/kg tyrosine subjects maintained higher pulse pressures. As a result, tyrosine appears to reverse blood pressure alterations and may therefore be protective when cardiovascular parameters in humans are disturbed.

21.3.9 L-TYROSINE AND SKIN PIGMENTATION

L-Tyrosine is one of the agents stimulating natural pigmentation because it is not only the precursor of brain catecholamines but also the precursor compound for the melanogenic pathway. This pathway leads to the formation of melanin, a dark-brown to black pigment that occurs in the skin, hair, pigmented coat of the retina, and medulla and zona reticularis of the adrenal gland. The rate-limiting enzyme in melanogenesis is tyrosinase, which catalyzes the conversion of tyrosine to L-dopa and then dopaquinone. Dopaquinone is then converted either to brownish-black eumelanins or reddish-yellow pheomelanins. Eumelanins are more photoprotective than pheomelanins.

It is assumed that L-tyrosine may induce melanogenesis because the primary substrate for tyrosinase may be limiting. Increasing L-tyrosine from 10 to 200 μM in the medium of melanoma cells results in a 10-fold increase of tyrosinase activity. In addition, 500 μM of L-tyrosine increased ca. 1.6-fold the melanin content of normal human epidermal melanocytes. These are pigment-producing cells located in the basal layer of the epidermis. L-Tyrosine at 1000 μM resulted in a 2.3-fold increase.

These findings suggest that L-tyrosine may be applied to induce tanning or to enhance ultraviolet B-induced tanning. Indeed, 48 mM L-tyrosine ethyl ester increased simulated-solar radiation (SSR)-induced pigmentation by 1.35-fold in rat skin. In humans, only19 mM L-tyrosine ethyl ester resulted in the same SSR increase as in the rats. However, no L-tyrosine derivatives or formulations have been demonstrated to induce tanning in animal or human skin without ultraviolet radiation (UVR).

Safety is a major concern when using L-tyrosine as a potential tanning agent. However, because natural pigmentation is associated with greatly reduced skin cancer risk, the use of pigmentation stimulators before sun exposure has the potential to reduce both solar radiation-induced damage and skin cancer incidence (Brown, 2001). The development of agents to increase pigmentation of the skin is still continuing. Whether L-tyrosine will become a safe and efficacious compound to enhance skin pigmentation depends on further research on the mechanisms that induce melanogenesis and the willingness of companies to produce and sell it as a tanning agent. The possible clinical use of L-tyrosine in treating psychiatric disorders, reversing behavioral deficits, or protecting against stress or sunburn can be inferred from the summary of studies on the topics in Table 21.1. From this table it is clear that most evidence is on L-tyrosine and the reversal of stress-induced behavioral deficits.

21.4 SAFETY OF TYROSINE AS A DIETARY SUPPLEMENT

21.4.1 RISK FACTORS

Excess L-tyrosine in low-protein diets may lead to a distinct syndrome of cataracts, skin lesions, and histopathological changes in rats. These changes are the consequence of providing L-tyrosine at levels of 3 to 5% in low-protein diets. The eye lesions appear to be related to the low solubility of the amino acid. These adverse effects of excess L-tyrosine appear to be intensified by adrenalectomy and hyperthyroidism. This syndrome has not been described in rats fed high-protein diets, even when as much as 12% L-tyrosine was added.

In humans, L-tyrosine has been studied extensively as a therapeutic agent in large doses. Doses of 70 to 150 mg/kg body weight are required because of the demonstration that this amount of precursor availability can influence the synthesis and release of brain catecholamines. Administration of 100 mg/kg body weight of L-tyrosine (7 g for a 70-kg individual) with or without food and as single or divided doses doubles the plasma tyrosine concentrations and increases the tyrosine to LNAAs ratio in humans (Cuche et al., 1985). Administration of this amount of L-tyrosine on a single day has not been associated with clinically significant changes in blood pressure, pulse rate, urinary volume, or abnormal neurological or psychological phenomena. Some studies have reported

TABLE 21.1
Studies on the Clinical Utility of Tyrosine

Disease/Condition	Type of Study	Daily Tyrosine Treatment	Number of Studies	Number and Type of Positive Outcomes	Number of Negative Outcomes
Stress-induced behavioral deficits	Animal	Tyrosine-enriched diet or 100–400 mg/kg IP tyrosine	5	5; reversion of behavioral deficits	0
	Human	85–170 mg/kg	7	7; reversion of behavioral deficits	0
High blood pressure	Animal	15 μg Tyr-containing dipeptides	2	2; blood pressure decrease	0
Low blood pressure	Animal	Tyr-containing dipeptides	1	1; blood pressure increase	0
	Human	100 mg/kg	1	1; blood pressure increase	0
Normal blood pressure	Human	100 mg/kg Tyr (2 g)-rich drink	2	2; decrease of blood pressure	0
Biochemical misbalance by weight loss	Animal	100 mg/kg	2	2; repletion DA and NE normalization adrenergic receptors	0
Physical condition	Human	20 g	1	0	1
Depression	Human	100 mg/kg	2	1; decrease in depression	1
Attention deficit disorder	Children	100 mg/kg	1	0	1
	Adults	150 mg/kg	3	0	3
Parkinson's disease	Human	1.8–4.7 g	2	1; less tremor, rigidity, akinesia	1
Skin pigmentation	Animal	48mM L-tyrosine ethyl ester	1	1; pigmentation increase	0
	Human	19 mM L-tyrosine ethyl ester	1	1; pigmentation increase	0

gastrointestinal side effects in some subjects receiving 100 mg/kg body weight of L-tyrosine given without food but not with food.

Administration of 100 mg/kg in divided doses for 4 weeks was associated with palpitations in 1 of 21 patients given L-tyrosine as a treatment for major depression. High plasma concentrations of tyrosine are associated with hepatic and renal failure in patients with tyrosinemia I, an autosomal recessive disorder. In tyrosinemia II, the enzyme tyrosine aminotransferase is defective, which can lead to mental retardation. Development of eye and skin lesions may occur in patients with tyrosinemia II. Transitory neonatal tyrosinemia (TNT) in premature and term infants has been associated with impaired intellectual abilities in childhood.

21.4.2 SAFE LEVELS OF HUMAN INTAKE

The usual daily L-tyrosine intake from the diet is about 2.2 g for an individual consuming 100 g protein per day. In one study in which the extremely high dose of 500 mg/kg body weight (35 g for a 70-kg individual) was given, no side effects were reported. Thus, from the information of one subject, L-tyrosine given at ca. 15 times the daily intake does not seem to induce side effects.

However, as noted previously, the occurrence of more severe effects of high doses of L-tyrosine is found in animals receiving low-protein diets. Therefore, humans with low intakes of protein may be more susceptible to possible adverse effects of chronic consumption of L-tyrosine as a dietary supplement.

Evidence of biochemical and behavioral effects in offspring of female rats given L-tyrosine during gestation suggests that women should not take L-tyrosine during pregnancy or lactation. High circulating levels of tyrosine during infancy, called transient neonatal tyrosinemia (TNT), resulting from the feeding of high-protein formulas (e.g., inappropriately diluted evaporated milk formulas) may be associated with impaired intellectual performance. Although some investigators have reported normal development of children with TNT, others have reported deficits in intellectual performance (Food and Drug Administration, 1992). Therefore, L-tyrosine should not be given to infants or children as a dietary supplement. Evidence that L-tyrosine increases the cytochrome P450 content in the liver of rats raises the possibility of interactions of tyrosine with some drugs. Likewise, evidence that compounds which stimulate the cytochrome P450-containing drug-metabolizing system also increase the acute toxicity of L-tyrosine in rats also suggests that L-tyrosine should not be used by persons taking pharmaceutical preparations, which stimulate this system.

The safety of continued ingestion of L-tyrosine as a dietary supplement by normal subjects cannot be determined from the data available. The relative insolubility of L-tyrosine raises concern about the possibility of localized adverse effects in the small intestine or in the lens with chronic exposure. Knowledge that greatly elevated plasma concentrations of tyrosine are associated with eye and skin lesions in persons with tyrosinemia II and that these lesions can be reversed by lowering plasma concentrations of L-tyrosine gives cause for concern about chronic high concentrations of plasma tyrosine. Likewise, demonstrations of pharmacological effects on catecholamine synthesis and release in stimulated catecholaminergic neurons in rats and on behavior of humans subjected to stressful situations by L-tyrosine given orally in single doses of 100 mg/kg body weight do not provide evidence of the safety of L-tyrosine at this level of intake.

Studies demonstrating evidence of safety, or lack thereof, with lower oral doses of L-tyrosine are not available. Amino acids used as dietary supplements can interact with prescription drugs such as monoamine oxidase inhibitors and other antidepressants, sympathomimetic amines, and opioids (Food and Drug Administration, 1992).

REFERENCES

Acworth, I. N., During, M. J. and Wurtman, R. J. (1988) Tyrosine: effects on catecholamine release. *Brain Research Bulletin,* 21, 473–477.

Avraham, Y., Bonne, O. and Berry, E. M. (1996) Behavioral and neurochemical alterations caused by diet restriction — the effect of tyrosine administration in mice. *Brain Research,* 732, 133–144.

Banderet, L. E. and Lieberman, H. R. (1989) Treatment with tyrosine, a neurotransmitter precursor, reduces environmental stress in humans. *Brain Research Bulletin,* 22, 759–762.

Basile-Filho, A., Beaumier, L., El Khoury, A. E., Yu, Y. M., Kenneway, M., Gleason, R. E. and Young, V. R. (1998) Twenty-four-hour L-[1-(13)C]tyrosine and L-[3,3-(2)H2]phenylalanine oral tracer studies at generous, intermediate, and low phenylalanine intakes to estimate aromatic amino acid requirements in adults. *American Journal of Clinical Nutrition,* 67, 640–659.

Bender, D. A. and Bender, A. E. (1997) *Nutrition. A Reference Handbook,* 1st ed. Oxford University Press, Oxford.

Brady, K., Brown, J. W. and Thurmond, J. B. (1980). Behavioral and neurochemical effects of dietary tyrosine in young and aged mice following cold-swim stress. *Pharmacology Biochemistry and Behavior,* 12, 667–674.

Brown, D. A. (2001) Skin pigmentation enhancers. *Journal of Photochemistry and Photobiology,* 63, 148–161.

Conlay, L. A., Maher, T. J. and Wurtman, R. J. (1981) Tyrosine increases blood pressure in hypotensive rats. *Science,* 212, 559–560.

Cotzias, G. C., Papavasiliou, P. S. and Mena, I. (1973) L-*m*-tyrosine and Parkinsonism. *Journal of the American Medical Association,* 223, 83.

Cuche, J. L., Prinseau, J., Selz, F., Ruget, G., Tual, J. L., Reingeissen, L., Devoisin, M., Baglin, A., Guéden and Frietel, D. (1985) Oral load of tyrosine or L-dopa and plasma levels of free and sulfoconjugated catecholamines in healthy men. *Hypertension,* 7, 81–89.

Deijen, J. B. and Orlebeke, J. F. (1994) Effect of tyrosine on cognitive function and blood pressure under stress. *Brain Research Bulletin,* 33, 319–323.

Deijen, J. B., Wientjes, C. J. E., Vullinghs, H. F. M., Cloin, P. A. and Langefeld, J. J. (1999) Tyrosine improves cognitive performance and reduces blood pressure in cadets after one week of a combat training course. *Brain Research Bulletin,* 48, 203–209.

Dollins, A. B., Krock, L. P., Storm, W. F., Wurtman, R. J. and Lieberman, H. R. (1995) L-tyrosine ameliorates some effects of lower body negative pressure stress. *Physiology and Behavior,* 57, 223–230.

During, M. J., Acworth, I. N. and Wurtman, R. J. (1988) Effects of systemic L-tyrosine on dopamine release from rat corpus striatum and nucleus accumbens. *Brain Research,* 452, 378–380.

Eisenberg, J., Asnis, G. M., van Praag, H. M. and Vela, R. M. (1988) Effect of tyrosine on attention deficit disorder with hyperactivity. *Journal of Clinical Psychiatry,* 49, 193–195.

Food and Drug Administration (FDA) (1992) Tyrosine. In *Safety of Amino Acids Used as Dietary Supplements* (Anderson, S.A. and Raiten, D.J., Eds.), pp. 178–192. Center for Food Safety and Applied Nutrition, Washington, D.C.

Fernstrom, M. H. and Fernstrom, J. D. (1995b) Acute tyrosine depletion reduces tyrosine hydroxylation rate in rat central nervous system. *Life Sciences,* 57, 97–102.

Fernstrom, M. H. and Fernstrom, J. D. (1995a) Effect of protein ingestion on rat central nervous system tyrosine levels and *in vivo* tyrosine hydroxylation rate. *Brain Research,* 672, 97–103.

Gelenberg, A. J., Wojcik, J. D., Falk, W. E., Baldessarini, R. J., Zeisel, S. H., Schoenfeld, D. et al. (1990) Tyrosine for depression: a double-blind trial. *Journal of Affective Disorders,* 19, 125–132.

Gelenberg, A. J., Wojcik, J. D., Growdon, J. H., Sved, A. F. and Wurtman, R. J. (1980) Tyrosine for the treatment of depression. *American Journal of Psychiatry,* 137, 622–623.

Gibson, C. J. (1988) Alterations in retinal tyrosine and dopamine levels in rats consuming protein or tyrosine-supplemented diets. *Journal of Neurochemistry,* 50, 1769–1774.

Growdon, J. H., Melamed, E., Logue, M., Hefti, F. and Wurtman, R. J. (1982) Effects of oral L-tyrosine administration on CSF tyrosine and homovanillic acid levels in patients with Parkinson's disease. *Life Sciences,* 30, 827–832.

Hao, S., Avraham, Y., Bonne, O. and Berry, E. M. (2001) Separation-induced body weight loss, impairment in alternation behavior, and autonomic tone: effects of tyrosine. *Pharmacology Biochemistry and Behavior,* 68, 273–281.

Harmer, C. J., McTavish, S. F. B., Clark, L., Goodwin, G. M. and Cowen, P. J. (2001) Tyrosine depletion attenuates dopamine function in healthy volunteers. *Psychopharmacology,* 154, 105–111.

Krause, M. V. and Mahan, L. K. (1984) *Food, Nutrition, and Diet Therapy. A Textbook of Nutritional Care,* 7th ed. W.B. Saunders, Philadelphia.

Lehnert, H., Reinstein, D. K., Strowbridge, B. W. and Wurtman, R. J. (1984) Neurochemical and behavioral consequences of acute, uncontrollable stress: effects of dietary tyrosine. *Brain Research,* 303, 215–223.

Lemoine, P., Robelin, N., Sebert, P. and Mouret, J. (1989) Medicine and therapeutics. *C.R. Académie des Sciences,* 309, 43–47.

Leyton, M., Young, S. N., Pihl, R.O., Etezadi, S., Lauze, R. N., Blier, P. et al. (2000) Effects on mood of acute phenylalanine/tyrosine depletion in healthy women. *Neuropsychopharmacology,* 22, 52–63.

Lieberman, H. R., Corkin, S., Spring, B. J., Growdon, J. H. and Wurtman, R. J. (1982) Mood, performance, and pain sensitivity: changes induced by food constituents. *Journal of Psychiatric Research,* 17, 135–145.

Maher, T. J., Kiritsy, P. J., Moya-Huff, F. A., Casacci, F., De Marchi, F. and Wurtman, R. J. (1990) Use of parenteral dipeptides to increase serum tyrosine levels and to enhance catecholamine-mediated neurotransmission. *Journal of Pharmaceutical Sciences,* 79, 685–687.

McTavish, S. F. B., Cowen, P. J. and Sharp, T. (1999) Effect of a tyrosine-free amino acid mixture on regional brain catecholamine synthesis and release. *Psychopharmacology,* 141, 182–188.

Melamed, E., Glaeser, B., Growdon, J. H. and Wurtman, R. J. (1980) Plasma tyrosine in normal humans: effects of oral tyrosine and protein-containing meals. *Journal of Neural Transmission,* 47, 299–306.

Möller, S. E., Mortensen, E. L., Breum, L., Alling, O. G., Larsen, O. G., Böge-Rasmussen, T., Jensen, C. and Bennicke, K. (1996) Aggression and personality: association with amino acids and monoamine metabolites. *Psychological Medicine,* 26, 323–331.

Neri, D. F., Wiegmann, D., Stanny, R. R., Shappell, S. A., McCardie, A. and McKay, D. L. (1995) The effects of tyrosine on cognitive performance during extended wakefulness. *Aviation, Space, and Environmental Medicine* 66, 313–319.

Owasoyo, J. O., Neri, D. F. and Lamberth, J. G. (1992) Tyrosine and its potential use as a countermeasure to performance decrement in military sustained operations. *Aviation Space and Environmental Medicine*, 63, 364–369.

Reimherr, F. W., Wender, P. H., Wood, D. R. and Ward, M. (1987) An open trial of L-tyrosine in the treatment of attention deficit disorder, residual type. *American Journal of Psychiatry*, 144, 1071–1073.

Reinstein, D. K., Lehnert, H., Scott, N. A. and Wurtman, R. J. (1984) Tyrosine prevents behavioral and neurochemical correlates of an acute stress in rats. *Life Science*, 34, 2225–2231.

Roberts, S. A., Thorpe, J. M., Ball, R. O. and Pencharz, P. B. (2001) Tyrosine requirement of healthy men receiving a fixed phenylalanine intake determined by using indicator amino acid oxidation. *American Journal of Clinical Nutrition*, 73, 276–282.

Salter, C. A. (1989) Dietary tyrosine as an aid to stress resistance among troops. *Military Medicine*, 154, 144–146.

Shukitt-Hale, B., Stillman, M. J. and Lieberman, H. R. (1996) Tyrosine administration prevents hypoxia-induced decrements in learning and memory. *Physiology and Behavior*, 59, 867–871.

Shurtleff, D., Thomas, J. R., Ahlers, S. T. and Schrot, J. (1993) Tyrosine ameliorates a cold-induced delayed matching-to-sample performance decrement in rats. *Psychopharmacology*, 112, 228–232.

Shurtleff, D., Thomas, J. R., Schrot, J., Kowalski, K. and Harford, R. (1994) Tyrosine reverses a cold-induced working memory deficit in humans. *Pharmacology Biochemistry and Behavior*, 47, 935–941.

Skeie, B., Kvetan, V., Gil, K. M., Rothkopf, M. M., Newsholme, E. A. and Askanazi, J. (1990) Branch-chain amino acids: their metabolism and clinical utility. *Critical Care Medicine*, 18, 549–571.

Smith, A. and Nutt, D. (1996) Noradrenaline and attention lapses. *Nature*, 380, 291.

Strüder, H. K., Hollmann, W., Platen, P., Donike, M., Gotzmann, A. and Weber, K. (1998) Influence of paroxetine, branched-chain amino acids and tyrosine on neuroendocrine system responses and fatigue in humans. *Hormone and Metabolic Research*, 30, 188–194.

Thomas, J. R., Lockwood, P. A., Sing, A. and Deuster, P. A. (1999) Tyrosine improves working memory in a multitasking environment. *Pharmacology Biochemistry and Behavior*, 64, 495–500.

Tohgi, H., Abe, T., Yamazaki, K., Murata, T., Ishizaki, E. and Isobe, C. (1999) Alterations of 3-nitrotyrosine concentration in the cerebrospinal fluid during aging and in patients with Alzheimer's disease. *Neuroscience Letters*, 269, 52–54.

Topall, G. and Laborit, H. (1989) Brain tyrosine increases after treating with prodrugs: comparison with tyrosine. *Journal of Pharmacy and Pharmacology*, 41, 789–791.

Tychsen, L. and Sitaram, N. (1989) Catecholamine depletion produces irrepressible saccadic eye movements in normal humans. *Annals of Neurology*, 25, 444–449.

Weiss, J. M. (1991) Stress-induced depression: critical neurochemical and electrophysiological changes. In *Neurobiology of Learning, Emotion and Affect* (Madden, J., Ed.), pp. 123–154. Raven Press, New York.

Wood, D. R., Reimherr, F. W. and Wender, P. H. (1985). Amino acid precursors for the treatment of attention deficit disorder, residual type. *Psychopharmacology Bulletin*, 21, 146–149.

Woods, S. K. and Meyer, J. S. (1991) Exogenous tyrosine potentiates the methylphenidate-induced increase in extracellular dopamine in the nucleus accumbens: a microdialysis study. *Brain Research*, 560, 97–105.

Wurtman, R. J. (1992) Effects of foods on the brain: possible implications for understanding and treating Tourette syndrome. In *Advances in Neurology*, Vol. 58 (Chase, N., Friedhoff, A. J. and Cohen, D. J., Eds.), pp. 293–301. Raven Press, New York.

Wurtman, R. J., Hefti, F. and Melamed, E. (1981) Precursor control of neurotransmitter synthesis. *Pharmacological Reviews*, 32, 315–335.

Yamori, Y., Fujiwara, M., Horie, R. and Lovenberg, W. (1980) The hypotensive effect of centrally administered tyrosine. *European Journal of Pharmacology*, 68, 201–204.

22 Popular Herbal Medicines Having Effects on the Central Nervous System

Edzard Ernst

CONTENTS

22.1 INTRODUCTION

According to the World Health Organization (WHO), ca. 80% of the world's population uses herbal medicines as its primary source of health care (Evans, 1998). In developing countries traditional herbal remedies have not yet been replaced by modern drugs, and in developed countries herbal medicine is experiencing a remarkable comeback (Eisenberg, 1998; Ernst and White, 2000). Many herbs have effects on the central nervous system (CNS), and some are specifically promoted for their CNS effects.

This chapter discusses herbal medicines that have been extensively researched for their CNS effects and are in popular use for such indications. In doing so, it briefly describes the mechanisms of action as well as summarizes the existing evidence on clinical efficacy and safety. To minimize random and selection biases, I refer (where possible) to systematic reviews of the published literature.

22.2 ST. JOHN'S WORT (*HYPERICUM PERFORATUM*)

22.2.1 MECHANISM OF ACTION

Tinctures or extracts of St. John's wort (SJW) applied topically or systemically have a wide range of traditional uses: for bronchitis, burns, cancer, enuresis, gastritis, hemorrhoids, hypothyroidism, insect bites, insomnia, kidney disease, scabies, and wound healing (Ernst et al., 2001). At present, SJW is employed almost exclusively to treat mild to moderate depression.

The constituents of SJW include anthraquinone derivatives (naphthodianthrones), including hypericin and pseudohypericin, flavonoids, prenylated phloroglucinols such as hyperforin and tautins, other phenols, and volatile oils (Barnes et al., 2002). Hypericin and hyperforin are prime candidates for the major pharmacologically active constituents but our knowledge on them is incomplete. The antidepressant mechanism of action has previously been assumed to consist of monoamine oxidase (MAO) inhibition (Blandt and Wagner, 1994). It is now believed to lie in a selective inhibition of serotonin, dopamine, and norepheniphrine reuptake in the CNS (Singer et al., 1999).

22.2.2 EFFICACY

A metaanalysis of good methodological quality included 27 double-blind randomized controlled trials (RCTs; Linde, 2000). The authors used the Jadad score (Jadad et al., 1996) for evaluating the methodological quality of the primary studies and found their average quality to be good. Seventeen RCTs were placebo-controlled and demonstrated efficacy in mild to moderate depression. Ten studies compared SJW extracts with standard reference medications and showed apparent equivalence to maprotiline, imipramine, bromazepam, amitriptyline, and diazepam. Two high-quality reviews applied stricter inclusion criteria (Gaster and Hoiroyd, 2000; Kim, 1999).Thus, only eight (Gaster and Hoiroyd, 2000) and six (Kim, 1999) double-blind RCTs were summarized. No formal assessment of the methodological quality of the primary studies was performed in either of these reviews. Their results confirmed that SJW is more effective than placebo in treating mild to moderate depression and similarly zaazin is as effective as low-dose tricyclic anti-depressants. The most critical metaanalysis (Williams et al., 2000) included only 14 RCTs. The authors made no formal assessment of trial quality. They confirmed that SJW extracts were significantly more effective to treat mild to moderate depression than was placebo (relative benefit 1.9, 95% CI: 1.2 to 2.8) and equivalent to tricyclic antidepressants (relative benefit 1.2, 95% CI: 1.0 to 1.4). There was some evidence of publication bias, which may have led to an overestimation of the effect.

Three recent three-armed RCTs (Philipp et al., 1999; Shelton et al., 2001; Hypericum Depression Trial Group, 2002) were not included in the reviews discussed previously. In Philipp et al. (1999), 263 patients with moderate depression received daily doses of 1050 mg standardized SJW extract or 100 mg imipramine or placebo for 8 weeks. The results show that SJW was more effective than placebo and equivalent to imipramine in reducing symptoms of depression as verified by Hamilton depression scores. The other two studies failed to demonstrate an antidepressant effect of SJW over and above that of placebo (Shelton et al., 2001; Hypericum Depression Trial Group, 2002). Both trials included patients with major depression. The most reasonable conclusion from data available to date is therefore that SJW is an effective treatment for mild to moderate but not for major or severe depression.

22.2.3 SAFETY

Taken as a monotherapy, SJW has an excellent safety profile that is clearly superior to that of conventional antidepressants (Ernst et al. 2001; Stevinson, 1999). The only potentially serious adverse effects relate to photosensitization, a complication of high-dose therapy that is extremely rare, and to the possibility that SJW might induce manic symptoms in predisposed patients (Schulz,

2000). Problems may, however, arise when patients take SJW with other medications. SJW acts as a hepatic enzyme inducer by activating the cytochrome P450 system (Ernst, 1999). In addiction, it activates the transporter protein P-glycoprotein, which further increases the elimination of synthetic drugs (Drewe, 2000). Through these mechanisms, it can decrease the plasma level of a range of prescribed drugs (e.g., anticoagulants, oral contraceptives, and antiviral agents). This interaction can have major clinical consequences (Piscitelli et al., 2000; Ruschitzka et al., 2000; Yue and Bergquist, 2000), of which endangering the success of organ transplantation through a lowering of plasma cyclosporine levels is perhaps the most serious (Ernst, 2002). Finally there is evidence that the combination of SJW with selective serotonin reuptake inhibitors can lead to serotonin overload or serotonin syndrome, particularly in elderly patients (Ernst, 1999).

22.3 GINKGO (*GINKGO BILOBA*)

22.3.1 Mechanism of Action

Various parts of the ginkgo tree have been used in traditional Chinese medicine for millennia, for example, for asthma and chilblains. At present, the ginkgo leaf is mostly used for medicinal purposes. It contains flavonoids and terpene lactones (ginkgolides and bilobalides), which are associated with diverse pharmacological actions. *In vitro* and *in vivo* experiments suggest that ginkgo has antiedema, antihypoxic, free radical-scavenging, antioxidant, metabolic, antiplatelet, hemorrheological, and microcirculatory actions (De Feudis, 1991; Johns and Cupp, 2000). Ginkgo extracts have been used experimentally for a wide range of indications: asthma, brain trauma, cochlea deafness, depression, free-radical damage to the retina, male impotence, reduction of myocardial reperfusion injury, and vertigo (Boon and Smith, 1999). In clinical practice, it is employed mostly for memory impairment, dementia, tinnitus, and intermittent claudication. In some European countries it is registered for these indications, whereas in the U.S. (like all other herbal medicines discussed in this chapter) it is marketed as a dietary supplement.

22.3.2 Efficacy

A systematic review of high methodological quality included 40 controlled clinical trials of ginkgo for cerebral insufficiency, that is, memory impairment but not dementia (Kleijnen and Knipschild, 1992). The authors used their own score to assess the methodological quality of the primary studies and only eight were judged to be rigorous. All but one of these eight showed positive effects of ginkgo on cognitive functions compared with placebo. In terms of global effectiveness ratings as well as single symptoms (e.g., forgetfulness, lack of concentration), the evidence was clearly in favor of ginkgo compared with placebo. However, the authors suspected publication bias because "there were no negative results reported in many trials of low quality" (Kleijnen and Knipschild, 1992). A further less rigorous metaanalysis focussed on one particular standardized extract (Lichtwer, Germany; Hopfenmüller, 1994). The author included no formal assessment of the methodological quality of the primary studies. Five of the 11 RCTs had not been included in the Kleijnen and Knipschild (1992) article. Again, the overall conclusions were positive.

Our recent systematic review addressed a slightly different research question: does ginkgo improve normal cognitive function in healthy volunteers below the age of 60 years free from any medical condition (Canter and Ernst, 2002)? Nine RCTs met our inclusion criteria. Their methodological quality was on average good, but none of the trials was entirely devoid of limitations. Collectively, the results of these studies did not show a consistent positive effect of ginkgo on any objective measure of cognitive function.

A further systematic review (Ernst and Pittler, 1999) included nine placebo-controlled double-blind RCTs of ginkgo for dementia. The methodological quality of the primary studies was assessed by the Jadad score (Jadad et al., 1996), and three of these studies achieved the highest possible rating. All these three and eight of the total nine studies suggested that ginkgo is significantly more effective than placebo in delaying the cognitive deterioration in dementia. By using stricter entry criteria, Oken et al. (1998) conducted a metaanalysis of high methodological quality of four placebo-controlled double-blind RCTs of ginkgo for Alzheimer's disease. All these studies were included in our (Ernst and Pittler, 1999) analysis. No formal assessment of the methodological quality of the primary studies was made. Again, the overall result was positive. The pooled effect size in terms of cognitive function was moderate and translated into 3% difference in the Alzheimer's Disease Assessment Scale — cognitive subtest. Thus the average size of the therapeutic effect associated with ginkgo seems to be modest but clinically relevant.

22.3.3 SAFETY

Adverse effects of ginkgo are usually mild, transient, and reversible. Potentially serious adverse effects are bleeding (e.g., subdural hematoma; De Feudis, 1991; Johns and Cupp, 2000) and seizures observed in children after excessive ingestion of seeds (Fetrow and Avila, 1999). Because of the antiplatelet activity of ginkgo (De Feudis, 1991) the possibility exists of interactions with anticoagulants (Ernst, 2000a, 2000b). Several such cases have been reported in the medical literature (Ernst, 2000b). In view of these findings, it might be wise to discontinue ginkgo several days before surgery.

22.4 KAVA (*PIPER METHYSTICUM*)

22.4.1 MECHANISM OF ACTION

Kava is made from the dried rhizome of the kava plant. It is traditionally used in the South Sea as a recreational drink. Kava has also been employed experimentally to attenuate seizures and to treat psychotic states (Fetrow and Avila, 1999). At present, it is mostly used for its anxiolytic effects (Ernst et al., 2001).

The active principle is constituted by a family of four pyrones (kavapyrones). Their main pharmacological properties include central muscle-relaxing and anticonvulsant actions (Schulz, 1999). The mechanism of the anxiolytic effect is still controversial; one theory is that kava pyrones enhance GABA binding in the amygdala without acting as direct antagonists at GABA receptors (Jussofie et al., 1994). Kava pyrones also are powerful strychnine antagonists (Schulz, 1999).

22.4.2 EFFICACY

A recent systematic review and metaanalysis (Pitter and Ernst, 2002) included seven placebo-controlled double-blind RCTs. The methodological quality of the primary studies was assessed by the Jadad score and found to be variable but good on average; four of the studies scored the maximum of 5 points on this scale. Three of the trials could be submitted to a metaanalysis. Its results demonstrated a reduction of the Hamilton Anxiety Score (HAS) in favor of kava with a weighted mean difference of 9.7 HAS units (95% CI: 3.5 to 15.8). One trial, not included in the analysis, compared kava with oxazepam (Lindenberg et al., 1990). Its results suggest that both medications are similarly efficacious anxiolytics. Collectively, these data leave little doubt that short-term administration of kava is efficacious in reducing anxiety in patients suffering from this condition.

22.4.3 SAFETY

In the previously mentioned RCTs, the incidence of adverse effects was similar in the experimental and placebo groups (Pittler and Ernst, 2000, 2002). Two postmarketing surveillance studies involving more than 6000 patients found adverse effects in 2.3 and 1.5% of patients taking 120 to 240 mg standardized kava extracts (Hofmann and Winter, 1996; Siegers et al., 1992). Several (in August 2002 the count was 68 worldwide) cases of toxic liver damage have been associated with kava (e.g., Stoller, 2000). So far the underlying mechanism is unclear.

These case reports are difficult to evaluate conclusively; in the vast majority of instances causality is a matter of debate not least because of concomitantly administered drugs, uncertainty of viral liver damage, and the role of alcohol abuse. Nevertheless, these cases were considered serious enough to warrant withdrawal of kava products from the markets of several countries.

When taken concomitantly with other medication acting on the CNS or with alcohol, there is a danger that the effects may be potentiated, leading to a temporal state of impaired vigilance or reduce consciousness. One such case report is on record (Almeida and Grimsley, 1996).

Long-term use of kava at high doses is associated with flaky, dry, and yellowish discoloring of the skin; ataxia; hair loss; partial loss of hearing; loss of appetite; and body weight reduction. The dermatological signs of kava abuse are known as kava dermopathy or kavaism (Norton and Ruse, 1994); usually they are reversible on discontinuation (Jappe et al., 1998). These phenomena have so far only been observed in inhabitants of the South Sea who took doses over prolonged periods of time, which were considerably higher than those recommended for therapeutic use (Schulz, 1999).

22.5 OTHER HERBAL MEDICINES

Numerous other herbal medicines have effects on the CNS. In many cases, they have not been studied in depth. Table 22.1 provides a list of some of these herbs based on data extracted from Ernst and White (2000) and Barnes et al. (2002).

22.6 COMMENTS

This brief overview is based (wherever possible) on systematic reviews of the literature. This approach minimized selection and random biases but is by no means free from drawbacks. The two main limitations are publication bias (i.e., the tendency for negative studies to remain unpublished) and the often low quality of the primary data that may render the conclusions of systematic reviews less convincing than one would hope. In addition, two further points are important for interpretating systematic reviews of herbal medicines. First, much of the (older) trial data were published in German or other non-English languages. It is thus important that systematic reviews (not only) in this area apply no language restrictions. Second, herbal medicinal products of one plant may vary hugely in terms of their chemical composition. Systematic reviews often pool data from extracts which differ significantly; for example, some may be of inferior quality. This creates the danger of generating false negative overall results with systematic reviews and stresses the importance of adequately standardizing and describing the extracts, particularly for clinical trials.

The evidence discussed previously demonstrates that many herbal medicines have CNS effects that differ in nature much as CNS effects of synthetic drugs would obviously differ. The important point is that generalizations across all herbal medicines, as they are often published in the lay press, are clearly nonsensical.

This overview also shows that herbal medicines have the potential to both benefit and harm patients, again much like synthetic drugs. One could argue therefore that they should be regulated in the same way as conventional medicines. In the current regulatory framework herbal medicines in the U.S. and U.K. are marketed as dietary supplements, and this is not in the best interests of consumer safety. Finally, if herbal and synthetic medicines are similar in crucial respects, they should be researched with the same rigor and this research should be funded with the same generosity.

TABLE 22.1
Examples of Other Herbal Medicines with Effects on the Central Nervous System

Latin (Common) Name	Main Constituents	Relevant Mechanism of Action	Efficacy/Indication[a] (Level of Evidence)	Major Safety Concerns	Comment
Acorus calamus (calamus)	Volatile oil	MAO inhibition, sedation	Sedation (+)	Genotoxic	May potentiate effects of barbiturates
Apium graveolens (celery)	Furanocuomarins	Sedation	Rheumatism, sedation, diuretic (+)	Photosensitivity	Fruits are used for medical purposes, stems for food
Calanitida (cola)	Caffeine	CNS stimulant	Depression, diuretic (+)	Sleeplessness, anxiety	Not for cardiovascular patients
Chamaemelum nobile (Roman chamomile)	Apigenin	Acts on central benzodiazepine receptors	Carminative, antiemetic (−)	Allergy	Should be avoided in hypersensitive individuals
Cinnamomium cassia (cassia)	Volatile oil	CNS stimulant	Range of gastrointestinal symptoms (+)	Allergy	Sometimes used as a substitute for true cinnamon
Crataegus (hawthorn)	Glycosides	CNS depressant	Congestive heart failure (+++)	Nausea, bradycardia	May potentiate action of other cardiac glycosides
Eleutherococcus senticosus (Siberian ginseng)	Eleutherosides	Sedation, CNS stimulant	General tonic (++)	Nerve irritation (prolonged use)	CNS stimulation/depression seem to depend on type of animal model
Ephedra sinica (ephedra/Ma huang)	Ephedrine	Sympathomimetic	Asthma (++)	Tachycardia, anxiety	Also promoted for weight reduction
Fumaria officinalis (fumitory)	Protopine (isoquinoline alkaloid)	Sedation	Diuretic, laxative (+)	Increase of intraocular pressure	Not for glaucoma or hypertensive patients
Humulus lupulus (hop)	Humulones, lupulones (deo-resin)	Sedation	Insomnia (+)	Allergy	Not for patients with depression
Inula helenium (elecampane)	Inulin	Sedation	Expectorant (+)	Allergy	None
Lobelia inflata (lobelia)	Lobeline (alkaloid)	CNS stimulant	Asthma (+)	Nausea, diarrhea	Actions are similar to those of nicotine
Matricaria recurita (German chamomile)	Apigenin	Acts on central benzodiazepine receptors	Wound healing (external use) (++)	Allergy	Should be avoided by hypersensitive individuals
Melissa officinalis (lemon balm)	Volatile oil	Sedation	Antiviral (++)	None reported	Cream against herpes labialis commercially available

Herb	Constituent	Mode of action	Uses[a]	Adverse effects	Remarks
Panax ginseng (American/Korean ginseng)	Ginsenoides	Inhibition of uptake of a range of neurotransmitters	Various, e.g., tonic, hypoglycemic (++)	Hypertension, diarrhea, insomnia	May potentiate action of MAO inhibitors
Passiflora incarnata (passionflower)	Harman	Sedation	Insomnia (+)	None reported	None
Pimpinella anisum (aniseed)	Transanethole (volatile oil)	Increases GABA in brain tissue of animals	Dyspepsia, catarrh (+)	Dermatitis, nausea vomiting	Traditionally used mostly for dyspepsia
Polygala senega (snake root)	Saponins	CNS depressant	Expectorant (+)	Gastric irritation	None
Salvia officinalis (sage)	Volatile oil	Anticholinesterase activity	Dyspepsia, pharyngitis (+)	Convulsions	Presently investigated as an antidementia agent
Sarothamus scoparius (broom)	Sparteine	CNS stimulant	Heart weakness, diuretic (+)	Bradycardia, respiratory arrest	Medical use not recommended
Scutellaria laterifolia (skullcap)	Scutellarin (glucoside)	Anticonvulsant, sedative	Epilepsy (+)	Hepatotoxicity	Medical use not recommended
Urtica dioica (stinging nettle)	Ligans	CNS depressant	Arthritis, benign prostatic hyperplasia (++)	Gastric irritation	Interactions with antidiabetic and antihypertensive drugs conceivable
Valeriana officinalis (valerian)	Valtrates	Affects GABA receptors	Insomnia (+++)	Hangover	May potentiate CNS depressants
Vitex agnus castus (chastburry)	Alkaloids (viticin)	Acts at dopamine D_2 receptors and affects prolactine release	Mastodynia, premenstrual syndrome (++)	Only minor adverse effects on record	Mostly used for its estrogenic effects

Note: + = mostly anecdotal evidence; ++ = several positive clinical trials; +++ = systematic review.

[a] Examples only.

Source: Extracted from Ernst, E. et al. (2001) *The Desktop Guide to Alternative and Complementary Medicine*, Mosby, Edinburgh, and Barnes et al. (2002) *Herbal Medicines: A Guide for Healthcare Professionals*, 2nd ed., Pharmaceutical Press, London. With permission.

REFERENCES

Almeida JC, Grimsley EW. (1996) Coma from the health food store: interaction between kava and alprazolam. *Ann Intern Med* 125:940–941.

Barnes J, Anderson LA, Phillipson JD. (2002) *Herbal Medicines. A Guide for Healthcare Professionals,* 2nd ed. Pharmaceutical Press, London.

Blandt S, Wagner H. (1994) Inhibition of MAO by fractions and constituents of *Hypericum* extract. *J Geriatr Psychiatr Neurol* 7:S57–S59.

Boon H, Smith M. (1999) *The Botanical Pharmacy.* Quarry Press, Ontario.

Canter P, Ernst E. (2002) *Ginkgo biloba*: a smart drug? *Psychopharmacol Bull* 36(3):108–123.

De Feudis FV. (1991) *Ginkgo biloba* extract: pharmacological activities and clinical applications. Elsevier, Amsterdam.

Drewe J. (2000) Mechanismen der Interaktionen mit Johanniskraut [Abstract] In *Phytopharmaka VI Forschung und Klinische Anwendung*, Vol. 14. Proceedings of the German Society for Clinical Pharmacological Therapy, June 16–17, 2000, Berlin.

Eisenberg DM, David RB, Ettner SL, Appel S, Wilkey S, Van Rompay M, Kessler RC. (1998) Trends in alternative medicine use in the United States. *JAMA* 280:1569–1575.

Ernst E. (1999) Second thoughts about safety of St. John's wort. *Lancet* 345:2014–2016.

Ernst E. (2000a) Possible interactions between synthetic and herbal medicinal products. Part 1: a systematic review of the indirect evidence. *Perfusion* 13:4–6,8.

Ernst E. (2000) Interactions between synthetic and herbal medicinal products. Part 2. A systematic review of the indirect evidence. *Perfusion* 13:60–70.

Ernst E. (2002) St. John's wort supplements endanger the success of organ transplantation. *Arch Surg* 137:316–319.

Ernst E, Pittler MH. (1999) *Ginkgo biloba* for dementia: a systematic review of double-blind placebo-controlled trials. *Clin Drug Invest* 17:301–308.

Ernst E, Pittler MH, Stevinson C, White AR, Eisenberg D. (2001)*The Desktop Guide to Complementary and Alternative Medicine.* Mosby, Edinburgh.

Ernst E, White AR. (2000) The BBC survey of complementary medicine use in the UK. *Complement Ther Med* 8:32–36.

Evans WC. (1998) *Trease and Evans' Pharmacognosy,* 14th ed. W.B. Saunders, London.

Fetrow CW, Avila JR. (1999) *Complementary and Alternative Medicines,* Springhouse, Philadelphia, PA.

Gaster B, Hoiroyd J. (2000) St. John's Wort for depression. *Arch Intern Med* 160:152–156.

Hofmann R, Winter U. (1996) Therapeutische Möglichkeiten met Kava-Kava bei Angsterkrankungen. *Psycho* 22(Suppl.):51–53.

Hopfenmüller W. (1994) Nachweis der therapeutischen Wirksamkeit eines Ginkgo biloba-spezialextraktes. *Arzneim.-Forsch* 44(II):1005–1013.

Hypericum Depression Trial Group. (2002) Effect of *Hypericum perforatum* (St. John's wort) in major depressive disorder: a randomized controlled trial. *JAMA* 287:1807–1814.

Jadad AR, Moore RA, Carrol D, Jenkinson C, Reynolds DJM, Gavaghan DJ, McQuay, DM. Assessing the quality of reports of randomised clinical trials is blinding necessary. *Contr Clin Trials* 17:1–12.

Jappe U, Franke I, Reinhold D, Gollnick HPM. (1998) Sebotropic drug interaction resulting from kava-kava extract therapy: a new entity? *J Am Acad Dermatol* 38:104–106.

Johns M, Cupp J. (2000) Toxicology and clinical pharmacology of herbal products. Humana Press, Totowa, NJ.

Jussofie A, Schmiz A, Hiemke C. (1994) Kavapyrone enriched extract from *Piper methysticum* as a modulator of GABA binding site in different regions of the rat brain. *Psychopharmacology* 116:469–474.

Kim HL. (1999) SJ. St. John's wort for depression. A meta-analysis of well-defined clinical trials. *J Nervous Mental Dis* 187:532–538.

Kleijnen J, Knipschild P. (1992) *Ginkgo biloba* for cerebral insufficiency. *Br J Clin Pharmacol* 34:352–358.

Linde K. (2000) St. John's wort for depression. In *The Cochrane Library,* Issue 3, pp.1–17. John Wiley & Sons, Chichester, U.K.

Lindenberg D, Pitule-Schödel H. (1990) D,L-Kavain im Vergleich zu Oxazepam bei Angstzuständen. *Fortschr Med* 108:31–34.

Norton S, Ruse P. (1994) Kava dermopathy. *J Am Acad Dermatol* 31:89–97.

Oken BS, Storzbach DM, Kaye JA. (1998) The efficacy of *Ginkgo biloba* on cognitive function in Alzheimer disease. *Arch Neurol* 55:1409–1415.

Philipp M, Kohen R, Hiller K-O. (1999) *Hypericum* extract versus imipramine or placebo in patients with moderate depression: a randomized, multicentre study of treatment for eight weeks. *Br Med J* 319:1534–1539.

Piscitelli SC, Burstein AH, Chaitt D, Alfaro RM, Falloon J. (2000) Indinavir concentrations and St. John's wort. *Lancet* 355:547–548.

Pittler MH, Ernst E. (2000) Efficacy of kava extract for treating anxiety: systematic review and meta-analysis. *J Clin Psychopharmacol* 20:84–89.

Pittler MH, Ernst E. (2002) Kava extract for treating anxiety. *The Cochrane Library*, Issue 3. John Wiley & Sons, Chichester, U.K.

Ruschitzka F, Meier PJ, Turina M, Lüscher TF, Noll G. (2000) Acute heart transplant rejection due to St. John's Wort. *Lancet* 355:548–549.

Schulz V. (1999) *Rationale Phytotherapie.* Springer, Berlin.

Schulz V. (2000) Incidence and clinical relevance of the interactions and side effects of hypericum preparations. *Perfusion* 13:486–444.

Shelton RC, Keller MB, Gelenberg A, Dunner DL, Hirschfeld R, Thase ME, Russell J, Lydiard RB, Crits-Cristoph P, Gallop R, Todd L, Hellerstein D, Goodnick P, Keitner G, Stahl SM, Halbreich U. (2001) Effectiveness of St. John's wort in major depression: a randomised controlled trial. *JAMA* 285:1978–1986.

Siegers C-P, Honold E, Krall B, Meng G, Habs M. (1992) Ergebnisse einer Anwendungsbeobachtung L1090 mit Laitan(r) Kapseln. *Ärztliche Forschung* 39:7–11.

Singer A, Wonneman M, Muller WE. (1999) Hyperforin, a major antidepressant constituent of St. John's Wort, inhibits serotonin uptake by elevating free intracellular Na^+. *J Pharmacol Exp Ther* 290:1363–1368.

Stevinson C, Ernst E. (1999) Safety of *Hypericum* in patients with depression. *CNS Drugs* 11:125–132.

Stoller A. (2000) Leberschädigungen unter Kava-Extrakten. *Schweizerische Ärztezeitung* 133:5–36.

Williams JW, Mulrow CD, Chiquette E, Noel PH, Aguilar C, Cornell J. (2000) A systematic review of newer pharmacotherapies for depression in adults: evidence report summary. *Ann Int Med* 132:743–756.

Yue QY, Bergquist GB. (2000)Safety of St. John's Wort (*Hypericum perforatum*). *Lancet* 355:576–577.

23 Phytochemicals in Foods and Beverages: Effects on the Central Nervous System

Barbara Shukitt-Hale, Amanda Carey, and James A. Joseph

CONTENTS

23.1 INTRODUCTION

Increased susceptibility to the long-term effects of oxidative stress (OS) and inflammatory insults is thought to be a contributing factor to the decrements in cognitive and motor performance seen in aging and neurodegenerative diseases. Deficits in brain functions due to OS may be due, in part, to a decline in the endogenous antioxidant defense mechanisms (Halliwell, 1994; Harman, 1981; Yu, 1994) and to the vulnerability of the brain to the deleterious effects of oxidative damage (Olanow, 1993). With respect to inflammation, increases in inflammatory mediators (e.g., cytokines), as well as increased mobilization and infiltration of peripheral inflammatory cells into the brain, produce deficits in behavior similar to those observed during aging (Hauss-Wegrzyniak et al., 2000).

Furthermore, age-related changes in brain vulnerability to OS and inflammation may be the result of membrane changes and differential receptor sensitivity (Joseph et al., 2001).

There has been a growing interest in a number of pharmacological approaches to help slow the rate of both cognitive and functional declines associated with these increases in OS and inflammation, with a view to maintaining a positive quality of life and reducing healthcare costs. These include antioxidants, monoamine oxidase inhibitors, antiinflammatory agents, cholinergic agents, estrogens, or neurotrophic factors. However, although scientists and pharmaceutical companies continue to invest tremendous resources into identifying agents that could be used to alleviate debilitating neurodegenerative disorders that continue to afflict numerous people around the world, an existing source of potentially beneficial agents, namely, phytochemicals, appears to have significant benefits that have yet to be fully exploited.

One class of phytochemicals is polyphenolics (also known as flavonoids), of which more than 4000 different structures have been identified (Macheix et al., 1990), and these occur ubiquitously in foods of plant origin (e.g., fruits, vegetables, nuts, seeds, grains, tea, and wine). Examples of polyphenolic families include flavonols, flavonals, procyanidins, and anthocyanins. Anthocyanins are the natural colorants responsible for the attractive pink, red, purple, and blue colors of flowers, leaves, fruits, and vegetables. Polyphenolics play a definite role in attracting animals as pollination and seed dispersal factors. They may also have a role in the resistance of plants to insect attack. Polyphenolic compounds have been recognized to possess many biological properties, including antioxidant, antiallergic, antiinflammatory, antiviral, antiproliferative, antitumorigenic, antianxiety, and anticarcinogenic ones (Eastwood, 1999; Hollman and Katan, 1999; Middleton, 1998). However, past interest in these compounds, with respect to their health-promoting benefits, has primarily focused on examining their roles in protection against the incidence and mortality rates of cancer and ischemic heart disease, by reducing risk factors associated with these diseases (Hollman and Katan, 1999). Consequently, few studies have investigated their role with regard to the central nervous system (CNS), such as their effects on brain function and behavior.

The phytochemicals that are most familiar to the general public are Chinese herbal remedies such as *Ginkgo biloba* (EGb 761) and ginseng. While continued research is being undertaken to further understand the biological actions of these extracts, the beneficial effects of phytochemicals from dietary sources are only now beginning to receive increased attention. This chapter focuses on nutritional interventions and the role played by phytochemical components found in foods in reducing the deleterious effects of OS and inflammation in the CNS. Recent reviews have highlighted the neuroprotective functions of vitamins E and C (Cantuti-Castelvetri et al., 2000; Martin et al., 2002), α-lipoic acid (Cantuti-Castelvetri et al., 2000; Lynch, 2001), Chinese remedies including *Ginkgo biloba* and ginseng (Cantuti-Castelvetri et al., 2000; Gold et al., 2002; Youdim and Joseph, 2001), caloric restriction (Casadesus et al., 2002a), dietary fatty acids (Youdim et al., 2000a), and other nutrients (McDaniel et al., 2002), and therefore are not covered in this chapter.

23.2 POLYPHENOLICS IN FRUITS AND VEGETABLES

Fruits and vegetables are known to contain numerous phytochemicals, and until recently their beneficial health effects on brain function had not been scientifically studied. The few studies in the literature on the role of dietary polyphenolics on brain function have found positive benefits following consumption. Although initially it was assumed that the vitamin component of fruits and vegetables was the primary source of dietary antioxidants, it is now well established that the phytochemical components also contribute substantially to the overall dietary antioxidant intake (Greenwood and Winocur, 1999). Fruit and vegetable extracts that have high levels of polyphenolics also display high total antioxidant activity, as assessed by the oxygen radical absorbance capacity (ORAC) assay (Cao et al., 1995); these include spinach and strawberries (Cao et al., 1996; Wang et al., 1996) and blueberries (Prior et al., 1998). Given that fruits and vegetables have antioxidant properties, and that increases in oxidative stress and declines in antioxidant defense mechanisms have been postulated as causative factors in age-related decrements in behavior, our laboratory, together with colleagues at the University of Colorado, decided to study the effects of fruit and vegetable supplementation for their ability to forestall or reverse age-related changes in performance (Bickford et al., 1999, 2000; Joseph et al., 1998, 1999; Youdim et al., 2000b).

23.2.1 SPINACH, STRAWBERRIES, AND BLUEBERRIES

In our first study, Fischer 344 (F344) rats underwent long-term feeding from adulthood (6 months) to middle age (15 months) with a control diet (AIN-93) or diets supplemented vitamin E (500 IU/kg diet) or with extracts of strawberry or spinach that contained identical antioxidant content (based upon mmol Trolox equivalents) to determine whether the feeding would prevent age-related decrements in motor and cognitive behavior as well as brain function (Joseph et al., 1998). A number of different parameters known to be sensitive to oxidative stress were prevented by the antioxidant diets: (1) receptor sensitivity, that is, oxotremorine-enhanced dopamine (DA) release in isolated striatal slices and cerebellar Purkinje cell activity; (2) calcium-buffering capacity, i.e., the ability of striatal synaptosomes to extrude calcium following depolarization, deficits of which ultimately result in reduced cellular signaling and eventually cell death; (3) changes in signal transduction assessed by carbachol-stimulated GTPase coupling/uncoupling in striatal membranes; and (4) cognition (spatial learning and memory) as measured by Morris water maze (MWM) performance (Joseph et al., 1998). Spinach-fed rats demonstrated the greatest retardation of age effects on all parameters except GTPase activity, in which strawberry had the greatest effect; strawberry and vitamin E showed significant but equal protection against these age-induced deficits on the other parameters.

Our subsequent experiments (Bickford et al., 1999, 2000; Joseph et al., 1999) found that dietary supplementations (for 8 weeks) with spinach, strawberry, or blueberry extracts in a control diet (AIN-93) were also effective in reversing age-related deficits in brain and behavioral function in aged (19 months) Fischer 344 rats. Blueberries showed the greatest increases in motor performance, carbachol-stimulated GTPase activity, and oxotremorine-enhanced DA release; in addition, the blueberry-fed group showed no decrements in calcium recovery following exposure to an oxidative stressor (Joseph et al., 1999). Interestingly, blueberries were added to this study because they were found to have the highest antioxidant capacity (ORAC) among all fruits and vegetables tested (Prior et al., 1998). Yet, even though these diets were supplemented based on equal antioxidant activity, as determined by the ORAC assay (Cao et al., 1996), they were not equally effective in preventing or reversing age-related changes. Therefore, it seems that antioxidant activity alone was not predictive in assessing the potency of these compounds at being beneficial against certain disorders affected by aging. In fact, oxidative stress parameters (as measured by DCF fluorescence and glutathione levels in the brain) were only modestly reduced by the diets (Joseph et al., 1999), suggesting that these fruit and vegetable polyphenolics possess a multiplicity of actions, aside from antioxidative, and differences in the polyphenolic composition of these extracts could account for the positive effects observed.

Based on this idea, findings from a recent study showed that two different cultivars of blueberries (that have different phytochemical compositions), even though they were supplemented at the same concentration (2% of the diet), afforded different degrees of protection against memory and learning declines in aging rats (Youdim et al., 2000b). Furthermore, the beneficial effects of blueberries were seen even when superimposed on an already well-balanced, healthy rodent diet (chow), which was more representative of a balanced human diet (Youdim et al., 2000b).

Some additional actions or properties elicited by polyphenolics that may be contributing factors in their protection against impairments in brain functions have been suggested by other examinations in our laboratory. Preliminary findings indicate that blueberry supplementation may (1) increase adult neurogenesis in the hippocampus (Galli et al., 2000; Casadesus et al., 2002b); (2) increase the protection against the *in vitro* application of the inflammatory agent tumor necrosis factor α (TNFα), on the expression of heat shock protein-70 (HSP-70; Galli et al., 2001); and (3) restore the ability of old rat hippocampal cells to respond to an acute *in vitro* inflammatory challenge with a large production of HSP-70 (Galli et al., 2001, in press). Moreover, blueberry supplementation from 4 to 12 months of age in mice transgenic for amyloid precursor protein (APP) and presenilin-1 (PS-1) mutations reduced deficits in Y-maze performance (at 12 mo of age), with no alterations in Aβ burden (Joseph et al., 2004). In addition, several parameters of neuronal signaling (e.g., ERK activity) showed parallel positive changes with the behavior in the blueberry-supplemented transgenic mice (Joseph et al., 2004).

The mechanisms involved, as well as the polyphenolic "families" that produce the most beneficial effects on neuronal aging and behavior, are being investigated further in our laboratory and other laboratories, but it is too early to identify the particular properties of the polyphenolics contained in blueberries that may be involved in these effects. However, we have found that several blueberry anthocyanins and flavonols can cross the blood-brain barrier and localize in various brain regions important for learning and memory (e.g., cerebellum, striatum, and hippocampus), suggesting that polyphenolic compounds may deliver their antioxidant and signaling modifying capabilities centrally (Shukitt-Hale et al., 2002).

23.2.2 SPINACH, SPIRULINA, APPLES, AND CUCUMBERS

Further research by Bickford and colleagues (Gemma et al., 2002) demonstrated that diets enriched with high antioxidant activity reversed age-induced decreases in cerebellar β-adrenergic function in a manner that corresponded to ORAC dose. Aged (18 months) male F344 rats were fed spirulina (a microalga with high ORAC), apple (intermediate ORAC), or cucumber (low ORAC) for 14 days. Spirulina reversed the age-related decrease in β-adrenergic receptor function, whereas apple had an intermediate and cucumber no effect on this parameter. However, both spirulina and apple significantly downregulated the age-related increase in proinflammatory cytokines and decreased malondialdehyde (a marker of oxidative damage) levels in the cerebellum, whereas cucumber had no effect (Gemma et al., 2002). These results suggest that, as discussed previously, one mechanism of action of the polyphenolics in fruits and vegetables may be antiinflammatory. In addition, even though there was some dose relationship between ORAC value and effectiveness, this relationship needs to be examined further because spirulina and apple were equally effective in reversing the inflammatory measures, even though spirulina has an ORAC that is 300-fold higher than that of apple. Therefore, it is possible that the different phytochemicals present in these foods might account for the observed effects in this study.

In another study by this research group, 6 weeks of a spinach-enriched diet ameliorated deficits in cerebellar-dependent delay classical eyeblink learning and reduced expression of the proinflammatory cytokines, TNF α and β, in the cerebellum of 18-month-old rats (Cartford et al., 2002). The authors believe the results indicate that there may be a link between the reduction of inflammatory processes in the brain and improvement in learning in older animals (Cartford et al., 2002) and lend further support to the idea that polyphenolics in fruits and vegetables have antiinflammatory properties.

Besides these studies from our groups, very few comprehensive studies have investigated nutritional supplementation from fruits and vegetables and their protection against OS and inflammation with regard to their prevention against age-related memory and cognitive impairments. However, in a manner similar to that discussed previously, several investigations have been conducted with other fruits and vegetables, which have reported similar benefits on either behavior or reduction of OS in aging, and these are discussed next.

23.2.3 AGED GARLIC EXTRACT

Aged garlic extract (*Allium sativum,* which contains *S*-allycysteine, *S*-allymercaptocysteine, allicin, and diallosulfides) administered at 2% (w/w) of the diet exhibits beneficial effects toward cognitive impairments in a novel strain of senescence-accelerated mouse (SAM; Moriguchi et al., 1994, 1997b; Nishiyama et al., 1997; Zhang et al., 1998). The garlic-supplemented SAM animals also show increased survival, less atrophy in the frontal brain areas, and better performance in both passive and active avoidance, which the authors attribute to the antioxidant properties of the aged garlic extract, although motor activity is not affected by the treatment. Although only tested *in vitro*, garlic extract has also been shown to protect against OS-induced increases in thiobarbituric acid-reactive substances (TBARS; Horie et al., 1989, 1992) and to promote the survival of neurons derived from various regions of the neonatal brain (i.e., increase neurogenesis; Moriguchi et al., 1997a). Until these effects are tested in the whole animal and until the aged garlic extract components are actually isolated and identified within the brain, the mechanisms involved in its protective effects are difficult to determine.

23.2.4 RED BELL PEPPER

Red bell pepper (*Capsicum annuum* L.) has also been reported to ameliorate the learning impairment in SAM (Suganuma et al., 1999). Following supplementation of an experimental diet that contained 20% (w/w) lyophilized powder of red bell pepper, SAMP8 mice showed greater acquisition in passive avoidance tasks compared with a control group given a nonsupplemented diet. However, the authors did not examine any other parameters, which makes it difficult to begin to delineate possible mechanisms by which red bell pepper has beneficial effects on learning in SAM.

23.2.5 TOMATOES

One study that investigated the relationship of plasma antioxidants to reduced functional capacity in the elderly found a strong positive association between plasma lycopene and the ability of the elderly nuns to perform self-care (e.g., dressing, feeding; Snowdon et al., 1996). Lycopene is a red pigment found in a small number of vegetables such as tomatoes, watermelons, and pink grapefruit, with the majority of lycopene being consumed in tomatoes and tomato-based products such as pizza.

23.2.6 BEANS

To study the effect of a radical scavenger on cellular aging, a water-soluble protein was isolated from broad beans (*Vicia faba*) and examined for its antioxidant properties in young and old human fibroblasts (Okada and Okada, 2000). The authors found that cells treated with this water-soluble protein showed increased endogenous activity of catalase as well as increased glutathione concentration, which they speculate may be related to the delay of cellular aging-dependent degeneration.

23.2.7 PERSIMMON AND GRAPE SEED

One laboratory found that persimmon and grape seed extracts have strong radical scavenging ability, as measured by decreases in lipid peroxidation and increases in antioxidant enzyme activity in the liver of rats administered these compounds (Ahn et al., 2002). The authors attributed the protective qualities of these compounds to the high tannin concentrations of these fruits.

23.2.8 CITRUS AND BLUEBERRIES

Tangeretin, a citrus flavonoid, when given orally to rats, exhibited neuroprotective properties in a lesion model of Parkinson's disease (Dada et al., 2001). This study showed that tangeretin was able to cross the blood-brain barrier to maintain the integrity and functionality of nigrostriatal dopaminergic pathways following 6-hydroxydopamine (6-OHDA) injections.

Neural grafting, a surgical technique for Parkinson's disease in which the lost DA neurons are replaced with embryonic neurons, shows promise as a treatment for this disease. Some transplant recipients have exhibited significant clinical benefit after transplantation of embryonic tissue. However, transplanted cells survive poorly, necessitating discovery of techniques for improving the survival of grafted DA neurons. One approach to improve survival of grafted tissue appears to be feeding a blueberry supplement to the transplant recipient for several weeks before surgery. In a rat model of Parkinson's disease, supplementation of the diet with blueberry extract improved the survival and functional ability of transplanted DA neurons by increasing the number of DA neurons surviving the grafting procedure, with consequent improvement in parkinsonian motor deficits (McGuire et al., 2002).

23.2.9 BIOAVAILABILITY OF FLAVONOIDS IN HUMANS

The previous findings suggest that, in addition to their known beneficial effects on cancer and heart disease, polyphenolics present in fruits and vegetables may be beneficial in reversing the course of brain and behavioral aging. However, one important question that needs to be addressed concerns the bioavailability of these compounds.

A recent investigation by Milbury et al. (2002) highlights the need for more information regarding the bioavailability and efficacy of plant polyphenols, particularly because of the renewed interest by industrial countries in traditional herbal medicines and the development of so-called functional foods. Therefore, Milbury et al. (2002) studied the bioavailability and pharmacokinetics of anthocyanins in humans and found that they were detected as glycosides in both plasma and urine. In addition, an earlier study found that consumption of strawberries, spinach, or red wine increased the serum antioxidant capacity in humans (Cao et al., 1998). The important finding in these and other similar studies is that flavonoids as a group are absorbed and may play a significant role in improving antioxidant status, even though their *in vivo* efficacy and bioavailability are less documented than their *in vitro* antioxidant capabilities (Milbury et al., 2002).

23.3 POLYPHENOLICS IN BEVERAGES

Phytochemicals are also found in beverages such as wine and tea. Epidemiological studies have demonstrated that the French population, despite a diet high in cholesterol, has lower mortality from coronary heart disease. This paradoxical finding, dubbed the *French Paradox*, is attributed to regular consumption of red wine, which is rich in polyphenols (Iijima et al., 2002). As with fruits and vegetables, the studies investigating the beneficial health effects of beverages rich in polyphenolics on brain function are limited and are presented next.

23.3.1 WINE AND GRAPE JUICE

People who are moderate wine drinkers (three to four glasses a day) have a significantly lower incidence of dementia and AD compared to nondrinkers (Launer et al., 1996; Letenneur et al., 1993; Orgogozo et al., 1997). The possible mechanisms for these protective effects include antioxidant and antiinflammatory properties as well as elevation of plasma apolipoprotein E levels (Orgogozo et al., 1997). Therefore, moderate wine drinking is seen as a possible preventive measure against senile dementia or AD, even though a direct demonstration of the protective effects of wine or a delineation of its mechanisms has not been shown.

Calling for a better understanding of the molecular mechanisms of action of polyphenolic compounds and their application to human health, one research group has been investigating the ability of grape polyphenols, particularly resveratrol (a major component in wine and the skin of grapes), to ameliorate neuronal damages due to chronic ethanol consumption in an animal model (Sun et al., 2002). Another group found that chronic treatment with *trans*-resveratrol (10 and 20 mg/kg, IP, for 21 days) prevented cognitive impairment in the passive avoidance and elevated plus maze tasks and reduced oxidative stress in rats given intracerebroventricular streptozotocin (a model of sporadic dementia of Alzheimer's type in rats; Sharma and Gupta, 2002). In addition, these authors point out that resveratrol possesses antiinflammatory properties (Sharma and Gupta, 2002).

Grape juice is also a rich source of flavonoids, and, in healthy adults, 10 ml of Concord grape juice for 2 weeks increased serum antioxidant capacity and protected low-density lipoprotein (LDL) against oxidation to an extent similar to that obtained with 400 IU α-tocopheral/day, but decreased native plasma protein oxidation more than vitamin E did (O'Byrne et al., 2002). Red grape juice concentrate (125 ml daily for 7 days) had similar protective effects on serum antioxidant capacity and LDL (Day et al., 1997), showing that nonalcoholic beverages also contain polyphenolics that can reduce oxidative stress, similar to the beneficial effects of red wine.

23.3.2 TEA

Polyphenolics from tea, particularly the catechins [i.e., catechin, epicatechin, and epigallocatechin gallate (EGCG)], have been suggested to elicit potential beneficial effects toward improving brain function. Tea catechins were shown to protect mice against the deleterious effects on memory induced by injection of glucose oxidase or cerebral ischemia, showing their radical scavenging abilities (Matsuoka et al., 1995). Levites et al. (2001) used the mouse N-methyl-4-phenyl-1,2,3,6-tetrahydropyridine (MPTP) model of Parkinson's disease to study potential protective activities of green tea extract and EGCG. MPTP is a neurotoxin that produces a pattern of neurodegeneration and a neuropathology similar to those seen in parkinsonian brains. Pretreatment with green tea extract (0.5 and 1 mg/kg) or EGCG (2 and 10 mg/kg) protected against MPTP-induced decreases in striatal DA content, demonstrating the neuroprotective activity of green tea and its individual polyphenol EGCG (Levites et al., 2001). Dietary treatment with green tea also prevented exercised-induced accumulation of lipid peroxidation products in kidney and, to a lesser extent, in liver, showing oxidative stress protection of EGCG (Alessio et al., 2002).

A recent pilot study showed that kombucha, a lightly fermented tea beverage, significantly inhibited weight gain, increased environmental awareness and responsiveness, and prolonged life in mice (Hartmann et al., 2000). However, there is a health safety issue with drinking kombucha because it has been linked to health problems and fatalities in humans, and chronic drinking by mice in the current study contributed to longer spleens and enlarged livers (Hartmann et al., 2000).

Furthermore, black tea ingestion in young adults was associated with rapid increases in alertness and information-processing capacity, and tea drinking throughout the day largely prevented diurnal patterns of performance decrements (Hindmarch et al., 1998). The authors concluded that the effects of tea consumption could not be entirely explained as the acute effects of caffeine ingestion, and felt that other factors, particularly the polyphenolics in the tea, might play a significant role in mediating these responses.

23.4 POLYPHENOLICS IN OTHER FOODS

23.4.1 MANDA

Manda, a natural product prepared by yeast fermentation of several fruits and black sugar, is high in flavonoids and has radical scavenging abilities; therefore, the effects of long-term administration of Manda on age-dependent changes in lipid peroxidation in the senescent rat brain were tested by Kawai and colleagues (1998). The addition of Manda to brain homogenates of adult rats, incubation of brain homogenates with Manda for 2 and 3 h, and oral administration of Manda suppressed age-related increases in lipid peroxidation in areas such as the hippocampus and striatum, regions predominantly responsible for controlling behaviors that seem to decline during aging (Kawai et al., 1998).

23.4.2 CURCUMIN

Curcumin, the yellow curry spice, is found in tumeric and is a polyphenolic with both antioxidant and antiinflammatory activities (Frautschy et al., 2001). Dietary curcumin (2000 ppm), but not ibuprofen, suppressed oxidative damage (as measured by isoprostane levels), synaptophysin loss, and reduced microgliosis in old (22 month) Sprague-Dawley (SD) rats given lipoprotein carrier-mediated intracerebroventricular infusions of amyloid-β (Frautschy et al., 2001). In a second study, curcumin (500 ppm) prevented amyloid-β-infusion-induced spatial memory deficits (as measured in the Morris water maze), postsynaptic density loss, and reduced amyloid-β deposits in middle-aged SD female rats, leading the authors to conclude that curcumin may be beneficial in preventing Alzheimer's disease (Frautschy et al., 2001).

Another investigation by this same group tested the effects of curcumin in an animal model for Alzheimer's disease, the Tg2576 APPSw transgenic mouse (Lim et al., 2001). Low (160 ppm) and high (5000 ppm) doses of curcumin significantly lowered brain levels of oxidized proteins and the inflammatory cytokine interleukin-1β. Only the low-dose curcumin also reduced the atrocytic inflammatory marker GFAP, insoluble and soluble amyloid, and plaque burden by 43 to 50%. Therefore, curcumin was able to suppress indices of inflammation and oxidative stress in the brains of APPSw mice, factors that have been implicated in Alzheimer's disease pathogenesis, and decrease levels of amyloid and plaque burden in different brain regions (Lim et al., 2001).

23.4.3 CHOCOLATE

Cocoa and chocolate are rich sources of flavonoids, including, but not limited to, epicatechins and catechins (Zhu et al., 2002). In male SD rats given a 100-mg intragastric dose of cocoa extract, plasma antioxidant capacity was increased between 30 and 240 min following feeding. Furthermore, erythrocytes obtained from the cocoa extract-fed animals showed an enhanced resistance to hemolysis (i.e., a reduction in the susceptibility of erythrocyte membranes to oxidation), showing that cocoa flavanols and procyanidins can provide membrane-protective effects (Zhu et al., 2002).

23.4.4 SOY

Soy and soy food products possess isoflavone phytoestrogens, which have also been studied for their effects on memory and age-related diseases (File et al., 2001; Lephart et al., 2002). A high soy diet (100 mg total isoflavones/day) fed to young healthy adults for 10 weeks improved short-term and long-term memory and mental flexibility, showing that cognitive improvements can arise from a brief soy intervention. Furthermore, these improvements were not restricted to women or to verbal tasks (File et al., 2001). A review by Lephart et al. (2002) suggests that dietary soy-derived phytoestrogens can significantly influence brain and behavior parameters in animals, and these results have important implications for brain and neural disorders such as Alzheimer's disease.

23.5 CONCLUSIONS

Numerous studies have shown that there is a significant relationship among fruit and vegetable intake, cardiovascular disease, and cancer. In comparison, the findings reviewed here with respect to the brain suggest that, although the studies are few and more work needs to be done, the polyphenolics present in antioxidant-rich foods and beverages can reduce the deleterious effects of oxidative stress and inflammation in the CNS. These reductions may, in turn, contribute to forestalling or reversing the course of neuronal and behavioral aging. Furthermore, the present information suggests that factors involving mechanisms and bioavailability remain to be determined and, in addition, the polyphenolic families involved in producing the positive effects in aging and behavior should be investigated further, with a view to increasing the protection in both neurodegenerative disease and aging. At present, the world population comprising people more than 65 years of age represents more than 50% of all those who have ever lived to attain this age, and the determinations of whether these dietary supplementations might be effective in aged organisms are of paramount importance.

REFERENCES

Ahn, H.S., Jeon, T.I., Lee, J.Y., Hwang, S.G., Lim, Y. and Park, D.K. (2002) Antioxidative activity of persimmon and grape seed extract: *in vitro* and *in vivo*. *Nutrition Research,* 22, 1265–1273.

Alessio, H.M., Hagerman, A.E., Romanello, M., Carando, S., Threlkeld, M.S., Rogers, J., Dimitrova, Y., Muhammed, S. and Wiley, R.L. (2002) Consumption of green tea protects rats from exercise-induced oxidative stress in kidney and liver. *Nutrition Research,* 22, 1177–1188.

Bickford, P.C., Gould, T., Briederick, L., Chadman, K., Pollock, A., Young, D. et al. (2000) Antioxidant-rich diets improve cerebellar physiology and motor learning in aged rats. *Brain Research,* 866, 211–217.

Bickford, P.C., Shukitt-Hale, B. and Joseph, J. (1999) Effects of aging on cerebellar noradrenergic function and motor learning: nutritional interventions. *Mechanisms of Ageing and Development,* 111, 141–154.

Cantuti-Castelvetri, I., Shukitt-Hale, B. and Joseph, J.A. (2000) Neurobehavioral aspects of antioxidants in aging. *International Journal of Developmental Neuroscience,* 18, 367–381.

Cao, G., Russel, R.M., Lischner, N. and Prior, R.L. (1998) Serum antioxidant capacity is increased by consumption of strawberries, spinach, red wine, vitamin C in elderly women. *Journal of Nutrition,* 128(12), 2383–2390.

Cao, G., Sofic, E. and Prior, R.L. (1996) Antioxidant capacity of tea and common vegetables. *Journal of Agriculture and Food Chemistry,* 44, 3426–3431.

Cao, G., Verdon, C.P., Wu, A.H.B., Wang, H. and Prior, R.L. (1995) Automated assay of oxygen radical absorbance capacity with the COBAS FARA II. *Clinical Chemistry,* 41, 1738–1744.

Cartford, C., Gemma, C. and Bickford, P. (2002) Eighteen-month-old Fischer 344 rats fed a spinach-enriched diet show improved delay classical eyeblink conditioning and reduced expression of tumor necrosis factor alpha (TNF α) and TNF β in the cerebellum. *Journal of Neuroscience,* 22(14), 5813–5816.

Casadesus, G., Shukitt-Hale, B. and Joseph, J.A. (2002a) Qualitative versus quantitative calorie intake: are they equivalent paths to successful aging? *Neurobiology of Aging,* 23(5), 747–769.

Casadesus, G., Stellwagen, H., Szprengiel, A., Galli, R.L., Smith, M.A., Shukitt-Hale, B. and Joseph, J.A. (2002b) Modulation of hippocampal neurogenesis and cognitive performance in the aged rat: the blueberry effect. *Society for Neuroscience Abstracts,* 28, 294.1.

Dada, K.P., Christidou, M., Widmer, W.W., Rooprai, H.K. and Dexter, D.T. (2001) Tissue distribution and neuroprotective effects of citrus flavonoid tangeretin in a rat model of Parkinson's disease. *Neuropharmacology and Neurotoxicology,* 12(17), 3871–3875.

Day, A.P., Kemp, H.J., Bolton, C., Hartog, M. and Stansbie, D. (1997) Effect of concentrated red grape juice consumption on serum antioxidant capacity and low-density lipoprotein oxidation. *Annals of Nutrition and Metabolism,* 41(6), 353–357.

Eastwood, M.A. (1999) Interaction of dietary antioxidants *in vivo*: how fruit and vegetables prevent disease? *Quarterly Journal of Medicine,* 92, 527–530.

File, S.E., Jarrett, N., Fluck, E., Duffy, R., Casey, K. and Wiseman, H. (2001) Eating soya improves human memory. *Psychopharmacology,* 157, 430–436.

Frautschy, S.A., Hu, W., Kim, P., Miller, S.A., Chu, T., Harris-White, M.E. and Cole, G.M. (2001) Phenolic anti-inflammatory antioxidant reversal of AB-induced cognitive deficits and neuropathology. *Neurobiology of Aging,* 22, 993–1005.

Galli, R.L., Bielinski, D., Szprengiel, A., Shukitt-Hale, B. and Joseph, J.A. (2001) Brain regional assessments of inflammatory markers in young and senescent rats. *Society for Neuroscience Abstracts,* 27, 861.1.

Galli, R.L., Bielinski, D.F., Szprengiel, A., Shukitt-Hale, B. and Joseph, J.A. (in press) Blueberry supplemented diet reverses age-related decline in hippocampal HSP 70 neuroprotection. *Neurobiology of Aging.*

Galli, R.L., Casadesus, G., Rottkamp, C., Shukitt-Hale, B., Denisova, N.A., Smith, M.A. et al. (2000) Immunocytochemical effects in the brains of blueberry supplemented rats showing reversals of age-related cognitive and motor deficits. *Society for Neuroscience Abstracts,* 26, 2078.

Gemma, C., Mesches, M.H., Sepesi, B., Choo, K., Holmes, D.B. and Bickford, P. (2002) Diets enriched in foods with high antioxidant activity reverse age-induced decreases in cerebellar B-adrenergic function and increases in proinflammatory cytokines. *Journal of Neuroscience,* 22(14), 6114–6120.

Gold, P.E., Cahill, L. and Wenk, G.L. (2002) *Ginkgo biloba*: A cognitive enhancer? *Psychological Science in the Public Interest,* 3, 2–11.

Greenwood, C.E. and Winocur, G. (1999) Decline in cognitive function with aging: impact of diet. *Mature Medicine,* 2, 205–209.

Halliwell, B. (1994) Free radicals and antioxidants: a personal view. *Nutrition Reviews,* 52, 253–265.

Harman, D. (1981) The aging process. *Proceedings of the National Academy Science of the United States of America,* 78, 7124–7128.

Hartmann, A.M., Burleson, L.E., Holmes, A.K. and Geist, C.R. (2000) Effects of chronic kombucha ingestion on open-field behaviors, longevity, appetitive behaviors, and organs in c57-bl/6 mice: a pilot study. *Nutrition,* 16, 755–761.

Hauss-Wegrzyniak, B., Vannucchi, M.G. and Wenk, G.L. (2000) Behavioral and ultrastructural changes induced by chronic neuroinflammation in young rats. *Brain Research,* 859, 157–166.

Hindmarch, I., Quinlan, P.T., Moore, K.L. and Parkin, C. (1998) The effects of black tea and other beverages on aspects of cognition and psychomotor performance. *Psychopharmacology,* 139, 230– 238.

Hollman, P.C. and Katan, M.B. (1999) Health effects and bioavailability of dietary flavonols. *Free Radical Research,* 31, S75–S80.

Horie, T., Awazu, S., Itakura, Y. and Fuwa, T. (1992) Identified diallyl polysulfides from an aged garlic extract which protects the membranes from lipid peroxidation. *Planta Medica,* 58, 468–469.

Horie, T., Murayama, T., Mishima, T., Itoh, F., Minamide, Y., Fuwa, T. and Awazu, S. (1989) Protection of liver microsomal membranes from lipid peroxidation by garlic extract. *Planta Medica,* 55, 506–508.

Iijima, K., Yoshizumi, M. and Ouchi, Y. (2002) Effect of red wine polyphenolics on vascular smooth muscle cell function-molecular mechanism of the "French paradox." *Mechanisms of Ageing and Development,* 123, 1033–1039.

Joseph, J.A., Arendash, G., Gordon, M., Diamond, D., Shukitt-Hale, B. et al. (2003) Blueberry supplementation enhances signaling and prevents behavioral deficits in an Alzheimer disease model. *Nutritional Neuroscience,* 6: 153–162.

Joseph, J.A., Denisova, N.A., Youdim, K.A, Bielinski, D., Fisher, D. and Shukitt-Hale, B. (2001) Neuronal environment and age-related neurodegenerative disease: nutritional modification. In *Annual Review of Gerontology and Geriatrics, Focus on Modern Topics in the Biology of Aging,* Vol. 21 (Cristofalo, V.J. and Adelman, D., Eds.), pp. 195–235. Springer, New York.

Joseph, J.A., Shukitt-Hale, B., Denisova, N.A., Bielinski, D., Martin, A., McEwen, J.J. and Bickford, P.C. (1999) Reversals of age-related declines in neuronal signal transduction, cognitive, and motor behavioral deficits with blueberry, spinach, or strawberry dietary supplementation. *Journal of Neuroscience,* 19, 8114–8121.

Joseph, J.A., Shukitt-Hale, B., Denisova, N.A., Prior, R.L., Cao, G., Martin, A., Taglialatela, G. and Bickford, P.C. (1998) Long-term dietary strawberry, spinach, or vitamin E supplementation retards the onset of age-related neuronal signal-transduction and cognitive behavioral deficits. *Journal of Neuroscience,* 18, 8047–8055.

Kawai, M., Matsuura, S., Asanuma, M. and Ogawa, N. (1998) Manda, a fermented natural food, suppresses lipid peroxidation in the senescent rat brain. *Neurochemical Research,* 23(4), 455–461.

Launer, L.J., Feskens, E.J., Kalmijn, S. and Kromhout, D. (1996) Smoking, drinking, and thinking. The Zutphen Elderly Study. *American Journal of Epidemiology,* 143, 219–227.

Lephart, E.D., West, T.W., Weber, K.S., Rhees, R.W., Setchell, K.D.R., Adlercreutz, H. and Lund, T.D. (2002) Neurobehavioral effects of dietary soy phytoestrogens. *Neurotoxicology and Teretology,* 24, 5–16.

Letenneur, L., Dartigues, J.F. and Orgogozo, J.M. (1993) Wine consumption in the elderly. *Annals of Internal Medicine,* 118, 317–318.

Levites, Y., Weinreb, O., Maor, G., Youdim, M.B. and Mandel, S. (2001) Green tea polyphenol (–)-epigallo-catechin-3-gallate prevents *N*-methyl-4-phenyl-1,2,3,6-tetrahydropyridine-induced dopaminergic neurodegeneration. *Journal of Neurochemistry,* 78, 1073–1082.

Lim, G.P., Chu, T., Yang, F., Beech, W. Frautschy, S.A. and Cole, G.M. (2001) The curry spice curcumin reduces oxidative damage and amyloid pathology in an Alzheimer transgenic mouse. *Journal of Neuroscience,* 21(21), 8370–8377.

Lynch, M.A. (2001) Lipoic acid confers protection against oxidative injury in non-neuronal and neuronal tissue. *Nutritional Neuroscience,* 4, 419–438.

Macheix, J.-J., Fleuriet, A. and Billot, J. (1990) *Fruit Phenolics.* CRC Press, Boca Raton, FL.

Martin, A., Youdim, K., Szprengiel, A., Shukitt-Hale, B. and Joseph, J. (2002) Roles of Vitamins E and C on neurodegenerative diseases and cognitive performance. *Nutrition Reviews,* 60, 308–334.

Matsuoka, Y., Hasegawa, H., Okuda, S., Muraki, T., Uruno, T. and Kubota, K. (1995) Ameliorative effects of tea catechins on active oxygen-related nerve cell injuries. *Journal of Pharmacology and Experimental Therapeutics,* 274, 602–608.

McDaniel, M.A., Maier, S.F. and Einstein, G.O. (2002) "Brain-specific" nutrients: a memory cure? *Psychological Science in the Public Interest,* 3, 12–38.

McGuire, S.O, Shukitt-Hale, B., Joseph, J.A., Sortwell, C.E., Fleming, M.A., Marchionini, D.M. et al. (2002) Dietary supplementation with blueberry extracts improves the survival and function of grafted embryonic dopamine neurons in rats. *Society for Neuroscience Abstracts,* 28, 787.7.

Middleton, E., Jr. (1998) Effect of plant flavonoids on immune and inflammatory cell function. *Advances in Experimental Medicine and Biology,* 439, 175–182.

Milbury, P.E., Guohua, C., Prior, R.L. and Blumberg, J. (2002) Bioavailability of elderberry anthocyanins. *Mechanisms of Ageing and Development,* 123, 997–1006.

Moriguchi, T., Matsuura, H., Itakura, Y., Katsuki, H., Saito, H. and Nishiyama, N. (1997a) Allixin, a phytoalexin produced by garlic, and its analogues as novel exogenous substances with neurotrophic activity. *Life Sciences,* 61, 1413–1420.

Moriguchi, T., Saito, H. and Nishiyama, N. (1997b) Anti-ageing effect of aged garlic extract in the inbred brain atrophy mouse model. *Clinical Experiments in Pharmacology and Physiology,* 24, 235–242.

Moriguchi, T., Takashina, K., Chu, P.J., Saito, H. and Nishiyama, N. (1994) Prolongation of life span and improved learning in the senescence accelerated mouse produced by aged garlic extract. *Biological and Pharmaceutical Bulletin,* 17, 1589–1594.

Nishiyama, N., Moriguchi, T. and Saito, H. (1997) Beneficial effects of aged garlic extract on learning and memory impairment in the senescence-accelerated mouse. *Experimental Gerontology,* 32,149–160.

O'Byrne, D.J., Devaraj, S., Grundy, S.M. and Jialal, I. (2002) Comparison of the antioxidant effects of Concord grape juice flavonoids alpha-tocopherol on markers of oxidative stress in healthy adults. *American Journal of Clinical Nutrition,* 76(6), 1367–1374.

Okada, Y. and Okada, M. (2000) Effect of a radical scavenger "water soluble protein" from broad beans (*Vicia faba*) on antioxidative enzyme activity in cellular aging. *Journal of Nutritional Science and Vitaminology,* 46, 1–6.

Olanow, C.W. (1993) A radical hypothesis for neurodegeneration. *Trends in Neuroscience,* 16, 439–444.

Orgogozo, J.M., Dartigues, J.F., Lafont, S., Letenneur, L., Commenges, D., Salamon, R., Renaud, S. and Breteler, M.B. (1997) Wine consumption and dementia in the elderly: a prospective community study in the Bordeaux area. *Review of Neurology (Paris),* 153, 185–192.

Prior, R.L., Cao, G., Martin, A., Sofic, E., McEwen, J., O'Brien, C. et al. (1998) Antioxidant capacity as influenced by total phenolic and anthocyanin content, maturity and variety of *Vaccinium* species. *Journal of Agriculture and Food Chemistry,* 46, 2586–2593.

Sharma, M. and Gupta, Y.K. (2002) Chronic treatment with trans resveratrol prevents intracerebroventricular streptozotocin induced cognitive impairment and oxidatve stress in rats. *Life Sciences,* 71, 2489–2498.

Shukitt-Hale, B., Andres-Lacueva, C., Galli, R.L., Jauregui, O., Lamuela-Raventos, R.M. and Joseph, J.A. (2002) Polyphenolics can localize within the brains of aged rats fed a blueberry-supplemented diet. *Society for Neuroscience Abstracts,* 28, 294.6.

Snowdon, D.A., Gross, M.D. and Butler, S.M. (1996) Antioxidants and reduced functional capacity in the elderly: findings from the Nun Study. *Journal of Gerontology Series A: Biological Sciences and Medical Sciences,* 51, M10–M16.

Suganuma, H., Hirano, T. and Inakuma, T. (1999) Amelioratory effect of dietary ingestion with red bell pepper on learning impairment in senescence-accelerated mice (SAMP8). *Journal of Nutritional Science and Vitaminology,* 45, 143–149.

Sun, A.Y., Simonyl, A. and Sun, G.Y. (2002) The "French paradox" and beyond: neuroprotective effects of polyphenols. *Free Radical Biology and Medicine,* 32(4), 314–318.

Wang, H., Cao, G. and Prior, R. L. (1996) Total antioxidant capacity of fruits. *Journal of Agriculture and Food Chemistry,* 44, 701–705.

Youdim, K.A. and Joseph, J.A. (2001) A possible emerging role of phytochemicals in improving age-related neurological dysfunctions: a multiplicity of effects. *Free Radical Biology and Medicine,* 30, 583–594.

Youdim, K.A., Martin, A. and Joseph, J.A. (2000a) Essential fatty acids and the brain: possible health implications. *International Journal of Developmental Neuroscience,* 18, 383–399.

Youdim, K.A., Shukitt-Hale, B., Martin, A., Wang, H., Denisova, N., Bickford, P.C. and Joseph, J.A. (2000b) Short-term dietary supplementation of blueberry polyphenolics: beneficial effects on aging brain performance and peripheral tissue function. *Nutritional Neuroscience,* 3, 383–397.

Yu, B.P. (1994) Cellular defenses against damage from reactive oxygen species. *Physiological Reviews,* 74, 139–162.

Zhang, Y., Moriguchi, T., Saito, H. and Nishiyama, N. (1998) Functional relationship between age-related immunodeficiency and learning deterioration. *European Journal of Neuroscience,* 10, 3869–3875.

Zhu, Q.Y., Holt, R.R., Lazarus, S.A., Orozco, T.J. and Keen, C.L. (2002) Inhibitory effects of cocoa flavanols and procyanidin oligomers on free radical-induced erythrocyte hemolysis. *Experimental Biology and Medicine,* 227(5), 321–329.

Index